T0329322

FOOD INDUSTRY WASTES
ASSESSMENT AND RECUPERATION
OF COMMODITIES

Food Science and Technology
International Series

A complete list of books in this series appears at the end of this volume.

FOOD INDUSTRY WASTES

ASSESSMENT AND RECUPERATION OF COMMODITIES

Edited by

MARIA R. KOSSEVA
Chemical and Environmental Engineering, Faculty of Science and Engineering
University of Nottingham Ningbo Campus, China
Expert on the European Commission LIFE Sciences Panel, Belgium

COLIN WEBB
School of Chemical Engineering & Analytical Science
University of Manchester, Manchester, UK

ELSEVIER

AMSTERDAM • BOSTON • HEIDELBERG • LONDON
NEW YORK • OXFORD • PARIS • SAN DIEGO
SAN FRANCISCO • SINGAPORE • SYDNEY • TOKYO
Academic Press is an imprint of Elsevier

Academic Press is an imprint of Elsevier
32 Jamestown Road, London NW1 7BY, UK
225 Wyman Street, Waltham, MA 02451, USA
525 B Street, Suite 1800, San Diego, CA 92101-4495, USA

First edition 2013

Notice
No responsibility is assumed by the publisher for any injury and/or damage to
persons or property as a matter of products liability, negligence or otherwise, or from
any use or operation of any methods, products, instructions or ideas contained in the
material herein.

Because of rapid advances in the medical sciences, in particular, independent
verification of diagnoses and drug dosages should be made

British Library Cataloguing-in-Publication Data
A catalogue record for this book is available from the British Library

Library of Congress Cataloging-in-Publication Data
A catalog record for this book is available from the Library of Congress

ISBN: 978-0-12-391921-2

For information on all Academic Press publications
visit our website at elsevierdirect.com

Typeset by MPS Limited, Chennai, India
www.adi-mps.com

Printed and bound in United States of America

13 14 15 16   10 9 8 7 6 5 4 3 2 1

Working together to grow
libraries in developing countries

www.elsevier.com | www.bookaid.org | www.sabre.org

ELSEVIER   BOOK AID
            International   Sabre Foundation

I dedicate this book to my family

Contents

III

IMPROVED BIOCATALYSTS AND INNOVATIVE BIOREACTORS FOR ENHANCED BIOPROCESSING OF LIQUID FOOD WASTES

IV

ASSESSMENT OF WATER AND CARBON FOOTPRINTS AND REHABILITATION OF FOOD INDUSTRY WASTEWATER

V

ASSESSMENT OF ENVIRONMENTAL IMPACT OF FOOD PRODUCTION AND CONSUMPTION

Contributors

Sheela Berchmans Central Electrochemical Research Institute, Karaikudi, Tamilnadu, India

Ashok K. Chapagain Freshwater Programmes, WWF-UK, Godalming, UK

Mario Díaz Department of Chemical Engineering and Environmental Technology, University of Oviedo, Spain

Mónica Herrero Department of Chemical Engineering and Environmental Technology, University of Oviedo, Spain

Keith James WRAP, Banbury, UK

R. Karthikeyan Central Electrochemical Research Institute, Karaikudi, Tamilnadu, India

Christopher A. Kent School of Chemical Engineering, University of Birmingham, UK

Maria R. Kosseva Chemical and Environmental Engineering, Faculty of Science and Engineering, University of Nottingham Ningbo Campus, China; Expert on the European Commission LIFE Sciences Panel, Belgium

Adriana Laca Department of Chemical Engineering and Environmental Technology, University of Oviedo, Spain

Wen-Wei Li Department of Chemistry, University of Science and Technology of China, Hefei, China

Mehmet Melikoglu Department of Energy Systems Engineering, Atilim University, Ankara, Turkey

Yutaka Nakashimada Department of Molecular Biotechnology, Graduate School of Advanced Sciences of Matter, Hiroshima University, Japan

Naomichi Nishio Department of Molecular Biotechnology, Graduate School of Advanced Sciences of Matter, Hiroshima University, Japan

A. Palaniappan Central Electrochemical Research Institute, Karaikudi, Tamilnadu, India

Monika Schröder School of Arts, Social Sciences & Management, Queen Margaret University, Edinburgh, UK

Guo-Ping Sheng Department of Chemistry, University of Science and Technology of China, Hefei, China

Colin Webb School of Chemical Engineering and Analytical Science, University of Manchester, UK

Han-Qing Yu School of Earth and Space Sciences, University of Science and Technology of China, Hefei, China

Preface

Large quantities of food waste are generated all over the world. They are generated largely by the fruit-and-vegetable, vegetable oil, fermentation, dairy, meat, and seafood industries. In the world's leading economies, such as the USA, United Kingdom, and Japan, approximately 30% to 40% of the food 'consumed' is actually discarded. Waste is generated at every stage in the process chain: agriculture (pre- and postharvest), food manufacturing, packaging, retail/catering, and consumer/household; up to 50% of produce can be wasted only along the supply chain. Food industry waste is therefore a significant global problem, with impacts on economic, environmental, and food security systems. In addition, wasting food, while millions around the world suffer from hunger, raises questions of ethics and morality and could well lead to future food crises. There are also environmental impacts associated with the inefficient use of associated natural resources, such as water, energy, and land. The disposal of food wastes to landfill causes pollution and produces methane, a powerful greenhouse gas. Government efforts are currently focused on diverting such wastes away from landfill through regulation, taxation, and public awareness. However, rather than dwelling entirely on the environmental challenges posed, governments would do well to realise that such food waste streams represent considerable amounts of potentially reusable materials and energy.

This book is a concise presentation of a variety of important aspects involved in dealing with food wastes. The main aim is to emphasize trends in food waste management techniques and processing technologies. Providing a number of case studies and examples, some emerging environmental technologies suitable for development towards a sustainable society are illustrated. The book consists of 5 major commodity-oriented sections. The first looks at the *Problems and Opportunities* associated with food wastes and considers current waste regulations, the variability of food wastes, and green production strategies. Next, *Treatment of Solid Wastes* is presented, with chapters considering waste bread, fruit-and-vegetable wastes, and others, for the production of fermentation products, functional foods, biogas,

and fertilizer. We then look at improved biocatalysts and innovative bioreactors for *Enhanced Bioprocessing of Liquid Wastes* including a mathematical modeling approach and a case study of thermophilic aerobic digestion. In the fourth section, *Impact Assessment of Water Footprint and Rehabilitation of Food Industry Wastewater*, addressing water conservation/use/reuse/waste, is introduced along with the benefits of electricity generation from wastewater. Finally, food chain management and the *Assessment of Environmental Impact of Food Production and Consumption* are considered through the application of life cycle assessment (LCA). The food industry uses LCA to identify the steps in the food chain that have the largest impact on the environment in order to target improvement efforts. It is then used to choose among alternatives in the selection of raw materials, packaging material, and other inputs as well as waste management strategies.

Key features of the book are that it provides guidance on current food process waste regulations and disposal practices and understanding of waste beneficial reuse and bio-processing. It is written by experts from around the globe, providing the latest information on international research and development of novel green strategies and technologies for coping with food wastes. The book includes both theoretical and practical information providing, we hope, inspiration for additional research and applications to recover energy and niche coproducts including water use and reuse. Food intake is a vital source of energy for human beings. In the same way, food wastes should be seen as a vital source of energy and a feedstock for novel manufacturing processes. We have therefore provided a strong focus on environmental and bioprocess engineering methodology for the simultaneous treatment of food wastes, reduction of water footprints, and production of valuable products. We are sure that the book will raise awareness of sustainable food waste management techniques and their appraisal *via* Life Cycle Assessment. Finally, the book will contribute to the state of the art in waste management and valorisation of food by-products, providing novel concepts in the conversion of waste to resource.

We would like to take this opportunity to acknowledge and thank the contributors to this book for their excellent collaboration in bringing a comprehensive range of topics together in a single volume. We would also like to thank Nancy Maragioglio, the senior acquiring editor for the Food Science and Technology book program at Academic Press, and the production team at Elsevier, in particular Carrie Bolger and Colin Williams for their helpful assistance throughout this project and for keeping us to time (almost) and ensuring everything came together properly.

At last but not least, we are grateful to our families for their current and continued support.

Maria R. Kosseva
Colin Webb

Introduction: Causes and Challenges of Food Wastage

Maria R. Kosseva

1. SUSTAINABILITY OF THE FOOD SUPPLY CHAIN

At the global level, there appears to be sufficient food available to feed the world's population. This total, however, hides a wide distribution of food consumption that stretches from acute hunger crises to excess food consumption and large quantities of food waste.

Food insecurity is a harsh reality for one billion people. At the other end of the spectrum, overeating and food waste is common among more than one billion people, too (Lundqvist, 2010). The problem of being overweight and obesity in adolescents and children mainly reflects increased energy intake. Long-term trends indicate marked increases in availability of added oils, meat, cheese, frozen dairy products, sweeteners, fruit, fruit juices, and vegetables, which may have influenced the prevalence of childhood obesity. Combined per capita chicken and turkey availability increased more than six-fold overall, from 5.1 kg/year in 1909 to 33.5 kg/year in 2007. Meat, poultry, and fish availability exceeded 90 kg/year in 2002 and in subsequent years, which represented a 60% increase over values from early in the 20th century. Estimated losses due to spoilage, waste, and cooking processes are as high as 57% for the meat food group (Barnard, 2010). In a loss-adjusted analysis, total meat, poultry, and fish availability rose from 48.3 kg/year in 1970 to 54.4 kg/year in 2007. According to USDA estimates, these data correspond to an increase in per capita energy availability from red meat, poultry, and fish, adjusted for losses, from 367 kcal/day in 1970 to 387 kcal/day in 2007 (US Department of Agriculture, 2007).

Lack of availability of empirical data hampers the analysis of the low efficiency in the food chain. Lundqvist (2010) compares the variation and trends in the food supply and norms of food intake requirements for *an active and healthy life*. For food supply reference is made to the international norm, which is usually set at 2,700–2,800 kcal/day per person (Molden, 2007). It was found that in many countries the food supply is much higher than the international norm and very much higher than the food intake requirements. If a comparison is made between the amount of food produced in the field and food intake requirements, the gap is even wider, since losses and conversions are substantial (Lundqvist, 2010). In parallel, calculations imply a steady increase in body weight among US adults over the past 30 years and a progressive increase in food waste, from 900 to 1,400 kcal/day per person between 1974 and 2003 (Hall et al., 2009). When supply of food increases and food is perceived as relatively cheap and easily accessible, the risk for a dual problem increases: the public health situation deteriorates and the waste increases, with negative repercussions on resource pressure, environment, and productivity in society (Lundqvist, 2010).

One-third of all food produced for human consumption on the planet, about 1.3 billion tonnes, is lost or wasted each year, according to the Food and Agriculture Organization report of the United Nations prepared by the Swedish Institute for Food and Biotechnology (Gustavsson et al., 2011). Food is wasted throughout the food supply chain (FSC), from initial agricultural production down to final household consumption. In medium- and high-income countries, food is to a great extent wasted, meaning that it is thrown away even if it is still suitable for human consumption. Significant food loss and waste do, however, also occur early in the FSC. It has been estimated that between 25% and 50% of food produce is wasted along the supply chain. In low-income countries, food is mainly lost during the early and middle stages of the FSC; much less food is wasted at the consumer level. Food losses represent a waste of resources used in production, such as land, water, energy, and other inputs. Producing food that will not be consumed leads to unnecessary CO_2 emissions in addition to loss of economic value of the food produced.

How much food is lost and wasted in the world today and how can we prevent food losses? It is impossible to give precise answers to these questions, and there is not much ongoing research in the area.

This is quite surprising as forecasts suggest that food production must increase significantly to meet future global demand. Insufficient attention appears to be paid to current global FSC losses (Gustavsson et al., 2011).

Historically, people secured food through two methods: hunting and gathering, and agriculture. Today, most of the food energy consumed by the world population is supplied by the food industry, which is operated by multinational corporations that use intensive farming techniques and industrial agriculture to maximize system output. While the European food system is undergoing remarkable change—spreading eastwards, concentrating, globalizing, altering internal relations—evidence for the food industry's impact on environment, health, and social inequalities has mounted. The two EU discourses—one on economic efficiency and high technological innovation (competitiveness) and the other on environmental and social progress (sustainable development)—are now in a state of tension. At the member state and EU level, there is recognition that both goals will either have to be addressed by the powerful industrial and retail conglomerates or those combines will themselves become policy targets (Rayner et al., 2008).

One feasible frontier in this "food crisis" has been identified as the environment. The shift in demand from local and seasonal toward imported, non-seasonal fruit and vegetables increases transportation, cooling, and freezing inputs, with a corresponding increase in energy. Greater processing of food leads to increased energy and material input and associated packaging waste. While energy in producing food has decreased, the environmental cost of acquiring food has risen with greater use of cars required to transport foods from supermarkets.

The second frontier is cultural: the impact on European food traditions and consciousness about food. A study across 15 European countries has suggested that three core attributes, or types of approach, guide Europeans in the selection of food products (Rayner et al., 2008):

- Food as a source of pleasure and sensations. Products are judged by taste, sight, smell, point of origin, trustworthiness of producer/retailer, etc.
- Food as a matter of price, convenience, or ease of use.
- Food as a consideration for health.

If the environment is not nurtured, it cannot yield wholesome food. On the other hand, if the food is not produced, processed, and distributed equitably, and if food cultures are irreversibly damaged by product marketing, it becomes a vehicle for social conflict, inequality, and worsening patterns of health. Europe's

dilemma is all too common: how to balance food production for large populations accustomed to unparalleled choice and cheapness with sustainability in both natural and human-ecological terms—managing supply chains in a manner that enables both them and the earth to sustain future generations (Rayner et al., 2008).

As a result, food waste is a significant global problem for economic, environmental, and food security reasons. The experts argue that, unless more sustainable and intelligent management of production and consumption are undertaken, food prices could indeed become more volatile and high in a world of seven billion people, rising to over nine billion by 2050, as a result of escalating environmental degradation. Up to 25% of the world food production may become 'lost' during this century as a result of climate change, water scarcity, invasive pests, and land degradation. These are environmental impacts associated with the inefficient use of natural resources such as water, energy, and land (e.g., causing deforestation and land degradation). World food production has already risen substantially in the 20th century, primarily as a result of increasing yields due to irrigation and fertilizer use as well as agricultural expansion into new lands, with little consideration of food energy efficiency. At the same time the world price of food is estimated to become 30–50% higher in the coming decades and to show greater volatility. Increased food prices have had a dramatic impact on the lives and livelihoods of those already undernourished or living in poverty and spending 70–80% of their daily income on food. Key causes of the current food crisis are the combined effects of speculation in food stocks, extreme weather events, low cereal stocks, growth in biofuels competing for cropland, and high oil prices (Nellemann et al., 2009).

The recommendations in the United Nations Environmental Programme were to capture and recycle postharvest losses/waste and to develop new technologies, thereby increasing food energy efficiency by 30–50% at current production levels. New strategies are needed that respond to the intimidating challenges posed by climate change mitigation and adaptation, water scarcity, the decline of petroleum-based energy, biodiversity loss, and persistent food insecurity in growing populations. There is also an economic impact of throwing food away, which ultimately affects all the organizations and individuals involved in the supply chain, including the final consumer (Ventour, 2008). Rather than focusing solely on increasing production, we can increase food security by enhancing supply through the optimization of food energy efficiency. Food energy efficiency is our ability to minimize the loss of energy in food from harvest potential, through processing, to actual consumption and recycling. By optimizing this chain, food supply can increase with much less damage to the environment, in a manner similar to improvements in efficiency in the traditional energy sector (Nellemann et al., 2009).

2. QUANTITY OF FOOD WASTES

Large quantities of food waste are generated all over the world. One of the key findings is that industrialized and developing countries dissipate roughly the same quantities of food. In developing countries more than 40% of the food losses occur at postharvest and processing levels, while in industrialized countries, more than 40% of the food losses occur at retail and consumer levels. Food waste at consumer level in industrialized countries (222 million tonnes) is almost as high as the total net food production in sub-Saharan Africa (230 million tonnes) (Gustavsson et al., 2011). The study also claims that fruits and vegetables, as well as roots and tubers, have the highest wastage rates.

In the UK, 8.3 million tonnes of food and drink are thrown away every year with a carbon impact exceeding 20 million tonnes of CO_2 equivalent emissions (WRAP, 2010). Most of this is avoidable and could have been eaten if only we had planned, stored, and managed it better. Less than a fifth is truly unavoidable—things like bones, cores, and peelings, which can be used as resources for other manufactured goods. The amount of food wasted per year in UK households is 25% of that purchased (by weight). The avoidable food and drink wastes are thrown away for two main reasons: 2.2 million tonnes is thrown away due to cooking, preparing, or serving too much; and a further 2.9 million tonnes because it was not used in time. For example, the avoidable food and drink waste consists of:

- 860,000 tonnes of fresh vegetables and salads
- 870,000 tonnes of drink
- 680,000 tonnes of bakery
- 660,000 tonnes of home made and pre-prepared meals
- 500,000 tonnes of fresh fruit
- 290,000 tonnes of meat and fish
- 530,000 tonnes of dairy and eggs
- 190,000 tonnes of cakes and desserts
- 67,000 tonnes of confectionery and snacks.

All this wasted food is costly; in the UK people spend £12 billion every year buying and then throwing away good food. This works out to £480 for the average UK household, increasing to £680 a year for households with children—an average of just over £50 a month.

By analogy with the UK, in Japan approximately 20 million tonnes food garbage is generated every year (Minowa et al., 2005). This means that as much as ¥11 trillion worth of food is lost to waste annually. In 2008, 70% of the wasted food in Japan was recycled, half of which was turned to animal feed, 30% converted to fertilizer, and 5% to methane. The rest of the food waste was mostly incinerated or sent to landfills (Sugiura et al., 2009). In Taiwan, approximately 16.5 million tonnes of food waste is produced annually (Mao et al., 2006). In the Republic of Korea, more than 22% of the municipal solid waste is reported to be food waste, and the generation rate for food waste is around 0.24 kg/person/day (Kim et al., 2008). Annually this is equivalent to 4.3 million tonnes of food waste.

The amount of food wasted in the USA is staggering. According to the US Environmental Protection Agency, the USA generates more than 34 million tonnes of food waste each year. Food waste is more than 14% of the total municipal solid waste stream. Less than 3% of the 34 million tonnes of food waste generated in 2009 was recovered and recycled. The rest—33 million tonnes—was thrown away. Food waste now represents the single largest component of municipal solid waste reaching landfills and incinerators. Currently in the USA, over 97% of food waste is estimated to be buried in landfills. When food is disposed in a landfill it quickly rots and becomes a significant source of methane—a potent greenhouse gas with 21 times the global warming potential of carbon dioxide. Landfills are a major source of human-related methane in the USA, accounting for more than 20% of all methane emissions, which can be used as an energy source. There is nonetheless interest in strategies to divert this waste from landfills as evidenced by a number of programs and policies at the local and state levels, including collection programs for source-separated organic wastes. Jones has estimated that overall food losses in the USA amount to US$90−100 billion a year, of which households throw away US$48.3 billion worth of food each year (Jones, 2006). The amount of food loss at the household level in the USA was estimated to be 14%, costing a family of four at least US$589.76 annually (Jones 2004).

Studies made in other OECD countries show broadly similar figures, but also that the magnitude of waste varies significantly. Norway has about the same level of waste as the UK (i.e., 71 kg/year per person) (Hanssen and Olsen, 2008). For Holland, Thoenissen (2009) reports that 43−60 kg of edible food is wasted per household annually. The Netherlands is throwing away €2.4 billion per year on food waste, representing

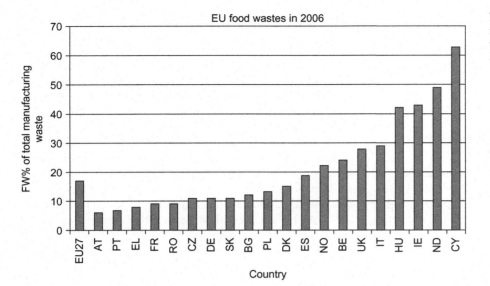

FIGURE 1 Waste generated by food industry in Europe in 2006 (percent of total waste generated in manufacturing). *Data from EUROSTAT (2009).*

more than 20% of the total food in the market (EUROSTAT, 2009). In the USA, the estimates vary significantly from about 25% to 50% (Jones, 2006; Schiller, 2009), and in Australia an average household annually throws out an estimated AUD$ 239 per person, or US$ 222 (Baker et al., 2009). The surprisingly large differences between societies that apparently have similar socioeconomic and cultural characteristics may partly depend on methodological differences and difficulties in defining the boundaries for the measurements (Lundqvist, 2010).

In Europe an estimated amount of approximately 50% of the food produced is wasted (EUROSTAT, 2009). This varies from country to country and from sector to sector, but in the best case approximately 20% of our food ends up as waste. At the same time more than 50 million Europeans are at risk of relative poverty. This is simply unacceptable from social, economic, and environmental points of view. In 2006 the manufacture of food accounted for 17% of total waste generated in manufacturing in the European Union (EU27) and over 40% in Cyprus, the Netherlands, Ireland, and Hungary (Figure 1). In Sweden, an average household is estimated to throw away 25% of food purchased. An average Danish family with 2 adults and 2 children wastes food worth €1341 a year (€2.15 billion for the whole country). Each French citizen throws away 7 kg of food still in the original package every year, when in the same country 8 million people are at risk of poverty.

Avoidable food waste in Finnish households is about 20–30 kg/person/year. This accounts for about 120,000 to 160,000 tonne/year for all households, or about 5% of all purchased food. In food service institutions, on average about 20% of prepared food is wasted, which amounts to about 75,000 to 85,000

tonne/year of avoidable food waste. In retail about 65,000 to 75,000 tonne/year of food products are discarded (due to the study methods, the result includes also some inedible parts like peels and bones). Households appear to be the biggest source of avoidable food waste in Finland. Edible food wasted in all households per year is worth about 500 million Euro (Koivupuro, 2011).

The waste produced by households ranged from 181 kg per capita in Poland to 576 kg per capita in the Netherlands in 2006, with an average of 423 kg per capita in the European Union (EUROSTAT, 2009) as shown in Boxes 1 and 2.

Food wastage depends largely on the society in which it was grown and consumed (Figure 2). In poor countries most food is lost at the producers' end: food gets lost in the fields or due to lack of storage and cooling systems or poor transport mechanisms. In the developing world, lack of infrastructure and associated technical and managerial skills in food production and postharvest processing have been identified as key drivers in the creation of food waste, both now and over the near future (WFP, 2009). For example, in India, it is estimated that 35% to 40% of fresh produce is lost because neither wholesale nor retail outlets have cold storage (Nellemann et al., 2009).

In developed countries, most food waste continues to be generated postconsumer, driven by the low price of food relative to disposable income, consumers' high expectations of food cosmetic standards, and the increasing disconnection between consumers and how food is produced. Similarly, the increasing urbanization within transitioning countries will potentially disconnect those populations from the sources of food, which is likely to further increase food waste generation.

BOX 1

EU WASTE GENERATED BY HOUSEHOLD IN 2006

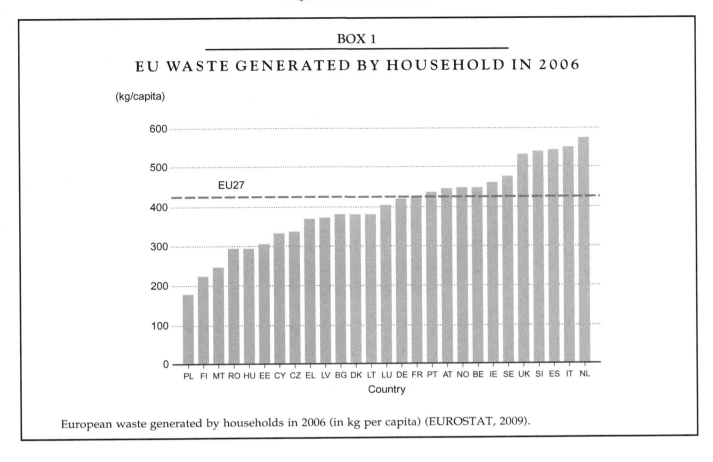

European waste generated by households in 2006 (in kg per capita) (EUROSTAT, 2009).

The lack of infrastructure in many developing countries and poor harvesting/growing techniques are likely to remain major elements in the generation of food waste. Less than 5% of the funding for agricultural research is allocated to postharvest systems (Kader, 2003), and yet reduction of these losses is recognized as an important component of improved food security (Nellemann et al., 2009). Irrespective of the global region, there is a need for successful introduction of culture-specific innovations and technologies across the FSC to reduce losses.

There are clearly fundamental factors affecting postconsumer food waste worldwide, some of which may require solutions that involve direct communication and awareness raising among consumers of the importance of reducing food waste. Others require government interventions and the support and cooperation of the food industry itself, such as improving the clarity of food date labeling and advice on food storage or ensuring that an appropriate range of pack or portion sizes is available that meets the needs of different households (Parfitt et al. 2010).

Undoubtedly, agricultural and food production losses are particularly high between field and market in developing countries, and wastage (i.e., excess caloric intake and obesity) is highest in the more industrialized nations. The loss of, or reduction in, other primary ecosystem services (e.g., soil structure and fertility; biodiversity, particularly pollinator species; and genetic diversity for future agriculture improvements) and the production of greenhouse gases (notably methane) by decomposition of the discarded food are just as important to long-term agricultural sustainability the world over (Nellemann et al., 2009).

The qualitative approach of Mena et al. (2011) helped to identify the main root causes of waste in the supplier—retailer interface, which were categorized into three groups: (1) mega trends in the market place, (2) natural causes related to the products and processes, and (3) management root causes on which practitioners have a direct impact. The results revealed that levels of waste are, to a large extent, dependent on the natural characteristics of the product, such as shelf life, temperature regime, and demand variability, and on mega trends in the markets, such as the increasing demand for fresh products and products out of season. Despite the natural constraints, it was found that there are many opportunities for reducing waste by

BOX 2

PERCENTAGE BREAKDOWN OF EU27 FOOD WASTE GENERATED BY MANUFACTURING, HOUSEHOLDS, WHOLESALE/RETAIL, AND FOOD SERVICE/CATERING SECTORS (BEST ESTIMATE)

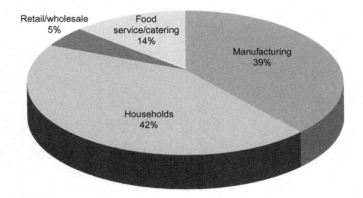

The main estimate of this study (EC DG ENV, 2011) relies more heavily on EUROSTAT data to estimate manufacturing, household, and "other sector" food waste: households produce the largest fraction of EU food waste among the four sectors considered, at about 42% of the total or about 38 Mt; manufacturing food waste was estimated at almost 35 Mt per year in the EU27 (70 kg per capita). The wholesale/retail sector accounts for close to 8 kg per capita (with an important discrepancy between Member States) representing around 4.4 Mt for the EU27; the food service sector accounted for an average of 25 kg per capita for the EU27, at 12.3 Mt for the EU27 overall. There is a notable divergence between the EU15 at 28 kg per capita (due to a higher trend of food waste in the restaurant and catering sector) and 12 kg per capita in the EU12. *Source: 2006 EUROSTAT data (EWC_09_NOT_093), various national sources.*

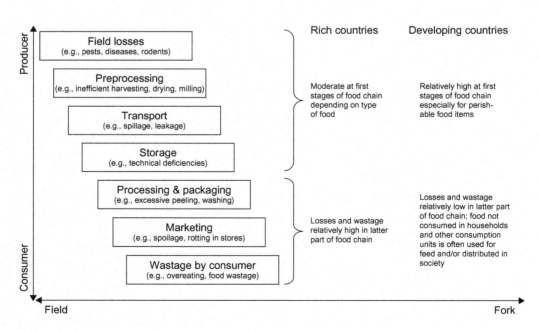

FIGURE 2   **Main types of food losses and wastage.** *From Lundqvist (2008).*

addressing the nine management root causes identified by the study. These root causes are: information sharing, forecasting and ordering, performance measurement, cold chain management, training, quality management, waste management responsibilities, promotions management, and packaging. This study was restricted to two countries (UK and Spain) and a limited range of products. It, however, identifies a number of common efficiency lapses at the supply-chain and management level, which can apply across the countries in the developed world. Despite the limitations, it could serve as a stepping-stone for future research trying to address the problem of food waste.

Ultimately, the most important reason for food waste at the consumption level in rich countries is that people simply can afford to waste food (Stuart, 2009).

3. WATER WASTE

Food waste is also water waste, as large quantities of water are used to produce the lost food. From the environmental perspective food waste accounts for more than one quarter of the total consumptive use of finite and vulnerable freshwater and more than 300 million barrels of oil per year. Globally, the amount of water withdrawn every year to produce the lost and wasted food could fill a lake of $1,300 \text{ km}^3$, about half the volume of Lake Victoria. In the US, annual food production consumes about 120 km^3 of irrigation water. People throw away 30% of this food, which corresponds to 40 billion liters of irrigation water. That is enough water to meet the household needs of 500 million people. Today, to meet global food demand some 7,100 billion cubic meters of water, equivalent to more than 3,000 liters per person per day, are used during crop production through evaporation and transpiration. In arid and semiarid countries, water is already a limiting factor in agricultural production. About 1.2 billion people, one-fifth of the world's population, live in basins where water is running out (Lundqvist, 2008).

Water losses accumulate as food is wasted before and after it reaches the consumer. In poorer countries, most uneaten food is lost before it has a chance to be consumed. Depending on the crop, an estimated 15–35% of food may be lost in the field. Another 10–15% is discarded during processing, transport, and storage. In richer countries, production is more efficient but waste is greater: people throw away much of the food they buy, and all the resources used to grow, ship, and produce the food are thrown away along with it.

The world water requirements for food production from 1960 to 2002 and its projection to 2050 are depicted in Box 3 (Nordpil, 2009).

By 2050, the demand for water for food production is predicted to double in order to cope with the needs of the growing human population (Rockström et al., 2005). The global need for energy production—and therefore water—is also projected to rise by 57% by the year 2030 (Hightower and Pierce, 2008). Clearly the time has come to address the central question: *Is there enough water to sustain our wasteful lifestyle?* (Cominelli et al., 2009). Therefore, the water footprint concept has been developed in order to have an indicator of water use in relation to consumption by people. The water footprint is more accurate and provides a more useful assessment of the water demands of a country than do the national figures for water consumption (Chapagain and Hoekstra, 2004).

4. ENVIRONMENTAL EFFECT OF FOOD WASTE

As carbon becomes the prominent form of measuring environmental impact, much attention is being given to determining the carbon impact of the UK food and drink retail supply chain. WRAP assessed the greenhouse gas emissions associated with food and packaging waste (Figure 3). WRAP's estimate of the carbon impact of the UK food and drink supply chain and household waste was performed to calculate the greenhouse gas emissions associated with grocery retail food and drink waste. Conversion factors were developed to convert the quantities of waste to quantities of CO_2 equivalent at each stage in the supply chain (a total conversion factor is 3.8 tonnes of CO_2 eq per tonne of waste) (WRAP, 2010). The greenhouse gas emissions of avoidable food waste from Finnish households roughly correspond to the emissions of 100,000 cars (Koivupuro, 2011).

The combined carbon impacts in the UK of food and packaging waste in the supply chain totals 10 million tonnes of CO_2 eq, and in the household 26 million tonnes CO_2 eq. In addition, the greenhouse gas impact associated with by-product going to animal feed is 3.7 million tonnes of CO_2 eq. For household and manufacturing, most of the impact comes from food waste, whereas for distribution and retail most comes from packaging waste.

5. CONCLUSIONS

Roughly 30–40% of food in both the developed and developing worlds is lost to waste, though the causes are very different. In the developing world losses are mainly attributable to the absence of food-chain infrastructure and the lack of knowledge or investment in storage technologies on the farm, although data are

BOX 3

HISTORIC AND PROJECTED CHANGES IN WATER CONSUMPTION FOR FOOD PRODUCTION, 1960-2050

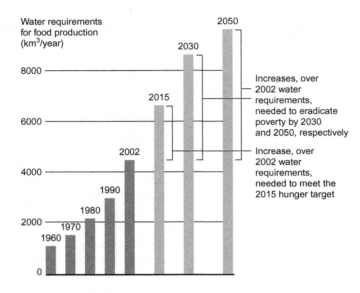

The requirements for water in agriculture will need to increase in order to meet the Millennium Development Goal of ending hunger. Boosting of water requirements is needed to meet this supply. *From Nordpil (2009)*

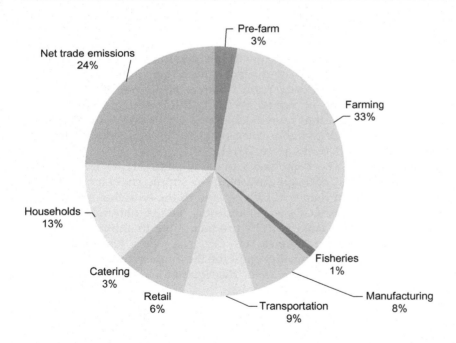

FIGURE 3 Greenhouse gas emissions associated with the UK food chain by sector in 2006. *From WRAP (2010).*

scarce. In contrast, in the developed world pre-retail losses are much lower but those arising at the retail, food service, and home stages of the food chain have grown dramatically in recent years (Nellemann et al., 2009; Stuart, 2009). Simultaneously, studies show that some Western countries, such as the USA, France, Sweden, and Brazil, are consuming every day a surplus of 1,400 calories per person for a total of 150 trillion calories a year. So, apart from the waste in the food supply chain, overeating is gradually becoming a serious public health issue in a growing number of countries (Lundqvist, 2010).

Godfray et al. (2010) propose strategies required to tackle the two types of waste. In developing countries, public investment in transport infrastructure would reduce the opportunities for spoilage, whereas better-functioning markets and the availability of capital would increase the efficiency of the food chain, for example, by allowing the introduction of cold storage (though this has implications for greenhouse gas emissions). Existing technologies and best practices need to be spread by education and extension services, and market and finance mechanisms are required to protect farmers from having to sell at peak supply, leading to gluts and wastage.

Reducing developed-country food waste is particularly challenging, as it is so closely linked to individual behavior and cultural attitudes toward food. Waste may be reduced by alerting consumers to the scale of the issue as well as to domestic strategies for reducing food loss. Advocacy, education, and possibly legislation may also reduce waste in the food service and retail sectors. Legislation such as that on sell-by dates and swill that has inadvertently increased food waste should be reexamined within a more inclusive competing-risks framework (Godfray et al., 2010). The need for a more sophisticated understanding of what causes household food waste should be pursued so that behavioural innovations have a better chance of bearing fruit. Consumers also have an important role as adopters of product innovation. But it is for the food industry to ensure that the consumer voice is accurately captured and translated when product developments are undertaken (Schröder, 2003).

Gustavsson et al. (2011) also propose that further research in the area is urgent, especially considering that food security is a major concern in large parts of the developing world. While increasing primary food production is paramount to meet the future increase in final demand, tensions between production and access to food can also be reduced by tapping into the potential to reduce food losses. Efficient solutions exist, along the whole food chain, for reducing total amounts of food lost and wasted. Actions should not be confined to isolated parts of the chain, since what is done (or not done) in one part has effects in others.

Food industry waste is a significant global problem for economic, environmental, and food security reasons. The disposal of waste to landfill causes pollution and produces methane, which is a harsh greenhouse gas. Therefore, government efforts have focused on diverting waste away from landfill through regulation, taxation, and public awareness. According to the European Landfill Directive (1999/31/EC), the amount of biodegradable waste sent to landfills in member countries by 2016 must reach 35% of the levels reached in 1995. Thus, the European food-processing industry operations have to comply with increasingly stringent European Union (EU) environmental regulations related to disposal or utilization of by-products and wastes. These include growing restrictions on land spraying with agro-industrial wastes, disposal within landfill operations, and the requirements to produce end products that are stabilized and hygienic. Unless suitable technologies are found for the processing and utilization of waste by-products, large numbers of food-processing operations will be under threat.

One alternative for the diversion of waste from landfills is to increase the quantity of food waste that is treated biologically, either by aerobic composting or anaerobic digestion. While programs and facilities to manage yard waste are well established, the management of food waste in composting facilities is less well developed and perhaps only in its infancy. There is nonetheless considerable interest in food waste composting, and the desire to increase food waste diversion is likely to increase (Levis et al., 2010). There are many regional factors that will influence technology selection, including the cost of competing waste management alternatives, local regulations, population density, and emissions standards. However, efforts to understand why food waste occurs have been limited, and detailed investigations are therefore necessary in the field.

Taking into consideration the above concluding remarks, this book is a concise presentation of important aspects of dealing with food system sustainability and sustainable consumption as well as the latest developments in the area of food industry waste reduction. The first and most important is to reduce the quantity of waste produced by the food industry, then to develop methods to valorise unused coproducts and improve the management of wastes that cannot be reused or recycled.

The objectives of this book are the following:

1. To emphasize trends in food waste management techniques and processing technologies.
2. To illustrate and provide inspiration for developing advanced methods to recycle and upgrade food industry by-products into high added-value food and feed commodities.
3. To encourage the recovery of energy and niche coproducts, including water use and reuse.
4. To focus on environmental and bio/ chemical engineering methods for treatment, which will reduce water footprints simultaneously.

5. To raise awareness of sustainable food chain management and sustainable consumption, applying the Life Cycle Assessment (LCA) principle. The food industry uses the LCA to identify the steps in the food chain that have the largest impact on the environment in order to target improvement efforts. It is then used to choose among alternatives in the selection of raw materials, packaging material, and other inputs as well as waste management strategies.
6. Knowledge of the principles of industrial ecology is essential for development of life cycle thinking and green technologies. We will describe *green production principles* and criteria, in order to distinguish key steps to *sustainability*, as well as discussing a *holistic approach* and *upgrading concept* in food production.

References

Baker, D., Fear, J., Denniss, R., 2009. What a waste. An analysis of household expenditure on food. Policy brief, 6 November. The Australian Institute, Canberra.

Barnard, N.D, 2010. Trends in food availability, 1909–2007. Am. J. Clin. Nutr. 91 (suppl), 1530S–1536S.

Chapagain, A. and Hoekstra, A.Y. 2004. Water footprints of nations. Research Report Series No.16. UNESCO-IHE, Delft, The Netherlands.

Cominelli, E., Galbiati, M., Tonell, C., Bowler, C., 2009. Water: the invisible problem. EMBO Rep. 10 (7), 671–676.

Europe In Figures Yearbook 2009. EC Eurostat Statistical books, Madrid Network, 2010.

Godfray, H.C.J., Beddington, J.R., Crute, I.R., Haddad, L., Lawrence, D., Muir, J.F., et al., 2010. Food security: the challenge of feeding 9 billion people. Science 327, 812–818.

Gustavsson, J., Cederberg, C., Sonesson, U., van Otterdijk, R., Meybeck, A. 2011. Global Food Losses and Food Wastes: Extent, Causes and Prevention. Food and Agriculture Organization of the United Nations.

Hall, K.D., Guo, J., Dore, M., Chow, C.C., 2009. The progressive increase of food waste in America and its environmental impact. PLoS ONE 4 (11), e7940.

Hanssen, O.J., Olsen, A. 2008. Kartlegging av matavfall. Forprosjekt for NorgesGruppen. Östfoldforskning. Norway. (Mimeo).

Hightower, M., Pierce, S.A., 2008. The energy challenge. Nature 452, 285–286.

Jones, T. 2004 The value of food loss in the American Household, Bureau of Applied Research in Anthropology, A Report to Tilia Corporation, San Francisco, CA, USA.

Jones, T. 2006 Addressing food wastage in the US. Interview: The Science Show.

Kader, A.A., 2003. Perspective on postharvest horticulture. HortScience 38, 1004–1008.

Kim, J.-D., Park, J.-S., In, B.-H., Kim, D., Namkoong, W., 2008. Evaluation of pilot-scale in-vessel composting for food waste treatment. J. Hazard. Mater. 154, 272–277.

Koivupuro, H.-K., 2011. FOODSPILL—Food Wastage and Environmental Impacts. Henvi Seminar Series, Food and Environment—Sustainable Food Cycle. MTT Agrifood Research, Finland.

Levis, J.W., Barlaz, M.A., Themelis, N.J., Ulloa, P., 2010. Assessment of the state of food waste treatment in the United States and Canada. Waste Manage. 30 (8–9), 1486–1494.

Lundqvist, J. 2008. In SIWI report "Saving Water: From Field to Fork - Curbing Losses and Wastage in the Food Chain," 16th Session of the United Nations Commission on Sustainable Development.

Lundqvist, J., 2010. Producing more or wasting less. Bracing the food security challenge of unpredictable rainfall. In: Martínez-Cortina, L., Garrido, G., López-Gunn, L. (Eds.), Re-thinking Water and Food Security: Fourth Marcelino Botín Foundation Water Workshop. Taylor & Francis Group, London, UK.

Mao, I.F., Tsai, C.-J., Shen, S.-H., Lin, T.-F., Chen, W.-K., Chen, M.-L., 2006. Critical components of odors in evaluating the performance of food waste composting plants. Sci. Total Environ. 370, 323–329.

Mena, C, Adenso-Diaz, B., Yurt, O., 2011. The causes of food waste in the supplier–retailer interface: evidences from the UK and Spain. Resour. Conserv. Recy. 55 (6), 648–658.

Minowa, T., Kojima, T., Matsuoka, Y., 2005. Study for utilization of municipal residues as bioenergy resource in Japan. Biomass Bioenergy 29, 360–366.

Molden, D. (Ed.), 2007. Water for Food, Water for Life: A Comprehensive Assessment of Water Management in Agriculture. Earthscan Publications, IWMI, London, UK and Colombo, Sri Lanka.

Nellemann, C., MacDevette, M., Manders, T., Eickhout, B., Svihus, B., Prins, A.G., 2009. The Environmental Food Crisis—The Environment's Role in Averting Future Food Crises. United Nations Environment Programme (UNEP), Norway.

Nordpil, H.A. 2009. Water requirements for food production 1960–2050. The Environmental Food Crisis—The Environment's Role in Averting Future Food Crises. Stockholm Environment Institute. Sustainable Pathways to Attain the Millennium Development Goals—Assessing the Key Role of Water, Energy and Sanitation.

Parfitt, J., Barthel, M., Macnaughton, S., 2010. Food waste within food supply chains: quantification and potential for change to 2050. Philos. Trans. R. Soc. B. 365, 3065–3081.

Rayner, G., Barling, D., Lang, T., 2008. Sustainable food systems in Europe: policies, realities and futures. J. Hung. Environ. Nutr. 3 (2–3), 145–166.

Rockström, J., Axberg, G.N., Falkenmark, M., Lannerstad, M., Rosemarin, A., Caldwell, I., et al., 2005. Sustainable Pathways to Attain the Millennium Development Goals: Assessing the Key Role of Water, Energy and Sanitation. Stockholm Environment Institute, Stockholm, Sweden.

Schiller, M. (2009). Dear EarthTalk: Food waste has become major issue in United States.

Schröder, M.J.A., 2003. Food Quality and Consumer Value. Delivering Food that Satisfies. Springer, Berlin.

Sugiura, K., Yamatani, S., Watahara, M., Onodera, T., 2009. Ecofeed, animal feed produced from recycled food waste. Vet. Ital. 45, 397–404.

Stuart, T., 2009. Waste: Uncovering the Global Food Scandal. Penguin Books, London.

Thoenissen, R., 2009. Fact Sheet: FoodWaste in the Netherlands. Ministry of Agriculture, Nature and Food Quality. November.

US Department of Agriculture, 2007. School Nutrition Dietary Assessment Study-III. Food and Nutrition Service, Office of Research, Nutrition, and Analysis, Alexandria, VA.

Ventour L. (2008) The Food We Waste: Food Waste Report. v2. WRAP.

WFP 2009. United Nations World Food Programme, 2009 Annual Report.

WRAP (2010). A review of waste arisings in the supply of food and drink to UK households in the UK. Banbury, UK.

Abbreviations and Glossary

ABPR	Animal By-Products Regulation	**FAO**	Food and Agriculture Organization
AD	Anaerobic Digestion	**FDA**	Food and Drug Administration
AI	Artificial Intelligence	**FIW**	Food Industry Wastes
ANN	Artificial Neural Network	**FPU**	Filter Paper Units
ASBR	Anaerobic Sequence Batch Reactor	**FOG**	Fats, Oil, and Grease
ASM	Activated Sludge Model	**FOS**	Fructo-oligosaccharides
ATAD	Autothermal Thermophilic Aerobic Digestion	**FSC**	Food supply chains
		FV	Fruit-and-vegetables
ATTD	Apparent Total Tract Digestibility	**FVW**	Fruit-and-vegetable wastes
ATP	Adenosine Triphosphate	**FW**	Food Waste
BAT	Best Available Technique	**GC**	Gas Chromatography
BOD	Biochemical Oxygen Demand	**GC-FID**	Gas Chromatography with Flame Ionization Detector
BSE	Bovine Spongiform Encephalopathy		
BSG	Brewers' Spent Grain	**GDP**	Gross Domestic Product
BTU	British Thermal Units - unit of energy equal to about 1,055 joules	**GHG**	Greenhouse Gas
		HPLC	High Pressure Liquid Chromatography
CE	Coulombic efficiency	**HRT**	Hydraulic Retention Time
CEM	Cation Exchange Membrane	**IAWQ**	International Association on Water Quality
CFU	Colony Forming Units		
CHP	Combined Heat and Power	**IE**	Industrial Ecology
COD	Chemical Oxygen Demand	**IPCC**	Intergovernmental Panel on Climate Change
CPC	Carotenoid-protein Cake		
CSR	Cellulase-to-solid Ratio	**IPPC**	Integrated Pollution Prevention and Control
CSTR	Continuous Stirred Tank Reactor		
CW	Chicken Waste	**ISO**	International Organization for Standardization
DAF	Dissolved Air Flotation Sludge		
DDGS	Distiller's Dried Grains with Solubles	**KBCS**	Knowledge Based Control Strategies
DF	Dietary Fibre	**LAB**	Lactic Acid Bacteria
DM	Dry Matter	**LCA**	Life Cycle Assessment
DO	Dissolved Oxygen	**LCFA**	Long-Chain Fatty Acids
DWAS	Dehydrated Waste-activated Sludge	**LCI**	Life Cycle Impact
EAMs	Electrochemical Active Microorganisms	**LCT**	Life Cycle Thinking
		LKF	Loquat Kernel Flour
EC	European Commission	**LSR**	Liquor-to-solid Ratio
EC DG ENV	Directorate-General for the Environment of the European Commission	**MBR**	Membrane Bioreactor
		MFC	Microbial Fuel Cells
		MS	Member States
ECN	European Compost Network	**MSW**	Municipal Solid Waste
EEA	European Environment Agency	**NAD**	Nicotinamide Adenine Dinucleotide
EGSB	Expanded Granular Sludge Bed	**NSC**	Nonstructural Carbohydrates
EMS	Environmental Management Systems	**OECD**	Organization for Economic Co-operation and Development
EP	European Parliament		
EPA	Environmental Protection Agency	**OFMSW**	Organic Fraction of Municipal Solid Waste
EU	European Union		
FAE	Feruloyl Esterase	**OLR**	Organic Loading Rate

xxvi ABBREVIATIONS AND GLOSSARY

OMWW	Olive-mill Wastewater
PAT	Process Analytical Technology
PD	Power Density
PEM	Proton Exchange Membrane
PoD	Point of Digestion Concept
PUFA	Poly-unsaturated Fatty Acids
PWM	Plant-wide Modeling Methodology
RBC	Rotating Biological Contactor
RDC	Retail Distribution Centre
SCP	Single Cell Protein
SI	Supporting Information
SL	Sustainable Livelihood
SmF	Submerged Fermentation
SOHO	Self-organizing Hierarchical Open System
SRT	Sludge Retention Time
SS	Suspended Solids
SSF	Solid State Fermentation
TAD	Thermophilic Aerobic Digestion
TAN	Total Ammonia Nitrogen
TKN	Total Kjehldahl Nitrogen

TOC	Total Organic Carbon
TOS	Theory of Sampling
TP	Total Phosphorous
TS	Total Solids
UAF	Upflow Anaerobic Filter
UASB	Upflow Anaerobic Sludge Blanket Reactor
UCO	Used Cooking Oil
UF	Ultrafiltration
UNDP	United Nations Development Programme
UNU	United Nations University
USDA	U.S. Department of Agriculture
VCR	Volumetric Concentration Ratio
VFA	Volatile Fatty Acids
VS	Volatile Solids
WB	Wheat Bran
WFD	Waste Framework Directive
WRAP	Waste and Resources Action Programme, UK
WWTP	Wastewater Treatment Plant

FOOD INDUSTRY WASTES: PROBLEMS AND OPPORTUNITIES

CHAPTER

1

Recent European Legislation on Management of Wastes in the Food Industry

Maria R. Kosseva

1. INTRODUCTION

The European Union (EU) 6th Environmental Action Programme (6th EAP)[1] provides the framework for environmental policymaking in the EU for the period 2002 to 2012 and outlines actions that need to be taken to achieve them. It identifies four environmental issues: climate change; nature and biodiversity; environment and health; and natural resources and waste. These are the priority issues of current European strategic policies. The 6th EAP also promotes the full integration and provides the environmental component of the community's strategy for sustainable development.

The analyses below indicate that today's understanding and perception of environmental challenges are changing—no longer can they be seen as independent, simple, and specific issues. The challenges are increasingly broad ranging and complex, part of a web of linked and interdependent functions provided by different natural and social systems. This implies an increased degree of complexity in the way we understand and respond to environmental challenges (European Environmental Agency [EEA], 2010).

In parallel, existing European environmental policies present a robust basis on which to build new approaches that balance economic, social, and environmental considerations. Future actions can draw on a set of key principles that have been established at the European level: the integration of environmental considerations into other measures; precaution and prevention; rectification of damage at source; and the polluter-pays principle. Waste policies can essentially reduce three types of environmental pressures: emissions

from waste treatment installations such as methane from landfill; impacts from primary raw material extraction; and air pollution and greenhouse gas emissions from energy use in production processes. The use of resources, water, energy, and the generation of waste are all driven by our patterns of consumption and production. Eating, drinking, and mobility are the areas of household consumption with the highest pressure intensities and the largest environmental impact. European policy has only recently begun to address the challenge of the growing use of resources and unsustainable consumption patterns. European policies such as the Integrated Product Policy and Directive on Ecodesign focused on reducing the environmental impacts of products, including their energy consumption, throughout their entire life-cycle. It is estimated that over 80% of all product-related environmental impacts are determined during the design phase of a product (EEA, 2010).

In this chapter, food wastes generated along the food supply chain are defined, and various legal aspects of this waste are presented. Best Available Technique candidates for the food and drink sector are evaluated as a reference point in the environmental permit regulation for industrial installations in EU Member States. This method is used to implement the Integrated Pollution Prevention and Control (IPPC) Directive (96/61/EC). Other EU documents, which we address here, are the Packaging Waste Directive (1994), the Animal By-products Directive (2000), and the Waste Framework Directive (2008).

1.1 Definitions of Food Industry Waste (FIW)

Different definitions of food waste with respect to the complexities of food supply chains (FSCs) exist.

[1]The 6th EAP (2002) is a decision of the European Parliament and Council adopted on 22 July 2002.

Food Industry Wastes.
DOI: http://dx.doi.org/10.1016/B978-0-12-391921-2.00001-9

3

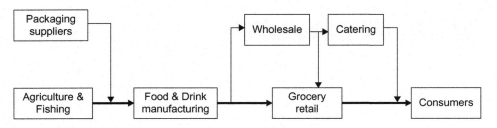

FIGURE 1.1 **The food and drink supply chain.** *Adapted from Mena et al. (2010).*

Food waste occurs at different points in the FSC, although it is most readily defined at the retail and consumer stages, where outputs of the agricultural system are self-evidently food for human consumption. In contrast to most other commodity flows, food is biological material subject to degradation, and different foodstuffs have different nutritional values. Below are five definitions referred to herein:

1. Wholesome edible material intended for human consumption, arising at any point in the FSC that is instead discarded, lost, degraded, or consumed by pests (FAO, 1981).
2. As (1), but including edible material that is intentionally fed to animals or is a by-product of food processing diverted away from the human food (Stuart, 2009).
3. Waste is "any substance or object the holder discards, intends to discard or is required to discard". "Products whose date for appropriate use has expired" targets food waste, with "date" referring to the *expiry date* of a food. This definition is from the EU Council Directive Waste 75/442/EEC [91/156/EEC] (EU, 1991a, b). Clearly any produce that does end up in landfill is a waste and can be quantified accordingly by the tonnage. But quantifying waste in farm-to-retailer supply chains is more difficult because rejection does not necessarily trigger disposal, but redirection to other markets.
4. "Uneaten food and food preparation wastes from residences and commercial establishments such as grocery stores, restaurants, and produce stands, institutional cafeterias and kitchens, and industrial sources like employee lunchrooms" as defined by the United States Environmental Protection Agency (2006).
5. Food or drink products that are disposed of (includes all waste disposal and treatment methods) by manufacturers, packers/fillers, distributors, retailers and consumers as a result of being damaged, reaching their end-of-life, are off cuts, or deformed (outgraded) (WRAP, 2010).

1.2 Waste Streams Considered in This Book

Within the literature, food waste postharvest is likely to be referred to as "food losses" and "spoilage".

Food loss refers to a decrease in food quantity or quality, which makes it unfit for human consumption (Grolleaud, 2002). At later stages of the FSC, the term food waste is applied and generally relates to behavioral issues. Food losses/spoilage, conversely, relate to systems that require investment in infrastructure. In this work, we refer to both food losses and food waste generated along the food and drink supply chain (Figure 1.1) as food industry waste, considering the first two definitions to be most relevant. The method of measuring the quantity of food waste is usually by weight, although other units of measure include calorific value, quantification of greenhouse gas impacts, and lost inputs (e.g., nutrients and water).

2. VARIOUS LEGAL ASPECTS OF FOOD WASTE

Legislation has been used around the world to prevent, reduce, and manage waste (e.g., promoting recycling and energy recovery). In the EU, the Council Directive on Waste (1991), originally introduced in 1975 and revised in 1991, deals with the regulatory framework for the implementation of the European Commission's Waste Management Strategy of 1989. It covers waste avoidance, disposal, and management. The EU Council directive on Hazardous Waste (1991) was introduced to align management of these materials across Member States (MS). Other documents related to food waste include the EU Council Directives on Packaging Waste (1994), Integrated Pollution Prevention and Control (1996), Landfill of Waste (1999), and Animal By-products (2000).

EU Directive 94/62/EC aimed to harmonize national measures concerning the management of packaging and packaging waste in order to prevent any impact thereof on the environment of all MS as well as of third countries or to reduce such impact. MS may encourage a system of reuse of packaging in an environmentally sound manner. Moreover, they should encourage the use of materials obtained from recycled packaging waste for the manufacture of packaging and other products (Arvanitoyannis, 2008).

The EU Council Directive on Animal By-products (2000) categorizes waste into three sections:

- Category 1: High risk, to be incinerated;
- Category 2: Materials unfit for human consumption; most types of this material must be incinerated or rendered;
- Category 3: Material which is fit for but not destined for human consumption.

The UK has its own order for animal by-products introduced in 1999 and amended in 2001 and again in 2003 (Statutory Instrument 2003 No. 1484) (OPSI, 2003), which aims to minimize disease transmission such as bovine spongiform encephalopathy (BSE). The current legislation requires the prevention of feeding livestock with catering waste that has been in contact with animal carcasses or material presenting similar hazards.

2.1 Selecting Best Available Technique Candidates for the Food and Drink Sector

Best Available Techniques (BATs) are an important reference point in the environmental permit regulation for industrial installations in the EU MS, which have to implement the IPPC. BATs correspond to the techniques and organizational measures with the best overall environmental performance that can be introduced at a reasonable cost (Derden et al., 2002). Central to this approach are scores given on technical feasibility, on cross media environmental performances, and on economic feasibility. The approach was tested in the fruit and vegetable processing industry. Their recommendation to map the sector from a technical and economic point of view, in order to understand its structure and financial capabilities, as well as to be able to assess sustainability of decisions taken, was adopted in a study by Midžić-Kurtagić et al. (2010).

Having in mind differences in the technological structure and the environmental priorities between countries, Schollenberger et al. (2008) propose a consistent and flexible assessment method for the evaluation of process improvements based on resource efficiency. They suggest that determination of candidate BATs requires the assessment of parameters from the three pillars of sustainability: economic, ecological, and social. Their logical application indicates that BAT candidate selection should be performed based on ecologic, economic, and social criteria. In this context, a concept of sustainable development can be understood as proposed by Strange and Baley (2008):

- *A conceptual framework:* a way of changing the predominant world view to one that is more holistic and balanced;

- *A process:* a way of applying the principles of integration, across space and time and to all decisions;
- *A goal:* identifying and solving specific problems of resource depletion, healthcare, social exclusion, poverty, etc.

Therefore, the selected method for assessing BAT sustainability should offer the criteria for analysis of relations between different issues and propose adequate solutions. Problems related to the over-exploitation of resources, environment, and human health are interconnected from the point of view of cause and effect, and the solutions should be pursued in technical, institutional, economic, and legal measures, as a multi-criteria procedure.

LaForest and Bertheas (2004) carried out a study to define BAT selection methods. This study revealed a great number of redundancies and heterogeneity in the considerations. They proposed a set of six objectives to which a technology must comply, if selected as a BAT. Those were: (i) limitation of environmental impact, (ii) economy of raw materials and energy, (iii) improvement of safety and risk minimization, (iv) valorization, (v) benchmarking, and (vi) innovation. Regardless of the objectives, the indicators of an existing state must be set up first. Thus, in accordance with Guidance on the Selection and Use of Environmental Performance Indicators (EC Recommendation No 2003/532/EC, 2003) and Practical Guide for the Implementation of an Environmental Management Scheme (Masoliver Jordana, 2001), the indicators of environmental performance, particularly input–output operational performance indicators, were selected as the most suitable set of indicators for the purpose of development of national reference documents on BATs (Midžić-Kurtagić et al., 2010). The environmental performance of food and beverage companies was assessed using input–output operational performance indicators (Masoliver Jordana, 2001; Strange and Baley, 2008), focusing on resource consumption and emissions generated. The available sources of information were: (i) Environmental monitoring reports for individual companies, and (ii) Activity Plans for Reduction of Emissions and Compliance with BAT for individual companies, prepared for the purpose of environmental permission procedure. Twenty-two companies from seven subsectors, including brewery, dairy, fish farming, fish processing, fruit and vegetable processing, meat processing, and slaughterhouses, expressed their willingness to voluntarily participate in the study and become subject to an environmental audit.

The information requested in the environmental auditing questionnaire included (Midžić-Kurtagić et al. 2010):

- *Basic data on facility*, including annual production capacity, number of employees, etc.
- *Description of the facility*, including information on equipment for pollution control, methods of maintenance and cleaning of the equipment and facility, and description of activities and production process.
- *Data on consumption* of raw materials, water, and energy.
- *Current environmental status* at the facility site, including wastewater, solid waste, and air emissions. For each waste flow a study was made of (i) the quantity generated, (ii) the process where it is generated, (iii) the environmental impact, and (iv) management of the waste flows.

The mapping activities revealed some important findings about production and environmental performance in the food and beverage sector in Bosnia and Herzegovina, as well as the suitability of the investigative method applied. Separation of waste streams is implemented in most sectors. Dairies close to rural areas separate whey and sell it to farmers as animal feed. Two large-scale dairies separate whey and make a new range of natural and aromatized whey products.

Only one fruit and vegetable processing industry implemented cleaner production measures aimed at separating the organic solid waste and recycling of packaging waste. Organic waste is given to farmers for composting while a small amount is used for animal feed. Implementation of cleaner production measures reduced the quantity of solid waste being disposed of by 534 tonnes of organic and 51 tonnes of packaging waste per year. The investment payback period was 12 months, with total saving of €9,963 per year (Silvenius and Grönroos, 2003). Slaughterhouses are using water extensively; there is little use of pressured hoses or triggers, especially in small traditional slaughterhouses and meat-processing companies. There is no reuse or recirculation of water.

In order to optimize energy consumption, most of the breweries, fruit and vegetable processors, and large-scale dairies use separate temperature control devices in cooling chambers. Production processes are almost completely automated and heating and cooling processes automatically programmed. However, there are a large number of small-scale dairies that use equipment, including milk pumps, motors, and heating and cooling equipment (pasteurization and drying equipment, refrigerators), that is not optimized to use energy rationally. Experience has shown that energy consumption in dairies can be reduced by 10–30% by employing and improving equipment and procedures with better energy efficiency and less heat waste, with drying air, speed control pumps, etc. Large-scale dairies use pasteurizers in the form of plate heat exchangers with high heat recovery. In the milk drying process, the energy consumption is reduced in the vaporization processes by use of secondary vapour. Large-scale dairies also apply pipeline and equipment insulation (Midžić-Kurtagić et al., 2010),

The mapping of environmental performance, carried out by Midžić-Kurtagić et al. (2010), revealed that the beer production subsector is more environmentally advanced than other subsectors analyzed. The reason may lie in the fact that major beer production companies have Environmental Management Systems in place, which oblige them to prevent pollution as well as to introduce environmental-friendly procedures and train employees to act responsibly in the production process. On the other hand, the slaughtering subsector, which in fact has the highest environmental impacts concerning type of solid waste produced and wastewater loads that can be expected, seems to be the least environmentally friendly and requires significant improvement in that sense.

Midžić-Kurtagić et al. (2010) concluded that the Best Available Techniques that will certainly get on the BAT candidate list will include wastewater stream separation, including economically feasible pollution prevention measures. Strong enforcement of law on waste, in terms of keeping records on waste generation and waste selection/separation at source, must be a priority. This will in the long-run result in an improvement in the waste recycling system and decrease waste quantities to be disposed of at municipal landfills.

3. EFFECTIVENESS OF WASTE MANAGEMENT POLICIES IN THE EUROPEAN UNION

3.1 Adoption of a "Recycling Society" in the EU

The waste management hierarchy is one of the guiding principles of zero-waste practice around the globe (Box 1.1).

In general, recycling levels may have increased across Europe, but there is a long way to go before the EU fully embeds the "recycling society" mentally—not only avoiding producing waste but using it as a resource. Such findings were outlined in a recent European Commission (EC) report (2011)—Supporting the Thematic Strategy on Waste Prevention and Recycling—on member states' performance in the prevention and recycling of waste. According to the

BOX 1.1

THE WASTE MANAGEMENT HIERARCHY

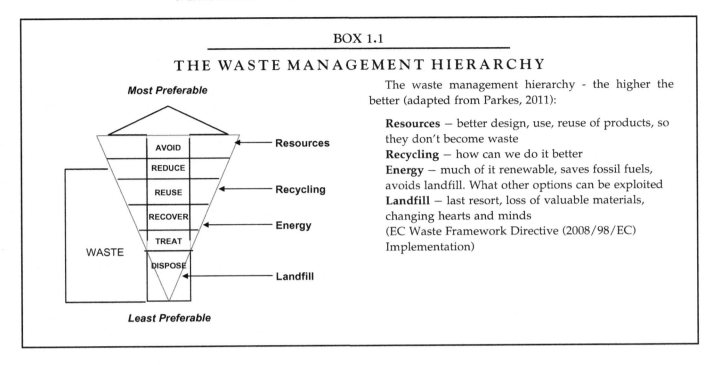

The waste management hierarchy - the higher the better (adapted from Parkes, 2011):

Resources — better design, use, reuse of products, so they don't become waste
Recycling — how can we do it better
Energy — much of it renewable, saves fossil fuels, avoids landfill. What other options can be exploited
Landfill — last resort, loss of valuable materials, changing hearts and minds
(EC Waste Framework Directive (2008/98/EC) Implementation)

report, for most MS, overall waste generation seems to be increasing, or at best stabilizing—but at a lower rate than economic growth. For example, in 2008 waste recycling was estimated at 38%, an increase of 5% compared with 2005 and 18% compared with 1995. For municipal waste, 40% was recycled or composted in 2008, an improvement of 11.4% between 2005 and 2008, with significant disparities between MS: from a few percent up to 70%. In addition, the report found that more consistency between product design and waste policies is needed to further boost recycling. According to the report, future trends in waste generation and treatment show that, without additional waste prevention policies, waste generation is expected to increase by 7% from 2008 to 2020.

3.2 Main Stipulations of the Landfill Directive 1999/31/EC

Diverting waste from landfill is an important element in EU policy on improving the use of resources and reducing the environmental impacts of waste management, in particular, in pursuance of Directive 1999/31/EC on landfill of waste (hereinafter referred to as the Landfill Directive). According to this Directive, MS must reduce the amount of biodegradable municipal waste going to landfill:

- to 75% of the total amount of biodegradable municipal waste generated in 1995 by 2006;
- to 50% of 1995 levels by 2010;
- to 35% of 1995 levels by 2016.

The Waste Framework Directive was revised and the new directive (2008/98/EC) was issued in November 2008. Several of the new provisions in the directive aim to reduce landfilling. Key issues are the introduction of quantitative targets on recycling of selected waste materials from households and other origins. It provided for the development of waste prevention and decoupling objectives for 2020.

The EC has published a green paper on the management of biowaste in the EU (EC, 2008b). Biodegradable waste means any waste that is capable of undergoing anaerobic or aerobic decomposition, such as food and garden waste and paper and paperboard (see the Landfill Directive). In this report, only the biodegradable waste included in municipal waste is addressed. Biowaste means biodegradable garden and park waste, food and kitchen waste from households, restaurants, caterers and retail premises, and comparable waste from food processing plants (see the Waste Framework Directive (2008/98/EC)). It sets out several options to improve biowaste management, including standards for composts, specific biowaste prevention measures, and tighter targets for biodegradable municipal waste sent to landfill. Greenhouse gas emissions are also becoming more and more relevant in waste management planning. Landfilled biodegradable waste produces methane many years after the waste has been deposited. Countries with high dependence on landfill can take positive action against climate change by landfilling less biodegradable waste. Likewise, in countries that have very low landfill rates, waste recycling and energy recovery can help avoid greenhouse gas emissions from the production of virgin material or energy (European Environment Agency, 2008).

Another Directive, 2001/77/EC, on the promotion of electricity produced from renewable energy sources in the internal electricity market, may stimulate waste incineration with energy recovery. The biodegradable fraction of industrial and municipal waste is defined in the directive as a renewable, nonfossil energy source. Production of electricity from incineration of municipal waste contributes to meeting the EU renewable energy target of 12% of total energy supply by 2010. Individual targets have been set for each Member State. According to the EC's 2008 integrated climate change and energy package (EC, 2008a) and the proposed directive on renewable energy sources (EC, 2008c), MS are expected to define ambitious new targets for generating electricity and heat from waste to help achieve the EU's goal of generating 20% of energy from renewable sources by 2020.

3.2.1 The European Environment Agency Report No 7/2009

3.2.1.1 AIMS

Waste policies must be seen in the broader life-cycle perspective of resource use, consumption, and production; prevention and recycling of waste are important elements in this life cycle. There are different routes to divert waste from landfill, including prevention and recycling, other material and energy recovery, and pretreatment. Not all of them are used by all MS. The EEA report (European Environment Agency, 2009) focused on why specific sets of measures were chosen and evaluated, which measures worked well and why, and it explored success factors and reasons for unsatisfactory results.

3.2.1.2 INDICATOR-BASED ANALYSIS

The methods employing favoring and hindering factors and the evaluation of each country/region are presented in detail in a series of background papers given in the EEA report. Individual country/region papers present the objectives, the policy instruments introduced to meet these objectives, and the waste management scene at the time of the transposition of the Landfill Directive. Further, these papers include an evaluation of the implemented policy of that Directive, which is a driver for landfill diversion (http://waste.eionet.europa.eu/publications).

3.2.1.3 INTERVIEWS WITH KEY STAKEHOLDERS

One way of analyzing the process of policy design and implementation is to review the course of actions taken regarding the policy process and objectives (upstream from the policy in place in Figure 1.2) and regarding the implementation of the policy and the outcomes (downstream from the policy in place in Figure 1.2). By describing changes in waste management in terms of a series of actions over time, it is possible to focus on the real actions and therefore choices made by authorities and other stakeholders, thus going beyond declarations of intent.

3.2.1.4 POLICY INSTRUMENTS

The following two case studies illustrate some of the instruments in control of wastes and waste management. A variety of waste management techniques have been adopted across the EU with mixed effects.

GERMAN CASE STUDY The German strategy on biodegradable waste has focused on separate collection and recycling of secondary raw materials (paper and biowaste), mechanical-biological treatment, dedicated incineration with energy recovery of mixed household waste, and banning the landfill of waste with organic content of more than 3%. Separate collection schemes have been successful in achieving very high recycling rates. A landfill ban was adopted in 1993, but due to several loopholes it was not implemented properly. The loopholes were closed with the Waste Landfilling Ordinance (2001), which confirmed the deadline of 1 June 2005 for implementing the landfill ban and included special provisions for landfilling residues from mechanical-biological treatment. Since the

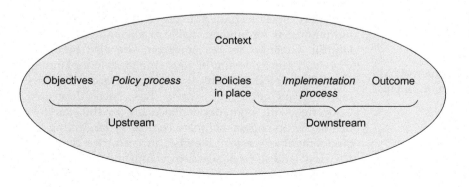

FIGURE 1.2 Policy effectiveness — from objectives to outcome (EEA Report No 7/2009).

deadline, the amount of municipal waste landfilled has fallen to 1%.

HUNGARIAN CASE STUDY The Hungarian waste strategy has focused on building capacity and setting up schemes for separate collection, mainly for packaging waste. An eco-taxation system of product charges has been in operation since 1995. A product charge is levied on certain products that have an impact on the environment, such as packaging materials including beverage packaging for commercial use. If a producer or importer meets the recycling or recovery targets, charges are returned. In practice, therefore, the product charge aims to ensure that recycling targets are met. The charge must be paid by the producer (or importer) and can be passed on to the consumers. Exemptions or discounts apply in the case of eco-labeled products. The Ministry of Environment and Water collects a share of the revenue from the charge and earmarks it for waste recovery and other environmental projects. Since 2003, landfilling of organic wastes has been partially banned. The amount permitted is gradually reducing in line with the interim targets for Biodegradable Municipal Waste (BMW). The National Biowaste Program 2005 includes initiatives for extending separate collection to include garden waste, green waste from public parks, organic kitchen waste, and paper by 2008.

3.2.1.5 LANDFILL TAXES AND GATE FEES

In general, it appears that a combination of policy instruments is required to divert waste from landfills effectively. Economic instruments such as user charges for the management of municipal waste (e.g., "pay-as-you-throw" schemes), landfill tax, and product charges can have a significant role if designed to regulate the behavior of households, waste companies, and producers. For a landfill tax to be effective, the tax level should be relatively high, although public perceptions of the tax burden are arguably as important as the tax rate.

The Landfill Directive provides that MS must ensure that all costs involved in setting up and operating a landfill site, as well as the estimated costs of the closure and aftercare of the site for a period of at least 30 years, are covered by the gate fee. The Waste Incineration Directive sets emission limits and monitoring requirements for pollutants entering air and water, and many plants also have to apply best available techniques according to the Integrated Pollution Prevention and Control Directive.

In 2004, Germany and Italy had the highest gate fees for landfilling at €80–90 per tonne at 2005 prices. Costs were lower in the Flemish Region of Belgium and in Finland at €47–60 per tonne. Hungary and Estonia had the lowest gate fees at €30–36 per tonne.

Reviewing gate fee growth in the decade to 2006, it is interesting to note that fees have rocketed in Estonia by 700%. Finland has experienced a similar change as fees have risen by almost 300%. The increase has been more moderate in the Flemish Region in the last ten years, with a rise of 40%. It seems reasonable to attribute these cost increases to implementation of the Landfill Directive—and anticipation of it. In the Flemish Region, Germany, and Italy, incineration prices are 30–70% higher than landfill gate fees, whereas the price in Finland was lower until 2006, when it rose to 25% higher than landfill. The price increase is the result of increasingly strict environmental standards, for example, investments to abate dioxin and NO_X emissions.

3.2.1.6 PUBLIC ACCEPTANCE

Public acceptance is absolutely crucial in determining what alternatives to landfilling are politically feasible. Communication and information programs therefore clearly have an important role to play in explaining to the general public the true costs and benefits of alternative waste management (and energy generation) strategies.

3.3 European Waste Framework Directive (WFD)

One promising development of the waste-management policies in the European Union is the new WFD, which should have been transposed by December 2010. It has still not passed into national law in many EU countries. MS had a transitional period of 2 years to put the necessary measures in place to comply with the new Directive. The new Directive modernizes and simplifies the approach to waste policy around the concept of "life cycle thinking", and introduces a binding waste hierarchy defining the order of priority for treating waste. The Directive obliges MS to modernize their waste management at the national level, including the requirements of the new WFD. But it will also seek to develop support for MS in designing appropriate strategies and policies upstream. With full implementation of existing acquis, recycling could increase from 40% in 2008 to 49% in 2020, according to the Commission. The Directive obliges MS to modernize their waste management plans and to set up waste prevention programs by 2013 (Europa Press releases RAPID, 2011).

The impact of waste policy (namely the Landfill Directive and the updated WFD), as well as the recommendations contained in the EC communication on future steps in biowaste management in the EU, on food waste generation is neutral. In other words it has no impact on the actual amount of food waste being generated. Waste policy does, however, have a

TABLE 1.1   Estimated Total Impact of Policies on Food Waste Tonnages Going to Landfill in the European Union (Million Tonnes)[1]

	2006	2010	2013	2020
EU12	7.5	5.6	3.0	0.8
EU15	32.7	24.5	13.1	3.2
EU27	40.2	30.1	16.1	4.0

Source: EUROSTAT data.
[1]*Based on 2006 figures, not taking into account socioeconomic changes.*

considerable impact on the treatment of food waste once it has been generated, and this section looks briefly at the potential impacts of likely treatment scenarios. Thus, the combined impact of waste diversion policies on the quantity of food waste going to landfill is estimated as (shown in Table 1.1):

- 25% *reduction in food waste going to landfill by 2010,* in comparison with that produced in 2006 (based on Landfill Directive targets);
- 60% *reduction in food waste going to landfill by 2013,* in comparison with that produced in 2006 (based on Landfill Directive [50%] and WFD [10%] targets);
- 90% *reduction in food waste going to landfill by 2020,* in comparison with that produced in 2006 (based on Landfill Directive [65%], WFD [15%], and future biowaste legislation following from the EC communication on future steps in biowaste management in the EU [10%]).

4. BIOWASTE MANAGEMENT POLICY UPDATES

As the result of a large impact assessment, the EC Directorate General (DG) for the Environment published a "Communication on Biowaste" in May 2010. The conclusion was that there is no major legal obstacle preventing MS from starting biowaste recycling. By contrast, a few months later in July, the European Parliament (EP) and the EP's Environment Committee voted by a vast majority for specific legislation on biowaste, which included quality assurance by the end of 2010. The parliament stated that the rules on the management of biowaste are fragmented and the current legislative instruments are not sufficient to achieve the overall objectives of sustainable management of biowaste. The European Compost Network (ECN), together with a large number of European stakeholders, is fully in line with the parliament's view. One of the potential benefits offered by the optimization of biowaste management is the potential saving of 10–50 million tonnes CO_2. In addition, 3–7% of agricultural soils could be improved. Additionally, optimization could even help meet up to 7% of the 2020

renewable energy, and 42% of the biofuel production targets, if biowaste is processed via anaerobic digestion and the resulting biowaste is used as a biofuel. The situation could be addressed further by the definition of end-of-waste criteria for compost/digestates to meet the legal status of a "product" in the context of the WFD in 2011. This way the compost will be seen as a high quality product fit for use and tradable across EU borders (Barth and Siebert, 2011).

4.1 Landfill Bans on Food Waste

Landfill bans on food waste have been introduced in certain European countries as well as across the Atlantic with mixed results. It remains to be seen, however, if closing the door on landfilled food wastes will provide feedstock to anaerobic digestion (AD) (Burrow, 2011).

Many countries are looking to the AD pacesetters like Germany and Sweden for inspiration. As a result, the idea of landfill bans on food waste has taken a central place. The most successful countries in terms of recycling are those that have introduced landfill bans. These include Germany, the Netherlands, Austria, Sweden, Denmark, France, Norway, Belgium, and various US states and Canadian provinces (Hyder Consulting Pty Ltd, 2010).

It has also sparked some interest among policymakers. If food waste is banned from landfill, there is more chance of meeting the targets under Europe's Landfill Directive, while also securing more renewable energy as the waste is diverted to anaerobic treatment plants.

Dr Hogg (a director at Eunomia, a waste management consultancy) thinks landfill bans are blunt instruments, and care has to be taken to ensure that they don't simply lead to a switch from landfill to incineration. A well-implemented "requirement to sort" organics, accompanied by "end of waste standards", is likely to do more to foster AD than is a ban (Burrow, 2011).

4.1.1 Introduction of New Regulations and the Right Policies

Austria, Germany, and the Holland region of the Netherlands all require the sorting of organic waste by household. Denmark provides an example of where the ban approach may not deliver much in the way of food-sourced AD. There is little source separation of food, and most food waste is incinerated in Denmark. Much also depends upon the relative costs of the alternative treatment routes. Landfill bans have been implemented by many countries in Europe and municipalities in North America and Canada. All have the overall aim of moving waste treatment and management up the hierarchy (to focus on prevention, reuse, and recycling; Box 1.1). To that end, the bans are usually implemented within a framework of existing policy measures, such as landfill taxes. The cost implication to any producer of waste will be the difference between treatment costs and landfill costs, and the latter is largely a function of landfill tax rates. Higher landfill taxes encourage diversion, and thus alternative treatments are more attractive. Therefore, the treatment of commercial and industrial waste has been and will continue to be driven commercially.

The British Government has already declared that it will not introduce wholesale landfill bans anytime soon. Instead it has declared its intention to divert waste from landfill, using the escalation of landfill tax, which will eventually make diversion and recycling a better financial option for the producers. Using landfill tax as a financial driver also ensures that the development of treatment capacity keeps pace with demand. Indeed, landfill bans cannot be introduced overnight; the infrastructure needs to be in place to take the waste for alternative treatment.

Germany is an example of how long a time is required for the implementation of a landfill ban. Gunnel Klingberg (secretary general at Municipal Waste Europe) suggests "the timing between introduction and enforcement of this action may be as long as a decade".

When the Flemish Region of Belgium implemented its landfill ban, legislation allowed landfill operators to apply for exemptions where the required alternative treatment infrastructure was not in place. This has had the effect of slowing down the development of new infrastructure needed to meet the requirements of the ban.

> Different country policies regarding waste management charges may result in market distortions. If landfill bans are enforced before sufficient infrastructure is constructed, that would impede the efficient market disposal of waste. For example, heightened gate fees charged at existing waste treatment facilities lead to an increase in the export of waste to those countries that can dispose of it in a more economic way, illustrated by the transport of waste between Germany and the Netherlands. *Tolvik Consulting Director, Adrian Judge, speaking to Burrow (2011).*

If a comprehensive treatment network is not in place, then countries also run the risk of biowaste being sent for incineration rather than composting or AD because of a lack of capacity, the costs involved with bulking and transportation, or local policy (Burrow, 2011).

In order to develop the infrastructure required, treatment facilities need to be confident that feedstock will be available and at a price that works economically. "Market forces are another complication in the landfill ban tax—and another reason why any ban needs to be part of a wider waste, environmental, and energy policy", suggests Municipal Waste Europe's

Klingberg (Burrow, 2011). "One needs to think about the social, technical, environmental, and social reasons for any decisions. This means that for AD one can't just think about what goes in—it is very important also to think about what comes out, and whether you are producing something the market wants and can afford." When it comes to AD, there are two main outputs: digestate and energy. These can both offer drivers for AD in their own right. Supporting the generation of renewable energy, and the market development of digestate, would make AD "competitive", particularly if combined with taxes or other measures to make landfilling or incineration more expensive.

4.2 Selection of Measures

Many experts, some quoted above, argue strongly that landfill bans must be part of a *balanced* portfolio of policy measures. The countries that have made bans work have installed a policy mix of measures; the ban is just one. In line with that they have developed composting or AD alternatives, supported the use of and the market for compost, prevented the generation of biowaste at source, limited cross-border movements, used fiscal drivers, and supported home composting. This requires joined-up policies through government too, including departments interested in soil quality, waste collection, energy, heat, and resource self-sufficiency. Fiscal, legislative, and social measures are all important in helping shift behavior and ensuring that organic waste is treated correctly and utilized more as a resource and not merely as waste.

In summary, in order to divert food waste from landfill, focusing on its prevention, reuse, and recycling, the following regulations and policies have to be combined:

- Sort-separated collection of organic/food waste
- Stipulation of end-of-waste criteria, and development of end-of-waste standards
- Introduction of landfill tax
- Establishment of balanced gate fees
- Establishment of infrastructure
- Establishment of a comprehensive treatment network
- Use of market forces to develop AD
- Landfill bans
- Others (Burrow, 2011).

4.3 Example of Application of Waste Management Legislation in Ireland

The amount of biodegradable municipal waste disposed of to landfill increased by 5%, or 1.5 million tonnes, in 2007, leaving Ireland in "danger" of missing

---

BOX 1.2
US EPA FOOD WASTE RECOVERY HIERARCHY

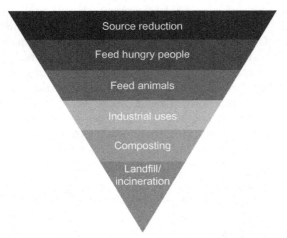

A hierarchy for food waste prevention has been developed by the US Environmental Protection Agency, following the spirit of the EU waste hierarchy as presented in the 2008 WFD. It prioritizes reduction at source and presents a list of preference for use, reuse, recycling, and waste treatment. While this study does not include composting, it should be noted that approximately one third of all food waste is inedible (WRAP, 2009), and thus options such as diversion to animal feed, industrial uses of food waste (e.g., cooking oils), and composting will usually be the environmentally preferable choice. Energy recovery can be another acceptable option where justified by a life cycle thinking approach. The US EPA hierarchy does not differentiate between waste treatment options; anaerobic digestion is likely to be environmentally preferable to incineration and landfilling (US EPA).

WRAP (2009) Household food and drink waste in the UK
Source: US EPA, www.epa.gov/epawaste/conserve/materials/organics/food/fd-gener.htm

---

its EU targets; only ~30% of the waste was recycled. Furthermore, 433,600 tonnes of food waste are sent to landfill each year, at a cost of approximately €1 billion to the country; of this, the catering sector accounts for over 100,000 tonnes of food waste, at a cost of more than €200 million wasted (Statutory Instrument No 508, 2009). Consequently, new regulations were designed to promote the segregation and beneficial use of food arising from the commercial sector and to reduce the amount of biodegradable waste going to landfill. Diversion of this waste type from landfill will help Ireland to achieve targets set down in the EU Landfill Directive 1999/31/EC and form part of the commitments of the National Strategy on Biodegradable Waste 2006. The Waste Management (Food Waste) Regulations 2009 (SI No 508) require that commercial premises:

- segregate and separately store all food waste arising on their premises for
- separate collection by an authorised waste collector.

Ultimately, the Commission intends to improve the implementation of legislation in the waste sector during the coming years. Looking at the current development, the European Parliament's critique of fragmented legislation can be fully supported. An optimized way of meeting the EU landfill directive, an all-in-one biowaste directive, including targets, would be far the best driver to force sustainable biowaste management across MS (Barth and Siebert, 2011).

4.4 Waste Management for the Food Industries in the USA and Canada

The main Directives of the EU and Acts of the USA and Canada on waste management for the food industries are provided by Arvanitoyannis (2008). He stated that the EU legislation is much more flexible and changeable (many amendments in a short period) than the respective US and Canadian legislation. The US Congress puts together the environmental laws. For such legislation to be enacted, lawmakers must perceive that environmental regulation benefits society; only after that will such laws be passed. In 1970, Congress created the US Environmental Protection Agency (EPA). Since then, the EPA has been responsible for enforcing applicable federal laws. In many cases, the laws allow the states to adopt and enforce the federal laws. The US EPA proposed the food waste recovery hierarchy (Box 1.2) based on the EU waste hierarchy presented in WRD 2008.

5. POLICY RECOMMENDATIONS IDENTIFIED FOR THEIR PREVENTION POTENTIAL

BIO Intelligence Service carried out a preparatory study on FW across EU27 (EC DG ENV, 2011). Recently they published a technical reports identifying five policy options for implementation at EU level to strengthen existing efforts to prevent food waste. The following five policy options were examined.

1. *EU food waste data reporting requirements.* EUROSTAT reporting requirements for MS on food waste and standardization methodologies for calculating food waste quantities at MS level to ensure comparability.
2. *Date labeling coherence.* The clarification and standardization of current food date labels, such as "best before", "sell by", and "display until" dates, and the dissemination of this information to the public to increase awareness of food edibility criteria, thereby reducing food waste due to date label confusion or perceived inedibility.
3. *EU targets for food waste prevention.* The creation of specific food waste prevention targets for MS as part of the waste prevention targets for MS by 2014, as recommended by the WFD 2008.
4. *Recommendation and subsidy on the separate collection of food waste in the MS.* Recommendation of MS adoption of separate collection of food waste or biodegradable waste for the household and/or food service sector. Subsidy for the development of separate collection and treatment infrastructure.
5. *Targeted awareness campaigns.* Targeted awareness campaigns, aimed at the household sector and the general public, to raise awareness on food waste generation, environmental and other impacts of biodegradable waste, prevention methods, and practical tips to encourage behavior change and a long-term reduction in food waste generation.

The concluding impact analysis of the EC DG ENV (2011) confirmed that the three priority options are data reporting requirements (1), date labeling coherence (2), and targeted awareness campaigns (5).

6. ENVIRONMENTAL MANAGEMENT STANDARDS AND THEIR APPLICATION IN THE FOOD INDUSTRY

The ISO 14000 environmental management family standards exist to help organizations to minimize the negative effect of their operations on the environment (which may cause adverse changes to air, water, or land), to comply with applicable laws, regulations, and other environmentally oriented requirements, and continually to improve on the above.

ISO 14000 is a series of standards, and guideline reference documents, which cover the following:

- Environmental Management Systems
- Environmental Auditing
- Eco Labeling
- Life Cycle Assessment
- Environmental Aspects in Product Standards
- Environmental Performance Evaluation.

ISO 14000 is similar to quality management, as it is related to the comprehensive outcome of how a product is produced rather than to the product itself. The overall idea is to establish an organized approach to systematically reduce the impact of the environmental aspects that an organization can control. Effective tools for the analysis of environmental aspects of an organization and for the generation of options for improvement are provided by the concept of Cleaner Production.

ISO 14001 is the standard against which organizations are assessed. The contents of this standard and how the food scientists and engineers can use these regulations in order to comply with them have been described elsewhere by Gekas and Nikolopoulou (2007).

The food industry has lagged behind other businesses in implementation of the ISO 14000 series. The main environmental challenges of food companies are water availability, wastewater discharge, air emissions, by-product utilization, solid waste disposal, and food packaging materials. These food industry management problems can be solved through ISO 14000. Laboratories working on this implementation will be strongly encouraged to register to ISO 14000 standards in order to enhance their business opportunities by performing life-cycle assessment among other tests.

Although the food industry was not one of the first industrial sectors to implement EMS/ISO 14000, a decade ago more than one thousand food companies in industrialized countries worldwide had already applied the ISO 14001 (Boudouropoulos and Arvanitoyannis, 2000).

Another standard of the ISO 14000 family, ISO 14040.2, has defined the Life Cycle Assessment (LCA) principles and guidelines. The LCA exercise, with focus on FIW, is presented in Chapter 15.

7. CONCLUSIONS

The goal of the European waste-related legislation is to protect public health and the environment, nature, and biodiversity and to mitigate climate change. Scientific methods of waste disposal or reuse should have a central place in environmental legislation. The legislator must first determine whether there are defensible scientific grounds for asserting that an environmental problem exists and then defend specific policy choices reflected in a bill that proposes to address the problem (Arvanitoyannis, 2008). However, government spending on environment and energy research and development typically remains at less than 4% of total government spending on research and development, which has declined dramatically since 1980. At the same time, research and development expenditure in the EU, at 1.9% of GDP, lags way behind the Lisbon strategy target of 3% by 2010 and behind major competitors in green technologies such as the USA, Japan, and, recently, China and India (European Environment Agency, 2010). By contrast, in many areas (such as air pollution reduction, water and waste management, eco-efficient technologies, green infrastructure, eco-tourism, and others), Europe already has first-mover advantages. The EU has more patents related to air/water pollution and waste than any other economic competitor (OECD, 2010). EU legislation is often replicated in China, India, California (USA), and elsewhere, highlighting further the multiple benefits of well-designed policies in the globalized economy.

Past experience shows that it often takes about 20–30 years or more from the framing of an environmental problem and scientifically based early warnings to a first full understanding of impacts. Interconnected policies that take the long-term view are monitored based on risk and uncertainty and have built-in interim steps for review and evaluation. This can help to manage the trade-offs between the need for long-term coherent action and the time it takes to put measures into place.

The main EU Directives considered in this chapter are related to diverting FIW from landfill, to minimization of packaging waste, and to integrated pollution prevention and control. The waste policy has no impact on the actual amount of food waste being generated, as it is not designed to function as a preventive measure but rather as an incentive mechanism for disposal and reuse. Consequently, waste policy has a considerable impact on the treatment of food waste once it has been generated. The EU Landfill Directive is one of the stakeholder key drivers for waste minimization, management, and co-product recovery in food processing (Waldron, 2007).

There are large differences between MS. Recycling rates vary from a few percent up to 70%. In some MS landfilling has virtually disappeared; in others more than 90% of waste is still buried in the ground. This shows a significant margin for progress beyond the current EU minimum collection and recycling targets. The introduction of a combination of economic and legal instruments used by the best performing MS should be encouraged, including landfill bans and applying the producer responsibility concept to additional waste streams across the EU. More consistency between product design and waste policies is needed to further boost recycling. The new WFD modernises and simplifies our approach to waste policy around the concept of "life cycle thinking". The Directive introduces a binding waste hierarchy defining the order of priority for treating waste. Top of the list is waste prevention, followed by reuse, recycling, and other recovery operations, with disposal such as landfill used only as the last resort. Waste is therefore increasingly also seen as a production resource and a source of energy. The EU brings together waste and resource use policies through the Thematic Strategy on the prevention and recycling of waste and the Thematic Strategy on the sustainable use of natural resources. The Directive obliges MS to modernize their waste management plans and to set up waste prevention programmes by 2013. They must also recycle 50% of their municipal waste by 2020 (Europa Press releases RAPID, 2011).

The next steps of the Commission are to continue to monitor the implementation and enforcement of waste legislation at national level, including the requirements of the new WFD. But it will also seek to develop support for MS in designing appropriate strategies and policies upstream. To further consolidate its waste policies, the Commission will make further proposals in 2012 including setting out the steps it will take in order to move closer to an EU resource-efficient recycling society (Europa Press releases RAPID, 2011), given that:

- Before implementations, likely scientific as well as economic effects of directives must be assessed.
- Mixed country policies distort the market and prevent economic efficiency gains in waste disposal.
- EU superstructures are in a good position to coordinate assessment studies and push countries towards a unified, European-wide policy of waste disposal, which takes into account the different disposal infrastructures available across MS and likely time lags between implementation.
- More specifically, efforts to push legislation with a likely strong impact, such as landfill bans, must be properly assessed and implemented according to MS capabilities to scientifically process waste into renewable energy and fine chemicals. It may be necessary to encourage or even subsidize the construction of recycling infrastructure.

References

Arvanitoyannis, I.S., 2008. Waste management for the food industries. Food Science and Technology, International Series. Academic Press, Elsevier, USA.

Barth, J., Siebert, S., 2011. Biowaste management returns to EU policy agenda. Waste Management World, 19–21.

Boudouropoulos, I.D., Arvanitoyannis, I.S., 2000. Potential and perspectives for application of Environmental Management System (EMS) and ISO 14000 to food industries. Food Rev. Int. 16 (2), 177–237.

Burrow, D., 2011. Landfill bans: handle with care. Waste Management World, 22–26. Available on-line at <www.waste-management-world.com>.

Derden, A., Vercaemst, P., Dijkmans, R., 2002. Best Available Techniques (BAT) for the fruit and vegetable processing industry. Resour. Conserv. Recycl. 34 (4), 261–271.

Directive 2008/98/EC of the European Parliament and of the Council of 19 November 2008 on waste and repealing certain Directives. Official Journal of the European Union, L 312/3. 22.11.2008.

EC, 2008a. Communication from the Commission to the European Parliament, the Council, the European Economic and Social Committee and the Committee of the Regions — 20 20 by 2020 — Europe's climate change opportunity. COM, 30 final. Commission of the European Communities, Brussels.

EC, 2008b. Green paper on the management of biowaste in the European Union. COM, 811 final. Commission of the European Communities, Brussels.

EC, 2008c. Proposal for a directive of the European Parliament and of the Council on the promotion of the use of energy from renewable sources. COM, 19 final. Commission of the European Communities, Brussels.

EC DG ENV, 2011. Preparatory study on food waste across EU 27.

EC Recommendation No 2003/532/EC, 2003. On guidance in the selection and use of environmental performance indicators in EMAS, Official Journal of the European Union 184 (25.6.2003), 19.

EU Council Directive 91/156/EEC, amending Directive 75/442/EEC, 1991a. Available at: <http://eur-lex.europa.eu/LexUriServ/LexUriServ.do?uri=CELEX:31991L0689:EN:HTML> [last visited 27/10/2009].

EU Council Directive on Animal By-Products (2000/76/EC), 2000. Available at: <http://www.central2013.eu/fileadmin/user_upload/Downloads/DocumentCentre/OP_Resources/Incineration.Directive_2000_76.pdf> [last visited 29/10/2009].

EU Council Directive on Hazardous Waste (91/689/EEC), 1991b. Available at: <http://eur-lex.europa.eu/LexUriServ/LexUriServ.do?uri=CELEX:31991L0689:EN:HTML> [last visited 29/10/2009].

EU Council Directive on Integrated Pollution Prevention and Control (96/61/EC), 1996. Available at: <http://eur-lex.europa.eu/LexUriServ/LexUriServ.do?uri=CELEX:31996L0061:EN:HTML> [last visited 29/10/2009].

EU Council Directive on Landfill (1999/31/EC), 1999. Available at: <http://eurlex.europa.eu/LexUriServ/LexUriServ.do?uri=OJ:L:1999:182:0001:0019:EN:PDF> [last visited 29/10/2009].

EU Council Directive on Packaging and Packaging Waste, 1994. Available at: <http://eur-lex.europa.eu/LexUriServ/site/en/consleg/1994/L/01994L0062-20050405-en.pdf> [last visited 29/10/2009].

Europa Press releases RAPID, 2011. EU moving towards 'recycling society' but room for progress remains. Brussels, Belgium. Available at: <http://ec.europa.eu/environment/waste/strategy.htm>

European Environment Agency, 2008. Better management of municipal waste will reduce greenhouse gas emissions, EEA Briefing 1/2008, EEA, Copenhagen, Denmark.

European Environment Agency, 2009. Report No.7/2009. Diverting waste from landfill. Effectiveness of waste-management policies in the European Union. Copenhagen, Denmark.

European Environment Agency, 2010. Report. The European Environment. State and Outlook 2010 Synthesis. EEA, Copenhagen, Denmark.

FAO, 1981. Food loss prevention in perishable crops. FAO Agricultural Service Bulletin, No. 43, FAO Statistics Division.

Gekas and Nikolopoulou, 2007. Introduction to food waste treatment: the 14001 standards. In: Oreopoulou, V., Russ, W. (Eds.), Utilization of by-Products and Treatment of Waste in the Food Industry. Springer.

Grolleaud, M., 2002. Post-harvest losses: discovering the full story. Overview of the Phenomenon of Losses During the Post-Harvest System. FAO, Agro Industries and Post-Harvest Management Service, Rome, Italy.

Hyder Consulting Pty Ltd, 2010. Landfill Ban Investigation. Report for the Department of Sustainability, Environment, Water, Population and Communities. Available at: <http://www.environment.gov.au/wastepolicy/publications/pubs/landfill-ban.pdf>

LaForest, V., Bertheas, R., 2004. Integrated environmental regulation–how to define Best Available Techniques? 9th European Roundtable on Sustainable Consumption and Production, Bilbao 12–14 May, 2004.

Masoliver Jordana, D., 2001. Practical Guide for the Implementation of an Environmental Management Scheme, Eco-management Manuals: 2, Generalitat de Catalunya Departament de Medi Ambient.

Mena, C., Adenso-Diaz, B., Yurt, O., 2010. The causes of food waste in the supplier–retailer interface: evidences from the UK and Spain. Resour. Conserv. Recycl. doi: 10.1016/j.resconrec.2010.09.006.

Midžić-Kurtagić, S., Silajdžić, I., Kupusović, T., 2010. Mapping of environmental and technological performance of food and beverage sector in Bosnia and Herzegovina. J. Clean. Prod. 18 (15), 1535–1544.

OECD, 2010. OECD Science, Technology and Industry Outlook, OECD Publishing.

OPSI, 2003. The Animal By-Products (Identification) (Amendment) (England) Regulations. Statutory Instrument 2003 No. 1484. Available at: <http://www.opsi.gov.uk/si/si2003/20031484.htm> [last visited 28/10/09].

Parkes, L., 2011. WFD Implementation, Waste & Resource Management, Environment Agency. Awareness Raising Events concerning the application and enforcement of the "new" Waste Framework. Directive 2008/98/EC - UK 2011.

Schollenberger, H., Treiza, M., Geldermann, J., 2008. Adapting the European approach of best available techniques: case studies from Chile and China. J. Clean. Prod. 16 (17), 1856–1864.

Silvenius, F., Grönroos, J., 2003. Fish Farming and the Environment, Results of Inventory Analysis, Finnish Environment Institute, Helsinki. Finland.

Statutory Instruments (S.I. No. 508), 2009. Waste management (food waste) regulations 2009.

Strange, T., Baley, A., 2008. Sustainable Development, Linking Economy Society and Environment. OECD, Paris.

Stuart, T., 2009. Waste: Uncovering The Global Food Scandal. Penguin Books, London.

United States Environmental Protection Agency, 2006. Available at: <http://www.epa.gov/OCEPAterms/fterms.html>. (Retrieved 2009-08-20).

Waldron, K.W., 2007. Handbook of waste management and co-product recovery in food processing (Volume 1). Woodhead Publishing Series in Food ScienceTechnology and Nutrition No. 141.

WRAP, 2009. Household food and drink waste in the UK. Banbury, UK.

WRAP, 2010. A review of waste arisings in the supply of food and drink to UK households in the UK. Banbury, UK.

Development of Green Production Strategies

Maria R. Kosseva

1. INTRODUCTION

This chapter focuses on the development of green food production strategies, which use a holistic approach while applying principles of industrial ecology and maintaining the integrity of the biosphere. The principles of industrial ecology are essential for development of life cycle thinking and green technologies. Prior to stepping out on a "sustainable Green Economy path", we need to introduce ecosystems as self-organizing systems to define sustainable livelihood and ecological integrity. The goal of green production is to fulfill our requirement for products in a *sustainable way*; consequently, we have to provide green production principles and criteria, distinguish key steps to sustainability, and describe a *holistic approach* in food production. Utilizing waste as a resource for recovery of energy and other materials, one can introduce a promising *upgrading concept*. For instance, the upgrading of vegetable residues creates a secondary use for the "waste products". An example of application of the principles of industrial ecology is integrated processing in *biorefineries*, which is rapidly gaining interest.

The European Commission estimates that about one third of the EU's 2020 target for renewable energy in transport could be met by biogas from biowaste. Thus, an *anaerobic digestion* (AD) strategy for treatment of biodegradable waste has shown both environmental and socioeconomic benefits with successful examples of application all over the globe. We use biogas production technology to illustrate energy conversion in the treatment of food wastes, including food and farm co-digestion, as an example of sustainable biowaste management in Europe. Investigation of the energy efficiency in the food supply chain (FSC) is discussed in Case Study 1. It provides calculations of energy embedded in the food production/consumption system in the USA and energy locked in household/retail food waste in the UK.

2. ENGINEERING DESIGN PRINCIPLES FOR INDUSTRIAL ECOLOGY

2.1 History and Definitions of Industrial Ecology

Erkman and O'Rourke et al. traced the roots of industrial ecology (IE) back to the early 1970s (O'Rourke et al., 1996; Erkman, 1997). Perhaps the first use of the term IE was by Japanese research and planning groups studying how to reduce their country's dependence on resources (Watanabe 1972). During the same period, an ecological scientist and physicist, James J. Kay, proposed to start a "new branch of engineering. This branch aims to bring together ecology, economics, engineering design systems theory and thermodynamics. It is responsible for providing engineers in the field with the methodology necessary for designing, implementing, and maintaining eco-compatible systems" (Kay, 1977). One strand of development of IE can be traced to the Institute of Electrical and Electronic Engineers (IEEE), Systems, Man, Cybernetics Society. IEEE has defined industrial ecology as "... the objective, multidisciplinary study of industrial and economic systems, and their linkages with fundamental natural systems" (Kay, 2002).

Frosch and Gallopoulos (1989) popularized industrial ecology in a *Scientific American* article. They expressed a vision that an industrial system behaves like an ecosystem, where the wastes of a species may be a resource to another species, and the outputs of one industry can be used as the inputs of another, thus reducing use of raw materials, pollution, and saving on waste treatment. Central to this field were the ideas of Ayres, which, using a related but different biological

Food Industry Wastes.
DOI: http://dx.doi.org/10.1016/B978-0-12-391921-2.00002-0

metaphor, he had come to call industrial metabolism (Ayres, 1989). One of the principles of IE is the view that societal and technological systems are bounded within the biosphere and do not exist outside of it. Ecology is used as a metaphor due to the observation that natural systems reuse materials and have a largely closed-loop cycling of nutrients. IE approaches problems with the hypothesis that by using principles similar to those of natural systems, industrial systems can be improved to reduce their impact on the natural environment. The metabolism of the industrial system would be also described through detailed material balances, which could be compiled for a production unit, such as a factory, or a geographic unit (Duchin and Hertwich, 2003).

In 1991, the US National Academy of Science's colloquium on IE constituted a defining moment in the development of IE as a field of study. Since the colloquium, members of industry, academia, and government have sought to further characterize and apply it. In early 1994, The National Academy of Engineering published *The Greening of Industrial Ecosystems* (B. Allenby and D. Richards, eds.). The book brings together many earlier initiatives and efforts to use systems analysis to solve environmental problems. It identifies tools of IE, such as design for the environment, life cycle design, and environmental accounting. It also discusses the interactions between IE and other disciplines such as law, economics, and public policy. A series of papers were also published after the colloquium, which enclosed the following definition:

> Industrial ecology is a new approach to the industrial design of products and processes and the implementation of sustainable manufacturing strategies. It is a concept in which an industrial system is viewed not in isolation from its surrounding systems but in concert with them. Industrial ecology seeks to optimize the total materials cycle from virgin material to finished material, to component, to product, to waste product, and to ultimate disposal .... Characteristics are: (1) proactive not reactive, (2) designed in not added on, (3) flexible not rigid, and (4) encompassing not insular. *Jelinski et al. (1992).*

Advances in the understanding of ecology have been made by researchers such as Kay (2002) and Nielsen (2007) based on *complexity* science. For IE, this may mean a shift from a more mechanistic view of systems to one where sustainability is viewed as an emergent property of a complex system (Ehrenfeld, 2004). Kay's description of IE is also based on an ecosystem approach:

> Industrial ecology is the activity of designing and managing human production-consumption systems, so that they interact with natural systems, to form an integrated ecosystem, which has ecological integrity, and provide humans with a sustainable livelihood. *Kay (2002).*

Essentially, it is about human transformation of mass and energy (i.e., industrial activity) from an ecosystem perspective. Let us explore an ecosystem proposal for IE as described by Kay in his chapter on *Complexity theory, exergy and industrial ecology* published in 2002. For this purpose, descriptions of several theoretical issues about ecosystems, sustainability, and complexity follow.

2.2 Complex Adaptive Self-Organizing Hierarchical Open (SOHO) System

Important for the understanding of the issues of complexity and complex systems thinking is to introduce the term *complex adaptive self-organizing hierarchical open (SOHO) system* (Kay et al., 1999).

Spontaneous and coherent behaviour and organization occurs in open systems (such as natural ecosystems and human systems). Open systems are processing a continuing flow of high quality energy—exergy. In these circumstances, coherent behaviour appears in systems for varying periods of time. This behaviour can change suddenly whenever the system reaches a catastrophe threshold and spins into a new coherent behavioural state (Nicolis and Prigogine, 1977). A "catastrophe threshold" is a point of discontinuity at which a continuous change of some variables generates sudden discontinuous responses. An example is the vortex that spontaneously appears in water from draining a bathtub.

Self-organizing dissipative processes emerge whenever sufficient exergy is available to support them. Once a dissipative process emerges and becomes established, it manifests itself as a structure. These structures provide a new context, nested within which new processes can emerge, which in turn cause new structures, nested within which .... Thus emerges a SOHO system, a nested collection of self-organizing dissipative processes/structures organized about a particular set of sources of exergy, materials, and information embedded in a physical environment that give rise to coherent self-perpetuating behaviours. A common example is the emergence of vortex in bathtub water as it drains (Kay, 2002).

2.2.1 Ecosystems as Self-Organizing Systems

Ecosystems can be viewed as the biotic, physical, and chemical components of nature acting together as a nonequilibrium self-organizing dissipative system. As ecosystems develop or mature, they should develop more complex structures and processes with greater diversity, more cycling, and more hierarchical levels, all to abet exergy degradation. Species that survive in an ecosystem are those that funnel energy in their own production and reproduction and contribute to auto-catalytic processes, which increase the total exergy degradation of the ecosystem. In short, ecosystems develop in a way that systematically increases their ability to degrade the incoming solar exergy (Kay and Schneider, 1992). Energy input and output will always balance according to the First Law of Thermodynamics, or the energy conservation principle. Exergy output will not balance the exergy input for real processes, since a part of the exergy input is always destroyed, according to the Second Law of Thermodynamics for real processes.

Keeping in mind that the more processes and reactions of material and energy take place within a system (a metabolism, cycling, building higher trophic levels), the more the possibility of exergy degradation exists, which attributes to maturing of the ecosystem (Kay, 2002).

2.3 Sustainable Livelihood (SL)

A broad definition of SL is given by the United Nations SL programme: sustainable livelihood is the capability of people to make a living and improve their quality of life without jeopardizing the livelihood options of others, either now or in the future.

Sustainability can be defined as:

- Economic efficiency, or use of minimal inputs to generate a given amount of outputs;
- Ecological integrity, ensuring that livelihood activities do not irreversibly degrade natural resources within a given ecosystem; and
- Social equity, which suggests that promotion of livelihood opportunities for one group should not foreclose options for other groups, either now or in the future (UNDP, 1998).

Sustainable livelihood is a socioeconomic impetus behind industrial ecology (Kay, 2002).

2.4 Ecological Integrity

Ecological integrity is a biophysical purpose of industrial ecology. Ecological integrity is about three aspects of the self-organization of an ecological system: (1) current well-being; (2) resiliency; and (3) capacity to develop, regenerate, and evolve (Kay and Regier, 2000).

Together the philosophy of ecological integrity and sustainable livelihood form the normative basis for industrial ecology.

2.4.1 A Conceptual Model of Industrial Ecology

The SOHO system model provides a conceptual basis for discussing ecological integrity and human sustainability; see Figure 2.1 (Kay, 1996). It furnishes us with an integrated ecosystem description of the relationship between natural and human systems. As such, it can serve as a basis in industrial ecology for scrutinizing these relationships.

Each self-organizing entity resides in an environment that provides: (a) the biophysical surroundings in which the entity exists; and (b) the flow of exergy, material, and information upon which the entity depends for the continuation of the self-organizing processes that maintain its structure. The biophysical surroundings, in conjunction with the flows into the system, constitute the context for the self-organizing entity. Referring to Figure 2.1, the relationships between societal and ecological systems are threefold:

- Ecological systems provide the context for societal systems.
- Societal systems can alter the structures in ecological systems.
- Societal systems can alter the context for the self-organizing processes of ecological systems (Kay, 2002).

FIGURE 2.1 Conceptual model of the ecological–societal system interface—a single horizontal level. *Source: Kay (1996).*

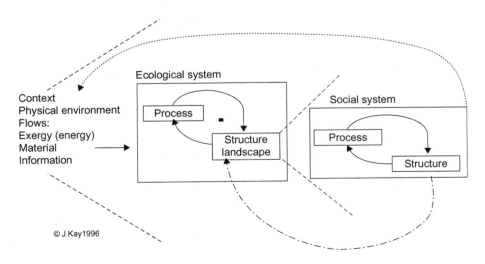

2.5 Design Principles and Tools for Industrial Ecology

Ehrenfeld (1997), van Berkel et al. (1997), and Allenby (1999) proposed the following design principles and tools, derived from the systematic applications of systems theory. They explicitly deal with the implications of the second law of thermodynamics, hierarchy, and attractors.

The design of the production–consumption system should take account of the following considerations:

- The problem of *interfacing*—the interface between societal system and natural ecosystem reflects the limited ability of natural ecosystems to provide energy and absorb waste before their survival potential is significantly altered, and the survival potential of a natural ecosystem must be maintained.
- *Mimicry*—the behavior and structure of large-scale societal systems should be as similar as possible to those exhibited by natural systems.
- Using *appropriate biotechnology*—whenever feasible the function of a component of a societal system should be carried out by a subsystem of the natural biosphere.
- Nonrenewable resources should be used only as capital expenditures to bring renewable resources on line.

2.5.1 Interfacing

The problem of interfacing is a precautionary principle. Whenever possible we should limit the effluent from societal systems (both waste materials and energy) flowing across the interface into natural systems. We should minimize the displacement on the landscape of natural systems by societal systems. Given that human society is appropriating more than half the photosynthetic capacity of the biosphere, human systems must decrease their use of energy. In short, adoption of the precautionary principle mandates that our designs minimize their ecological footprints.

Effectiveness measures how much exergy is used in the system relative to a theoretical best scenario (it is about the quality). The theoretical best case, according to the second law of thermodynamics, is when all processes are performed reversibly and all the available work (exergy) is extracted from the energy. Efficiency is about how well the quantity of the physical flow is used.

2.5.1.1 FOCUS ON SUBOPTIMIZATION AND EXAMPLE WITH A STUDENT RESIDENCE CAFETERIA

The assumption that, if individual processes and subsystems are made efficient, then the overall system will be efficient, is valid only for linear systems. This is

rarely true in a real physical system. For example, the task was undertaken to change a student residence cafeteria so that less waste would be produced. Observation of the food that remained on plates after students had finished their meals revealed a large amount of untouched food and unopened packages, which ended up in the garbage. Surveys of students revealed that this was because food was served at a fixed price. A redesign of the cafeteria was undertaken so that students paid for what they put on their plates. This reduced the waste from plates after meals by 72%. The redesign seemed to be a great success. However, if one monitored all the waste generated, from the time the food entered the university, through all the processing steps, until it was consumed or disposed of, the changes in the overall food system required by the redesign actually increased the waste generated in some subsystems. When all the waste generated was taken into account, a 45% decrease was observed.

The point is that any time one part of a system is optimized in isolation, another part will be displaced further from its optimum in order to accommodate the change. Generally, when a system is optimal, its components are themselves run in a suboptimal way (Kay, 2002).

Allenby (1999) took this observation one step further. He noted that sustainability, like efficiency, must be a property of the overall nested system, not of each of the subsystems and components.

2.5.2 Mimicry of Natural Ecosystems

Nature is the undisputed master of complex systems, and in our design of a global industrial system we could learn much about how the global natural ecosystem functions (Tibbs, 1992).

Ecosystems are complex, adaptive, self-organizing, hierarchical systems. Their dynamics are largely a function of positive and negative feedback loops.

Mimicry of natural resources should be tempered by an appreciation that humans have a set of priorities, which will cause them to find a different balance between the need to make good use of the resources while coping with a changing environment.

2.5.3 Using Appropriate Biotechnology

Composting and AD have been actively promoted by regional government, thus diverting biodegradable solid wastes from local landfills. Wetlands (both existing and man-made) have been used for sewage treatment plants and for remediation of mine tailing ponds. The experience with appropriate biotechnology has been that it saves much money, both in capital and operating costs.

2.5.4 *Renewable Resources*

When a resource is used at a rate that is less than the rate at which it can be replenished by natural systems, then the use of the resource is renewable. Recycled materials are a renewable resource in so far as the cost of recycling is borne by renewable resource consumption. The human population must be such that it can be supported by renewable resources (Kay, 2002).

Human society must fit in with nature (McHarg, 1999). Humans must understand that the integrity of human societal ecosystems is inextricably linked to the integrity of natural ecosystems. Maintaining the integrity of the biosphere is necessary for the continuation of our society. This means that we must design our physical systems so as to maintain the context for the integrity of the self-organizing processes of natural ecosystems, which are necessary for the continued existence on this planet of self-organizing human ecosystems. This is the task that IE must accomplish, to design an intertwined ecological–social system, which is emerging on our planet (Kay, 2002).

At present, there is a considerable gap between theoretical approaches to IE and what is being implemented in a world in which the value chain of manufacturing companies is increasingly globalized. However, some applications of IE have been attempted through the establishment of "eco-industrial parks" (e.g., in Kalundborg, Denmark). These parks comprise a cluster of companies that seek to harness industrial symbioses through close cooperation with each other, and with the local community, by sharing resources to improve economic performance while minimizing waste and pollution. This idea is also promoted by the United Nations University (UNU) Zero Emissions Forum, which is establishing pilot eco-park projects as well as researching industrial synergies and sustainable transactions (Kuehr, 2007).

3. BARRIERS TO ADOPTION OF INDUSTRIAL ECOLOGY AND DRIVERS OF CHANGE

3.1 Constraints and Incentives for Industrial Ecology

While IE represents many points of view and many types of contributions, nonetheless variants of material balances, namely substance flow analysis and material flow analysis, provide the unifying conceptual and methodological core for IE (Duchin and Hertwich, 2003). The global IE can be modeled as a network of industrial processes that extract resources from the Earth and transform (manufacture) those resources

into commodities, which can be bought and sold to meet the needs of humanity (customers). To be ultimately sustainable, this model should have a cyclic character, as waste to one component of the system represents resources to another; in addition, a healthy recycling and remanufacturing entity is needed. Limited resources are available as an external input flow, which is much smaller than the flows within the system. Thus, IE cannot be studied and optimized in isolation from the human institutions of various kinds that promote or constrain the cyclic materials or energy flows within the entire industrial ecosystem, indicated in Box 2.1. In this context, Jelinski et al. (1992) considered the following constraints and incentives for IE, presented at the colloquium of the National Academy of Sciences, USA:

1. Engineering excellence can often promote cyclic behavior within the manufacturing node by designing processes to promote materials reuse.
2. The desire to avoid toxic wastes may promote process changes to reduce the quantity of wastes or (better) to substitute materials or components that result in less toxic or nontoxic wastes.
3. The economic system may make it difficult to raise capital to alter a process and render it more efficient, that is, to improve its cyclic nature.
4. Taxation may promote raw material flows or import/export flows that are contrary to cyclization of the industrial ecosystem.
5. Government regulations may make reuse of materials so difficult that enhanced waste flow is de facto encouraged.
6. The price system, by failing to include relevant externalities in prices and costs, may preclude adoption of IE by manufacturers and producers.
7. The standard of living of the consumer may encourage long product use or, alternatively, may promote early product disposal.
8. The rapid rate of technological evolution and obsolescence contributes to an enhanced waste stream.

Regulations that are based on incentives rather than on sanctions are those that have proven to be the most effective (Henrichs, 1992). Regulatory matters, from the point of view of risk assessment and public perceptions, constitute an important perspective for IE (Pariza, 1992). Processes and products can be designed to avoid the use of materials with known ecological consequences but cannot be designed to avoid the use of materials whose use might be prescribed by irrational legislation. In this connection, a good case can be made for close cooperation among industry, government, and the public, if real environmental progress is to be achieved.

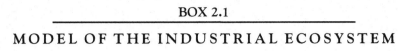

BOX 2.1

MODEL OF THE INDUSTRIAL ECOSYSTEM

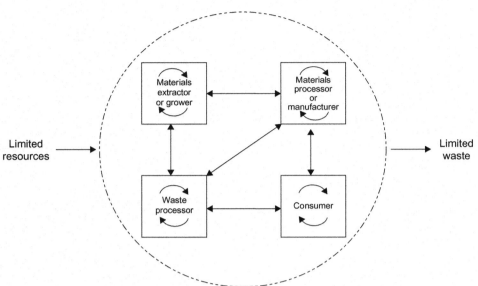

The central domain of industrial ecology is conveniently pictured with four central nodes: the materials extractor or grower, the materials processor or manufacturer, the consumer, and the waste processor. To the extent that they perform operations within the nodes in a cyclic manner or organize to encourage cyclic flow of materials within the entire industrial ecosystem, they evolve into modes of operation that are more efficient and have less disruptive impact on external support systems. Examples range from the recycling of iron scrap by the heavy metals industry to the popularity of garage sales for the reuse of products within the consumer domain. Note that the flows within the nodes and within the industrial ecological system as a whole are much larger than the external resource and waste flows (Jelinski *et al.*, 1992).

The development and implementation of sustainable systems (closed-loop production systems) necessitates a multidisciplinary and multi-organizational approach, in which stakeholders from various industrial sectors, areas of society, and disciplines engage in intelligent and cooperative partnership. Thus no company can become sustainable on its own (OECD, 2009).

3.2 Eco-Innovation as a Driver of Sustainable Manufacturing

To meet the growing environmental challenges, much attention has been paid to innovation as a way of developing sustainable solutions, also known as eco-innovation (OECD, 2009). Eco-innovation is defined as "new products and processes that provide customer and business value, while significantly decreasing environmental impacts" (Charter and Clark, 2007). This concept is

gaining ground in industry and policy making as a way to facilitate the more radical and systematic improvements in corporate environmental performance that are increasingly needed. This has led to understanding eco-innovation in the sense that solutions concern not only technological developments but also non-technological changes such as those in consumer behavior, social norms, cultural values, and formal institutional frameworks. Changes across all these areas cannot be achieved by a single company (Reid and Miedzinski, 2008).

Eco-innovation plays a key role in driving manufacturing industries toward sustainable production. Every shift in environmental initiatives—from traditional pollution control to cleaner production and the establishment of eco-industrial parks—can be characterized as shifts facilitated by eco-innovation. The concept of eco-innovation can help companies and governments to consider and make these shifts through technological advances, changes in management tools, social acceptance of new products and procedures, as well as changes in institutional frameworks for facilitating progressive change (OECD, 2009).

3.3 Drivers of and Barriers to Eco-Innovation

Rennings and Zwick (2003) define five drivers of eco-innovation: regulation, demand from user, capturing new markets, cost reduction, and firm's image. On the other hand, the EU Environmental Technologies Action Plan refers to the following barriers to the introduction and dissemination of environmental technologies: economic barriers, regulations and standards, insufficient research efforts, inadequate availability of risk capital, and lack of market demand. Ashford (1994) provides the following additions to the list of barriers: consumer-related, supplier-related, managerial, and labor force-related barriers.

4. EDUCATING INDUSTRIAL ECOLOGISTS

Education and training programs are critical for eco-innovation. They develop the human capital needed to deliver the eco-innovative solutions and create a potential labor force. Education and training would also be relevant to the demand-side policy, as it builds public concern for environmental challenges and helps shift consumer behavior to a more sustainable mode (OECD, 2009). For example, the US EPA has organized a wide range of programs of environmental education and training. The Green Engineering Program developed a textbook entitled *Green Engineering* to promote green thinking in chemical engineering processes and applications (Allen and Shonnard, 2002).

IE can be taught as a separate course or incorporated into existing courses in schools of engineering, business, public health, and natural resources. Due to the multidisciplinary nature of environmental problems, the course can also be a multidisciplinary offering. Degrees in IE might be awarded by universities in the future (Garner and Keoleian, 1995).

Starr (1992) has written of the need for schools of engineering to lead the way in integrating an interdisciplinary approach to environmental problems in the future. This would entail educating engineers so that they could incorporate social, political, environmental, and economic factors into their decisions about the uses of technology. Current research in environmental education attempts to integrate pollution prevention, sustainable development, and other concepts and strategies into the curriculum. Examples include environmental accounting, strategic environmental management, and

environmental law. It can be argued that the crucial item in educating industrial ecologists is bridging the traditional separation between the study of technology and society.

Industrial Ecology by Graedel and Allenby (1994), the first university textbook on the topic, provides a well-organized introduction to and overview of IE as a field of study. Another good textbook is *Pollution Prevention: Homework and Design Problems for Engineering Curricula* by Allen et al. (1993) (Garner and Keoleian, 1995).

5. GREEN PRODUCTION

In the middle of the 1990s a Greenpeace briefing defined green production systems using the principles of IE (Kruszewska and Thorpe, 1995):

The goal of Green Production is to fulfill our need for products in a sustainable way, i.e. using renewable, non-hazardous materials and energy efficiently while conserving biodiversity. Green Production systems are circular and use fewer materials and less water and energy. Resources flow through the production-consumption cycle at slower rates. In the first place, a Green Production approach questions the very need for the product or looks at how else that need could be satisfied or reduced.

Green Production implements the Precautionary Principle—it is a new holistic and integrated approach to environmental issues centered around the product. This approach recognizes that most of our environmental problems—for example global warming, toxic pollution, loss of biodiversity—are caused by the way and rate at which we produce and consume resources. It also acknowledges the need for public participation in political and economic decision making.

Lately a definition of Green Production with a focus on hazardous substances was provided by Thorpe (2009):

Green production is any practice, which eliminates at source the use or formation of hazardous substances through the use of non-hazardous chemicals in production processes, or through product or process redesign, and thereby prevents releases of hazardous substances into the environment by all routes, directly or indirectly.

5.1 Principles of Green Production

According to Thorpe, in its complete form, Green Production must integrate four underlying principles or key elements:

1. **The Precautionary Principle**—The precautionary principle requires that action should be taken as far as possible to avoid damage to the environment before it occurs and recognizes that there are limitations and uncertainties to scientific

knowledge. For example, information should be acquired in advance and a company should demonstrate the safety of their discharge, rather than require regulators or the surrounding community to prove that the discharge could be harmful.

2. **The Preventive Principle**—It is cheaper and more effective to prevent environmental damage than to attempt to manage or "cure" it.

3. **The Public Participation Principle**—Public access to information about emissions and releases of hazardous chemicals from manufacturing facilities, the amounts and types of chemicals and materials used in production processes, and the chemical ingredients in products is necessary to move to safer alternatives and can hasten the adoption of clean production.

4. **The Holistic Principle**—Green Production is an integrated approach to production, constantly asking what happens throughout the life cycle of the food product. It is necessary to think in terms of integrated systems, which is how the living world functions.

Other elements of a complete clean production system include that it:

- is energy efficient and aims for 100% renewable energy;
- conserves water and other raw materials;
- evaluates the function of the product and seeks non-material ways of fulfilling the product's service (e.g., eliminating hazardous pesticides with organic agricultural techniques);
- recirculates ecologically safe wastes and materials back into the production process;

- reduces consumption in current material-intensive economies while maintaining quality of life and materials;
- protects biological and social diversity.

5.2 Green Production Criteria

Green Production systems for food and manufactured products (Kruszewska and Thorpe, 1995):

- are nontoxic;
- are energy efficient;
- use renewable materials that are routinely replenished and extracted in a manner that maintains the viability of the ecosystem and community from which they were taken, or
- use nonrenewable materials previously extracted but able to be reprocessed in an energy efficient and nontoxic manner.

The products are:

- durable and reusable;
- easy to dismantle, repair, and rebuild;
- minimally and appropriately packaged for distribution using reusable or recycled and recyclable materials.

Above all, Green Production systems:

- are nonpolluting throughout their entire life cycle;
- preserve diversity in nature and culture;
- support the ability of future generations to meet their needs.

The life cycle includes:

- the product/technology design phase;

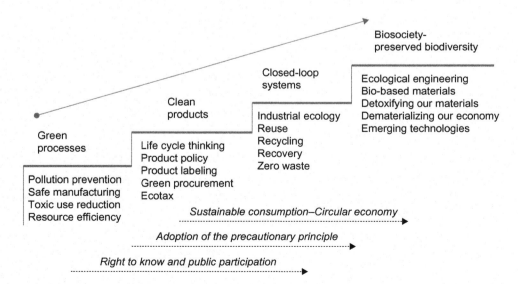

FIGURE 2.2   Steps on the journey to Green Production. *Adapted from Thorpe (2009).*

- the raw material selection and production phase;
- the product manufacture and assembly phase;
- the consumer use of the product phase;
- the societal management of the materials at the end of the useful life of the product.

The complete journey to Green Production can be thought of in four main steps as shown in Figure 2.2 (Thorpe, 2009). The focus of the zero discharges/emissions and losses of hazardous substances is concentrated on the first two steps: Green Production and Clean Products. These are necessary building blocks for a sustainable closed-loop economy wherein all materials are reused and recycled safely and for a bio-society, where ecological protection and economy are given equal weight and are fully integrated.

CHANGING THE PRODUCTION PROCESS Green Production is both a process and a goal. The first steps toward this goal are changes in the production process. These include improvements in housekeeping (preventing leaks and spills), reduction of toxin use, and the introduction of in-house recycling systems to reuse waste waters or heat that would otherwise be dissipated. These initial measures can be undertaken with no or low cost investment and considerable savings.

CHANGING THE PRODUCT While changing production processes, the transition to Green Production also requires examining the product. Whereas traditionally the technical design of a product was aimed at minimizing production costs, today society must move to full cost accounting as a way to understand the environmental, social, and monetary costs of resource depletion and waste generation.

6. SUSTAINABILITY IN THE GLOBAL FOOD AND DRINK INDUSTRY

In the face of growing pressures upon the global population and the world's natural resources, many leaders in the food and drinks industry have been implementing a number of major sustainability initiatives. Sustainability is increasingly considered to be a key strategic priority, and many companies are setting sustainability targets hand in hand with the delivery of commercial objectives (e.g., improving energy efficiency and reducing carbon emissions). The initiatives includes effort (1) to reduce carbon and water footprints, (2) to make innovations in the health and well-being sector, (3) to reduce and recycle packaging, and (4) to make improvements to the supply chain.

Leaders of several companies within the global food and drinks industry are currently pursuing sustainability strategies. These companies include The Coca-Cola Company, ConAgra Foods Inc., Danisco, General Mills, Heineken, Groupe Danone, Kirin Holdings, Kraft Foods, Nestlé, Unilever, United Biscuits, etc. Additionally, sustainability strategies are currently being carried out by a number of the world's leading food retailers, such as Tesco, Wal-Mart, and Carrefour (www.just-food.com).

7. HOLISTIC APPROACH IN FOOD PRODUCTION

When discussing the environmental impact of food production it is important to use a holistic approach. This approach tries to connect different goals, such as highest product quality and safety, highest production efficiency, and the integration of environmental aspects into product development and food production. Within the concept every factor and aspect should be taken into account in a coherent manner. Present R&D in food technology is unthinkable without taking environmental aspects into account. A responsible management of inadequate resources is needed, especially in view of tighter living spaces. Based on these considerations, the holistic concept of food production, shown in Figure 2.3, has been developed (Laufenberg et al., 2003).

This approach tries to connect differing goals, such as highest product quality and safety, highest production efficiency, and the integration of environmental

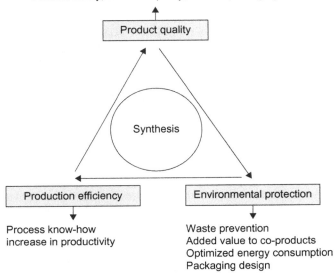

FIGURE 2.3 The holistic concept of food production. *Source: Laufenberg et al. (2003).*

aspects into product development and food production. The recycling of residues is important to every manufacturing branch and includes high developing potential. A systematic reduction of product losses and emissions is profitable under both economical and ecological aspects. Concepts like the differentiation and separate treatment of wastewater streams and task-oriented by-product management support this trend; in this connection special attention is drawn to the recovery of valuable substances or product losses and internal process water recycling (Laufenberg et al., 2003).

7.1 Development of Green Production Strategy

Green or clean production can be considered so far as a strategic element in manufacturing technology for present and future products in several industrial branches. Demand is focused on the development of cost effective technology, the optimization of processes, including separation steps, alternative processes for the reduction of wastes, optimization of the use of resources, and improvement in production efficiency (Paul and Ohlrogge, 1998). Hence current industrial waste management techniques can be classified into three options: (1) source reduction via in-plant modification, (2) waste recovery/recycle, and (3) waste treatment by detoxifying, neutralizing, or destroying the undesirable compounds. The first two options, plant modification and waste recovery/recycle, represent the most promising waste management strategies. Indeed, waste recovery is a particularly attractive option. Significant environmental and economic benefits can accrue from separating industrial wastes with the objective of recycling/reusing these valuable components and/or the bulk of water. Considering the vegetable industry the mentioned goals could be fulfilled by the usual approaches, such as minimization, disposal, feeding, fertilization/composting, closed-loop production, or conversion. Promising concepts include the upgrading of vegetable residues to create a secondary use for the "waste products" (Laufenberg et al., 1999).

7.2 The Upgrading Concept

An important factor for the upgrading process is the development of a procedure using technical standard equipment. The goal of the upgrading is a product with desired, reproducible properties designed under economical and ecological conditions (Laufenberg et al., 2003).

Most vegetable residues consist mainly of water and cellulose and have a poor microbiological quality because of numerous spoilage bacteria on the surface, particularly if stored in the production unit prior to use; thus they quickly decompose in an uncontrolled way. A pretreatment step in the form of inoculation with lactic acid bacteria may produce a more stable substrate, which should be dried to further enhance shelf and storage life. An alternative to fermentation is acidification by acids such as citric, acetic, or ascorbic acid. For sensory reasons and because of the influence on color stability, ascorbic acid is likely to be most useful for food applications. Hence almost any recycling process will start with the steps pretreatment (ensiling), drying, size reduction, and fractionation. The overall recycling strategy, described in Figure 2.4, is designed in a modular manner, subdivided into (1) substance characterization, (2) definition of objectives, (3) product and process design, (4) application, and (5) optimization phases. The result is a final product that is optimized, in regard to the requested product properties, in the demonstrated way as a multifunctional food ingredient (Laufenberg, 2003).

(1) The first phase is mainly the *substance characterization*; based on these data the optimal recycling and application areas and possibilities are worked out. Particle classification, chemical analysis, and physical-chemical properties are the important steps.

(2) Following this, the *definition of objectives* will capture the desired properties of the future food ingredient as well as the food to which it is to be applied. At this point a decision has to be made about use in theory. Based on these "key properties", advantages will arise for technological benefit, health, or taste of a product.

(3) *Product and process design* covers product and dispersion properties as well as how they vary with process parameters. Obvious examples are desirable or undesirable interactions between the food ingredients in general or during processing and interactions with surrounding and processing factors. The range of possible interactions is enormous; thus, concentration on the valuable ingredients as well as on the desired technological, sensorial, and physiological properties is useful. Continuous control and improvement of the upgrading process and product can be achieved by prototype development, definitions of partial qualities, and incorporation of feedback loops.

(4) At the *application phase*, food products and newly designed food ingredients will be combined. At this interaction point the estimated use and practical application in a real food system meet each other. Quality-related properties of the new product have to be assessed and compared with similar products already on the market. Hence a successful launch may be forecast. The sensorial quality is the most important

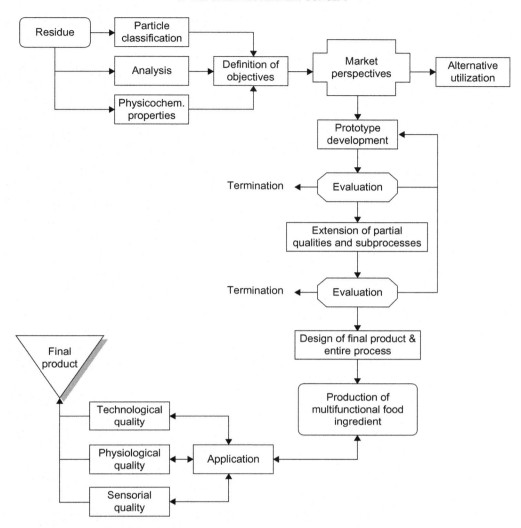

FIGURE 2.4 **Modular strategy for the development of multifunctional food ingredients made of vegetable residues—the upgrading concept.** *Source: Laufenberg (2003).*

criteria for a multifunctional food ingredient applied in a new product.

(5) Since sensorial, technological, and nutritional quality of the new product is compared with a so-called "gold standard", the *optimization* is nearly complete. Final investigations into product properties will answer questions that are useful in establishing benefit to consumers or even a unique selling point. The latter is often science based and hence measurable. Instead of producing a multifunctional food ingredient the goal could alternatively be bioconversion into a food flavor or the development of an operational supply like a bioadsorbent (Laufenberg, 2003).

Different objectives will affect the product and process conception and the application phase. In the following sections the theoretical description of the upgrading concept will be verified by several implementation examples.

8. THE GREEN BIOREFINERY CONCEPT

One key approach, which has applied principles of IE, is integrated processing in *biorefineries*. Analogous to petrochemical refineries producing wide ranges of products from crude oil, the same principles and innovative biotechnology processes can be applied to convert plant molecular structures to a variety of products needed in society. Biorefineries have many advantages over petrochemical refineries and are rapidly gaining interest (Kamm et al., 2006; Lasure and Zhang, 2007). A *biorefinery* can be defined as a processing facility that converts biomass in a highly integrated mode into an assortment of valuable products like food, feed, fuels, fibers for building and insulation purposes, and last but not least organic fertilizers.

An important key phrase here is *process integration*. A biorefinery has the capability of utilizing multiple biological feedstocks or food wastes depending on their availability and market conditions. Furthermore, organic by-products generated in one part of the biorefinery often serve as substrates in another. This is apparent when integrating production of liquid biofuels with biogas. The anaerobically digested biogas substrate can be separated into a liquid, as highly valuable organic fertilizer, and a solid lignocellulose-rich fiber fraction suitable for further enzymatic hydrolysis and subsequent bioethanol fermentation or other purposes. The term *waste products* is more or less non-existent in the context of biorefining.

In this new biotechnological approach, fundamental advantages exist; low process temperature, low energy consumption, and high product specificity are some of the most important factors (Thomsen, 2005). Still, the overall energy efficiency of a biorefinery has to be optimized to ensure economical and environmental feasibility. Applied synergistic and integrated processes can solve many practical problems, provided that the solutions are implemented in the planning phase of the biorefinery concepts (Holm-Nielsen, 2008). An example of a green biorefinery flow diagram showing cereal (and possibly food waste) as input resources, several biotechnological process steps, and the final end product portfolio is illustrated in Chapter 4.

9. ANAEROBIC DIGESTION AND BIOGAS PRODUCTION TECHNOLOGY

Food wastes are a plentiful source of organic material for use as feedstock in anaerobic digestion. By definition, *anaerobic digestion* (AD) is the use of mixed consortia of microorganisms in the absence of oxygen for the stabilization of organic material, by which four main steps of conversion processes simultaneously take place. During these stages, organic substrates are degraded into the end product, biogas. Biogas is a mixture comprising major components CH_4 (50—70%) and CO_2 (30—50%), as well as traces of H_2S and NH_3. Hydrogen sulfide (H_2S), ammonia (NH_3), and other trace gases are present at the level of parts per million (ppm). The exact composition of biogas depends on the feedstock composition (e.g., digestion of carbohydrates results in about 50% CH_4 and 50% CO_2; digestion of pure proteins and fats, around 70% CH_4 and 30% CO_2). Each consortium of microorganisms thrives optimally at a given set of biological, chemical, and physical conditions and with various conversion rates. The various steps of anaerobic digestion are highly complex but take place simultaneously when well managed at biogas plants.

Biogas technology can be applied for many purposes, including production of renewable energy, sustainable utilization of agricultural biomass resources, and environmental protection through treatment of various food wastes and wastewater. In the context of renewable energy production, organic substrates are degraded in the absence of oxygen, producing biogas in the gaseous phase and a nutrient-rich liquid fertilizer applicable for crop cultivation in the liquid phase. Biogas plants are a new class of versatile bioconversion processing plant that can be implemented and optimized for various purposes. For energy production, biogas plants are capable of treating organic wastes, energy crops, and agricultural residues. In the context of wastewater treatment, biogas technology can be applied for the removal of persistent organic pollutants and thus to clean up and/or secure the water environment (Holm-Nielsen, 2008). Applying anaerobic thermophilic digestion to food waste (FW), one can combine the advantages to pasteurize the effluent material with high rates of biodegradation. Such processes provide a method of bacterial sanitation without preceding pasteurization of the incoming organic waste and produce a high-value fertilizer. The most common waste categories used in European biogas production are listed in Box 2.2.

The first full-scale application of AD was in the 1890s, when the city of Exeter in the UK used this technology to treat wastewater. From there, it continued to be widely applied as a way to stabilize sewage sludge, as it is today. The first systems were large, unheated and unmixed tanks with significant operational problems due to solid settling and scum formation. These frequent system disturbances limited the adoption of the technology until the twentieth century (Lusk, 1999). Examples of modern commercial AD technology and available designs are Kompogas, Dranco, Linde-BRV, and Valorga processes (Vandeviviere et al., 2002). European nations have led the way in expanding AD to be a significant part of the organic fraction of municipal solid waste (OFMSW) management. Over 50 plants process MSW either alone or with sewage in Germany, Denmark, France, Spain, Austria, Holland, Belgium, UK, and other European nations. Several types of digesters process between 50,000 and 80,000 tonnes of organic wastes (e.g., source-separated biowastes) per year, with the largest treating 100,000 tonnes annually (De Baere, 2000; van Lier et al., 2001). Some plants accept mixed MSW, for example the Vagron plant, which treats 232,000 tonnes of mixed waste per year, 92,000 tonnes of which are organics (Grontmij, 2004). While anaerobic digestion of OFMSW is relatively well established in other nations, especially in Europe, it remains an undeveloped or developing technique in the USA (van Opstal, 2006). Feedstock security, public perception, and finance are the three interlinked foundations that are key to AD achieving its full potential in the market (see Chapter 7).

BOX 2.2

THE MOST COMMON WASTE CATEGORY USED IN EUROPEAN BIOGAS PRODUCTION

Biological waste waste code*	Waste description	Subcategories
02 00 00	Waste from agriculture, horticulture, aquaculture, forestry, hunting and fishing, food preparation and processing	Waste from agriculture, horticulture, aquaculture, forestry, hunting, and fishing Waste from the preparation and processing of meat, fish, and other foods of animal origin Wastes from fruit, vegetables, cereals, edible oils, cocoa, tea, and tobacco preparation and processing; conserve production; yeast and yeast extract production; molasses preparation and fermentation Wastes from sugar processing Wastes from the dairy products industry Wastes from the baking and confectionery industry Wastes from the production of alcoholic and nonalcoholic beverages (except coffee, tea, and cocoa)

* *The 6-digit code refers to the corresponding entry in the European Waste Catalogue (EWC) adopted by the European Commission.*

Some biowastes, suitable for biological treatment and AD, according to European Waste Catalogue (2007)

10. ENERGY GENERATED BY FOOD AND FARM CO-DIGESTION

The European Commission (EC) estimates that about one third of the EU's 2020 target for renewable energy in transport could be met by biogas from biowaste. Co-digestion of FW with animal slurries could resolve the dilemmas of biodegradation of FW, reducing greenhouse gas (GHG) emissions simultaneously. The digestion of cattle slurry can make a major contribution to targets for reduction of GHG emissions, which in the EU27 in 2008 were estimated to be 50.26 million tonnes CO_2 equivalent from manure management alone. If the process is carried out entirely on the farm, the digestion plant would have to meet the requirements of the Animal By-products Regulation (ABPR) (EC 1774/2002). Under the ABPR, either the feedstock or digestate must be heat treated; this reduces the net energy gained from the process and tends to favor the use of an on-site combined heat and power (CHP) plant, where the waste heat generated can be used for this purpose.

Banks (2011) proposes a solution in the Hub or "point of digestion" (PoD) concept. The basis of this idea is that source-separated FW collected from households is taken to a centralized processing facility (the Hub) where it is homogenized, blended, and pasteurized to ensure it is safe. This could be an existing site such as a landfill or waste-to-energy plant that already has facilities for waste handling and may also have a CHP plant, which produces spare heat for pasteurization. Pasteurized FW is then transported by tanker to a farm, the PoD, where the material is used as feedstock for the digester. Biogas produced is used to meet farm energy needs, with any excess exported. The digestate provides a valuable organic fertilizer with a nutrient balance similar to that required for crop production so that it can be used to replace mineral fertilizers. The Hub and PoD system has both environmental and economic benefits. If managed correctly, the nutrients returned in FW imported onto the farm will balance the fertilizer requirements for crop production, giving farmers enough nitrogen, phosphate, and potash to replace the amounts exported in their produce. This will close the nutrient cycle between towns and countryside.

A dairy farm of about 300 cows would need about two tanker loads of pasteurized FW per week, with no significant effect on traffic movements to and from the farm. Production of biogas from the FW provides a renewable energy source and allows effective recovery of the energy in the manure, in both cases replacing an equivalent amount of fossil fuel and helping to reduce our reliance on centralized energy production. Combining 5 million tonnes of the UK's FW with 40 million tonnes of manures would allow the generation of 3,541 GWh of electricity—enough to supply 913,000 households and to save 1.8 million tonnes of CO_2 equivalent GHG from grid-based electricity production (Banks, 2011).

In economic terms, making on-farm digestion feasible and efficient allows the farmer to make the necessary capital investment in infrastructure and reduces the capital costs to local government and ratepayers, as large centralized plants are not required. In their place is a more cost-effective array of Hubs, which only carry out pretreatment to ensure biosecurity and nutrient balance. The Hub and PoD system could be managed as a separate entity—an "association". This involves waste disposal authorities working in long-term and sustainable routes for recovery of FW with a single point of contact, rather than having to deal with large numbers of individual farmers. The "association"

could reduce management costs and ensure security of the disposal route, while potentially leading to reduced gate fees (Figure 2.5) (Banks, 2011).

One example of co-digestion of manure with source-separated OFMSW is Holsworthy, UK (Farmatic), which co-digests manures and household wastes. The manure is collected from 25—30 local farms within a 5—10 mile radius. The food waste is collected from food processors in the area southwest of Devon in the UK. The total annual inputs to the Holsworthy plant are projected to be 160,000 tonnes of food and animal waste. About 440 tonnes/day of feedstock is added, resulting in a daily biogas production of about 17,840 m^3. This corresponds to a biogas yield of around 40.35 m^3 per tonne of waste input to the plant. The biogas will be used to generate electricity and recover heat from two engines with a total power capacity of approximately 2.1 MW. Expected power production is around 14.4 million kWh/year. Recovered heat is expected to be sold for use in a new district heating system.

Case Study 1: Energy Lost in Food Waste

The Energy Embedded in Food Waste in the USA

Cuellar and Webber (2010) calculated the energy required to produce food in the USA. They organized their estimate for the energy by the following food

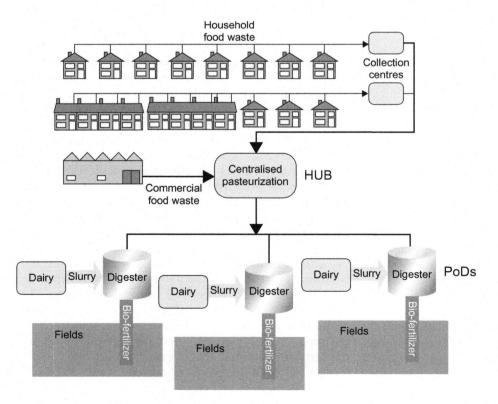

FIGURE 2.5 The Hub and PoD system could form a separate entity—an "association". *Source: Banks (2011).*

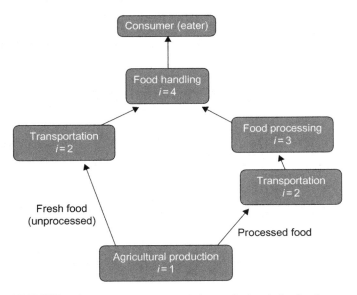

FIGURE 2.6 The pathways used for analysis of the food production system in the USA. *Source: Cuellar and Webber (2010).*

production steps (as shown in Figure 2.6): agricultural production, transportation, food processing, and food handling.

To calculate the energy embedded in wasted food one can calculate the energy required at each of the four food production steps (shown in Table 2.2 and Figure 2.6 as $i = 1$ to $i = 4$) to produce food in the ten categories (shown in Table 2.2 as $j = 1$ to $j = 10$) and then use the food loss percentages (f_j from Table 2.2) to calculate the energy embedded in wasted food. First one defines the energy consumed annually for food production ($E\ tot$) in Eq (1).

$$E_{tot} = \sum_i E_i \qquad (1)$$

Equation (1) states that the energy consumed to produce food is equal to the sum of the energy required for each production step (Ei), i, shown in Figure 2.6 and listed in Table 2.1. For Eq (1) the energy intensity (and therefore embedded energy in wasted food) also varies by food category. Consequently, Eq (1) is rewritten, to account for the differences in energy intensity between food categories, as Eq (2).

$$E_{tot} = \sum_i \sum_j E_{ij} \qquad (2)$$

In Eq (2) the total energy for food production is the sum of the energy required to produce each food category, j (listed in Table 2.2), at each production step, i. However, values for E_{ij} are not available in the literature, and thus they must be deduced. Consequently, E_{ij} is replaced with the total energy required for each

production step, i, and the relative energy intensity for food category j and production step i, A_{ij}, as shown in Eq (3).

$$E_{tot} = \sum_i \sum_j E_i A_{ij} \qquad (3)$$

When one includes the fraction of food lost in each category (f_j) in Eq (3), an estimate is obtained for the energy embedded in wasted food (E_{loss}) as shown in Eq (4).

$$E_{loss} = \sum_i \sum_j E_i A_{ij} f_j \qquad (4)$$

In Eq (4), E_i and f_j can be determined by normalizing and scaling values published in the literature, as shown in Tables 2.1 and 2.2. In this section, Cuellar and Webber (2010) developed reasonable estimates for A_{ij}, which then were used to calculate the total energy embedded in wasted food. Data for the energy consumed in food production are mostly from the year 2002, whereas the available data on food loss are from 1995, and food quantities are given for 2004. In order to minimize error, the energy values for food production were determined for 2002 and then scaled to estimate 2007 values. For the estimate of the energy required to produce food consumed in the USA, data from various sources, including government reports and scientific literature, were compiled (Canning et al., 2010; Heller and Keoleian, 2000; Pimentel et al., 2008; Steinhart and Steinhart, 1974).

The total uncertainty estimated in the energy consumption in 2002 is ± 730 trillion BTU after propagating the calculated and 20% error values throughout all calculations. Total uncertainty was estimated by the following relationship: $U_{tot} = (\Sigma u_i^2)^{1/2}$, where U_{tot} is the total uncertainty and u_i is the uncertainty for the steps listed in Table 2.1. The energy estimate for food production scaled to 2007 energy values is 8080 ± 760 trillion BTU (EIA, 2007). The total food consumed in the USA includes imported food and excludes food exports. To calculate the energy required to produce food consumed domestically, Cuellar and Webber (2010) assumed that the energy intensity of the agricultural production of each food category is the same in the USA and importing countries. Components used to calculate the energy for agriculture production consist of energy production for agriculture chemicals, fuel, fertilizers, electricity, fisheries, as well as domestic and imported agriculture (shown in Table 2.1). Further, the food handling process was split to include the following categories: food services, packaging, and residential energy consumption. In

TABLE 2.1 Energy Necessary for Food Production Steps in the USA and Calculated Energy Lost in Food Waste in 2002, 2004, and 2007

Food production steps	Energy production 2002 (10^{12} BTU)	Energy production 2004 (10^{12} BTU)	Energy production 2004 (%)	Energy production 2007 (10^{12} BTU)	Energy lost 2004 (10^{12} BTU)	Energy lost 2007 (%)	Energy lost 2007 (10^{12} BTU)	Percentage lost of total energy
Agricultural chemicals, fuel, electricity	1160							
Fisheries	18							
Aquaculture domestic	8.8							
Aquaculture imported	55.8							
Agriculture total	**1240**	**1270**	**15.9**	**1285**	**275**	**14**	**284**	**3.5**
Transportation total	**1650**	**1690**	**21.1**	**1705**	**440**	**22**	**447**	**5.5**
Food processing total	**1120**	**1150**	**14.4**	**1165**	**280**	**14**	**284**	**3.5**
Food services	1530	1569	19.6	1585	421		421	
Packaging	684	701	9	708	184		184	
Residential energy consumption	1570	1610	20	1627	410		410	
Total food handling	**3780**	**3880**	**48.6**	**3920**	**1015**	**50**	**1015**	**13**
Total energy	**7790**	**7980**	**100**	**8080**	**2010**		**2030**	**25.5**

Source: Cuellar and Webber (2010).

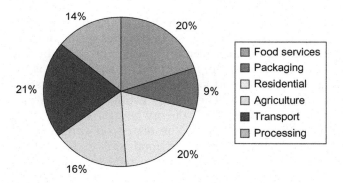

FIGURE 2.7 Distribution of energy used for food production in the USA in 2004.

Table 2.1, the greatest contributor to the total energy required to produce food consumed in the USA is the food handling step ($i = 4$). The greatest single component of the food production step in the same table is energy consumption in residences for food preservation, preparation, and clean up. This result is especially useful because it suggests that consumers can have a direct effect on 20% of the energy required for food production, as shown in Figure 2.7. The USDA estimates that 27% of edible food was wasted in 1995 with fresh foods. A similar result for total energy lost ($\sim 26\%$) in 2007 was obtained in Table 2.1. Because of unaccounted food losses and the potential for increased waste due to economic conditions, the results in the present analysis represent a lower bound of the actual values.

To calculate the energy embedded in wasted food, Cuellar and Webber (2010) used 1995 food loss data provided by the USDA for ten food categories, shown in Table 2.2. These data show that fresh vegetables (23%wt), dairy (20%wt) and grain (14%wt) products, as well as fruits (14%wt) are proportionally the most wasted foods, as depicted in Figure 2.8. The index j is used as an index for the ten different food categories for the calculations using Eqs (2)–(4). The term fj denotes the fraction of the total production for food category j that is wasted.

TABLE 2.2 The Main Food Categories in the USA, Their Energy of Production, and Energy Loss in 2004

Food category, j index $= j$	Energy production total (10^{12} BTU)	Mass food (10^6 t)	Mass loss (10^6 t)	Commodity fraction (%), fj	Mass loss (% weight)	Total energy embedded (10^9 MJ)	Energy loss/food mass (MJ/t)
Grains	837.40	28.30	9.06	32.00	13.45	0.89	98.37
Vegs	1573.50	62.20	15.74	25.30	23.37	1.67	106.37
Fruits	1037.90	41.20	9.64	23.40	14.32	1.10	114.53
Dairy	1363.00	41.60	13.31	32.00	19.77	1.45	108.92
M&P&F	1952.00	43.30	6.93	16.00	10.29	2.08	299.74
Eggs	229.10	4.90	1.54	31.40	2.29	0.24	158.41
Dry beans	35.39	1.00	0.16	15.90	0.24	0.04	236.79
Peanuts	41.13	1.50	0.24	15.90	0.35	0.04	183.46
Sweeteners	566.00	20.90	6.37	30.50	9.47	0.60	94.46
Fat & Oil	353.40	13.00	4.34	33.40	6.45	0.38	86.59
Sum	7988.82	257.90	67.33		100.00	8.50	Av. = 148.76

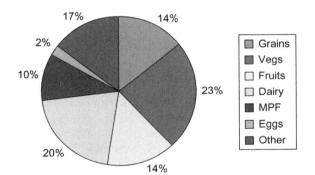

FIGURE 2.8 Distribution of food waste in the USA separated by category (weight %).

Cuellar and Webber (2010) calculated A_{ij} in three different ways for the different production steps. In agriculture some products are far more energy intensive than others. For instance, the production of animal products requires energy to grow the animal's feed and must account for efficiency losses in the animal when converting feed to edible mass. Data from Pimentel (Pimentel, 2007; Pimentel et al., 2008) on the amount of energy necessary to produce a kcal of protein energy were used for subcategories in eight different food categories (grains, fruits, vegetables, meat, dairy, eggs, dry beans/peas, and tree nuts/peanuts), and food mass data obtained from the USDA Economic Research Service (USDA, 2009) were used to calculate the relative energy intensity of each category. To calculate the energy intensity of the food categories listed in this paper, first the energy intensities of

subcategories per mass were calculated using the Eshel−Martin methodology (Eshel and Martin, 2006), and then a weighted average of the energy intensities of the subcategories was calculated to represent the average energy intensity of the eight food categories. A detailed account of the methods for calculating the relative energy intensity of each food category for the agriculture production step is given in the Supporting Information (SI, Cuellar and Webber, 2010) available at http://pubs.acs.org.

The Energy Estimated in UK Retail and Household Food Waste in 2008

The energy embedded in the FW generated in the UK can be approximately estimated using the approach of Cuellar and Webber (2010) and data from the Waste & Resources Action Programme (WRAP). WRAP (2010) reported a total quantity of wastes (18.4 million tonnes) created by the UK food and drink supply chain via their stages: manufacture, distribution, and retail, including household waste. The quantity of food waste, reported in 2008, comprises approximately 11.3 million tonnes, and its distribution through the supply chain is shown in Figure 2.9.

We decided to compare the energy embedded in food wasted through the USA food production scheme (Figure 2.6) with the energy of FW generated in the UK (Figure 2.9), because both countries are economically well developed. By analogy with the USA food handling step, which is the biggest contributor to the total wasted energy (approximately 50%), retail and household waste contribute 72.3% to the total FW in the UK supply chain. Assuming that the energy used

TABLE 2.3 The Main Food Categories in the UK Household and Retail Sectors as Well as Their Energy Loss in 2008

Household food categories UK	Household mass FW (10^6 tonnes)	Household FW energy (10^6 MJ)	Retail food categories UK	Retail mass FW (10^6 tonnes)	Retail FW energy (10^6 MJ)	Total retail & household energy of FW UK (10^6 MJ)
Bakery	1.00	98.37	Bakery	0.0417	4.10	102.47
Vegetables	1.90	202.11				202.11
Fruit	1.30	148.89				148.89
Dairy & egg	0.58	160.40	Dairy	0.0336	3.66	158.71
Meat & fish	0.61	185.84	Meat and fish	0.0286	8.57	191.41
			Meal & other			154.99
			Dry goods	0.0382	9.05	9.05
Drinks	1.30	122.80	Drinks			122.80
Meal & other	1.69	155.86	Produce	0.0901	13.40	13.40
Sum	**8.38**	**1074.25**	**Sum**	**0.2322**	**38.78**	**1103.83**

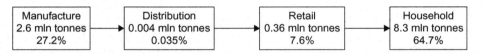

FIGURE 2.9 **Total food waste generated by the UK food and drink supply chain and households in 2008.** *Source: WRAP (2010).*

to produce 1 tonne of similar food category consumed in both countries is comparable, we subsequently compared the following food categories and took the energy embedded in 1 tonne of the wasted food categories as follows:

- Bakery waste (UK) with grain waste (USA)
- Dairy and egg wastes (UK and USA)
- Fruit and vegetable wastes (UK and USA)
- Meat & fish (UK) with meat, poultry, & fish (USA)
- Drinks (UK) with sweeteners (USA)
- Meals & other (UK) with fat & oils (USA)
- Dry goods (UK) with dry beans/peas, etc. (USA)
- Produce (UK) with the average energy value per tonne of all USA FW.

We assume that the energy intensity of the agricultural production of each food category is similar in both countries. The results are shown in Table 2.3, where the values of FW generated in the UK are taken from WRAP (2010). For our estimate, we used a breakdown of total food waste created by the UK grocery retail store and sent to landfill in 2008.

Although the data used were incomplete and likely representing a lower bound on the actual values, we found that the total energy embedded in the UK household and retail FW is around 1.1 billion MJ ($\sim 10^{12}$ BTU), or the equivalent of approximately 307 billion kWh. If we apply AD for treatment of the same

quantity of FW, the volume of biogas produced can generate about 30 million kWh energy, taking into account that one tonne of FW (VS = 80% DM) can produce biogas (61% methane) containing $\sim 3000-3700$ kWh energy. The UK food categories with the greatest embedded energy in their wastes are fresh vegetables, meat/fish, dairy, and fruits. Using the same approach, we estimated that total energy embedded in the UK's FW (11.3 million tonnes) is approximately 40 GWh.

A £12 million anaerobic digestion (AD) facility that will handle 45,000 tonnes of food waste produced by local businesses has recently been opened in Doncaster, England. It will eventually provide enough power for 5,000 homes, while generating 40,000 tonnes of nutrient-rich fertilizer, which will be available for farmers across South Yorkshire.

To summarize, the food category that requires the greatest energy to produce is meat, poultry, and fish. Nonetheless, the food categories with the greatest embedded energy in their waste are dairy and vegetables in the USA. This discrepancy results from the greater proportional waste of dairy and vegetables—32% and 25.3%, respectively (Kantor and Lipton, 1997)—as compared with meat—16% wasted annually (Kantor and Lipton, 1997)—in addition to their high energy requirements. Cuellar and Webber (2010) stated that the data used were incomplete and out of date, likely representing a lower bound on the actual value. The energy embedded in wasted food represents approximately 2% of annual energy consumption in the USA. The energy discarded in the USA's wasted food is more than the energy available from many popular efficiency and energy procurement strategies: for example, the annual production of ethanol from grains (RFA, 2009). The results estimated in the USA suggest that consumers can save up to 20% of the energy required for food production, which they use for food preservation, preparation, and clean up. In the same way, in the UK the food categories with the greatest embedded energy in their waste are fresh vegetables, meat/fish, and dairy products, which are mainly generated by households. The total energy embedded in the UK household and retail FW is around 1.1 billion MJ ($\sim 10^{12}$ BTU).

11. CONCLUSIONS

Industrial ecology is a comparatively new approach to the industrial design of products and processes and the implementation of sustainable manufacturing strategies. The main design principles of the production–consumption system in IE are: *interfacing* between societal system and natural ecosystem, *mimicry* of natural ecosystems, using *appropriate biotechnology*, and *renewable resources* (Kay, 2002). Sustainability is increasingly considered to be a key strategic priority in the food and drink industry. The application of the eco-innovation concept can offer a promising way to move industrial production towards true sustainability, which requires manufacturing industry to integrate and apply the concept in a more holistic way. It involves a conscious reassessment of each phase of the production system, identifying areas in which to apply innovative solutions. It also includes development of new knowledge networks and partnerships that can function as co-creative processes. Government policy initiatives and programs that promote eco-innovation are diverse and include both supply-side and demand-side measures (OECD, 2009).

Using a Green Production strategy (Kosseva, 2011), a bio-based economy combines cost-effective societal and ecological progress, offering lower production costs, considerable reduction of CO_2 emissions, and a new market for agricultural commodities. Integrated processing in biorefineries and anaerobic digestion serve as examples of *appropriate biotechnology* applications. As noted in the recent EC communication on biowaste management (COM(2010)235), producing renewable energy, AD offers a means of realizing a wide range of environmental benefits in a cost-effective manner (Banks, 2011). On-farm co-digestion of FW and animal slurries is a prime example of maximizing these benefits by exploiting the synergies between a centralized processing facility and "point of digestion", which can be managed as a joint "association" for waste treatment. Thus, energy locked in FW can be further recovered, producing methane gas available for heat and power generation.

References

Al Seadi, T., Rutz, D., Prassl, H., Köttner, M, Finsterwalder, T., Volk, S., et al., 2008. In: Al Seadi, T. (Ed.), Biogas Handbook. Published by University of Southern Denmark Esbjerg, Denmark, ISBN 978-87-992962-0-0.

Allen, D.T., Bakshani, N., Rosselo, K.S.T., 1993. Pollution Prevention: Homework and Design Problems for Engineering Curricula. American Institute of Chemical Engineers, American Institute for Pollution Prevention, and the Center for Waste Reduction Technologies, Los Angeles.

Allen, D.T, Shonnard, D.R., 2002. Green Engineering. Prentice Hall, New Jersey.

Allenby, B.R., 1999. Industrial Ecology: Policy Framework and Implementation. Prentice Hall, New Jersey.

Ashford, N.A., 1994. Government Strategies and Policies for Cleaner Production. UNEP, Nairobi.

Ayres, R., 1989. Industrial Metabolism. In: Ausubel, J.H., Sladovich, H.E. (Eds.), Technology and Environment. National Academy Press, Washington, DC.

Banks, C.J., 2011. Food and farm co-digestion. Waste Management World 12, 3.

Canning, P., Charles, A., Huang, S., Polenske, K.R., Waters, A., 2010. Energy Use in the U.S. Food System. U.S. Department of Agriculture, Economic Research Service.

Charter, M., Clark, T., 2007. Sustainable Innovation: Key Conclusions from *Sustainable Innovation Conferences 2003-06*. Center for Sustainable Design, Farnham, UK.

Cuellar, A.D., Webber, M.E., 2010. Wastes food, wasted energy: the embedded energy in food waste. Environ. Sci. Technol. 44, 6464–6469.

De Baere, L., 2000. Anaerobic digestion of solid waste: state-of-the-art. Wat. Sci. Technol. 41, 283–290.

Duchin, F., Hertwich, E., 2003. Industrial Ecology. International Society for Ecological Economics, Online Encyclopaedia of Ecological Economics.

Ehrenfeld, J., 1997. Industrial ecology: a framework for product and processes design. J. Clean. Prod. 5, 85–87.

Ehrenfeld, J., 2004. Industrial ecology: a new field or only a metaphor? Applications of Industrial Ecology. J. Clean Prod. 12, 825–831.

EIA, 2007. Annual Energy Review 2007. Energy Information Administration.

Erkman, S., 1997. Industrial ecology: a historical view. J. Clean. Prod. 5, 1–10.

Eshel, G., Martin, P.A., 2006. Diet, energy, and global warming. Earth Interact. 10, 1–17.

Frosch, R., Gallopoulos, N., 1989. Strategies for manufacturing. Sci. Am. 261, 144–152.

Garner, A., Keoleian, G.A., 1995. Industrial ecology: an introduction. Pollution Prevention and Industrial Ecology. National Pollution Prevention Center for Higher Education, University of Michigan, pp. 1–31.

Graedel, T., Allenby, B., 1994. Industrial Ecology. Prentice Hall, New York.

Grontmij, 2004. Municipal Solid Waste Mechanical Separation and Biological Treatment Plant. VAGRON, Netherlands.

Heller, M.C., Keoleian, G.A., 2000. Life Cycle-Based Sustainability Indicators for Assessment of the U.S. Food System. University of Michigan, Ann Arbor.

Henrichs, R. 1992. Proc. Nat. Acad. Sci. USA 89, 856—859.

Holm-Nielsen, J.B., 2008. Process Analytical Technologies for Anaerobic Digestion Systems. Esbjerg Institute of Technology, Aalborg University, Denmark, PhD Thesis.

Jelinski, L.W., Graedel, T.E., Laudise, R.A., McCall, D.W., Patel, C.K. N., 1992. Industrial ecology: concepts and approaches. Proc. Nati. Acad. Sci. USA 89, 793—797.

Kamm, B., Gruber, P.R., Kamm, M. (Eds.), 2006. Biorefineries—Industrial Processes and Products. Status Quo and Future Directions, 1. Wiley-VCH Verlag.

Kantor, L.S., Lipton, K., 1997. Estimating and addressing America's food losses. Food Rev. 20, 2.

Kay, J., 1977. An investigation into engineering design principles for a conserver society. Systems Design Engineering. University of Waterloo thesis, Waterloo, Ontario, Canada.

Kay, J.J., 1996. Some notes on the ecosystem approach, ecosystems as complex systems. In: Tamsyn, Gallopin (Eds.), Integrated Conceptual Framework for Tropical Agroecosystem Research Based on Complex Systems Theories. SIAT, Cali, Colombia, pp. 69—98.

Kay, J.J., 2002. On complexity theory, exergy and industrial ecology. In: Kibert, C., Sendzimir, J., Guy, B. (Eds.), Construction Ecology: Nature as the Basis for Green Buildings. Spon Press, London, UK, pp. 72—107.

Kay, J.J., Regier, H., 1996, 2000. Uncertainty, complexity, and ecological integrity: insights from an ecosystem approach. In: Crabbe, P. et al., (Eds.), Implementing Ecological Integrity: Restoring Regional and Global Environmental And Human Health. Kluwer, The Netherlands, pp. 121—156.

Kay, J., Raiger, H., Boyle, M., Francis, G., 1999. An ecosystem approach to sustainability: addressing the challenges of complexity. Futures 31, 721—742.

Kay, J.J., Schneider, E.D., 2002. Thermodynamics and measures of ecosystem integrity. In: McKenzie, D.H., Hyatt, D.E., MacDonald, V.J. (Eds.), Ecological Indicators. Amsterdam, Elsevier, Fort Lauderdale Florida, pp. 159—182.

Kosseva, M.R., 2011. Management and processing of food wastes. In: Moo-Young, M. (Ed.), second ed. Comprehensive Biotechnology, 6. Elsevier B.V, Netherlands, Amsterdam, pp. 557—593.

Kruszewska, I., Thorpe, B., 1995. Strategies to promote Clean Production. Briefing. Greenpeace International.

Kuehr, R., 2007. Toward a sustainable society: United Nations Zero Emission's approach. J. Clean. prod. 15, 1198—1204.

Lasure, L.L., Zhang, M., 2007. Bioconversion and biorefineries of the future. United States Department of Energy, Pacific Northwest National Laboratory. Available at: <http://www.pnl.gov/bio-based/docs/biorefineries.pdf>.

Laufenberg, G., Hausmanns, S., Kunz, B., Nystroem, M., 1999. Green productivity concept for the utilisation of residual products from food industry—trends and performance. In: Lecture at the 2nd Asia-Pacific Cleaner Production Roundtable and Trade Expo, Brisbane.

Laufenberg, G., Kunz, B., Nystroem, M., 2003. Transformation of vegetable waste into value added products: (A) the upgrading concept; (B) practical implementations. Bioresour. Technol. 87, 167—198.

Lusk, P., 1999. Latest progress in anaerobic digestion. Biocycle 40, 7.

McHarg, I., 1999. Architecture in an ecological view of the world. In: McHarg, I., Steiner, F.R. (Eds.), To Heal the Earth. Selected writings of Ian L. McHarg, 1999. Island Press, Washington, DC, pp. 175—185.

Nicolis, G., Prigogine, I., 1977. Self-Organization in Non-Equilibrium Systems. J.Wiley & Sons, New York, USA.

Nielsen, S.N., 2007. What has modern ecosystem theory to offer to cleaner production, industrial ecology and society? The views of an ecologist. J. Clean. Prod. 15, 1639—1653.

O'Rourke, D., Connelly, L., Koshland, C., 1996. Industrial ecology: a critical review. Int. J. Environ. Pollut. 6, 89—112.

OECD, 2009. Eco-innovation in industry. Enabling green growth OECD Publishing, Paris, France.

Paul, D., Ohlrogge, K., 1998. Membrane separation processes for clean production. Environmental Progress 17, 137—141.

Pariza, M. 1992. Proc. Nat. Acad. Sci. USA 89, 860—861.

Pimentel, D., 2007. Food, Energy, and Society. Taylor & Francis Ltd., CRC Press, Boca Raton, Florida.

Pimentel, D., Williamson, S., Alexander, C., Gonzalez-Pagan, O., Kontak, C., Mulkey, S., 2008. Reducing energy inputs in the US food system. Hum. Ecol. 36, 459—471.

Reid, A., Miedzinski, M., 2008. Eco-innovation: Final Report for Sectoral Innovation Watch (Brighton: Technopolis Group). Available at: <www.technopolis-group.com/resources/downloads/661_report_final.pdf>.

Rennings, K., Zwick, T., 2003. Employment Impacts of Cleaner Production. ZEW Economic studies, Heidelberg.

RFA, 2009. Ethanol Industry Statistics. Available at: <http://www.ethanolrfa.org/industry/statistics/>.

Starr, C., 1992. Education for industrial ecology. Proc. Natl. Acad. Sci. U.S.A. 89, 868—869.

Steinhart, J.S., Steinhart, C.E., 1974. Energy use in the U.S. food system. Science 184, 307—316.

Thomsen, M.H., 2005. Complex media from processing of agricultural crops for microbial fermentation. Appl. Microbiol. Biotechnol. 68, 598—606.

Thorpe, B., 2009. Clean Production Action. Greenpeace International. Available at: <www.cleanproduction.org>.

Tibbs, B.C, 1992. Industrial ecology: an environmental agenda for industry. Whole Earth Rev., 4—19.

UNDP, 1998. Sustainable livelihood concept paper (web page).

USDA, 2009. Food Availability. U.S. Department of Agriculture.

van Lier, J.B., Tilche, A., Ahring, B.K., Macrie, H., Moletta, R., Dohanyos, M., et al., 2001. New perspectives in anaerobic digestion 43, 1—18.

van Opstal, B., 2006. Evaluating AD system performance for MSW organics. Biocycle 47, 35—39.

Vandeviviere, P., De Baere, L., Verstraete, W., 2002. Types of anaerobic digester for solid wastes. In: Mata-Alvarez, J. (Ed.), Biomethanization of the Organic Fraction of Municipal Solid Wastes. IWA Publishing, London, pp. 111—137.

Watanabe, C., 1972. Industrial-Ecology: Introduction of Ecology into Industrial Policy. Ministry of International Trade and Industry (MITI), Tokyo.

WRAP, 2010. A Review of Waste Arisings in The Supply of Food and Drink to UK Households in the UK. Banbury, UK.

CHAPTER

3

Sources, Characterization, and Composition of Food Industry Wastes

Maria R. Kosseva

1. INTRODUCTION

Food processing industries generate large volumes of mostly biodegradable wastes. Food industry wastes (FIW) are sorted mainly into three categories down the food and drink supply chain: (1) food/drink manufacturing, (2) grocery retail/catering business, and (3) consumer/household wastes. Fruit and vegetables are usually among the most-wasted items, followed by other perishables such as bakery/grains, dairy products, eggs, meat and fish, etc. There is often a large variation in the wastage rates for various food types and different countries. Some hazardous wastes are generated occasionally, when incidents occur or disease spreads. One example is pork meat contamination with carcinogenic chemicals (dioxin and polychlorinated biphenyls) in Ireland in 2008. Another example is microbial contamination of sprout beans by the rare strain *of Escherichia coli* known as *O104:H4* in Germany in 2011.

This chapter is a comprehensive presentation of environmental pollution effects of food wastes, their characteristics, and chemical composition. Important waste streams covered include: fruit and vegetables, apple, onion and potato processing co-products, dairies, meat, poultry and seafood processing by-products, and olive oil manufacturing waste. Knowledge of environmental quality parameters of wastes is required for development of treatment strategies; also knowledge of their chemical composition is essential for the design of novel products and food waste valorization. Therefore, methods used for the analysis of chemical and biochemical compounds of FIW are presented, including biochemical oxygen demand (BOD), chemical oxygen demand (COD), total organic carbon (TOC), total nitrogen and phosphorus, carbohydrates,

lipids, proteins, organic acids, and others. Furthermore, food processing requires a considerable amount of water, which in turn generates a large volume of wastewater containing substantial contaminants in soluble, colloidal, and particulate forms. The degree of contamination depends on the particular operations of the manufacturing process, therefore specific characteristics of dairy, meat, seafood, and distillery wastewater are evaluated in this chapter. Numerous illustrations of the best disposal routes of food waste and wastewater are also extensively discussed.

1.1 Sources of Food Wastes

Most studies of the identity of food types wasted find that the most perishable food items account for the highest proportion of FIW. Some of the most wasted foods in the USA include vegetables (~23 wt%), dairy products (~20 wt%), grain products (~14 wt%), fruits (~14 wt%), and total meat (~10 wt%) (Cuellar and Webber, 2010). Commonly, in the UK, fruit-and-vegetable (FV) wastes are among the most-wasted items, followed by other fresh products like bakery and dairy products, eggs, meat, and fish (Pekcan et al., 2006; WRAP, 2008; Morgan, 2009; Thonissen, 2009). There is often a large variation in the wastage rates for different food types: the Waste and Resources Action Programme (WRAP, 2009) found that over 50% of lettuce/leafy salads (by weight), 36% of bakery, and 7% of milk purchases are wasted. The main sources of FIW in the food supply chain are usually the food and drink manufacturing industry, grocery retailers, and the catering business; on the other hand, food is also wasted by consumers, known as household waste.

Food Industry Wastes.
DOI: http://dx.doi.org/10.1016/B978-0-12-391921-2.00003-2

1.1.1 Household Waste

Although food and drink categories are not fully consistent across studies, variation in household food waste composition in the UK, which represented 64.7% of total FIW in 2008, is shown in Figure 3.1 (WRAP, 2010). The extent to which such differences relate to consumption patterns or different wastage rates cannot be derived from these data alone. Nor do these compositional data distinguish between avoidable and unavoidable waste, the exception being the UK data shown in Table 3.1 (Parfitt et al., 2010).

A very different picture emerges if one considers the value of the waste rather than its weight. The largest contributors by value are "meals" and "meat and fish", due to their high price relative to other food groups. An alternative measure for consideration can be the "avoidable" household food and drink waste generated (see Table 3.1.).

The composition of the household food waste generated across five countries—UK, Netherlands, Austria, USA, and Turkey—is summarized in Box 3.1.

1.1.2 Retailer Wastes

Food, drink, and packaging waste can occur at all points within the retail supply chain, but it is a significant cost to retailers and so is minimized wherever possible. However, availability and having full shelves are key drivers for retailers, who want to ensure they are able to offer customers a full range of products. This not only maintains service levels for the customers but can also make financial sense to the company. However, in some instances this activity may create surplus product and waste back-of-store. A report from Economic Research Service found that annual supermarket losses for 2005 and 2006 averaged 11.4% for fresh fruit, 9.7% for fresh vegetables, and 4.5% for fresh meat, poultry, and seafood, for those participating in the study (WRAP, 2010). Results of this study are illustrated in Figure 3.2.

It is noticeable that grocery retailers together with households create a larger proportion of FIW in countries such as the UK and USA. In contrast, developing countries report a relatively high fraction of losses and spoilage in the first part of the food chain (Lundqvist, 2008).

In addition, every year the world's food-processing industry produces vast volumes of aqueous wastes. These include: fruit and vegetable residues and discarded items; molasses and bagasse from sugar refining; bones, flesh, and blood from meat and fish processing; stillage and other residues from wineries, distilleries, and breweries; dairy wastes such as

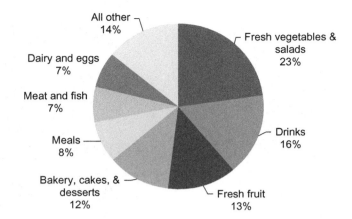

FIGURE 3.1 **Total food and drink waste generated by UK households by food group.** *Source: WRAP, (2008).*

TABLE 3.1 Estimated Food and Drink Waste Generated by UK Households by Food Group in 2008 (million tonnes) (WRAP, 2010)

Food group	Total	Avoidable	Unavoidable/possibly avoidable	Percentage avoidable
Fresh vegetables and salads	1.9	0.86	1.04	45%
Drinks	1.3	0.87	0.43	67%
Fresh fruit	1.3	0.50	0.60	46%
Bakery, cakes, and desserts	0.99	0.87	0.12	88%
Meals	0.69	0.66	0.30	96%
Meat and fish	0.61	0.29	0.32	48%
Dairy and eggs	0.58	0.53	0.50	91%
All other	1.1	0.75	0.36	67%
Total	**8.3**	**5.3**	**3.0**	**64%**

Source: Household food and drink waste in the UK, WRAP (2009).
Note: The figures are subject to rounding and therefore may not sum to their total.

cheese whey; and wastewaters from washing, blanching, and cooling operations (Arora et al., 2002). Many of these contain low levels of suspended solids and low concentrations of dissolved materials. Apart from the environmental challenges posed, such streams represent considerable amounts of potentially reusable materials and energy. Much of the material generated as wastes by the food-processing industries throughout Europe—and about to be generated within biofuels programs—contains components that

BOX 3.1

SUMMARY OF HOUSEHOLD FOOD WASTE COMPOSITION ACROSS FIVE COUNTRIES

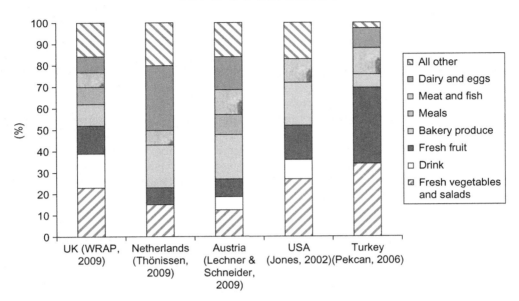

One can distinguish that the main components of household wastes are fruit and vegetables (FV), dairy products and eggs, bakery produce, as well as meat and fish. Thonissen (2009) found that dairy products accounted for an unusually high proportion of food waste in the Netherlands; by contrast, FV accounted for the highest proportion in the Turkish data (Pekcan et al., 2006). The composition of household wastes varies from country to country and depends on the diversity of national diets, complexity of food choice, as well as household size/composition, income, demographics, and culture.

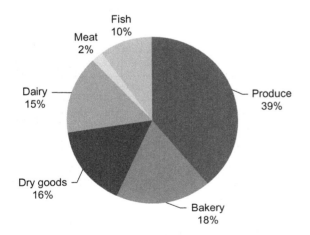

FIGURE 3.2 **Estimated breakdown of total UK grocery retail store food waste sent to landfill, 2008.** *Source: WRAP (2010).*

could be utilized as substrates and nutrients in a variety of microbial/enzymatic processes, to give rise to valuable products.

2. CHARACTERIZATION AND COMPOSITION OF FOOD WASTES

2.1 Fruit-and-Vegetable Wastes

Fruit-and-vegetable wastes (FVW) represent almost 20% to 50% of total household waste in several countries, as shown in Box 3.1. Pekcan et al. (2006) note the importance of fruits in the diets of households studied, so wasted FV accounted for the highest proportion in some countries (e.g., in Turkey ~70% of FVW, with a preponderant amount of fruit). The whole production of FVW collected from the market of Tunisia has been measured and estimated to be 180 tonnes per month (Bouallagui et al., 2003). Without proper storage and transport facilities, perishable food items are particularly vulnerable. The high losses in developing countries are mainly due to a lack of technology and infrastructure as well as other intrinsic and extrinsic factors such as high insect infestations, unwanted microbial growth, improper handling or transportation, and high temperatures and humidity during growth and harvesting (Buys and Nortje, 1997). For example, in India the spoilage can be as high as 30% in the case of vegetables and fruits (Swaminathan, 2006). Considering the predominant amount of FVW and their perishable nature, we will discuss experience of their management in a few countries with hot and humid climate.

Angulo et al. (2012) performed a quantitative and nutritional characterization of FVW from a main marketplace in Medellín (Colombia) during different periods of the year. The results indicate that the Minorista market produces approximately 8.35 tonne/day of FVW, which is composed of 43% fruit, 30% vegetables, and 27% leaves, stems, leaf wrappers, corncobs, roots, refuse, and others. The forty-two products included in FVW were classified into four groups, with the basic composition found in the products of groups 1 and 2 (oranges, cabbage, swiss chard, tomatoes, bananas, lettuce, lemons, papaw, stems, leaf wrappers, and corncobs), which changed according to factors such as the harvest period, the demand for the product, and its handling. In contrast, other market places in Colombia (e.g., Corabastos in Bogotá city) generate FVW containing a higher proportion of vegetable (67.6%) than fruit (14.1%) waste. In other countries, such as India, waste from marketplaces was composed of 85% vegetables (Mukherjee and Kumar, 2007). The evaluation of Angulo et al. (2010) was carried out 7 days per week during 4 periods of the year. On average, FVW contained 10% crude protein (CP), 36.6% neutral detergent fiber (NDF), 29.6% acid detergent fiber (ADF), 87.8% ruminal degradability at 24 h, 3657 kcal/kg, 0.59% calcium (Ca^{2+}), and 0.21% phosphorous (P). There were no statistical differences between days or between periods of evaluation ($p > 0.05$) for CP or for Ca^{2+}. As for NDF and ADF, there were statistically significant differences between periods but not between days. There are different aspects that could influence the kind of products present in FV, including the geographical location of the marketplace, the harvest period, the demand for the products, as well as characteristics of the products and their handling. For example, potatoes (group 4) had the biggest entries in the marketplace during all periods of evaluation but had a low constant presence in FV, explained by their high demand. The orange, which was the fruit most included in FV, had a constant harvest period during the evaluation and had a high demand in the marketplace, but oranges were the most damaged fruits reported in this study with the shortest shelf life (Angulo et al., 2010).

With regard to nutritional and the microbiological composition, FVW represents a potential resource that

might be used in bovine diets. According to Esteban et al. (2007), fiber percentage of organic waste was around 13%, which contradicts data found in the study of Angulo et al. (2010), where neutral detergent fiber (NDF) reached values between 22.6% and 47.7%, and acid detergent fiber (ADF) reached values between 19.8% and 37.1%, with high dry matter degradability in all periods. Thus, more than 50% of FV was degraded at 2 h, and more than 80% was degraded at 24 h for all periods, which demonstrates that there were components with a high digestibility during all evaluated periods. An increase was observed in the content of analyzed microorganisms, especially in mesophilic aerobic microorganisms, when the humidity of the samples was over 12% (Table 3.2.). In a general way, FV with a maximum humidity level of 12% should be used for animal feeding to avoid microorganism proliferation and to allow the best product preservation. The microbiological determinations indicated no apparent health threat to animals, which agreed with data found by Myer et al. (1999) and by Sancho et al. (2004). So recycling FV from market places for animal feeding is an important waste disposal alternative that might reduce the amount of biodegradable urban waste going to landfills and minimize the environmental impact.

Knowledge of environmental quality parameters of wastes is needed for the development of treatment strategies; also, knowledge of chemical composition is essential for designing novel products and food waste valorization. Environmental quality parameters such as COD, BOD, suspended solids (SS), volatile solids (VS), and total organic carbon (TOC)

TABLE 3.2　Microbiological Analysis of Fruit and Vegetables in the Minorista Market, Medellin, Colombia (Average/Period)

Analysis	Samples with more than 12% humidity	Samples with less than 12% humidity	Reference value in feed destined for animals[1]
Mesophylls aerobic microorganisms/g	761,900	27,775	<1,000,000
Total coliforms/g	<226	<3	<1000
Fecal coliforms/g	<5	Absent	Absent
Molds and yeasts	110,805	943	<100,000
Salmonella spp/25 g of food	Absent	Absent	Absent

Source: Angulo et al. (2010).
[1]*Laboratory of Microbiological analysis, Facultad de Ciencias Agrarias, Universidad de Antioquia.*

characterize the polluted environment and are used to indirectly measure the amount of organic compounds in the system. Other parameters of interest are concentration of total solids (TS), total nitrogen (TN), and total phosphorus contents (TP). According to Ruynal et al. (1998), the total initial solid concentration of FVW is between 8% and 18%, with a total VS content of about 87%. The organic fraction includes about 75% sugars and hemicellulose, 9% cellulose, and 5% lignin. In general, these wastes consist mainly of carbohydrates and relatively small amounts of proteins and fat and have a moisture content of 80–90% (Grobe, 1994). The wastewaters contain dissolved compounds, pesticides, herbicides, and cleaning chemicals. Largely, FV are rich in antioxidants, pectins, fibers, carbohydrates, mineral salts, food flavors, and colorants, and these substances can be utilized to integrate animal feedstuffs or to increase the quality of food products. Generally, the wastes from FV processing industries contain high values of BOD and COD, large amounts of SS, and are characterized by a variation in pH. Indicative values of BOD, COD, SS, and pH for the processing of some fruit and vegetables are summarized in Table 3.3. Apples demonstrate the highest values of BOD/COD and acidic pH among the FV items presented in this table. The high organic load of the apples reflects the high contents of carbohydrates, organic acids, fibers, and some proteins found in them.

2.1.1 Fruit Wastes

2.1.1.1 APPLE POMACE

Huge amounts of apple pomace (press residue) are produced worldwide, and its disposal represents a serious environmental problem. In Brazil alone, about 800,000 tonnes of apple pomace are produced every year, and it is mostly used as animal feed (Vendruscolo et al., 2008). This utilization is limited by a low protein and vitamin content and hence low nutritional value. Apple pomace is a natural source of pectic substances (~8%), being an important raw material for pectin production throughout the world. It has high water content and is mainly composed of insoluble carbohydrates such as cellulose, hemicellulose, and lignin. Being rich in different carbon sources including simple sugars, such as glucose, fructose, and sucrose (high contents of carbohydrates ~60%), it appears to be an excellent substrate for bioprocesses (Albuquerque et al., 2006). We will describe the biotechnological applications of this valuable resource in Chapter 5, emphasizing production of fine chemicals and nutrient supplements.

2.1.1.2 GRAPE POMACE

Grape (Vitis vinifera L.) is one of the world's largest fruit crops, in excess of 60 million tonnes (FAO Statistics, 2008). Grape pomace, the main by-product of wine production, consists of skins, seeds, and stalks, reaching an estimated amount of 13% by weight of processed grapes (Torres et al., 2002). It was calculated that grape pomace amounts to more than 9 million tonnes per year (Mazza and Miniatry, 1993). Grape pomace represents a rich source of various high-value products such as ethanol, tartrates and malates, citric acid, grape seed oil, hydrocolloids, and dietary fibre. The chemical composition is complex: alcohols, acids, aldehydes, esters, pectins, polyphenols, mineral substances, and sugars are the principal classes of compounds (Ruberto et al., 2008). Concentrations of compounds were determined from their peak areas in gas chromatograph with a flame ionization detector (GC-FID) profiles and gas-chromatography–mass spectrometry.

2.1.1.3 CITRUS POMACE

The family of citrus fruits consists of oranges, kinnow, khatta, lime, lemon, grapefruit, malta, sweet orange, etc. Orange production in 2010 is estimated to be approximately 66.4 million tonnes (Mamma et al., 2007). Citrus peels are the principal solid by-product of the citrus-processing industry and constitute about 50% of fresh fruit weight. The disposal of the fresh peels is becoming a major problem for many factories. Dry citrus peels are rich in pectin, cellulose, and hemicellulose. Various microbial transformations have been proposed for the use of this waste to produce valuable products like biogas, ethanol, citric acid, chemicals, various enzymes, volatile flavoring compounds, fatty acids, and microbial biomass (Dhillon et al., 2010).

TABLE 3.3 Fruit-and-Vegetable Waste Characteristics

Fruit/Vegetable	BOD (gO$_2$/L)	COD (gO$_2$/L)	SS (g/L)	pH
Apples	9.60	18.70	0.45	5.9
Carrots	1.35	2.30	4.12	8.7
Cherries	2.55	2.50	0.40	6.5
Corn	1.55	2.50	0.21	6.9
Grapefruit	1.00	1.90	0.25	7.4
Green peas	0.80	1.65	0.26	6.9
Tomatoes	1.025	1.50	0.95	7.9

Source: Thassitou and Arvanitoyannis (2001).
BOD, biochemical oxygen demand; COD, chemical oxygen demand; SS, suspended solids.

2.1.2 Vegetable Waste

2.1.2.1 ONION WASTES

Onions (*Allium cepa L.*) are the second most important horticultural crop worldwide with current annual production around 66 million tonnes (FAO Statistics, 2008). More than half a million tonnes of onion wastes are produced annually in the European Union, mainly from Spain, the UK, and Holland (Waldron, 2001). The main onion wastes include onion skins, two outer fleshy scales and roots generated during industrial peeling, and undersized, malformed, diseased, or damaged bulbs. Onion composition is variable and depends on cultivar, stage of maturation, environment, agronomic conditions, storage time, and bulb section (Abayomi and Terry, 2009). Water makes up most (80–95%) of the fresh weight of onion. Up to 65% or more of the dry weight may be in the form of non-structural carbohydrates (NSC), which include glucose, fructose, sucrose, and fructo-oligosaccharides (FOS) (Davis et al., 2007). Moreover, brown skin showed high concentrations of quercetin aglycone and calcium, and top/bottom showed high concentrations of magnesium, iron, zinc, and manganese. Onions are rich in several groups of plant compounds, such as dietary fiber (DF), FOS, flavonoids, and alk(en)yl cystein sulfoxides (ACSOs) that have perceived benefits to human health. Different flavonols were detected using HPLC, an aglycone, quercetin, and five flavonol glucosides. Therefore, the onion wastes can be used as a source of food ingredients. The main FOS in onion bulbs are kestose, nystose, and fructofuranosylnystose (Jaime et al., 2001). Benitez et al. (2011) described the content of minerals, DF, NSC, ACSOs, and flavonoids together with antioxidant activity in different onion wastes. They summarized the chemical analysis/parameters and mineral composition of onion produce by two Spanish cultivars and their industrial wastes. It is clear that valorization of industrial onion wastes will require exploitation of many components available in these by-products. Such information may be useful to food technologists for the appropriate exploitation of each industrial onion waste as a source of a specific functional compound (addressed in Chapter 6).

2.1.2.2 POTATO CO-PRODUCTS

Potato (*Solanum tuberosum*) co-products from processing for frozen food products generated around 4.3 million tonnes of co-product in the USA and Canada in 2008 (Nelson, 2010). Of the 21.8 million tonnes (fresh weight) of potatoes produced in the USA and Canada in 2008, less than 1% were fed, 7% were used for seed, 7% were shrinkage and loss, 28% were used as table stocks, and 57% (11.3 million tonnes) were processed (Agriculture and Agri-Food Canada, 2008;

USDA National Agriculture Statistics Service, 2008). Potatoes contain approximately 18% starch, 1% cellulose, and 81% water, plus dissolved organic compounds such as protein and carbohydrate. The major processes in all products are storage, washing, peeling, trimming, slicing, blanching, cooking, drying, etc. (Hung et al., 2006). There are four main types of potato processing co-products available. Some processors combine the types into one product (called slurry). The four types of potato co-products are: (1) potato peels; (2) screen solids, bits and pieces, white waste, or hopper box (small potatoes and pieces); (3) cooked product (fries, hash browns, crowns, batter, crumbles); and (4) material from water recovery systems (oxidation ditch, belt solids, filter cake) that varies from mostly microbial cells and solubles (oxidation ditch) to fine potato particles from clarifiers after drum or belt-type vacuum filtration (filter cake). The chemical composition of potato co-products varies, as do dry matter contents, which vary from 10% to 30%. Commercial potato starch granules are described as slowly degrading. Monteils et al. (2002) reported that *in sacco* potato peel starch disappearance rate was 5%/h compared with 34%/h for wheat starch. In contrast, Szasz et al. (2005) reported that ensiled potato slurry starch was 27% to 38% soluble, and the insoluble fraction disappeared *in vitro* at 14%/h regardless of pasteurization at 54°C. This slurry probably was mostly peels and cooked co-product, based on chemical composition, in which most of the starch would have been gelatinized.

Processing potato starch results in potato pulp as a major by-product, particularly in Europe. Research indicates that potato pulp can be fractionated to produce several commercially viable resources. Pectin and starch can be isolated, as well as cellulase enzyme prepared. Potato pulp may also have applications for reuse in the following industries: replacement of wood fiber in paper making and as a substrate for yeast and B_{12} production. Protein can also be isolated from the starch processing wastewater and sold as fractionated constituents (Kingspohn et al., 1993). Potato pulp isolated from potato starch production can be isolated and sold as pomace (Treadway, 1987).

Potato wastes have also been evaluated as a potential source for the production of acetone, butanol, and ethanol by fermentation techniques. This application of biotechnology resulted in a biofuel that utilizes potato wastes as a renewable resource (Grobben et al., 1993). Potato processing solid wastes are often applied to agricultural land as a disposal medium (Smith, 1986). Their composition percentage is shown in Table 3.4. The solid wastes are filtered or centrifuged from the primary clarifiers, then used as a nitrogen fertilizer for crops. Disposal routes of potato co-products include landfills, application to cropland, and composting.

TABLE 3.4 Composition Percentage of Potato Waste Solids

Component	Amount (%)
Total organic nitrogen as N	1.002
Carbon as C	42.200
Total phosphorus as P	0.038
Total sulfur as S	0.082
Volatile solid	95.2

Source: Pailthorp, (1987).

TABLE 3.5 Chemical Composition of Organic Fraction and Quality Characteristics of Liquid Olive Oil Waste

Component	Value (%)	Parameter	Value
Lipids	1.0−1.5	pH	3−5
Organic acids	0.5−1.55	SS	65.0 g/L
Pectin, colloids, tannins	1.0−1.5	TS	6.39%
Sugars	2.0−8.0	BOD	43.0 gO_2/L
Total nitrogen content	1.2−1.5	COD	100.0 gO_2/L

Source: Thassitou and Arvanitoyannis (2001).

Clearly the volumes of potato co-products are too large to be disposed of reasonably and economically through these methods. Potential difficulties with disposal include drainage from co-products; sprouting and regrowth of potatoes; insects, nematodes, and pathogen exposure to surrounding crops; and nuisances from smell and animals. Approximately two million tonnes per year of potato peels are produced from potato processing as by-products. Potato peels contain 40 g dietary fiber per 100 g dry matter, depending on the variety of potato and the method of peeling (Arora et al., 1993). Potato peels provide a good source of dietary fiber, particularly when processed by a lye-peeling technique (Smith, 1987).

Feeding potato co-products to ruminants can be the best disposal route, if these high moisture co-products are transported only short distances and are handled/fed properly. However, potato co-product identification, nutrient composition, anti-quality components and hygiene, storage and preservation, feeding value, and effects on meat quality must be considered for proper utilization in ruminant diets (Nelson, 2010).

2.2 Olive Oil Industry

Olive-mill technology generates a variety of waste in both energy and mass forms. At present, extra virgin olive oil can be obtained by two main milling processes: the three-phase system, which is widely used in Italy, Greece, and other Mediterranean countries, and the two-phase system, which is mainly used in Spain. In the first case, olive oil is separated from two other by-products, olive mill wastewater and a solid olive residue (pomace), whereas in the latter system, only a semisolid waste is obtained. The production of large quantities of these by-products, especially in the 3−4 months of intensive oil production, and the variability of the waste composition are the main problems in residue management (Aliakbarian et al., 2011).

Liquid waste is known as olive-mill wastewater (OMWW), which consists of substantial amounts of added water, olive juice combined with small amounts of unrecoverable oil, and fine olive pulp particles. From an environmental point of view, OMWW is the most critical waste emitted by olive mills in terms both of quantity and quality. The problems created in managing this waste have been extensively investigated during the last fifty years without finding a technically feasible, economically viable, and socially acceptable solution (Niaounakis and Halvadakis, 2006). Liquid waste from the olive oil industry is a dark-colored juice, which contains organic substances such as sugars, organic acids, polyalcohols, pectin, colloids, tannins, and lipids with low pH (Table 3.5). The difficulty of disposing of OMWW is mainly related to its high BOD and COD, as shown in the Table 3.5, and high concentration of organic substances (e.g., polyphenols), which make degradation a difficult and expensive task (Saez et al., 1992).

Olive pomace can constitute up to 30% of olive oil manufacturing output, depending on the milling process. After oil extraction, the pomace is generally distributed by means of controlled spreading on agricultural soil. However, a large quantity of olive-mill solid residue remains without actual application because only small amounts are used as natural fertilizers, combustible biomass, or additives in animal feeding (Pagnanelli et al., 2003). Many researchers have also studied the use of olive pomace in direct combustion (Khraisha et al., 1999), in the production of chemical compounds (Montané et al., 2002), as animal feed or soil conditioner (Mellouli et al., 1998), and as activated carbon (Baçaoui et al., 2001).

2.3 Fermentation Industry Wastes

The fermentation industry comprises three main categories: brewing, distilling, and wine manufacture. Each of these industries produces liquid waste with many common characteristics, such as high BODs and CODs, but differs in the concentration of the organic compounds such as tannins, phenols, and organic acid. The difficulty in dealing with fermentation wastewaters is in the flows and loads of the waste. Similar challenges face other activities that use bioprocessing on a large scale, a problem likely to grow rapidly with the production of wastes and by-products from bioethanol processes. Such production already occurs on a significant scale within the brewing and distilling industries but will increase in volume substantially as national "biofuels" programmes develop. They have to follow the European Parliament's Directive 2003/30/EC, which imposed upon Member States of the EU the task of introducing legislation that guarantees that, from the year 2010, the minimum proportion of biofuels in the total of fuels produced in each country would be 5.75%. Directive 2009/28/EC of the European Parliament and of the Council of 23 April 2009 on the promotion of the use of energy from renewable sources amended and subsequently repealed Directive 2003/30/EC. Accordingly, each Member State shall ensure that the share of energy from renewable sources in all forms of transport in 2020 is at least 10% of the final consumption of energy in transport. As a result, a series of "bioethanol" programmes was considered by the European Parliament to be one of the top priorities in establishing alternative fuel production. Global production of bioethanol increased from 17.25 billion liters in 2000 to over 46 billion liters in 2007. With all of the new government programs in America, Asia, and Europe in place, total global fuel bioethanol demand could grow to exceed 125 billion liters by 2020 (Balat and Balat, 2009). This can produce vast volumes of dilute aqueous by-product or waste streams similar to those from the brewing and distilling industries.

2.3.1 Quantities of Bioethanol Production

An initial substrate used for production of ethanol consists of three major groups of feedstock: (1) sucrose-containing feedstocks (e.g., sugar cane, sugar beet, sweet sorghum, and fruits), (2) starchy materials (e.g., corn, milo, wheat, rice, potatoes, cassava, sweet potatoes, and barley), and (3) lignocellulosic biomass (e.g., wood, straw, and grasses). In the short-term the production of bioethanol as a vehicular fuel is almost entirely dependent on starch and sugars from existing food crops (Enguídanos et al., 2002). More than 95% of the ethanol produced is obtained from agricultural or agriculture-related feedstocks. Of these, sugar-based feedstocks account for approximately 42%, and non-sugar feedstocks (mainly starch-based) for about 58% of the ethanol volume produced (Tolmasquim, 2007). Approximately 67% of the global ethanol volume, which was a total of 39 billion liters (REN21, 2008) in 2006, is used for fuel production.

The major sugar-based substrates that are used on a global scale for ethanol production include sugar cane, cane molasses, beet molasses, and sugar beets. Sugar cane is a basic substrate for ethanol production in Brazil, where approximately one third of the world's overall ethanol volume is being produced, and ~79% of it comes from sugar cane juice. Cane molasses, which ranks second as a substrate for ethanol production in Brazil, is practically the sole substrate for the manufacture of ethanol in India, a country that is the fourth largest ethyl alcohol producer in the world (Cardona and Sánchez, 2007). France, Europe's largest and the world's fifth largest producer of ethanol, uses predominantly sugar beets as the substrate (Bernard and Prieur, 2007). France also established an ambitious biofuels plan, with goals of 7% by 2010 and 10% by 2015. The USA is a global leader in the use of starch-based feedstocks for ethanol production. In 2006 US production accounted for 38% of the world's overall ethanol production (REN21, 2008). The US principal starch feedstock is corn. Besides corn, some other grains are in use as feedstocks, including wheat, sorghum, or rye (Tolmasquim, 2007). In 2005 in the USA, about 95 distilleries/factories were operated, with a total capacity of 16.4 billion liters per year. In mid-2006 there were 35 additional plants under construction, providing further capacity of 8 billion liters per year (FAO/IEA, 2007). In 2006, fuel ethanol production in the USA and Brazil amounted to 18.3 billion liters and 17.5 billion liters, respectively (REN21, 2008). In the USA over 99% of the ethanol produced is used for fuel production (Seelke and Yacobucci, 2007); in Brazil this proportion is 90%. Canada has an annual ethanol volume of 550 million liters produced in 2006 (FAO/FAO, 2007) and fuel ethanol production of 200 million liters (REN21, 2008). Starch feedstock (mainly maize grains) is used for ethanol production in Canada (Berg, 2004). Other widely accepted feedstocks include wheat, barley, triticale, and rye (Klein et al., 2008).

In 2006 the EU produced approximately 1.6 billion liters of ethanol for fuel (EUBIA, 2008). Most of this was distilled from cereals, accounting for 976 million liters or roughly 61% of total ethanol feedstock. Among the different grains used for ethanol production, wheat is the most important cereal, with a market share of 36%, followed by rye (15%), barley (7%), maize (only 2%), and triticale (marginal) (Biofuels International, 2007). In 2006 the largest starch-based bioethanol fuel producer in the EU was Germany with an overall output of 430 million liters (Biofuels International, 2008). The main feedstock used for the German ethanol production was rye (49.5%), followed by wheat (41.3%), barley (8%), and triticale (no precise data). Spain, as the second largest producer of ethanol in the EU (396 million liters in 2006), mainly processed wheat (57%) followed by barley (21%). The rest of the Spanish production came from wine alcohol. In 2007 France achieved the greatest production (578 million liters), almost double its previous highest annual figure (Biofuels International, 2008), but France was the only large-scale producer of bioethanol from sugar beet (around 81%) (Biofuels International, 2007). In 2007 the next largest producers were Germany (394 million liters) and Spain (348 million liters). In that year fuel ethanol production decreased compared with 2006, with the most significant decline in Sweden, where production dropped to 70 million liters. This was partly because of high cereal prices, which increased the costs of bioethanol production to a level that made it unprofitable (Biofuels International, 2008).

In a similar manner to the USA, Europe had invested in large distillery plants, whose overall ethanol production increased by 71% in 2007, reaching 2.9 billion liters. The potential demand for bioethanol as a transportation fuel in the EU countries, calculated on the basis of Directive 2003/30/EC, is estimated at about 12.6 billion liters in 2010 (Zarzyycki and Polska, 2007). As for Asia, the leadership in utilizing starch-based feedstocks for ethanol production belongs to China, which was ranked third as a global ethanol producer, with a yearly volume of approximately 4.1 billion liters in 2006 (Tolmasquim, 2007). Overall, global production of bioethanol increased from 17.25 billion liters in 2000 to over 46 billion liters in 2007 (REN21). Figure 3.3 shows global bioethanol production between 2000 and 2007.

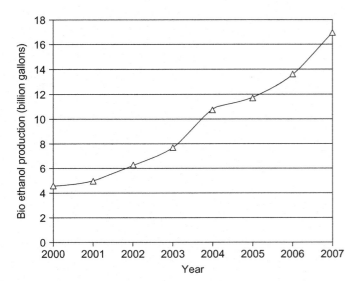

FIGURE 3.3 Global bioethanol production from 2000 to 2007.
Source: Balat and Balat (2009).

2.3.2 Composition of Distillery Wastes

2.3.2.1 SUGAR-BASED FEEDSTOCK

Stillage (distillery wastewater) is the main by-product from distilleries, and its volume is approximately ten times higher than that of the ethanol produced. In European countries, beet molasses is the most commonly used sucrose-containing feedstock. Sugar beet crops are grown in most of the EU countries and yield substantially more bioethanol per hectare than wheat (EUBIA, 2007). The by-product remaining after distillation of the sugar-based feedstock, liquid waste called vinasse, is characterized by a high concentration of organic acids and polyphenols. This is the reason for the relatively high values of COD typical of that type of waste. Regardless of whether it originated from sugar cane or sugar beets, vinasse is a high-strength effluent with COD values frequently exceeding $100 \, gO_2/L$ (Satyawali and Balakrishnan, 2008). Moreover, besides potassium ions, vinasse contains large quantities of other metal ions and sulfates, as well as phenol compounds, which may exert some toxic effect. Concentrated vinasse constitutes an environmental problem also because of its high organic load (BOD 61,000 to 70,000 mg/L) and high salinity (EC 250−300 dS/m, Na 28 g/kg) (Madejón et al., 2001). The main pollutant in beet molasses vinasse (not detected in cane-based vinasse) is betaine (N,N,N-trimethylglycine) (Cibis et al., 2011). Betaine is a soluble compound, which is not consumed to any significant extent and appears to largely pass through the subsequent processing stages, causing serious problems in the wastewater produced by these industries.

When used as fodder, vinasse has a lower nutritive value than does the stillage from starch-based feedstock. As a result, the use of vinasse as fodder, practiced in the countries of Western Europe (Krzywonos et al., 2009), is limited to a level lower than 10% of the diet in the case of ruminants and less than 2% of the diet in the case of pigs and poultry (Wilkie et al., 2000). Another application of vinasse in agriculture is its use as a fertilizer, which is common practice in Brazil and some European countries (Madejón et al., 2001; Tejada et al., 2009). Cibis et al. (2011) reported aerobic batch biodegradation of betaine in vinasse, utilized at 27−54°C and pH 8.0. Maximal COD removal (88.73%) was obtained at 36°C and pH 6.5, using a mixed population of *Bacillus spp*.

2.3.2.2 STARCH-BASED FEEDSTOCK

In contrast, distillery stillage from the fermentation of starch-based feedstock by yeasts contains not only some feedstock components (Davis et al., 2005; Sweeten et al., 1981−1982) but also degraded yeast cells (Sanchez et al., 1985). Many of those substances are characterized by a high nutritive value. They contain B vitamins, proteins rich in exogenous amino acids (Mustafa et al., 1999), and mineral components.

Because of the chemical composition of its dry matter content (Table 3.6), potato stillage has been regarded as valuable fodder (despite its high water content, 90–95.3%). As far as the stillage derived from wheat, rye, triticale, and barley is concerned, the lowest nutritive value (measured as crude protein and neutral detergent fiber fraction) is that of barley-derived stillage, both in its liquid and solid fraction (Mustafa et al., 2000). Unprocessed warm stillage has the highest feeding value but also a serious drawback: it cannot be stored over a long period because of its proneness to souring and mould growth. This means that the animals should be fed shortly after the stillage has been produced, which makes this method of stillage utilization really difficult. Feeding farm animals with "raw" stillage is cost-effective only if the users live in the close proximity of the distillery. Because of the high water content, the transport of the stillage over long distances is unprofitable. For example, in Germany the integration of distilleries with stock farms permits them to avoid troubles in the use of starch-based stillage. This is not so in the USA, where ethanol plants have difficulties with the profitability of stillage marketing or disposal. Most of the stillage produced is sold in the form of Distiller's Dried Grains with Solubles (DDGS).

Other methods of utilizing starch-based stillage include concentration and drying. Both are characterized by a high energy demand, which makes them applicable to very large distilleries only (Murphy and Power, 2008). The concentration and drying of starch stillage that is to be used as fodder have found wide acceptance in the USA and in Canada, where the processes are carried out in several variants (Akayezu

et al., 1998). One of these involves the separation of the solid fraction from the liquid fraction using sieves, centrifuges, or pressing devices (the product is referred to as Wet Distiller's Grain). The solid phase is then dried (Distiller's Dried Grain), while the liquid phase is concentrated until a 30–40% suspended solids content has been obtained (Condensed Distiller's Solubles). In some distilleries the concentrated liquid fraction is dried (Distiller's Dried Solubles) or mixed with the previously separated solid fraction, and the mixture obtained is dried to form DDGS. During the last decade, many new dry-grind fuel ethanol plants have been constructed, and there are now approximately 165 plants in the USA, projected to produce more than 18 million tonnes of DDGS in 2008 (Renewable Fuels Association, 2008).

Much research has been conducted during recent years to further evaluate the nutrient concentration, digestibility, feeding value, and unique properties associated with feeding DDGS to swine. Most of the starch in the grain is converted to ethanol during the fermentation process and only a small amount of starch is present in DDGS. However, the fiber in corn is not converted to ethanol, and as a result, DDGS contains ~35% insoluble and 6% soluble dietary fibers. The apparent total tract digestibility (ATTD) of dietary fiber is 43.7%, and this results in a reduced digestibility of DM. It is also the reason why the digestibility of energy in DDGS is less than with many other feed ingredients. The concentration of phosphorus (P) in DDGS ranges from 0.60% to 0.70%, and the ATTD of P in DDGS is about 59%. This value is much greater than in corn (Pedersen et al., 2007). The ATTD of P in DDGS corresponds to bioavailability values between 70% and 90% relative to P bioavailability in dicalcium phosphate (Whitney and Shurson, 2001). Therefore, when DDGS is included in diets fed to swine, the

TABLE 3.6 Chemical Composition of Dry Matter Content of Selected Starch Stillage (%)

Stillage	Dry matter	Crude protein	Fat	Crude fiber	Sugars	Starch	Ash	References
Grain sorghum	5.8	1.7	nd	1.51	2.6	1.01	3.77	Sweeten et al., 1981–82
Barley	5.97	2.21	0.76	2.35	2.14	0.04	0.58	Mustafa et al., 1999
Maize	6.2	1.3	1.3	0.1*	2.8	0.5	0.8	Kim et al., 1999
Maize	3.7	1.44	nd	1.81	0.97	0.56	0.27	Sweeten et al., 1981–82
Maize	7.5	2.3	nd	nd	0.5	nd	2.1	Maiorella et al., 1983
Potato	6.0	1.45	0.05	0.7	3.1	nd	0.7	Czupryński et al., 2000
Wheat	8.4	3.8	1.14	2.86	2.67	0.185	0.7	Mustafa et al., 1999
Wheat	12	3.8	2.3	0.12	6	nd	0.156	Davis et al., 2005

*Acid detergent fiber.
nd, no data available.

utilization of organic P will increase and the need for supplemental inorganic P will be reduced (Stein and Shurson, 2009).

Stillage is largely a high-strength effluent that, because of considerable organic matter content (Table 3.7), can neither be sent to the sewer system nor be discharged into water/soil; at least a certain portion of the COD load must be removed at source. Comparison of the stillage chemical composition (Table 3.7) shows that it varies considerably according to its properties. This means that the COD content is influenced by the technology of alcohol production and the method of feedstock and stillage storage (Krzywonos et al., 2009).

The diversity of the feedstocks used for ethanol production, and the fact that distillery plants differ widely in size/production capacity, are the contributory factors in the use of different methods for distillery wastewater utilization and biodegradation. There are reports on the use of concentrated liquid corn stillage fractions as the feedstock for the production of some cosmetics, including alternan (Leathers, 1998) and pullulan (West and Strohfus, 1996). Furthermore, corn stillage was used as the feedstock for the synthesis of

TABLE 3.7 Chemical Composition of Liquid Phase in Starch Stillage (g/L)*

	Type of stillage									
Parameter	Maize (Cibis et al., 2004)	Wheat (Nagano et al., 1992)	Wheat (Hutnan et al., 2003)	Barley (Kitamura et al., 1996)	Rye (Cibis et al., 2004)	Grain (Laubscher et al., 2001)	Awamori (Rice) (Tang et al., 2007)	Starch waste feedstock (Cibis et al., 2004)	Potato (Cibis et al., 2002)	Potato (Cibis et al., 2006)
pH	3.7	4.6	3.35	3.7–4.1	3.94	4.0–4.5	3.65	3.88	3.69	3.88
Density	2.9	–	–	–	3.1	–	–	12.2	4.6	7.9
SS	–	18.4–23.0	38.6	97	–	1.0**	–	–	–	–
COD	21.85	18.5–20.8	90.75	83	28.98	20–30	56	122.33	48.95	103.76
BOD$_5$	–	12.5–13.6	–	–	–	–	50	–	–	–
TOC	9.15	–	–	–	10.70	–	28.33	45.60	–	35.15
Reducing substances	4.05	–	–	–	11.81	–	–	37.06	10.47	37.44
Glycerol	3.95	–	–	–	3.22	–	–	3.91	3.04	5.96
Lactic acid	6.63	–	–	–	3.51	–	1.4	61.14	–	17.53
Propionic acid	0.21	08–1.24	–	–	0.12	–	0.623	2.77	–	2.64
Succinic acid	0.21	–	–	–	0.31	–	1.059	0.23	–	0.43
Acetic acid	0.44	2.1–6.6	–	–	0.27	–	0.132	4.14	–	2.10
Sum of organic acids	9.67	–	–	–	5.29	–	10.795[‡]	75.11	–	24.46
TN	0.67	1.5–1.6	4.09[†]	6.0	0.83	0.17–0.18[†]	2.18[†]	2.57	0.52	1.05
Ammonia N	0.096	0.5–0.6	–	–	0.19	–	0.052	0.361	2.375	0.308
TP	0.441	0.17–0.18	–	–	0.47	0.27–0.30	–	0.816	0.259	0.277
Phosphate P	0.363	–	0.4	–	0.28	–	0.004	0.588	0.167	0.165

Source: Krzywonos et al. (2009).
Except pH, density (oBlg), COD, and BOD$_5$ (gO$_2$/L);
[†]*Total Kjeldahl Nitrogen (TKN);*
**Total suspended solids (TSS);*
[‡]*Volatile fatty acids (VFA).*
BOD, biochemical oxygen demand; COD, chemical oxygen demand; SS, suspended solids; TN, total nitrogen; TOC, total organic carbon; TP, total phosphorus.

astaxanthin carotenoid by *Phaffia rhodozyma* (Leathers, 2003). Other starch stillage has been used for the synthesis of protease (Yang and Lin, 1998), chitosan (Yokoi et al., 1998), and biodegradable plastics (e.g., poly β-hydroxybutyrate (Khardenavis et al., 2007)).

Anaerobic biodegradation of stillage has been carried out on an industrial scale. Worldwide, at least 135 anaerobic bioreactors are operated; nine of these (with four operating in Germany) are used for the treatment of starch stillage, and the others are primarily used to treat vinasse (Wilkie et al., 2000).

In contrast to anaerobic methods, aerobic biodegradation of starch stillage with thermo- and mesophilic bacteria has not yet been conducted on an industrial scale. The results obtained during laboratory-scale aerobic thermo- and mesophilic biodegradation of distillery wastewater indicate that the effectiveness of this method (Anastassiadis and Rehm, 2006; Battestin and Macedo, 2007; Choorit and Wisarnwan, 2007) is influenced by the aerobic conditions, pH, and temperature (Cibis et al., 2002; Krzywonos et al., 2008).

2.4 Dairy Industry

The dairy industry is an important part of the food industry and contributes significant liquid wastes, whose disposal requires a large amount of capital investment. World milk production in 2009 reached approximately 701 million tonnes; in the EU it remained at ∼54 million tonnes, while in the USA output fell to ∼85.5 million tonnes (FAO, 2010). A sustained worldwide increase in the production of dairy products has led to the generation of additional vast amounts of cheese whey. As a general rule, to make 1 kg of cheese about 9 liters of whey are generated (Guimarães et al., 2010). The world whey production is over 160 million tonnes per year, showing a 1–2% annual growth rate (OECD-FAO, 2008). Whey exhibits a BOD of 30–50 g/L and a COD of 60–80 g/L. Lactose is largely responsible for the high BOD and COD. Protein recovery reduces the COD of whey by only about 10 g/L (Gonzalez-Siso, 1996). Despite its use in many food products, about half of world cheese whey production is not treated but discarded as effluent. Cheese whey has a high COD, mainly owing to its high lactose content (4.5–5%, w/v), soluble proteins (0.6–0.8%, w/v), lipids (0.4–0.5%, w/v) and mineral salts (8–10% of dried extract). Whey also contains appreciable quantities of other components, such as lactic (0.05% w/v) and citric acids, non-protein nitrogen compounds (urea and uric acid), and B group vitamins. A dairy farm processing 100 tonnes of milk per day produces approximately the same quantity of organic products in its effluent as would a town with 55,000 residents (Gonzalez-Siso, 1996).

The lack of affordable methods for COD elimination in whey still forces industries to dump large volumes into sewage lines or on to the land, a fact that poses a permanent hazard in terms of environmental pollution (Rubio-Texeira, 2006). However, legislative regulations for the dumping of whey are forcing industries to come up with alternatives to make this process of elimination environmentally safer. Dairy wastewater contains milk solids, detergents, sanitizers, milk wastes, and cleaning water. It is characterized by high concentrations of nutrients and organic and inorganic contents (USDA-SCS, 1992). Significant variations in COD ($80-95 \times 10^3$ mg/L) and BOD ($40-48 \times 10^3$ mg/L) have been reported by various investigators of dairy wastewater (Table 3.8). The total COD of dairy wastewater is mainly influenced by the milk, cream, or whey. The pH varies in the range 4.7–11 (Passeggi et al., 2009), whereas the concentration of suspended solids (SS) varies in the range 0.024–4.5 g/L. Significant amount of nutrients, 14–830 mg/L of total nitrogen (Rico Gutierrez et al., 1991), and 9–280 mg/L of total phosphorus are also found in dairy wastewater. In dairy wastewaters, nitrogen originates mainly from milk proteins and is either present in organic nitrogen form, such as proteins, urea, and nucleic acids, or as ions such as NH^{4+}, NO^{2-}, and NO^{3-}. Phosphorus is found mainly in inorganic forms such as orthophosphate (PO_3^{4-}) and polyphosphate ($P_2O_4^{7-}$) as well as in organic forms. Significant amounts of Na, Cl, K, Ca, Mg, Fe, CO, Ni, and Mn are also always present in dairy wastewater. The presence of high concentrations of Na and Cl is due to the use of large amounts of alkaline cleaners in dairy plants (Demirel et al., 2005). The quality and quantity of the product content in the dairy wastewater at a given time changes with the application of another technological cycle in the processing line (Janczukowicz et al., 2008). Because the dairy industry produces different products, such as milk, butter, yogurt, ice cream, and various types of desserts and cheese, the characteristics of these effluents also vary widely both in quantity and quality, depending on the type of system and the methods of operation used (Rico Gutierrez et al., 1991). Some of the integrated solutions for valorization of the cheese whey are provided in Chapter 8.

2.5 Meat and Poultry Industry Wastes

2.5.1 Meat Production Waste

Global meat production was estimated at approximately 280 million tonnes in 2008. Experts predict that by 2050 nearly twice as much meat will be produced as today, for a projected total of more than 465 million tonnes. Consumption of meat and other animal products also continues to grow. Currently nearly

TABLE 3.8 Characteristics of Dairy Waste Effluents

Waste type	COD (gO$_2$/L)	BOD (gO$_2$/L)	pH	SS (g/L)	VS (g/L)	FOG (g/L)	TKN (mg/L)	TP (mg/L)	References
Cheese industry, whey	0.377–2.214	0.189–6.219	5.2	0.188–2.330	–	–	13–172	0.2–48.0	Andreottola et al. (2002)
Cheese industry, effluent	1.0–7.0	0.588–5.0	5.5–9.5	0.5–2.5	–	–			Monroy et al. (1995)
Cheese industry, effluent	–	–	4.7	2.5	–	–	830	280	Gavala et al. (1999)
Cheese industry, effluent	–	2.83	4.99	–	–	0.32	102	45	Sparling et al. (2001)
Dairy industry, effluent	0.980–7.500	0.680–4.500	–	0.3	–	–	–	–	Kolarski and Nyhuis (1995)
Milk industry	–	0.713–1.410	7.1–8.1	0.36–0.92	0.54–0.55	–	–	–	Samkutty et al. (2002)
Milk industry, cream	1.2–4.0	2.0–6.0	8–11	0.35–1.0	–	3–5	50–60	–	Ince (1998a, b)
Raw cheese, whey	–	68.814	–	1.30	1.89	–	1462	379	Malaspina et al. (1996)
Blue Stilton cheese, whey	66.0–72.5	–	3.86–5.35	–	–	–	Protein 27006000	–	Kosseva et al. (2003)

Source: Britz et al. (2006).

42 kilograms of meat is produced per person per year worldwide, but meat consumption varies greatly by region and socio economic status. In the developing world, people eat about 30 kg of meat a year, while consumers in the industrial world eat more than 80 kg per person each year. Rising food prices are pushing consumers to choose cheaper cuts of meat, like chicken. Global poultry output in 2010 was expected to reach 94 million tonnes (FAO, 2008). The production of meat across the farm-to-fork chain makes not just meat for human consumption but also wastes. The nature and quantity of the waste varies at each stage but includes the carcasses of dead animals, parts of animals which are treated as inedible, bones, hides, and blood. The quantity of meat production waste is staggering (Ontario Report, 2005). Humans consume only a portion of a food animal. Approximately 50–54% of each cow, 52% of each sheep or goat, 60–62% of each pig, 68–72% of each chicken, and 78% of each turkey end up as meat consumed by humans, with the remainder becoming waste after processing (Scotland Regulations, 2003). Based on mortality rates and livestock statistics in Ontario, it has been estimated that the annual mass of deadstock alone is greater than 86,000 tonnes. The meat waste from federal and provincial slaughterhouses in Ontario is believed to be 333,000 tonnes each year. This does not take into account other waste from meat processing, which is also substantial. As a direct result of meat processing operations, a slaughterhouse (abattoir) generates waste comprising the animal parts that have no apparent value; it also generates wastewater as a result of washing and cleaning processes. The operations taking place within a slaughterhouse and the types of waste and products generated are summarized in Figure 3.4.

The enormous volume of the meat waste makes the issue of the safety risks associated with its disposal an immediate, ongoing, and serious one. The common methods for disposal of blood by meat processors are rendering, land application, composting, and transfer to a wastewater treatment plant. In the USA, the federal government provides guidance, while state governments regulate composting and land application. Rendering is defined as the process of breaking down, through heat application, blood, meat pieces, and other animal by-products to useful components. Unlike raw waste materials, the products derived from rendering can be stored for long periods of time. The temperature and length of the rendering process kills or inactivates traditional disease-causing organisms (Food Safety Network, 2003). The rendering industry in North America recycles over 20.8 million tonnes of perishable material generated by livestock and poultry/meat processing, food processing, grocery, and restaurant industries each year (National Renderers Association, 2004). Rendering plants have started charging a disposal fee for blood. For this reason,

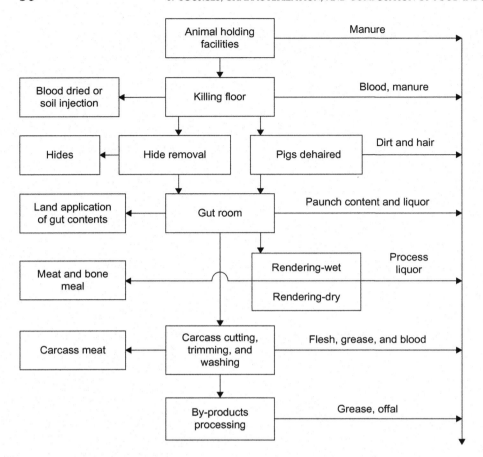

FIGURE 3.4 **Flow diagram indicating the products and sources of wastes from a slaughterhouse.** *Source: Banks and Wang (2006).*

rendering is now less attractive and less economical (Mittal, 2006).

In contrast, the EU deals with approximately 16.1 million tonnes of animal waste each year (Scotland Regulations, 2003). Meat, poultry, and fish industries produce the highest loads of waste within the food industry. Approximate composition of lean meat/ muscle and its range in meat products is presented in Table 3.10. The wastewater has a high strength, in terms of BOD, COD, suspended solids (SS), total nitrogen (TN), and phosphorus (TP), and is odorous. Table 3.9 shows the characteristics of meat-processing wastewater.

The mineral chemistry of the wastewater is influenced by the chemical composition of the slaughterhouse's treated water supply. Waste additions such as blood and manure can contribute to the heavy metal load in the form of copper, iron, manganese, and zinc in process plants. The wastewater contains a high density of total coliform, fecal coliform, and fecal *Streptococcus* groups of bacteria due to the presence of manure material and gut contents. Numbers are usually in the range of several million colony forming units (CFU) per 100 mL. Proper management is a prerequisite to ensure that potentially

high levels of pathogens are eliminated (Banks and Wang, 2006).

Slaughterhouses in Ontario and Quebec provinces of Canada generally discharge their wastewater in municipal sewers after some degree of primary or chemical pretreatment at the plant. Slaughterhouses are therefore required to pay a surcharge to dispose of their wastewater for further treatment at the municipal treatment plants (Masse and Masse, 2000). Except in Scandinavia, there are few waste treatment plants installed at slaughterhouses across Europe. In his review paper, Mittal (2006) summarized a variety of pretreatment and treatment methods for wastewater from slaughterhouses before land application. Recommendations on how to save water usage have also been made. Pretreatments are screening, catch basins, flotation, equalization, and settlers for recovering proteins and fats from abattoir wastewater. With chemical addition, dissolved air flotation (DAF) units can achieve COD reductions ranging from 32% to 90% and are capable of removing large amounts of nutrients. Anaerobic systems are lagoon, anaerobic contact (AC), up-flow anaerobic sludge blanket (UASB), anaerobic sequence batch reactor (ASBR),

TABLE 3.9 Characteristics of Meat Processing Wastewater

Type of meat	pH	COD (gO$_2$/L)	BOD (gO$_2$/L)	SS (g/L)	TN (mg/L)	TP (mg/L)	References
Cattle	6.7–9.3	3.0–12.873	0.900–4.62	–	–	93–148	Kostyshyn et al. (1987)
			7.237	3.574	378	79	USEPA (2002)
Hog	7.3	3.015	1.95	–	14.3	5.2	Gariepy et al. (1989)
			2.22	3.677	253	154	USEPA (2002)
Mixed	6.7	5.10	3.10	0.310	405	30	Borja et al. (1994)
	7.3–8.0	12.16–18.768	8.833–11.244	10.588–18.768	448–773	–	Arora & Routh (1980)
	7.0	0.583	0.404	0.20–1.00	152	–	Millamena (1992)

Source: Banks and Wang (2006).
BOD, biochemical oxygen demand; COD, chemical oxygen demand; SS, suspended solids; TN, total nitrogen; TP, total phosphorus.

TABLE 3.10 Composition and Range of Muscle in Meat and Fish Products

Characteristics	Typical lean meat (%)	Range in meat products (%)	Average fish muscle (%)
Water	70	22–80	75
Protein	20	9–34	15
Lipids (fats)	3	1.5–65	0.2–25
Ash (minerals)	1	1–12	5

Source: Montana Meat Processors Convention (2001).

and anaerobic filter (AF) processes. Typical reductions of up to 97% BOD, 95% SS, and 96% COD have been reported. UASB's average COD removal efficiencies are 80–85%. UASB seems to be a suitable process for the treatment of slaughterhouse wastewater, on account of its ability to maintain a sufficient amount of viable sludge.

The operation of a pilot-scale UASB treating the effluent from a beef slaughtering operation was reported by Torkian et al. (2003a). The researchers were able to obtain steady-state operation of the UASB at organic loading rates of 13–39 kg soluble COD per m^3 per day and HRTs of 2 to 7 hours under mesophilic conditions. SCOD removals of 75–90% were obtained at these loading rates with influent feed concentrations of 3.0–4.5 g soluble COD per liter. In a connected study, the effluent from the UASB was processed through a pilot-scale rotating biological contactor (RBC) to obtain additional organic load reduction (Torkian et al., 2003b). At organic loading rates of 5.3 g soluble BOD per m^3 per day, soluble BOD removals of 85% were obtained, with a vast majority occurring within the first half of the six-stage reactor. As part of a wastewater treatability study, Del Pozo and Diez (2003) noted that the high COD concentrations (7.23 g/L) warranted anaerobic treatment and conducted a series of anaerobic batch tests on a beef slaughterhouse's wastewater that yielded COD removals of 80%. Bohdziewicz et al. (2003) were successful in obtaining high contaminant removals from meat-processing wastewater using ultrafiltration followed by reverse osmosis. TP and TN were removed at greater than 98% efficiencies, and BOD and COD removals were greater than 99%. Membrane fouling has commonly been cited as the primary impediment to the use of filtration technologies in the meat-processing industry. Allie et al. (2003) reported on the use of lipases and proteases to effectively remove fouling proteins and lipids from flat-sheet polysulphone ultrafiltration membranes.

Luste et al. (2009) studied the effect of five pretreatments (thermal, ultrasound, acid, base, and liquid certizyme) on hydrolysis and methane production potentials of four by-products from the meat-processing industry. Liquid certizyme 5™ increased the soluble COD (COD$_{sol}$) of digestive tract content by 62% and drum sieve waste by 96%, compared with untreated waste. Ultrasound was the most effective in increasing the COD$_{sol}$ of dissolved air flotation (DAF) sludge (88%) and grease trap sludge. Thermal treatment in batch experiments increased methane production potential of drum sieve waste and acid production from grease trap sludge and all pretreatments of DAF sludge. Methane production potential decreased with thermal treatment of all other pretreated waste (compared with untreated waste). Methane production potentials (m^3 methane per total volatile solids added) from the untreated materials were digestive tract content (400 ± 50 m^3), drum sieve waste (230 ± 20 m^3), DAF sludge (2340 ± 17 m^3), and grease trap sludge (900 ± 44 m^3).

Pretreated abattoir wastewater was fed to an upflow anaerobic filter (UAF) at an HRT of 2 days under mesophilic (37°C) and thermophilic (55°C) conditions (Gannoun et al., 2009). The UAF was operated at organic loading rates (OLRs) of 0.9–6 g COD/L/d under mesophilic conditions and at OLRs of 0.9–9 g COD/L/d under thermophilic conditions. COD removal efficiencies of 80–90% were achieved for OLRs up to 4.5 g COD/L/d in mesophilic conditions, while the highest OLRs (i.e., 9 g COD/L/d) led to efficiencies of 70–72% in thermophilic conditions. The biogas yield in thermophilic conditions was about 0.32–0.45 L biogas per gram of COD removed for OLRs up to 4.5 g COD/L/d. Mesophilic anaerobic digestion has been shown to destroy pathogens partially, whereas the thermophilic process was more efficient in the removal of indicator microorganisms and pathogenic bacteria at different organic loading rates.

2.5.2 Poultry Wastes

Poultry wastes are as problematic as meat wastes. Poultry processing uses a relatively high amount of water, with an average consumption of 26.5 L per bird, and the wastewater contains proteins, fats, and carbohydrates from meat, blood, skin, and feathers, resulting in relatively high BOD and COD levels. Ultrafiltration (UF) membranes in membrane bioreactor (MBR) treatment systems allow the wastewater to be recycled, which reduces overall potable water demand (Avula et al., 2009). Valladao et al. (2009) evaluated the effect of pre-hydrolysis time and enzyme concentration on the anaerobic biodegradability of poultry slaughterhouse wastewaters. The effect of sludge reuse on treatment efficiency was also observed (Frenkel et al., 2010).

Starkley (2000) reviewed the considerations for selection of a treatment system for poultry processing wastewater, including land availability, previous site history, publicly owned treatment work discharge, conventional waste treatment systems, and land application systems. The performance of anaerobic treatment systems, including lagoons, contact processes, sludge beds, filters, packed beds, and hybrid reactors, were outlined (Ross and Valentine, 1992). In another study, anaerobic and aerobic fixed-film reactors in tandem were used for the combined treatment of poultry-processing wastewater (Del Pozo and Diez, 2003). COD removals of 92% were observed with system organic loading rates of 0.39 kg COD/m^3/d and 95% total Kjeldahl nitrogen (TKN) removals for applied N loads of 0.064 kg TKN/m^3/d. The authors reported the effects on nutrient and organic removals at varying recycle rates between the two reactors and varying reactor size. Pretreatment is also regarded as necessary for poultry waste to reduce the moisture and increase the porosity with the addition of bulking agents, which also increase the aeration and carbon level in wastewater. Proper treatment is needed to eliminate the pathogens. Typically, the thermophilic processes are more efficient in the removal of indicator microorganisms and pathogenic bacteria at various organic loading rates, achieving also higher biogas yields.

2.6 Seafood By-Products

One of the most important environmental problems, characteristic of coastline areas, is the large volume of waste generated by fishing, aquaculture, or foodstuff-processing industries. According to FAO (2008), global fish production, including capture fisheries, aquaculture, and trade volume, was about 140 million tonnes in 2007, forecasting ~144.1 million tonnes in 2009. About 116.6 million tonnes of this has been utilized as food/feed and other uses. Hence, global fish wasted annually reached roughly 27.5 million tonnes in 2009. Seafood processing involves the capture and preparation of fish, shellfish, marine plants and animals, as well as by-products such as fish meal and fish oil. The processes used in the seafood industry generally include harvesting, storing, receiving, eviscerating, precooking, picking or cleaning, preserving, and packaging (Carawan et al., 1979). Figure 3.5 shows a general process flow diagram for seafood processing. It is a summary of the processes typical to most seafood-processing operations; however, the actual process will vary depending on the product and the species being processed. There are several sources of wastewater, including:

- fish storage and transport;
- fish cleaning;
- fish freezing and thawing;
- preparation of brines;
- equipment sprays;
- offal transport;
- cooling water;
- steam generation;
- equipment and floor cleaning (Tay et al., 2006).

Seafood processed include also mollusks (oysters, clams, scallops), crustaceans (crabs and lobsters), and saltwater/freshwater fish products. Tuna, including yellowfin, skipjack, bluefin, albacore, and bigeye, is one of the worldwide favorite fish species (Aewsiri et al., 2008). The total catch of tuna in the world has increased continuously from 0.4 to 3.9 million metric tonnes from 1950 to 2000 (Miyake et al., 2004). Several new regulations for the global tuna industry came into force in 2010. Overhanging the sector, especially Japanese consumers, is the listing of Atlantic and Mediterranean bluefin on the Convention on International Trade in Endangered Species of Wild Fauna and Flora (CITES). In Thailand, tuna is usually processed as canned products, which are exported to many countries over the world. During the processing, a large amount of wastes involving skin, bone, and fin is generated (Shahidi, 1994). These wastes

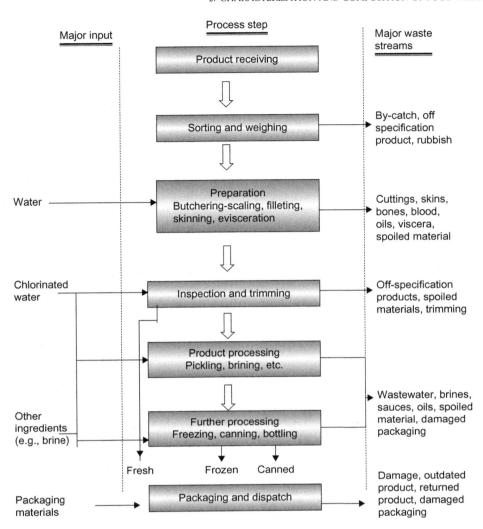

FIGURE 3.5 General process flow diagram for seafood processing operations. *Source: Tay et al. (2006).*

are commonly utilized as low value fish meal or fertilizer. So far, the utilization of fish-processing wastes has received increasing attention as a promising means to increase revenue for producers and to decrease the cost of disposal or management of those wastes. Fishery wastes can be used for enzyme recovery (Klomklao et al., 2005), protein hydrolysate production (Slizyte et al., 2005), collagen (Fernandez-Dıaz et al., 2001; Muyonga et al., 2004a), gelatin (Choi and Regenstein, 2000; Muyonga et al., 2004b), and lipids extraction (Kristinsson and Hultin, 2004). Seafood-processing operations produce wastewater containing substantial contaminants in soluble, colloidal, and particulate forms. Wastewater from seafood-processing operations can be very high in BOD, fat, oil and grease (FOG), and nitrogen content. Literature data for seafood-processing operations showed a BOD production of 1−72.5 kg per tonne of product. White fish filleting processes typically produce 12.5−37.5 kg of BOD for every tonne of product. BOD is derived mainly from the butchering process and general

cleaning, and nitrogen originates predominantly from blood in the wastewater stream (Environment Canada, 1994).

2.6.1 Chemical Composition of Fish Waste

As with many animal products, fish and fishery products contain water, proteins and other nitrogenous compounds, lipids, carbohydrates, minerals, and vitamins (Table 3.10). However, the chemical composition of fish varies greatly from one species and one individual fish to another depending on age, sex, environment, and season. Proteins and lipids are the major components, whereas carbohydrates are detected at very limited levels (<0.5%). Vitamin content is comparable to that of mammals except for vitamins A and D, which are found in large amounts in the meat of fatty species, especially in the liver of species such as cod and halibut. As for minerals, fish meat is a particularly valuable source of calcium and phosphorus as well as iron, copper, and selenium, which can create a problem when wasted. In addition, saltwater fish contain high levels of iodine. Fish lipids contrast greatly to mammalian lipids in that they include up to 40% of long-chain fatty acids (long-chain omega-3 polyunsaturated fatty acids, PUFA) that contain five or six double bonds. Esteban et al. (2007) evaluated fruit-and-vegetable and fish wastes as alternative feedstuffs in pig diets. They found that the fish waste contained 58% crude protein, 22% ash, 19% ether extract, and 1% crude fiber, whereas the fruit-and-vegetable waste contained 65% nitrogen-free extract, 13% crude fiber, 12% crude protein, 8% ash, and 2% ether extract.

2.6.2 Crustacean Wastes

Among seafood by-products, cooked cephalopod (particularly octopus) has higher commercial value and larger production of wastewater, particularly in NW Spain (Vazquez and Murado, 2008). These massive spills with high protein concentration generate a negative environmental impact on marine ecosystems. Over the past two decades the shellfish industry has also experienced a significant expansion, concentrating crustacean waste materials in some areas and in larger quantities. The crustacean species commercially harvested to the greatest extent are crab, shrimp, prawn, Antarctic krill, and crayfish. In 2006, 145.18 megatonnes of frozen shrimp was produced, and it can be estimated that nearly 150–175 megatonnes of shrimp waste per annum are generated from shrimp-processing companies in India (Babu et al., 2008). Use of these crustacean wastes has been of interest to researchers for two reasons: (1) these wastes are highly perishable and create environmental pollution (Tan and Lee, 2002); (2) they are rich sources of protein, chitin, and carotenoids. Proximate composition of shrimp shell and crab is presented in Table 3.11 (Santhosh and Mathew, 2008). These large quantities of waste materials are useful in production of chitin. Liming and Wenshui (2009) studied recovery of sodium hydroxide from alkaline wastewater in chitin processing using ultrafiltration and nanofiltration. Flux behaviors were observed with

respect to filtration time, volumetric concentration ratio (VCR), operating pressure, temperature, and cleaning. They found that the most suitable VCRs were ~50, resulting in a total sodium hydroxide recovery of 96% with removal of 94% COD and 100% SS.

A laboratory-scale sequencing batch reactor was successfully operated for shrimp aquaculture wastewater with high concentrations of carbon and nitrogen (Boopathy, 2009). Sequential operation of the reactor (aerobic and anoxic modes) yielded nitrification and denitrification as well as removal of carbon. Ammonia in the waste was nitrified within 4 days, and 100% denitrification was observed within 15 days of reactor operation.

Anaerobic treatment of food waste, including the treatment of fruits/vegetables, beverage, dairy, meat, poultry, and fish, recently has regained attention and proved to be an efficient method of waste valorization. Another advantage is the potential for stabilization of the solid residues. Co-digestion of waste material and energy production from anaerobic treatment processes has been often reported recently. The co-digestion plant is an environmentally friendly method of waste treatment and appears to have overcome the problems of trying to digest slaughterhouse and fish solid wastes in isolation (Neves et al., 2009; Martín-González et al., 2010).

3. BIOCHEMICAL/CHEMICAL ANALYTICAL METHODS

This section lists and describes those methods used for the analysis of chemical and biochemical constituents of waste and wastewaters, including BOD, COD, total organic carbon (TOC), total nitrogen and phosphorus, carbohydrates, proteins, lipids, organic acids, and other compounds.

BOD Traditionally, the biodegradable component of waste and wastewater is measured by the standard method of **biochemical oxygen demand (BOD$_5$)**. BOD$_5$ is a measure of the amount of dissolved oxygen required for the biochemical oxidation of the organic solutes in 5 days from the time when the test sample is seeded with a microbial system at 20°C. It is most commonly expressed in milligrams/grams of oxygen consumed per liter of sample. An alternative for the measurement of BOD is the development of biosensors, which are devices for the detection of an analyte that combines a biological component with a physico-chemical detector component. Biosensors can be used to indirectly measure BOD via a fast (<30 min) to-be-determined BOD substitute and a corresponding calibration curve method (pioneered by Karube et al., 1977). Consequently, biosensors are now commercially available, but they do have several limitations such as their high maintenance costs, limited run lengths due

TABLE 3.11 Proximate Composition of Shrimp Shell and Crab

Parameter	Shrimp shell (%)	Crab (%)
Moisture	75–80	70
Ash (dry basis)	30–35	45–50
Protein (dry basis)	35–40	30–35
Chitin (dry basis)	15–20	13–15
Fat (dry basis)	3–5	1.0–1.5

Source: Santhosh and Mathew (2008).

to the need for reactivation, and the inability to respond to changing quality characteristics as would normally occur in wastewater treatment streams as a result of diffusion processes.

COD The **chemical oxygen demand (COD)** test is commonly used to indirectly measure the amount of organic compounds in liquid waste. It is expressed in milligrams/grams per liter, which indicates the mass of oxygen consumed per liter of solution. Older references may express the units as parts per million (ppm). COD is determined using Hach tubes and a method based upon the reduction of (orange) potassium dichromate to (green) chromium salts at high temperature, followed by absorbance measurement using a dedicated colorimeter. The method has been chosen by the water industry as its standard, and is available from CHEMetrics® USA as a US EPA-approved dichromate reactor digestion method. The International Organization for Standardization describes a standard method for measuring COD in ISO 6060. Potassium dichromate is a strong oxidizing agent under acidic conditions. Acidity is usually achieved by the addition of sulfuric acid. In the process of oxidizing the organic substances found in the water sample, potassium dichromate is reduced, forming Cr^{3+}. The amount of Cr^{3+} is determined after oxidization is complete and is used as an indirect measure of the organic contents of the water sample. One of the main limitations of the COD test is its inability to differentiate between biodegradable and biologically inert organic matter on its own (Bourgeois et al., 2001). The use of chemicals such as chromium and strong acid produce liquid hazardous waste that requires disposal. To control for availability of organic matter accidentally added to the sample of water, a blank sample is usually created by adding all reagents to a volume of distilled water. COD is measured for both the water and blank samples, and the two are compared. The oxygen demand in the blank sample is subtracted from the COD for the original sample to ensure a true measurement of organic matter.

TOC An alternative for estimating the organic content is the **total organic carbon (TOC)** method, which is based on the combustion of organic matter (at 650–800°C) to carbon dioxide and water in a TOC-analyzer. After separation of water, the combustion gases are passed through an infrared analyzer and the response is recorded. The TOC-analyzer is gaining acceptance in some specific applications, as the test can be completed within a few minutes, provided that a correlation with the BOD_5 or COD contents has been established. An added advantage of the TOC test is that the analyzer can be mounted in the plant for online process control.

TS AND VS These are measured according to standard APHA methods (2005).

TOTAL NITROGEN This is determined, on pre-digested samples, using the **Kjeldahl method (TKN)**.

TOTAL PHOSPHORUS This is obtained, in pre-digested and neutralized samples, using a method involving the creation of phosphomolybdate blue.

HIGH-PRESSURE LIQUID CHROMATOGRAPHY (HPLC) This is a technique that can separate a mixture of compounds and is used for identification and quantification of carbohydrates and organic acids including VFA, glycerol, alcohols, etc. Chromatographic methods, commonly finding application in industry and research laboratories, are capable of separating the individual VFA acids and provide quantitative measures of their concentrations. Chromatographic separation is based on relative affinity between a mobile and a stationary phase in a separation column. Gas chromatographic separation of the VFA acids is feasible due to the relatively large differences in boiling points. Brondz (2002) presented a comprehensive review outlining the technological and methodological developments for quantification of fatty acids by liquid and gas chromatography. These techniques are routinely applied by professionals and require a great deal of expertise and sample preprocessing.

The Aminex HPX-87H column (Bio-Rad), using sulfuric acid elution for organic acids analysis, combined with a UV/VIS detector at 220 nm, and the Rezex RHM Monosaccharide column with RI detector, are applied for sugar and ethanol analysis.

GAS CHROMATOGRAPHY (GC) This is a common type of chromatography used in analytic chemistry for separating, identifying, and analyzing compounds that can be vaporized without decomposition. In gas chromatography, the mobile phase is a carrier gas, usually an inert gas such as helium or an unreactive gas such as nitrogen. The stationary phase is a microscopic layer of liquid or polymer on an inert solid support, inside metal tubing called a column. The gaseous compounds being analyzed interact with the walls of the column, which is coated with different stationary phases. This causes each compound to elute at a different time, known as the retention time of the compound. The comparison of retention times is what gives GC its analytical usefulness. GC can be also used for quantification of gases produced during anaerobic fermentation of food waste (e.g., methane (CH_4), carbon dioxide (CO_2) and hydrogen (H_2)). Neves et al. (2009) developed a method for long-chain fatty acids (LCFA) extraction, identification, and further quantification by GC with application to liquid and solid samples collected from anaerobic digesters. This method was applied to track conversion of oleic acid in a cow manure digester receiving pulses of an industrial effluent containing high lipid content from a a fish cannery.

LIQUID CHROMATOGRAPHY TANDEM COUPLED TO MASS SPECTROMETRY (LC-MS/MS) This proved to be a powerful tool to selectively screen phospholipids (major component of biological membranes) in extracts for the occurrence of structurally related substances and their concentrations (Islam et al., 2011). We separated the lipid molecules using a Fortis C_{18} analytical column (2.0 mm i.d. $\times$ 100 mm) temperated at 45°C. Elution was performed at a flow rate of 0.25 mL/min in a binary gradient mode. The mobile phase consists of 5 mM ammonium formate (pH 4.0)/methanol/tetrahydrofuran (5/2/3 → 1/2/7). The HPLC was coupled to an Agilent 6520 Q-TOF mass spectrometer equipped with an electrospray ion (ESI) source. ESI spectra were recorded in positive ionization for a mass range from m/z 100 to 1000. Capillary voltage was set to 4000. Nitrogen was used as the nebulizing gas at a flow rate and temperature of 8 L/min and 350°C, respectively.

PROTEIN ASSAY Protein concentrations measured using a Coomassie protein assay reagent (Pierce, USA) based on the Bradford method (Bradford, 1976).

4. CONCLUSIONS

This chapter explores the main FIW resources generated along the food and drink supply chain. Fruit-and-vegetable wastes are usually among the most-wasted items, followed by other perishables such as bakery and dairy products, meat, fish, and eggs. Important waste streams, covered by references published in 2010−11, include fruit and vegetables (apple and grape pomaces, onion and potato processing co-products), dairy, meat/poultry, and seafood processing by-products, as well as olive oil manufacturing and distillery wastes. This chapter also presents methods used for the analysis of chemical and biochemical components of FIW, including BOD, COD, TOC, total nitrogen and phosphorus, carbohydrates, lipids, proteins, organic acids, and others. We evaluated the effects of environmental pollution causes of food wastes, their characteristics, and chemical composition. The highest COD values were obtained for starch (122 gO_2/L) and potato (104 gO_2/L) stillages, followed by the COD of liquid olive oil waste (100 gO_2/L), wheat (90 gO_2/L), and barley (83 gO_2/L) stillages. The COD of Blue Stilton cheese whey was obtained in the range 66−72 gO_2/L. COD value of apples was about 19 gO_2/L, which was quite similar to the COD of mixed meat-processing wastewater.

Determining biowaste constituents is a first step to their valorization. Fruit and vegetables are rich in antioxidants, pectin, fibers, carbohydrates, organic acids, mineral salts, food flavors, colorants, etc. These compounds can be utilized as raw materials for secondary processes or as ingredients of new products. Fruits are rich in reducing sugars, therefore it will be interesting to produce first-generation bioethanol from these matrices (Joshi and Sandhu, 1996). Vegetable wastes are rich in cellulose, hemicellulose, and lignin, thus the production of second-generation bioethanol could be useful (Zheng et al., 2009). In general, FIW are characterized by heterogeneity of their sources, variation of their chemical composition and microbial instability, which is not easy to control. Hazardous wastes may be occasionally generated from discarded gourmet foods, depending on situations such as contamination by toxic chemicals or pathogens. Food-processing wastewater is mainly characterized by a high degree of organic pollution, sometimes containing polyphones, and aromatics forming inhibitor or toxic substances (e.g., liquid olive oil wastes), which constitute a serious environmental problem for soil, rivers, and groundwater. The great variety of components found in liquid and solid waste requires development of appropriate technologies to eliminate those that have harmful effects on the environment. In practical terms, it is imperative to develop cost-effective uses for the final waste product.

Food waste streams have a vast potential, which has been underestimated so far. In the following chapters we will discuss development of sustainable technologies, bioremediation strategies, production of animal feed, fertilizer, energy, and functional food ingredients, which can create new opportunities for the markets in agriculture, the food industry, fine chemistry, cosmetics, pharmaceuticals, and others.

References

Abayomi, L.A., Terry, L.A., 2009. Implications of spatial and temporal changes in concentration of pyruvate and glucose in onion (*Allium cepa L.*) bulbs during controlled atmosphere storage. J. Sci. Food Agric. 89, 683−687.

Aewsiri, T., Benjakul, S., Visessanguan, W., Tanaka, M., 2008. Chemical compositions and functional properties of gelatin from pre-cooked tuna fin. Int. J. Food Sci. Technol. 43, 685−693.

Agriculture and Agri-Food Canada. 2008. Canada's potato industry. Available at: <http://www.ats-sea.agr.gc.ca/pro/3639-eng.htm> (accessed 05.07.09.).

Akayezu, J.M., Linn, J.G., Harty, S.R., Cassady, J.M., 1998. Use of distillers grains and co-products in ruminant diets. Midwest Animal and Dairy Science Meeting.

Albuquerque, P.M., Koch, F., Trossini, T.G., Esposito, E., Ninow, J.L., 2006. Production of *Rhizopus oligosporus* protein by solid state fermentation of apple pomace. Braz. Arch. Biol. Technol. 49, 91−100.

Aliakbarian, B., Casazza, A.A., Perego, P., 2011. Valorization of olive oil solid waste using high pressure−high temperature reactor. Food Chem. 128, 704−710.

Allie, Z., Jacobs, E.P., Maartens, A., Swart, P., 2003. Enzymatic cleaning of ultrafiltration membranes fouled by abattoir effluent. J. Membr. Sci. 218, 107.

Anastassiadis, S., Rehm, H.-J., 2006. Oxygen and temperature effect on continuous citric acid secretion in *Candida oleophila*. Electron. J. Biotechnol. 9 (4), 414−423.

Angulo, J., Mahecha, L., Yepes, S.A., Yepes, A.M., Bustamante, G., Jaramillo, H., et al., 2012. Quantitative and nutritional characterization of fruit and vegetable waste from marketplace: a potential use as bovine feedstuff? J. Environ. Manage. 95, S203−209.

APHA, 2005. Standard methods for the examination of water and wastewater: contennial edition. Am. Public Health Assoc.

Arora, A., Jianxin, Z., Camire, M.E., 1993. Extruded potato peel functional properties affected by extrusion conditions. Food Sci. 58, 335−337.

Arora, J.K., Marwaha, S.S., Grover, R., 2002. In: Arora, J.K., Marwaha, S.S., Grover, R. (Eds.), Biotechnology in Agriculture and Environment. Asiatech Publishers Inc., New Delhi, pp. 129−149.

Avula, R.Y., Nelson, H.M., Singh, R.K., 2009. Recycling of poultry process wastewater by ultrafiltration. Inno. Food Sci. Emerg. Technol. 10, 1−8.

Babu, C.M., Chakrabarti, R., Sambasivarao, K.R.S., 2008. Enzymatic isolation of carotenoid-protein complex from shrimp head waste and its use as a source of carotenoids. LWT 41, 227−235.

Baçaoui, A., Yaacoubi, A., Dahbi, A., Bennouna, C., Phan Tan Luu, R., Maldonado-Hodar, F.J., 2001. Optimisation of conditions for the preparation of activated carbons from olive-waste cakes. Carbon 39, 425−432.

Balat, M., Balat, H., 2009. Recent trends in global production and utilization of bio-ethanol fuel. Appl. Energy 86, 2273−2282.

Banks, C.J., Wang, Z., 2006. Treatment of meat wastes. In: Wang, L. K., Hung, Y.-T., Lo, H.H., Yapijakis, K. (Eds.), Waste Treatment in Food Processing Industry. CRC Taylor & Fransis Group LLC, New York, pp. 67−100.

Battestin, V., Macedo, G.A., 2007. Effects of temperature, pH and additives on the activity of tannase produced by *Paecilomyces variotii*. Electron. J. Biotechnol. 10, 191−199.

Benítez, V., Mollá, E., Martín-Cabrejas, M.A., Aguilera, Y., López-Andréu, F.J., Cools, K., et al., 2011. Characterization of industrial onion wastes (*Allium cepa L.*): dietary fibre and bioactive compounds. Plant Foods Hum. Nutr. 66, 48−57.

Berg, C., 2004. World fuel ethanol—analysis and outlook.

Bernard, F., Prieur, A., 2007. Biofuel market and carbon modelling to analyse French biofuel policy. Energy Policy 35, 5991−6002.

Biofuels International, 2007. Regulations: Feedstocks of the future, 25th September 2007, vol. 1, no. 4.

Biofuels International, 2008. European ethanol production improves, News 10th April 2008.

Bohdziewicz, J., Sroka, E., Korus, I., 2003. Application of ultrafiltration and reverse osmosis to the treatment of the wastewater produced by the meat industry. J. Environ. Studies 12, 269.

Boopathy, R., 2009. Biological treatment of shrimp production wastewater. J. Ind. Microbiol. Biotechnol. 36, 989−992.

Bouallagui, H., Cheikh, R.B., Marouani, L., Hamdi, M., 2003. Mesophilic biogas production from fruit and vegetable waste in a tubular digester. Bioresour. Technol. 86, 85.

Bourgeois, W., Burgess, J.E., Stuetz, R.M., 2001. On-line monitoring of wastewater quality: a review. J. Chem. Technol. Biotechnol. 76, 337−348.

Bradford, M.M., 1976. A rapid and sensitive method for the quantification of microgram quantities of protein using the principle of protein-dye binding. Anal. Biochem. 72, 248−254.

Britz, T.J., van Schalkwyk, C., Hung, Y.-T., 2006. Treatment of dairy processing wastewaters. In: Wang, L.K., Hung, Y.-T., Lo, H.H., Yapijakis, K. (Eds.), Waste Treatment in Food Processing Industry. CRC Taylor & Francis Group LLC, Boca Raton, London, New York, pp. 1−28.

Brondz, I., 2002. Development of fatty acid analysis by high-performance liquid chromatography, gas chromatography, and related techniques. Anal. Chim. Acta 465, 1−37.

Buys, E.M., Nortje, G.L., 1997. HACCP and its Impact on Processing and Handling of Fresh Red Meats. Food Industries of South Africa, October Issue.

Carawan, R.E., Chambers, J.V., Zall, R.R., 1979. Seafood Water and Wastewater Management. The North Carolina, Agricultural Extension Service, U.S.A.

Cardona, C.A., Sánchez, O.J., 2007. Fuel ethanol production: process design trends and integration opportunities. Bioresour. Technol. 98, 2415−2457.

Choi, S.S., Regenstein, J.M., 2000. Physicochemical and sensory characteristics of fish gelatin. J. Food Sci. 65, 194−199.

Choorit, W., Wisarnwan, P., 2007. Effect of temperature on the anaerobic digestion of palm oil mill effluent. Electron. J. Biotechnol. 10, 376−385.

Cibis, E., Kent, C.A., Krzywonos, M., Garncarek, Z., Garncarek, B., Miśkiewicz, T., 2002. Biodegradation of potato slops from rural distillery by thermophilic aerobic bacteria. Bioresour. Technol. 85, 57−61.

Cibis, E., Krzywonos, M., Miśkiewicz, T., 2006. Aerobic biodegradation of potato slops under moderate thermophilic conditions: effect of pollution load. Bioresour. Technol. 97, 679−685.

Cibis, E., Ryznar-Luty, A., Krzywonos, M., Lutosawski, K., Miskiewicz, T., 2011. Betaine removal during thermo- and mesophilic aerobic batch biodegradation of beet molasses vinasse: influence of temperature and pH on the progress and efficiency of the process. J. Environ. Manage. 92, 1733−1739.

Cuellar, A.D., Webber, M.E., 2010. Wasted food, wasted energy: the embedded energy in food waste in the United States. Environ. Sci. Technol. 44, 6464−6469.

Czupryński, B., Kłosowski, G., Kotarska, K., Sadowska, J., 2000. Studies on utilization of potato slops in the production of rigid polyurethane-polyisocyanurate foams. Polymers 45, 439−441.

Davis, F., Terry, L.A., Chope, G.A., Faul, C.F.J., 2007. Effect of extraction procedure on measured sugar concentrations in onion (*Allium cepa L.*) bulbs. J. Agric. Food Chem. 55, 4299−4306.

Davis, L., Jeon, Y.J., Svenson, C., Rogers, P., Pearce, J., Peiris., P., 2005. Evaluation of wheat stillage for ethanol production by recombinant *Zymomonas mobilis*. Biomass Bioenergy 29, 49−59.

Del Pozo, R., Diez, V., 2003. Organic matter removal in combined anaerobic-aerobic fixed-film bioreactors. Water Res. 37, 3561.

Demirel, B., Yenigun, O., Onay, T.T., 2005. Anaerobic treatment of dairy wastewaters: a review. Process Biochem. 40, 2583−2595.

Dhillon, G.S., Brar, S.K., Verma, M., Tyagi, R.D., 2010. Recent advances in citric acid bio-production and recovery. Food Bioprocess Technol. 4, 505−529.

Directive 2009/28/EC of the European Parliament and of the Council. Promotion of the use of energy from renewable sources. Available at: <http://europa.eu/legislation_summaries/energy/renewable_energy/en0009_en.htm)>.

Environment Canada, 1994. Canadian biodiversity strategy: Canadian response to the convention on biological diversity, Report of the Federal Provincial Territorial Biodiversity Working Group; Environment Canada, Ottawa.

Esteban, M.B., Garcia, A.J., Ramos, P., Marquez, M.C., 2007. Evaluation of fruit, vegetable and fish wastes as alternative feedstuffs in pig diets. Waste Manag. 27, 193–200.

EUBIA, European Biomass Industry Association, 2007. Biofuels for Transportation. European Biomass Industry Association, Renewable Energy House, Brussels.

EUBIA, European Biomass Industry Association, 2008. Bioethanol.

FAO (U.N. Food and Agriculture Organization), 2008. Meat and meat products. Food Outlook.

FAO Statistics, 2008. Productions.

Fernandez-Dıaz, M.D., Montero, P., Gomez-Guillen, M.C., 2001. Gel properties of collagens from skins of cod (*Gadus morhua*) and hake (*Merluccius merluccius*) and their modification by the co-enhancers magnesium sulphate, glycerol and transglutaminase. Food Chem. 74, 161–167.

Food and Agriculture Organization of the United Nations for a world without hunger. Available at http://faostat.fao.org/site/339/default.aspx

Food Safety Network, 2003. Rendering Fact Sheet. University of Guelph.

Frenkel, V., Cummings, G., Scannell, D.E., Tang, W.Z., Maillacheruvu, K.Y., Treanor, P., 2010. Food processing wastes. Water Environ. Res. 82, 1468–1484.

Gannoun, H., Bouallagui, H., Okbi, A., Sayadi, S., Hamdi, M., 2009. Mesophilic and thermophilic anaerobic digestion of biologically pretreated abattoir wastewaters in an upflow anaerobic filter. J. Hazard. Mater. 170, 263–271.

Gonzalez-Siso, M.I., 1996. The biotechnological utilization of cheese whey: a review. Biores. Technol. 57, 1–11.

Grobben, N.G., Egglink, G., Cuperus, F.P., Huizing, H.J., 1993. Production of acetone, butanol and ethanol (ABE) from potato wastes: fermentation with integrated membrane extraction. Appl. Microbiol. Biotechnol. 39, 494–498.

Grobe, K., 1994. Composter links up with food processor. BioCycle 34, pp. 40, 42–43.

Guimarães, P.M.R., Teixeira, J.A., Domingues, L., 2010. Fermentation of lactose to bio-ethanol by yeasts as part of integrated solutions for the valorisation of cheese whey. Biotechnol. Adv. 28, 375–384.

Hung, Y.-T., Lo, H.H., Awad, A., Salman, H., 2006. Potato wastewater treatment. In: Wang, L., Hung, Y.-T., Lo, H.H., Yapijakis, C. (Eds.), Waste Treatment in the Food Processing Industry. CRC Taylor & Francis, Boca Raton, London, New York, pp. 193–254.

Hutnan, M., Hornak, M., Bodik, I., Hlavacka, V., 2003. Anaerobic treatment of wheat stillage. Chem. Biochem. Eng. Q. 17, 233–241.

International Energy Agency. 2007. Key World Energy Statistics 2007. OECD/IEA, Paris, France.

Islam, M.N., Chambers, J.P., Ng, C.K.Y., 2011. Lipid profiling of the model temperate grass. *Brachypodium distachyon*. Metabolomics (online).

Jaime, L., Martín-Cabrejas, M.A., Mollá, E., López-Andréu, F.J., Esteban, R.M., 2001. Effect of storage on fructan and fructooligosaccharide of onion (*Allium cepa L.*). J Agric Food Chem. 49, 982–988.

Janczukowicz, W., Zielinski, M., Debowski, M., 2008. Biodegradability evaluation of dairy effluents originated in selected sections of dairy production. Bioresour. Technol. 99, 4199–4205.

Joshi, V.K., Sandhu, D.K., 1996. Effect on type of alcohols in the distillates from the solid state fermentation of apple pomace by different yeasts. Natl. Acad. Sci Lett. 49, 219.

Karube, I., Matsunaga, T., Mitsuda, S., Suzuki, S., 1977. Microbial electrode BOD sensors. Biotechnol. Bioengng 19, 1535–1545.

Khardenavis, A.A., Kumar, S.M., Mudliar, S.N., Chakrabarti, T., 2007. Biotechnology conversion of agro-industrial wastewaters into biodegradable plastics, poly β-hydroxybutyrate. Bioresour. Technol. 98, 3579–3584.

Khraisha, Y.H., Hamdan, M.A., Qalalweh, H.S., 1999. Direct combustion of olive cake using fluidized bed combustor. Energy Sources 21, 319–327.

Kim, J.-S., Kim, B.-G., Lee, C.-H., 1999. Distillery waste recycle through membrane filtration in batch alcohol fermentation. Biotechnol. Lett. 21, 401–405.

Kingspohn, U., Bader, J., Kruse, B., Kishore, P.V., Schugerl, K., Kracke-Helm, H.A., et al., 1993. Utilization of potato pulp from potato starch processing. Proc. Biochem. 28, 91–98.

Kitamura, Y., Maekawa, T., Tagawa, A., Hayashi, H., Farrell-Poe, K.L., 1996. Treatment of strong organic, nitrogenous wastewater by an anaerobic contact process incorporating ultrafiltration. Appl. Eng. Agric. 12, 709–714.

Klein, K., Romain, R., Olar, M., Bergeron, N., 2008. Ethanol policies, programs and production in Canada.

Klomklao, S., Benjakul, S., Visessanguan, W., Simpson, B.K., Kishimura, H., 2005. Partitioning and recovery of proteinase from tuna spleen by aqueous two-phase systems. Process Biochem. 40, 3061–3067.

Kristinsson, H.G., Hultin, H.O., 2004. The effect of acid and alkali unfolding and subsequent refolding on the pro-oxidative activity of trout hemoglobin. J. Agric. Food Chem. 52, 5482–5490.

Krzywonos, M., Cibis, E., Miskiewicz, T., Kent, C.A., 2008. Effect of temperature on the efficiency of the thermo- and mesophilic aerobic batch biodegradation of high strength distillery wastewater (potato stillage). Bioresour. Technol. 99, 7816–7824.

Krzywonos, M., Cibis, E., Miśkiewicz, T., Ryznar-Luty, A., 2009. Utilization and biodegradation of starch stillage (distillery wastewater). Electron. J. Biotechnol. 12, 1–12.

Laubscher, A.C.J., Wentzel, M.C., Le Roux, J.M.W., Ekama, G.A., 2001. Treatment of grain distillation wastewater in an upflow anaerobic sludge bed (UASB) system. Water SA 27, 433–444.

Leathers, T.D., 1998. Utilization of fuel ethanol residues in production of the biopolymer alternan. Process Biochem. 33, 15–19.

Leathers, T.D., 2003. Bioconversions of maize residues to value-added coproducts using yeast-like fungi. FEMS Yeast Res. 3, 133–140.

Liming, Z., Wenshui, X., 2009. Stainless steel membrane UF coupled with NF process for the recovery of sodium hydroxide from alkaline wastewater in chitin processing. Desalination 249, 774–780.

Lundqvist, J., 2008. Saving Water: From Field to Fork—Curbing Losses and Wastage in the Food Chain. Stockholm International Water Institute, Stockholm, 16th Session of the United Nations Commission on Sustainable Development.

Luste, S., Luostarinen, S., Sillanpää, M., 2009. Effect of pretreatments on hydrolysis and methane production potentials of by-products from meat-processing industry. J. Hazard. Mater. 164, 247–255.

Madejón, E., López, R., Murillo, J.M., Cabrera, F., 2001. Agricultural use of three (sugar-beet) vinasse composts: effect on crops and chemical properties of a Cambisol soil in the Guadalquivir river valley (SW Spain). Agric. Ecosyst. Environ. 84, 55–65.

Mamma, D., Kourtoglou, E., Christakopoulos, P., 2007. Fungal multienzyme production on industrial by-products of the citrus-processing industry. Bioresour. Technol. 99, 2373–2383.

Martín-González, L.F., Colturato, X., Font, T., Vicent, 2010. Anaerobic co-digestion of the organic fraction of municipal solid waste with FOG waste from a sewage treatment plant. Waste Manag. 30, 1854–1859.

Masse, D.I., Masse, L., 2000. Treatment of slaughterhouse wastewater in anaerobic sequencing batch reactors. Can. Agric. Eng. 42, 131–137.

Mazza, G., Miniati, E., 1993. Grapes, Anthocyanins in Fruits, Vegetables, and Grains. CRC Press, Boca Raton, FL, 149–199.

Mellouli, H.J., Hartmann, R., Gabriels, D., Cornelis, W.M., 1998. The use of olive mill effluents as soil conditioner mulch to reduce evaporation losses. Soil Till. Res. 49, 85–91.

Mittal, G.S., 2006. Treatment of wastewater from abattoirs before land application—a review. Bioresour. Technol. 97, 1119–1135.

Miyake, M.P., Miyabe, N., Nakano, H., 2004. Historical trends of tuna catches in the world. In: FAO Fisheries Technical Paper. 467, 1–6.

Montané, D., Salvadó, J., Torras, C., Farriol, X., 2002. High temperature dilute-acid hydrolysis of olive stones for furfural production. Biomass Bioenergy 22, 295–304.

Monteils, V., Jurjanz, S., Colin-Schoellen, O., Blanchart, G., Laurent, F., 2002. Kinetics of ruminal degradation of wheat and potato starches in total mixed rations. J. Anim. Sci. 80, 235–241.

Morgan, E., 2009. Fruit and Vegetable Consumption and Waste in Australia. State Government of Victoria, Victorian Health Promotion Foundation, Victoria, Australia.

Mukherjee, S.N., Kumar, S., 2007. Leachate from market refuse and biomethanation study. J. Environ. Monit. Assess. 135, 49–53.

Murphy, J.D., Power, N.M., 2008. How can we improve the energy balance of ethanol production from wheat? Fuel 87, 1799–1806.

Mustafa, A.F., Mckinnon, J.J., Christensen, D.A., 1999. Chemical characterization and in vitro crude protein degradability of thin stillage derived from barley- and wheat-based ethanol production. Anim. Feed Sci. Technol. 80, 247–256.

Mustafa, A.F., Christensen, D.A., McKinnon, J.J. 2000. Effects of Pea, Barley, and Alfalfa Silage on Ruminal Nutrient Degradability and Performance of Dairy Cows. Journal of Dairy Science, 83(12), 2859–2865.

Muyonga, J.H., Cole, C.G.B., Duodu, K.G., 2004a. Characterization of acid soluble collagen from skins of young and adult Nile perch (Lates niloticus). Food Chem. 85, 81–89.

Muyonga, J.H., Cole, C.G.B., Duodu, K.G., 2004b. Extraction and physico-chemical characterization of Nile perch (Lates niloticus) skin and bone gelatin. Food Hydrocolloids 18, 581–592.

Myer, R.O., Brendemuhl, J.H., Johnson, D.D., 1999. Evaluation of dehydrated restaurant food waste products as feedstuffs for finishing pigs. J. Anim. Sci. 77, 685–692.

Nagano, A., Arikawa, E., Kobayashi, H., 1992. Treatment of liquor wastewater containing high-strength suspended solids by membrane bioreactor system. Water Sci. Technol. 26, 887–895.

Nelson, M.L., 2010. Utilization and application of wet potato processing co-products for finishing cattle. J. Anim. Sci. 88, E133–E142.

Neves, L., Pereira, M.A., Mota, M., Alves, M.M., 2009. Detection and quantification of long chain fatty acids in liquid and solid samples and its relevance to understand anaerobic digestion of lipids. Bioresour. Technol. 100, 91–96.

Niaounakis, M., Halvadakis, C.P., 2006. Olive Processing, Second Ed. Waste Management: Literature Review and Patent Survey, Waste Management, 5. Elsevier, Amsterdam.

Nord, M., Coleman-Jensen A., Andrews M., Carlson S. (2010) Household Food Security in the United States, 2009, Economic Research Report No. (ERR–108) 68.

OECD/FAO, 2007. Agricultural Outlook 2007–2016: Paris, France.

OECD/FAO, 2008. OECD-FAO Agricultural Outlook 2008-2017. Paris, France.

Ontario Report of the Meat Regulatory and Inspection Review, 2005. Disposal of meat production waste. Paris, France.

Pagnanelli, F., Mainelli, S., Vegliò, F., Toro, L., 2003. Heavy metal removal by olive pomace: biosorbent characterisation and equilibrium modeling. Chem. Eng. Sci. 58, 4709–4717.

Pailthorp, R. E., J. W. Filbert, and G. A. Richter. 1987. Treatment and disposal of potato wastes. In: Potato Processing. W. F. Talburt and O. Smith, (eds.) AVI Publ. Co., Westport, CT, pp. 747–788.

Parfitt, J., Mark Barthel, M., Macnaughton, S., 2010. Food waste within food supply chains: quantification and potential for change to 2050. Phil. Trans. R. Soc. B 365, 3065–3081.

Passeggi, M., Lopez, I., Borzacconi, L., 2009. Integrated anaerobic treatment of dairy industrial wastewater and sludge. Water Sci. Technol. 59, 501–506.

Pedersen, C., Boersma, M.G., Stein, H.H., 2007. Digestibility of energy and phosphorus in 10 samples of distillers dried grains with solubles fed to growing pigs. J. Anim. Sci. 85, 1168–1176.

Pekcan, G., Koksal, E., Kucukerdonmez, O., Ozel, H., 2006. Household Food Wastage in Turkey. FAO, Rome, Italy.

REN21 (Renewable Energy Network for the 21st Century), 2008. Renewable 2007 Global Status Report. Paris: REN21 Secretariat and Worldwatch Institute, Washington, DC.

Renewable Fuels Association., 2008. Renewable fuels standard. Washington, USA.

Rico Gutierrez, J.L., Garcia Encina, P.A., Fdz-Polanco, F., 1991. Anaerobic treatment of cheese-production wastewater using a UASB reactor. Bioresour. Technol. 37, 271–276.

Ross, C.C., and Valentine, G.E., 1992. Anaerobic treatment of poultry processing wastewaters. In: Proc. 1992 Natl. Poultry Waste Manage. Symp., Auburn, AL, p. 199.

Ruberto, G., Renda, A., Amico, V., Tringali, C., 2008. Volatile components of grape pomaces from different cultivars of Sicilian Vitis vinifera L. Bioresour. Technol. 99, 260–268.

Rubio-Texeira, M., 2006. Endless versatility in the biotechnological applications of Kluyveromyces LAC genes. Biotechnol. Adv. 24, 212–225.

Ruynal, J., Delgenes, J.P., Moletta, R., 1998. Two phase anaerobic digestion of solid waste by a multiple liquefaction reactors process. Bioresour. Technol. 65, 97–103.

Saez, L., Perez, J., Martinez, J., 1992. Low molecular weight phenolics attenuation during simulated treatment of wastewaters from olive oil mill in evaporation ponds. Water Res. 26, 1261–1266.

Sanchez, R.F.S., Cordoba, P., Sineriz, F., 1985. Use of the UASB reactor for the anaerobic treatment of stillage from sugarcane molasses. Biotechnol. Bioeng. 27, 1710–1716.

Sancho, P., Pinacho, A., Ramos, P., Tejedor, C., 2004. Microbiological characterization of food residues for animal feeding. Waste Manag. 24, 919–926.

Santhosh, S., Mathew, P.T., 2008. Preparation and properties of glucosamine and carboxymethylchitin from shrimp shell. J. Appl. Polym. Sci. 107, 280–285.

Satyawali, Y., Balakrishnan, M., 2008. Wastewater treatment in molasses based alcohol distilleries for COD and color removal: a review. J. Environ. Manage. 86, 481–497.

Scotland Regulations, 2003 The Animal By-Products. Training Seminar materials (Edinburgh, 4 November 2003); EU, Questions and Answers on Animal By-Products (Brussels, 6 May 2004).

Seelke, C.R. and Yacobucci, B.D., 2007. Ethanol and Other Biofuels: Potential for U.S.–Brazil Energy Cooperation. CSR Report for Congress.

Shahidi, F., 1994. Seafood processing by-products. In: Shahidi, F., Botta, J.R. (Eds.), Seafoods Chemistry, Processing, Technology and Quality. Chapman & Hall, London, pp. 320–334.

Slizyte, R., Dauksasa, E., Falch, E., Storro, I., Rustad, T., 2005. Yield and composition of different fractions obtained after enzymatic hydrolysis of cod (*Gadus morhua*) by-products. Process Biochem. 40, 1415—1424.

Smith, J.H., 1986. Decomposition of potato processing wastes in soil. Environ. Qual. 15, 13—15.

Smith, O., 1987. Potato chips. In: Talburt, W.F., Smith, O. (Eds.), Potato Processing. Van Nostrand Reinhold Co., New York, pp. 371—474.

Starkey J., 2000. U.S. Poultry & Egg Association, Tucker, Georgia. Personal Communication.

Stein, H.H., Shurson, G.C., 2009. Board-invited review: the use and application of distillers dried grains with solubles in swine diets. J. Anim. Sci. 87, 1292—1303.

Swaminathan, M.S., 2006. 2006-07: Year of agricultural renewal. 93 Indian Science Congress in Hyderabad, Public Lecture, January 4.

Sweeten, J.M., Lawhon, J.T., Schelling, G.T., Gillespie, T.R., Coble, C.G., 1981. Removal and utilization of ethanol stillage constituents. Energ. Agr. 1, 331—345.

Szasz, J.I., Hunt, C.W., Turgeon Jr., O.A., Szasz, P.A., Johnson, K.A., 2005. Effects of pasteurization of potato slurry by-product fed in corn- or barley-based beef finishing diets. J. Anim. Sci. 83, 2806—2814.

Tan, E.W.Y., Lee, V.R., 2002. Enzymatic Hydrolysis of Prawn Shell Waste for the Purification of Chitin. Department of chemical Engineering, Loughborough University, UK, Final report R&D project supervised by Hall GM.

Tang, Y.Q., Fujimura, Y., Shigematsu, T., Morimura, S., Kida, K., 2007. Anaerobic treatment performance and microbial population of thermophilic upflow anaerobic filter reactor treating *awamori* distillery wastewater. J. Biosci. Bioeng. 104, 281—287.

Tay, J.-H., Show, K.-Y., Hung, Y.-T., 2006. Seafood processing wastewater treatment. In: Wang, L.K., Hung, Y.-T., Lo, H.H., Yapijakis, K. (Eds.), Waste Treatment in Food Processing Industry. CRC Taylor & Francis Group LLC, New York, pp. 29—66.

Tejada, M., García-Martínez, A.M., Parrado, J., 2009. Effects of a vermicompost composted with beet vinasse on soil properties, soil losses and soil restoration. CATENA 77, 238—247.

Thassitou, P.K., Arvanitoyannis, I.S., 2001. Bioremediation: a novel approach to food waste management. Trends in Food Sci. Tech. 12, 185—196.

Thonissen, R., 2009. Food waste: The Netherlands. Presentation to the EU Presidency Climate Smart Food Conf., November 2009, Lund, Sweden.

Tolmasquim, M.T., 2007. Bioenergy for the future. In: Conference on Biofuels: an option for a less carbon intensive economy (December 2007) Sao Paulom, Brazil.

Torkian, A., Alinejad, K., Hashemian, S.J., 2003b. Posttreatment of upflow anaerobic sludge blanket-treated industrial wastewater by a rotating biological contactor. Water Env. Res. 75, 232.

Torkian, A., Eqbali, A., Hashemian, S.J., 2003a. The effect of organic loading rate on the performance of UASB Reactor Treating Slaughterhouse Effluent. Resour. Conserv. Recyc. 40, 1.

Torres, J.B., Varela, M.T., Garcia, J., et al., 2002. Valorization of grape (*Vitis vinifera*) by products. Antioxidant and biological properties of polyphenolic fractions differing in procyanidin composition and flavonol content. J. Agric. Food Chem. 50, 7548—7555.

Treadway, R.H., 1987. Potato starch. In: Talburt, W.F., Smith, O. (Eds.), Potato Processing. Van Nostrand Reinhold Co., New York, New York, pp. 647—666.

USDA National Agriculture Statistics Service., 2008. Potatoes 2007 Summary Cornell University, USA.

USDA-SCS (US Department of Agriculture-Soil Conservation Service), 1992. Agricultural Waste Management Field Handbook. Washington, DC.

Valladao, A.B.G., Sartore, P.E., Freire, D.M.G., Cammarota, M.C., 2009. Evaluation of different pre-hydrolysis times and enzyme pool concentrations on the biodegradability of poultry slaughterhouse wastewater with a high fat content. Water Sci. Technol. 60, 243—249.

Vazquez, J.A., Murado, M.M., 2008. Enzymatic hydrolysates from food wastewater as a source of peptones for lactic acid bacteria productions. Enzyme Microb. Technol. 43, 66—72.

Vendruscolo, F., Albuquerque, P.M., Streit, F., Esposito, E., Ninow, J.L., 2008. Apple pomace: a versatile substrate for biotechnological applications. Crit. Rev. Biotechnol. 28, 1—12.

Waldron, K.W., 2001. Useful ingredients from onion waste. Food Sci Technol. 15, 38—41.

West, T.P., Strohfus, B., 1996. Pullulan production by *Aureobasidium pullulans* grown on ethanol stillage as a nitrogen source. Microbios 88, 7—18.

Whitney, M.H., Shurson, G.C., 2001. Availability of phosphorus in distillers dried grains with solubles for growing swine. J. Anim. Sci. 79, 108.

Wilkie, A.C., Riesedel, K.J., Owens, J.M., 2000. Stillage characterization and anaerobic treatment of ethanol stillage from conventional and cellulosic feedstock. Biomass Bioenergy 19, 63—102.

Worldwatch Institute, 2008. Meat Production Continues to Rise.

WRAP, 2008. The Food We Waste. Banbury, UK.

WRAP, 2009. Household Food and Drink Waste in the UK. Banbury, UK.

WRAP, 2010. A Review of Waste Arisings in the Supply of Food and Drink to Households in the UK. Banbury, UK.

Yang, F.C., Lin, I.H., 1998. Production of acid protease using thin stillage from a rice-spirit distillery by *Aspergillus niger*. Enzyme Microb. Technol. 23, 397—402.

Yokoi, H., Aratake, T., Nishio, S., Hirose, J., Hayashi, S., Takasaki, Y., 1998. Chitosan production from shochu distillery wastewater by funguses. J. Ferment. Bioeng. 85, 246—249.

Zarzyycki A., Polska W., 2007. Bioethanol Production from Sugar Beet—European and Polish perspective. The first TOSSIE workshop on technology improvement opportunities in the European sugar industry, Ferrara, Italy.

Zheng, Y., Pan, Z., Zhang, R., 2009. Overview of biomass pretreatment for cellulosic ethanol production. Int J Agric Biol Eng. 2, 51—68.

PART II

TREATMENT OF SOLID FOOD WASTES

Use of Waste Bread to Produce Fermentation Products

Mehmet Melikoglu and Colin Webb

1. INTRODUCTION

In the 21st century, with the presence of supermarkets and fast food chains, people of the developed world are very different from their predecessors and even from the people of the developing world. Not only do they have much better access to food but they are also much wealthier and can therefore, unfortunately, afford to waste more food. With the convenience of excellent food packaging, people today are not always aware of purchasing more than they need and do not think about the food that they are wasting as much as they might have done in the past.

Consequently, large quantities of food of many different types are wasted globally. It is difficult to quantify them individually, but the top three major categories of food wastes are meat, fruit and vegetables, and bakery products. Most of the latter category is bread (Jones, 2006; Jones, 2007; Kantor et al., 1997; Parry, 2007). This chapter explores the potential use of waste bread as a renewable raw material for the production of value-added products using fermentation as the principal processing route.

2. BREAD AS A MAJOR DIETARY STAPLE

Bread has long been the staple food of the western world and, increasingly, it is becoming a major alternative to other cereal-based foods in the Far East. Our very civilization is based on the domestication and cultivation of cereals around 10,000 years ago in the so-called Fertile Crescent. It was this cultivation of cereals that enabled much larger populations to live in close-knit communities and begin to build cities. Nowadays, more than 2 billion tonnes of cereals are grown each year around the world and cereals account for more than half of all the food we eat or feed to animals.

Cereals cannot be eaten without first being processed. In the case of rice, which makes up approximately one third of the total cereal harvest, this involves polishing the grain, after which it can be cooked and eaten directly. Maize (corn), which also accounts for approximately one third of the harvest, is grown primarily as an animal feed, though it is increasingly being diverted into biofuels production. Arguably the most important of the cereals, wheat, also accounts for one third of the total cereal crop (all other cereals are produced in very minor quantities by comparison to the big three) and the majority of this is processed into bread of one sort or another. Bread making, in very crude terms, consists of grinding the wheat grain to produce flour, mixing this to a dough with water, allowing this to condition (often involving yeast, as a leavening agent), and then baking into large loaves or small rolls. The traditional process can take more than a day and even using modern rapid processing techniques requires several hours.

While other carbohydrate-rich food sources, such as rice and potatoes, are generally cooked as part of a meal in the required quantity, bread is traditionally baked separately and in relatively large quantities, regardless of how much is required. This, of course, is in part due to the comparatively lengthy process of bread making and also to the tradition of baking loaves rather than individual servings. As a consequence, a large proportion of the bread that is baked is not eaten immediately and, while baked loaves can be kept fresh for several days, much of the bread ends up as stale waste. Another key difference between wheat consumption and that of other carbohydrate sources is that much more of the processing (and cooking) is done outside of the domestic environment. Even in

Food Industry Wastes.
DOI: http://dx.doi.org/10.1016/B978-0-12-391921-2.00004-4

63

small rural communities it is common to have a bakery, while in large communities and cities there will usually be multiple, industrial scale, bread factories servicing local supermarkets. Thus further considerations for wastage are unsold fractions of production from the bakery along with those from the retailer.

2.1 Staling and Spoilage

While cereal grains in general are quite indigestible in their raw state, and therefore quite stable to microbial attack, bread is a highly nutritious food, not just for humans. Due to the combined effects of heat and moisture during baking, most of the starch in bread is gelatinized, making it much more readily digestible and therefore susceptible to microbial attack. Nutrient compositions of different types of bread vary but, typically, 100 grams of white bread contains around 50 g carbohydrate (where 47 g is in the form of starch), 37 g water, and about 8 g protein. This composition makes bread an excellent, near complete source of nutrition for many microorganisms as well as for its intended consumers.

2.1.1 Staling of Bread

Bread has a unique texture, and the baked loaf consists of two distinct layers. An outer, hard "crust" surrounds the inner, soft and porous "crumb". The crust dries faster than the interior during baking and becomes less porous as a result. This lack of porosity permits it to act as a protective layer following production and enables the crumb to remain moist. Even so, bread has a relatively short shelf life. Various physical and chemical changes occur during its storage. These include crumb firming, as well as changes in taste, aroma, water absorption capacity, crystallinity, opacity, and soluble starch content (Chen et al., 1997). The global effects of these individual changes cause the deterioration of bread (Mandala, 2005) known as staling. Stale bread loses its sensory qualities. This results in a negative perception from the consumer even though the product is still healthy and rich in nutrients (Ribotta and Bail, 2007). Consequently, large quantities of bread are discarded, which causes an important economic loss and makes waste bread a major component of the food waste problem.

Two main features of staling are the loss of moisture and retrogradation of starch (Hui, 2006). Moisture is lost due to migration from the crumb to the crust. During this process, the crumb becomes drier and the crust becomes soft and leathery. During baking, the temperature inside the crumb reaches 95–98°C (Wrigley and Walker, 2004). At temperatures above 70°C, starch readily absorbs water and gelatinizes. However, the gelatinized starch is no longer stable (the very reason for cooking it in the first place).

On cooling, starch crystallization or retrogradation occurs usually over a period of a few days (Morgan et al., 1992) and is believed to be the principal cause of staling. The two major starch components, amylose and amylopectin, act differently during retrogradation, with amylose retrograding during the first few hours following baking and amylopectin remaining stable for much longer (Ribotta and Bail, 2007).

Starch retrogradation is both time and temperature dependent. It proceeds fastest at low temperatures just above freezing (Hui, 2006) and so, contrary to common belief, refrigeration of bread (at 4°C) accelerates staling compared with storage at room temperature. Furthermore, bread stales rapidly at temperatures in the range −3°C to −10°C but hardly at all at −20°C or below (Ranken et al., 1997). Therefore, freezing and frozen storage of bread is excellent for preserving it as long as the temperature is sufficiently low. Such low-temperature freezing drastically reduces water activity, which retards staling by slowing down the rearrangement of the starch molecules. Staling can be therefore also be retarded by the addition of ingredients that lower water activity such as salt and sugar or bind water (e.g., hydrocolloids and proteins). Staling can also be slowed by the incorporation of surfactants, shortening, or heat-stable α-amylase into the bread mix (Hui, 2006).

2.1.2 Spoilage of Bread

Legan (1993) mentioned that the most important common factors for different breads are their high moisture content (about 40% wb) and water activity, a_w (0.94–0.97). While they make the bread attractive to consumers, these factors also make it susceptible to mold attack and thereby limit its shelf life to 3–7 days unless special steps are taken. Recipe formulation and storage conditions can be used for extending the shelf life of bread.

Spoilage of bread and the consequent waste problem cause large economic losses. Microbial spoilage is the major problem causing deterioration of bread products and accounts for the loss of 1–5% of product, depending on season, type of product being produced, and method of processing (Needham et al., 2005). Bakers use preservatives to reduce spoilage and ensure food safety (Suhr and Nielsen, 2004). However, consumers are not in favour of such preservatives and urge bakers to reduce the quantities. Reduction to subinhibitory levels might even stimulate the growth of fungi and, potentially, mycotoxin production. The most common food preservatives are:

- **Propionic acid**: This preservative inhibits molds and *Bacillus* spores but not yeasts to the same extent. Therefore, it has been used as the traditional additive for bread preservation.

- **Sorbic acid**: This preservative is more effective than propionic acid. It inhibits both molds and yeasts. It is used in a broad variety of food products, including fine bakery products, confectionery, and bread.
- **Benzoic acid**: This preservative is used in many types of acidic food products and bread. However, it is mainly associated with fruit preservation.

Even with the addition of preservatives, most bread is still susceptible to spoilage by microorganisms, including molds of the genera *Penicillium*, *Aspergillus*, *Cladosporium*, *Mucorales*, and *Neurospora* (Legan, 1993). Figure 4.1 shows a typical example of bread spoilage by molds.

3. THE SIZE OF THE BREAD WASTE PROBLEM

So, how much bread do we waste? Let's first consider the size of the market. In 2004, global bread and rolls production was around 90.8 million tonnes (Anon, 2011). It has since risen by around 10% because of increases in the world population and the trend for greater bread consumption in China and the Far East. In very round terms it is therefore about 100 million tonnes per annum. Despite recent trends, Europe still dominates the market, with more than 65% of sales value. The USA accounts for just 9%, similar in absolute terms to the Asia Pacific Region, though much higher on a per capita basis. The rest of the world consumes the remaining 17% (Anon, 2011). Artisan bakeries supply most of the bread market in Europe, especially in France, Spain, Greece, and Turkey. However, in countries such as the UK and Germany, large baking companies and in-store bakeries dominate the market.

FIGURE 4.1 Bread spoilage by filamentous fungi.

In the UK, the bread and rolls sector is one of the largest in the food industry and produces almost 3 million tonnes of product. Most of this is supplied as standard 800 g loaves and thus the equivalent of almost 10 million such loaves is produced daily. The average UK household buys over 86 loaves per year, the vast majority of which are produced by large plant bakeries owned by just three companies (Allied Bakeries, British Bakeries, and Warburtons). A relatively small but growing volume (ca. 15%) of bread is baked "in-store" at major supermarkets, and just 1% is produced by artisan "Master" bakers.

3.1 Estimated Wastage

It is possible to estimate roughly the size of losses, or wastage, of bread, and the following example considers the case of the UK. As mentioned, wastage occurs at bakeries, retail outlets, and consumer households. Although it is difficult to determine the actual amount of bread that is lost along its life cycle, the average amount of waste can be estimated from the results of surveys and statistical analysis. Fallows and Wheelock (1982) estimated that in the region of 2–5% of the bread production within the UK baking industry was wasted. They calculated that this waste was between 700 and 1,750 tonnes per week. Assuming that the percentage loss has not changed with time, and taking annual bread production as 3 million tonnes, this suggests that current wastage from the industry is around 58,000 to 145,000 tonnes annually. However, this does not include bread sold but subsequently wasted by the consumer.

In a recent report, published by the Prudential[1], it was stated that 60% of people surveyed throw away a loaf of bread every week. Assuming a standard loaf of 800 grams, annual bread production in the UK of 2.9 million tonnes, and an adult population of 46.7 million, it was calculated that approximately 1.2 million tonnes, or 40% of annual bread production, might be wasted by consumers in the UK (Melikoglu, 2008). Parry (2007) stated that avoidable bread and bakery waste in the UK is 407,000 tonnes annually. This is approximately equal to 15% of the annual bread production. It must be emphasized that this is just the avoidable fraction of bread waste in the UK, which suggests that the actual loss is much greater. According to a more recent study by *Delicious* magazine, it was found that more than 40% of British people throw out bread on a regular basis. The results from this survey match very well with the

[1]http://dl.dropbox.com/u/21130258/resources/Attitudes/soggy_lettuce_pru.pdf

calculation carried out above. Whatever the exact figure, it is clear that overall wastage is considerable.

Of course, the waste bread problem is not unique to the UK. It has been reported that in Turkey, almost every day 120 million loaves of bread, each weighing 200 grams, are produced. Around 10% of these are never consumed, causing an economic loss of approximately $700 million per year to the economy and producing around 0.9 million tonnes of waste. In Germany, 8–10% of bread on sale is returned to the manufacturers before the sell-by date in order to supply fresher bread for the customer to buy (Meuser, 1998). Processing this bread, referred to as bread returns, constitutes a major problem for most German industrial bakeries.

In Japan during the manufacturing process of bread, substandard breads, bread crusts removed to make sandwiches, and unsold bread from retail stores that is returned to the bakeries are constant by-products of the bread industry (Oda et al., 1997). Although it is difficult to calculate the total amount lost, Oda and coworkers mentioned that it might account for more than 1% of the flour consumed by the bread industry in Japan. It should be noted that this loss is just within the manufacturing process and does not include commercial, retail, or household wastage, where losses are much higher. Bread wastage is a worldwide problem. Tens, if not hundreds of thousands, of tonnes of bread are wasted daily around the world, and this valuable biomass constitutes a potential resource that could be utilized in productive processes rather than lost to the environment. Perhaps a fourth R, "Reprocess", could be added to the classical 3 Rs (reduce, reuse, and recycle) of waste management strategy.

4. UTILIZATION OF BREAD AND BAKERY WASTES

In common with all cereal products, bread is much less stable than the unprocessed grain from which it is produced, and so, once disposed of, it deteriorates rapidly. Its degradation is almost always biological and usually consists of a solid-state fermentation involving filamentous fungi, often species of Aspergilli. Hence, it should be possible to harness this natural fermentation process to produce potentially valuable products rather than leaving the bread to be transformed into greenhouse gas emissions in the form of carbon dioxide or, worse still, methane. Using data reported by Adhikari et al. (2006) for food waste in general, it can be estimated that each 800 g loaf of bread is responsible for around 100 L of biogas generation, of which 60–65% is methane and the rest is CO_2. According to

the EU Landfill Directive[2], biodegradable municipal waste disposed into landfills should be decreased to 35% of 1995 levels by 2020. Espinoza-Orias et al. (2011) determined the carbon footprint of a typical 800 g loaf of bread to range from 977 to 1,244 g CO_2 equivalent. It is likely, therefore, that realizing the potential of bread waste as a renewable raw material will be an attractive proposition to governments. One way in which this potential might be realized is through the production of value-added products based on fermentation.

In the Satake Centre for Grain Process Engineering, at the University of Manchester, research over many years has led to the development of processes for the production of generic fermentation feedstocks from a variety of raw materials (Webb and Wang, 1997; Webb et al., 2004; Koutinas et al., 2007; Wang et al., 2010). Such feedstocks provide the basis for the grain-based biorefinery concept (see Box 4.1). Bread is an ideal candidate for such processes, providing a well-balanced source of carbon, nitrogen, and other major and minor nutrients. In fact the main function of producing bread is to make cereal grains (usually wheat) more digestible to humans, and therefore more fermentable.

From a simple literature survey it is clear to see that little attention has been given to the problems, and potential utilization, of bakery wastes and waste bread. Searching the Web of Science[SM], spanning the years 1945 to 2012, yields just 59 results for "bread waste" or "waste bread". The Scopus database, covering a wider range of publication types, returns 96. These are very low counts for what are, after all, global databases. There is, however, a growing interest, with some 20 of those publications appearing within the last 2 years. Most recently, Kawa-Rygielska and Pietrzak (2011) have investigated the possibility of using bakery wastes (dough, bread, cakes) as raw materials for ethanol fermentation by Saccharomyces cerevisiae. This has been a theme of much of the work with such wastes and is an obvious target, since the yeast for ethanol production is essentially the same as the one used for leavening the bread. Leung et al. (2012) have reported successful production of succinic acid through the fermentation of waste bread pieces, while Doi et al. (2009) have investigated the feasibility of producing biohydrogen from waste bread using microorganisms isolated from rice. They showed that at a reactor temperature of 35°C, with a hydraulic retention time of 12–36 h they could produce up to 1.30 mol-H_2/mol-hexose consumed.

Early work on the utilization of bakery wastes was carried out by Nakano and Yoshida (1977), who patented a process to use crushed waste bread pieces

[2]http://www.defra.gov.uk/environment/waste/topics/landfill-dir/

BOX 4.1

THE BIOREFINERY CONCEPT

Most types of chemicals produced from petroleum can be produced from biomass.

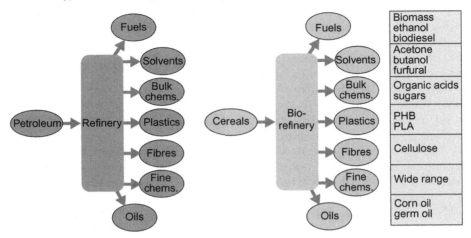

In the conventional refinery, petroleum is distilled to separate light and heavy fractions. Through processes such as cracking and reforming, large hydrocarbon molecules are broken down and used to synthesize new chemicals to form a wide range of products. Analogously, in the biorefinery, large macromolecules such as starch and proteins are broken down through hydrolysis to produce simple sugars and amino acids. These provide nutrients for diverse microorganisms to synthesize a wide range of functional products.

mixed with molasses, celluloytic, proteolytic, and saccharifying enzymes. The mix was incubated at 40–50°C for 70–80 hours to produce a syrup, which could be used as a sugar substitute. Martin (1984) tried to determine the possible yields of ethanol from the waste products of major bakeries in the Denver, Colorado, area by distillation in a laboratory-scale column. After a couple sets of experiments, he concluded that the production of alcohol from bakery wastes is feasible, and the average yield of ethanol from bread type products ranged up to 25%. According to patented research by Menge (1986), lactic and acetic acid-producing bacteria can be used to ferment a mixture of waste bread and flour at temperatures between 20°C and 30°C to produce an acidic dough for recycling into bread and other bakery products.

More recently, Berghofer et al. (1995) developed a process for syrup production from waste bread by using hot mash with malt and enzymatic hydrolysis processes. The optimum process parameters found in the laboratory were tested in pilot plant experiments and, using these results, calculations were carried out to determine overall economics of potential processes. Another patented process was registered by Maeda et al. (2004) for

the treatment and utilization of waste bread for ethanol and animal feed production. Their process involves four steps. First, any uncooked materials from the feedstock are removed. The feedstock is then hydrolyzed with water and commercial enzymes. The hydrolysate is fermented to produce alcohols, and then, finally, the alcohol solution is mixed with a portion of the original feedstock. They also stated that the product from this stream contains little water, so does not decay straightaway and could be stored or transported to another place for the production of feeding materials.

Oda and coworkers (1997) tried to optimize the production of lactic acid from bread crust, and the application of the culture filtrate obtained from the lactic acid fermentation was used in the bread-making process as an economical method of recycling bakery wastes. It is believed that the major breakthrough in this research is the conversion of the starch in bakery wastes to lactic acid without supplementing starch-degrading enzymes. Production of lactic acid from discarded bread was carried out using an amylolytic lactic acid bacterium, *Lactobacillus amylovorus*. Kumar and coworkers (1998) tried to utilize bakery wastes to produce ethanol. The bakery wastes included bread, biscuits, buns, cakes, and donuts, which were 2–3 weeks old. They ground the bakery wastes and mixed them with water and commercial α-amylase and glucoamylase to hydrolyze the starch found in the bakery wastes. Then the hydrolysate was used for ethanol fermentation using distiller's dried yeast and an ethanol tolerant *Saccharomyces cerevisiae*.

Daigle and colleagues (1999) also conducted fermentations on waste bread crumb but for the production of aroma compounds. Fermentations were carried out with 35% white bread crumb and 65% water, using *Geotrichum candidum* ATCC 62217 in Erlenmeyer flasks at 30°C and 300 rpm. Others, more recently Asghar et al. (2002), have tried to optimize the fermentation parameters for α-amylase production to establish the relationship between *Arachniotus* sp. and waste bread medium. The fermentations were carried out using shake flasks for different fermentation periods with varying levels of substrate, pH, temperature, $(NH_4)_2SO_4$, $MgSO_4.7H_2O$, $CaCl_2.2H_2O$, and KH_2PO_4. The experiments were carried out in such a way that the parameter optimized in one experiment was maintained in the subsequent investigation.

According to the patented research by Yahagi et al. (2003), starch-rich food wastes such as rice, noodle, bread, and draff (sediments) from the liquor industry can be used for the manufacturing of industrial starch. According to the manufacturing method, first the wastes are spread into a thin layer then rapidly heated at the same time for laying and drying, cooling, and crushing. They also proposed that a twin drum drying apparatus might be used for laying and drying. Yamashita and Miwa (2003) on the other hand suggested that food wastes such as bread and dough can be decomposed by hydrothermal reaction under supercritical water conditions and optionally subjected to wet oxidation or methane fermentation under anaerobic conditions. Their work was followed by that of Murase and Yoshino (2005), who stated in their patent that a sugar solution could be produced by hydrolyzing starchy waste materials such as waste bread, with selected enzymes such as amylase, protease, and lipase and incubation without agitation. They proposed that their method is easy to use and provides a cheap alternative, which does not require filtration and centrifugation, to other processes.

European workers tried to process bread returns in such a way as to obtain products that could all be used economically in the bread production process (Meuser, 1998). These products can be used either directly (liquid sour, baker's yeast, baking ingredients) or indirectly (ethanol, carbon dioxide) to manufacture bakery products. Amylolytic and proteolytic enzymes were added to the mash in order to degrade the carbohydrate fraction to glucose and the proteins to nitrogenous substances that can be digested by yeast. They pointed out that industrial bakeries have so far mainly disposed of the bread returns by selling them as a raw material for animal feed or for the manufacture of ethanol in Germany.

A summary of the various approaches adopted by those working in the area of waste bread utilization is given in Table 4.1.

4.1 Conceptualizing How Best to Utilize Waste Bread

Although there have been a number of studies involving the fermentation of waste bread to various products, there is no well-established bioprocess for the utilization of waste bread as a generic feedstock, such as there is for crop-based biorefineries. Yet bread is rich in starch and also contains proteins and other nutrients, so it should be suitable, after modification, as a fermentation substrate. The large macromolecules need reducing to smaller units to be accessible to microorganisms but have, at least, favourable elemental compositions. The modification necessary would therefore be the preparation of a hydrolysate rich in glucose, nitrogen, and minerals that could subsequently be converted to almost any desired product with the proper bioconversion.

There are essentially two ways to produce a nutrient rich hydrolysate from waste bread. The first involves the use of commercial enzymes. However, this can be costly, and commercial enzymes are sold in relatively pure form, so several different enzyme preparations— such as amylases and gluco-amylases for starch hydrolysis, proteases for protein hydrolysis, etc.—would be required in order to produce a complete nutrient rich hydrolysate from waste bread. Each commercial enzyme preparation might also require different operating conditions for the hydrolysis of its specific substrates, leading to lengthy sequential processing. The second option is to utilize a portion of the bread waste for a solid-state fungal fermentation in which the necessary cocktail of enzymes to effect the full hydrolysis is produced. Using these enzymes, the remaining portion of the waste bread could be hydrolyzed for the production of a nutrient rich hydrolysate. Finally, a value-added product or products could be produced from this hydrolysate as conceptualized in Figure 4.2. This concept was first discussed and investigated by Melikoglu (2008). A similar approach has recently been reported by Leung et al. (2012). In their process, the waste bread pieces were subjected to two solid-state fermentations: one to produce starch-degrading enzymes, the other to produce proteases. The enzymes were then used to hydrolyze the remaining fraction of the bread waste, as shown in Figure 4.3.

TABLE 4.1 Summary of Research into the Utilization of Waste Bread and Bakery Products

Publication	Drying	Enzymatic hydrolysis	Ethanol fermentation	Lactic acid fermentation	Other fermentations	End product
			Process			
Nakano and Yoshida, 1977		+				Glucose rich syrup
Martin, 1984		+	+			Ethanol
Menge, 1986				+	+	Lactic and acetic acid
Berghofer et al., 1995		+				Glucose rich syrup
Oda et al., 1997				+		Lactic acid and filtrate
Kumar et al., 1998		+	+			Ethanol
Meuser, 1998		+	+	+		Ethanol, liquid sour
Daigle et al., 1999					+	Aroma compounds
Asghar et al., 2002					+	α-amylase production
Yahagi et al., 2003	+					Starch substitute
Yamashita and Miwa, 2003					+	Methane
Maeda et al., 2004		+	+			Ethanol and animal feed
Murase and Yoshino, 2005		+				Sugar solution
Melikoglu, 2008		+	+		+	Enzymes and ethanol
Doi et al., 2009					+	Biohydrogen
Leung et al., 2012		+			+	Succinic acid

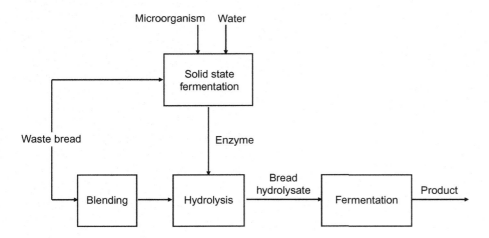

FIGURE 4.2 **Novel bioprocess for the utilization of waste bread based on multi-enzyme-producing solid-state fermentation.** *Source: Melikoglu (2008).*

5. SOLID-STATE FERMENTATION OF BREAD WASTE

Whereas the vast majority of commercial fermentation processes involve submerged culture, where the microorganism is suspended in an aqueous solution of nutrients, most microbial cultures occurring naturally do so on the surfaces of solid or semisolid substrates. Thus, solid-state fermentation (SSF) is defined as the growth of microorganisms on solid or semisolid substrates or supports (Rosales et al., 2007). This phenomenon is responsible for the large number of traditional solid-state

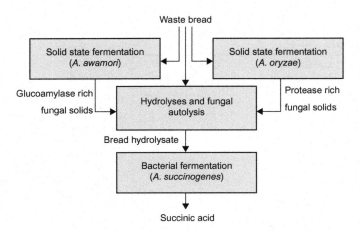

FIGURE 4.3 Novel bioprocess for the utilization of waste bread based on amylolytic and proteolytic solid-state fermentations. *Source: Leung et al. (2012).*

fermentations used particularly in the Far East and based largely on the Koji process, considered as one of the most important fields in microbiological studies (Fujikawa and Morozumi, 2005). Comparative studies between submerged (SmF) and solid-state fermentations (SSF) by various researchers have shown superior product yields (e.g., enzyme activities) for SSF (Nagel et al., 2001).

In recent years, SSF has attracted interest in western countries because of its advantages in the production of secondary metabolites and production of novel foods (Oostra et al., 2000). In addition, via SSF, solid wastes can be used as commercially desirable substrates (Barrios-González et al., 1993). Enzyme production by SSF is a growing field on account of simplicity, high productivity, and production of concentrated products (Castilho et al., 2000). Various raw materials are used for SSF but, because of their high nutrient composition and availability, cereals such as corn, wheat, and rice are the most common. However, utilization of such food grade raw materials carries economic and ethical problems. Instead, utilization of food wastes can be a synergistic solution to these problems, and waste bread is an ideal substrate for SSF. It is both nutritious and highly porous yet its use does not compete with food supplies.

A novel process (Figure 4.2), based on the production of hydrolytic enzymes from a small portion of waste bread, via SSF, and the subsequent use of these enzymes to hydrolyze the remaining portion for the production of a nutrient rich hydrolysate, has been developed at the University of Manchester. The nutrient rich hydrolysate product of the process can be converted into a range of desired products by subsequent fermentations. In the study reported by Melikoglu (2008), production of a multienzyme solution, rich in glucoamylase and protease, from waste bread pieces using the fungus

Aspergillus awamori, was optimized. Medium temperature, inoculum size, pH, particle size, initial moisture content, and duration are among the most important process parameters in simple SSF. Optimum values for medium temperature, 30°C (Pestana and Castillo, 1985); inoculum size, 1.0×10^6 spores/g dry substrate (Wang et al., 2007); and medium pH, natural (Legan, 1993) have been reported elsewhere (Wang et al., 2009); but the remaining key process parameters—particle size, initial moisture content, and fermentation time, all of which will affect enzyme production in SSF of waste bread pieces—have not previously been reported. These have now been studied and are reported below. First, initial estimates were taken from the literature, then individual parameters were studied stepwise. After each step, the best value found for the selected parameter was used as the set value for the next study.

One of the important aspects, and major advantages, of using bread waste is that it is a complete nutrient source and can therefore be used without supplementation. The process thus uses bread waste (typically white sliced bread) as the sole source of nutrients during fermentation. The typical composition of waste white sliced bread is presented in Table 4.2. Such slices must first be reduced in size before use and so a study based on cutting the bread into small pieces prior to solid state fermentation was carried out.

5.1 Optimum Particle Size

In solid-state fermentations, particle size of the substrate can be crucially important, as it determines the amount of void space (Pandey, 1991). This in turn will affect the rate of oxygen transfer and consequently microbial growth. Recently, Botella et al. (2009) suggested the term particulate bioprocessing to describe SSF in which the substrate is present as discrete particles. The fermentation of waste bread pieces would fit this definition. Particle size also influences degradation of the substrate during SSF by filamentous fungi (Zadrazil and Puniya, 1995). The smaller the particle size becomes, the larger the surface area for growth

TABLE 4.2 Composition of a Single Slice of White Bread Waste

Component	Weight (g)
Water	28.67
Starch	45.34
Nitrogen (N)	1.61
Protein (N × 5.7)	9.18
Phosphorus	0.10
Ash	2.26

and penetration by hyphae. However, reduction in particle size increases the packing density, which causes a reduction in the void space, as mentioned above. This leads to a concomitant reduction in the transfer of gases with the surrounding atmosphere, which results in reduction in microbial growth and enzyme production (Kumar et al., 2003). In solid-state systems there is thus a trade-off between small sizes providing large specific surface area and large sizes providing an open bed of particles. In studies of both glucoamylase and protease production, Melikoglu (2008) determined the optimum size for waste bread pieces to be around 20 mm.

5.2 Optimum Moisture Content

The most important difference between solid-state and submerged fermentation is the absence of free-flowing water in the growth environment. However, even in so-called solid-state fermentations, some water must be available if microbial growth and biochemical activity are to occur (Wiseman, 1985). In SSF, water content influences the physical state of the substrate, nutrient availability, diffusion of nutrients, and oxygen–carbon dioxide exchange in a complex way (Nishio et al., 1979; Ramesh and Lonsane, 1990). Water content is adjusted in terms of the initial moisture in most SSF. In a study involving a large number of individual fermentations, carried out at different initial moisture contents, the results presented in Figure 4.4 were obtained. The highest moisture content (around 350% on a dry basis)

represented the maximum water-carrying capacity of waste bread (i.e., around 3.5 times more water than bread). The figure clearly shows a significant influence of moisture on the success of the SSF in terms of enzyme production. The best conditions for the production of both enzymes were in the range 150–190% (db). These results are consistent with levels reported elsewhere in the literature (Malathi and Chakraborty, 1991) for *Aspergillus flavus* of 170% (db).

It is interesting to note that *Aspergillus awamori* appears to be influenced to different extents by initial moisture levels for glucoamylase and protease production. It has been reported that at both low and high initial moisture levels, metabolic activities of the culture and subsequent product synthesis are seriously affected in SSF (Sandhya et al., 2005). In fungal SSF, lower moisture leads to reduced solubility of the nutrients, a lower degree of substrate swelling, and higher water tension (Zadrazil and Brunert, 1981). Furthermore, higher moisture contents were reported to cause decreased porosity, loss of particle structure, development of stickiness, reduction in gas volume, and decrease in gas exchange (Lekha and Lonsane, 1994).

5.3 Optimum Duration for Solid-State Fermentation

During solid state fermentations, medium pH, nutrient concentration, temperature, moisture content, and physical structure of the raw materials are subject to continual change. All these parameters affect microbial

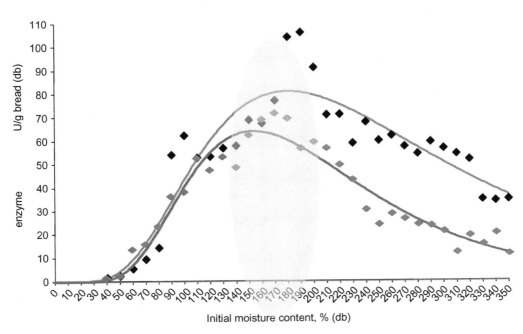

FIGURE 4.4 Effect of moisture content on the production of glucoamylase (upper line) and protease (lower line) in solid-state fermentation of waste bread by *Aspergillus awamori*.

growth and enzyme production. Moreover, changes in the medium temperature and pH directly affect the stability and activity of the enzymes produced *in situ*. There will naturally therefore be a profile of production and activity of the enzymes throughout a typical batch fermentation. It is worthwhile, then, looking at the various stages of the typical batch fermentation, which can be described under four different phases: Germination (or Lag), Growth, Stationary, and Death phases. In the following sections, the growth of *Aspergillus awamori* on waste bread is explained, based mainly on the analogy suggested by Auria et al. (1993), who studied the growth of *Aspergillus niger* during SSF.

5.3.1 Germination (Lag) Phase

During the first 18 hours of incubation, the only major activity is the swelling of the fungal spores prior to the onset of germination. Compared with submerged fermentations where fungal germination usually occurs within the first 6 hours, this is very slow. However, other researchers have observed similarly long lag times in solid SSF; for example Baldensperger et al. (1985) reported 24 hours for *Aspergillus niger* germination on banana wastes.

5.3.2 Growth Phase

During the first 24 hours of a typical waste bread fermentation, only 1% of the starch is hydrolyzed. This rises to 10% during the next 48 hours, during which significant quantities of hydrolytic enzymes are produced. Growth continues rapidly over several more days until the substrate is fully consumed by the fungus. Fungal mycelia completely cover both external and internal surfaces of the bread crumb.

5.3.3 Stationary Phase

After around 5 days of fermentation, fungal hyphae begin to deteriorate in certain regions of the substrate solids, there is a concurrent autolysis of cells, and spore production starts. However, SSF are very heterogeneous and so, in other parts of the substrate, fungal cells continue to grow well, consuming nutrients and producing metabolites. Approximately 95% of the starch has typically been consumed by 144 hours. Such depletion of nutrients is a common characteristic of the stationary phase in microbial growth. The highest glucoamylase and protease activities (see Figure 4.5) are measured during this period.

5.3.4 Death Phase

Once the activity of the enzymes has started to decline and the available starch has all but been consumed, the fermentation enters a death phase. The fungal cells continue to autolyze and disintegrate, and only spores remain intact. A small residue of starch is associated with regions of the solid substrate where the fungi could not penetrate.

5.3.5 Termination of the Fermentation

The kinetics of enzymatic starch hydrolysis by amylases shows that the rate of formation of glucose is retarded by product inhibition (Fujii and Kawamura, 1985). Therefore, a low concentration of glucose in the enzymatic extract is beneficial if the main aim of the solid-state fermentation is to produce hydrolytic enzymes. From the results shown in Figure 4.5, it is clear that for enzyme production the best time to terminate the fermentation is 144 hours. At this point in the fermentation, both glucoamylase and protease activities are at a maximum, pH is low, and glucose

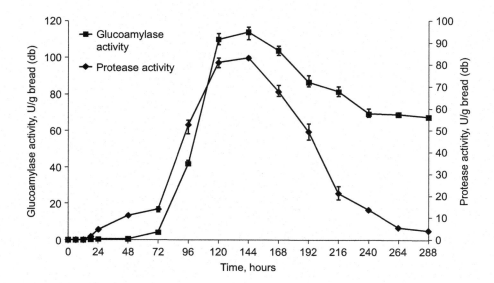

FIGURE 4.5   Time course of solid-state fermentations to produce glucoamylase and protease on waste bread.

concentration in the extract is insignificant. Enzyme activities fall dramatically after around 150 hours of fermentation such that, if the fermentation were to continue to 288 hours, the maximum duration observed, only 60% of the peak glucoamylase and just 5% of the protease activity would remain. These results demonstrate the importance of optimizing the duration of fermentation for enzyme production.

6. PROCESS DEVELOPMENT OPPORTUNITIES

In the study by Melikoglu (2008) reported above, optimum conditions for extracellular enzyme production via solid-state fermentation of waste bread pieces by *A. awamori* were 20 mm particles, 180% (db) initial moisture content, and 144 hours duration. Under these conditions, a very high-activity multienzyme solution was produced with activities, on a dry basis, of 114 U (glucoamylase)/g(bread) and 83.2 U(protease)/g(bread). In a previous study with the same fungus but using whole wheat as raw material, glucoamylase and protease activities were just 81.3 and 66.5 U/g (db) (Wang et al., 2010). Waste bread, therefore, represents an excellent alternative raw material for the development of a cereal-based biorefinery.

In Figure 4.6 a proposed biorefinery process for the utilization of waste bread is presented. Waste bread, from collection and storage, is first lightly crushed or chopped to form pieces of around 20 mm equivalent diameter. These are then sterilized if necessary, depending on the conditions during storage, before being subjected to a solid-state fermentation with filamentous fungus *Aspergillus awamori*. The fungus produces a tailored cocktail of enzymes to degrade the macromolecules in the bread and these can be stored, without separation, along with the dried fermentation solids. Melikoglu (2008) found that enzymes stored in this way had an increased shelf life compared with aqueous extracts and were also more thermally stable. The enzyme-carrying solids can subsequently be sold as a product for use in the direct hydrolysis of further bread waste or for wheat and corn flour hydrolysis. Alternatively, the enzyme cocktail can be extracted into warm water and stored as a concentrate, to be used in the same way as the solid, crude enzyme mix referred to above. The remaining soft solids and associated fungal spores can be recycled to the next SSF.

The lower half of the process shown in Figure 4.6 involves the direct hydrolysis of waste bread using the crude enzyme cocktail produced during the SSF. For this process, the waste bread is made into a paste rather than kept as a solid. Following hydrolysis, a

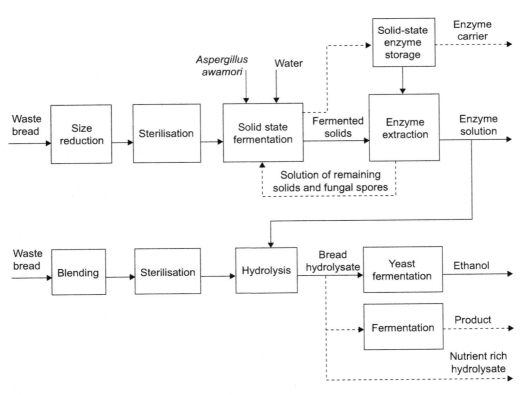

FIGURE 4.6 **Proposed bioprocess for the production of a generic fermentation feedstock from waste bread by solid-state fermentations.** Dotted lines indicate alternative steps in the process.

glucose and nitrogen rich medium is produced, which can be used for a wide variety of subsequent fermentations including, for example, ethanol production. In simple experiments to demonstrate the principle, Melikoglu (2008) was able to generate 0.2 kg ethanol per kg of bread waste, close to the theoretical maximum yield. The hydrolysate could readily be fermented to other products such as the succinic acid reported by Leung et al. (2012) or the range of biodegradable plastics based on poly-hydroxy-alkoanates (PHA). It could, of course, also be sold directly as a generic fermentation feedstock and nutrient source.

Collection of the waste bread will, of course, be a crucial issue, and obtaining it in relatively pure form would be a major concern. Waste bread from households is generally thrown away with other wastes and so collection could be problematic from these sources. However, waste bread from bakeries, supermarkets, and other retail stores could be collected much more readily. In the UK, waste from these sources could be between 2% and 15% of annual production (i.e., between 58,000 and 435,000 tonnes/y).

7. CONCLUSIONS

Industrial production of chemicals from sustainable feedstocks will require a balance to be achieved between the supply of food and the nonfood processing of bio-based raw materials. Research at the Satake Centre for Grain Process Engineering has resulted in the production of various generic fermentation feedstocks that can be converted into a range of platform chemicals, biodegradable polymers, and biofuels via microbial bioconversion from a range of starch-based raw materials. One of the major untapped resources for such processes is waste cereal-based foods and, in particular, bread. Bread possesses the characteristics of an ideal substrate for SSFs, and waste bread possesses little or no commercial value. However, when disposed of in landfill sites it is converted into methane, an extremely potent greenhouse gas, by anaerobic fermentation. Studies of a proposed bioprocess for the utilization of waste bread for the production of value-added products, in particular a multienzyme solution of amylolytic and proteolytic enzymes, show significant promise. Preliminary SSFs demonstrated that waste bread pieces are suitable for the growth of the filamentous fungus *Aspergillus awamori* and for the production of glucoamylase and protease at high concentrations.

Kinetic studies showed that the glucoamylase produced via SSF had a much longer shelf life than the same enzyme produced under submerged conditions. It appears that using SSF enhances the thermostability of the enzyme significantly. The multienzyme solution produced from waste bread was used in subsequent hydrolysis reactions to produce nutrient rich hydrolysates. These were then used for ethanol production. The yield for these fermentations was found to be equivalent to 255 liters of ethanol per tonne of waste bread. Consequently, 300 million liters of ethanol, with a current market value of over £100 million, could be produced annually in the UK alone if all of the waste bread could be recovered and processed in this way.

Wheat and corn prices have increased drastically in recent years due, in part, to the development of the bioethanol market. In order to make ethanol production fully commercially feasible the capital and operating costs of these processes should be decreased (Pasha et al., 2007). Raw materials such as lignocellulosic residues are cheap but their pretreatment costs are high. Waste bread, on the other hand, can offer a much cheaper alternative for bioethanol production. This approach would find a solution for the disposal of these wastes and at the same time reduce the environmental impact, in an eco-friendly and profitable manner.

References

Adhikari, B.K., Barrington, S., Martinez, J., 2006. Predicted growth of world urban food waste and methane production. Waste Manag. Res. 24, 421–433.

Anonymous, 2011. Global Bread & Rolls. Datamonitor Group, Reference Code: 0199–0020.

Asghar, M., Azhar, U., Rafiq, S., Sheikh, M.A., Asad, M.J., 2002. Production of α-amylase by *Arachniotus* sp. using waste bread medium. Int. J.Agric. Biol. 4, 26–28.

Auria, R., Morales, M., Villegas, E., Revah, S., 1993. Influence of mold growth on the pressure drop in aerated solid state fermentors. Biotechnol. Bioeng. 41, 1007–1013.

Baldensperger, J., Le Mer, J., Hannibal, L., Quinto, P.J., 1985. Solid state fermentation of banana wastes. Biotechnol. Lett. 7, 743–748.

Barrios-Gonzalez, J., Gonzalez, H., Mejia, A., 1993. Effect of particle size, packing density and agitation on penicillin production in solid state fermentation. Biotechnol. Adv. 11, 539–547.

Berghofer, E., Schleining, G., Matzner, G., 1995. Syrup production out of waste bread (Sirupherstellung aus Restbrot). Ernahr./Nutr. 19, 603–610.

Botella, C., Diaz, A.B., Wang, R., Koutinas, A., Webb, C., 2009. Particulate bioprocessing: a novel process strategy for biorefineries. Proc. Biochem. 44, 546–555.

Castilho, L.R., Medronho, R.A., Alves, T.L.M., 2000. Production and extraction of pectinases obtained by solid state fermentation of agroindustrial residues with *Aspergillus niger*. Bioresour. Technol. 71, 45–50.

Chen, P.L., Long, Z., Ruan, R., Labuza, T.P., 1997. Nuclear magnetic resonance studies of water mobility in bread during storage. Lebenson. Wiss. und-Technol. 30, 178–183.

Daigle, P., Gelinas, P., Leblanc, D., Morin, A., 1999. Production of aroma compounds by *Geotrichum candidum* on waste bread crumb. Food Microbiol. 16, 517–522.

Doi, T., Matsumoto, H., Abe, J., Morita, S., 2009. Feasibility study on the application of rhizosphere microflora of rice for the bio-hydrogen production from wasted bread. Int. J. Hydrogen Energy 34, 1735–1743.

Espinoza-Orias, N., Stichnothe, H., Azapagic, A., 2011. Carbon footprint of bread. Int. J. Life Cycle Anal. 16, 351–365.

Fallows, S.J., Wheelock, J.V., 1982. By-products from the U.K. food system 4. The cereals industries. Conserv. Recycling 5, 191–201.

Fujii, M., Kawamura, Y., 1985. Synergestic action of alpha-amylase and glucoamylase on hydrolysis of starch. Biotechnol. Bioeng. 27, 260–265.

Fujikawa, H., Morozumi, S., 2005. Modeling surface growth of *Escherichia coli* on agar plates. Appl. Environ. Microbiol. 71, 7920–7926.

Hui, Y.H., 2006. Food Biochemistry and Food Processing. Blackwell Publishing.

Jones, T.W., 2006. Food loss and the american household. Biocycle 47, 28.

Jones, T.W., 2007. Using contemporary archaeology and applied anthropology to understand food loss in the American food system. Available at: Communitycompost.com.

Kantor, L.S., Lipton, K., Manchester, A., Oliveira, V., 1997. Estimating and addressing America's food losses. Food Rev. 20, 2–12.

Kawa-Rygielska, J., Pietrzak, W., 2011. Studies on the suitability of the baking industry waste to produce ethanol fuel. Przemysl Chemiczny 90, 1269–1272.

Koutinas, A.A., Wang, R., Webb, C., 2007. The biochemurgist—Bioconversion of agricultural raw materials for chemical production. Biofuels, Bioprod. Biorefin. 1, 24–38.

Kumar, D., Jain, V.K., Shanker, G., Srivastava, A., 2003. Citric acid production by solid state fermentation using sugarcane bagasse. Proc. Biochem. 38, 1731–1738.

Kumar, J.V., Shahbazi, A., Mathew, R., 1998. Bioconversion of solid food wastes to ethanol. Analyst. 123, 497–502.

Legan, J.D., 1993. Mould spoilage of bread: the problem and some solutions. Int. Biodeterior. Biodegradation. 32, 33–53.

Lekha, P.K., Lonsane, B.K., 1994. Comparative titres, location and properties of tannin acyl hydrolase produced by *Aspergillus niger* PKL104 in solid state, liquid surface and submerged fermentation. Proc. Biochem. 29, 497–503.

Leung, C.C.J., Cheung, A.S.Y., Zhang, A.Y-Z., Lam, K.F., Lin, C.S.K., 2012. Utilisation of waste bread for fermentative succinic acid production. Biochem. Eng. J. In Press.

Maeda, S., Imashiro, S., Amagai, T., Harada, T., 2004. Feed production from wastes of bakery products. Kanagawa Prefecture, Japan, Ishikawajima Kensa Keisoku Co. Ltd, Japan. Jpn. Kokai Tokkyo Koho, 7 pp. CODEN: JKXXAF JP 2004065185 A 20040304.

Malathi, S., Chakraborty, R., 1991. Production of alkaline protease by a new *Aspergillus flavus* isolate under solid-state fermentation conditions for use as a depilation agent. Appl. Environ. Microbiol. 57, 712–716.

Mandala, I.G., 2005. Physical properties of fresh and frozen stored, microwave-reheated breads, containing hydrocolloids. J. Food Eng. 66, 291–300.

Martin, J.E., 1984. Conversion of bakery waste into ethanol. Energy Citations Database. Available at: <http://www.osti.gov/energy-citations/product.biblio.jsp?osti_id=6167186>.

Melikoglu, M. (2008). Production of sustainable alternatives to petrochemicals and fuels using waste bread as a raw material, PhD thesis. The University of Manchester, United Kingdom.

Menge W. (1986) Acidic fermenting dough for the preparation of bread and bakery products using waste bread. United States, CODEN: USXXAM US 4613506 A 19860923.

Meuser, F. (1998) Process for recycling bakery products, more specially rests of bread and bread remainders, *European Patent*, EP0821877.

Morgan, K.R., Fumeaux, R.H., Stanley, R.A., 1992. Observation by solid-state C CP MAS NMR spectroscopy of the transformations of wheat starch associated with the making and staling of bread. Carbohydr. Res. 235, 15–22.

Murase Y. and Yoshino A. (2005) Enzymic preparation of sugar solution from starch waste by decantation. Jpn. Kokai Tokkyo Koho, 17 pp. CODEN: JKXXAF JP 2005278588 A 20051013.

Nagel, F-J.J.I., Tramper, J., Bakker, M.S.N., Rinzema, A., 2001. Temperature control in a continuously mixed bioreactor for solid-state fermentation. Biotechnol. Bioeng. 72, 219–230.

Nakano M. and Yoshida S. (1977). Syrup Production from Bread Waste by Fermentation. Jpn. Kokai Tokkyo Koho, 1 pp. CODEN: JKXXAF JP 52117446 19771001.

Needham, R., Williams, J., Beales, N., Voysey, P., Magan, N., 2005. Early detection and differentiation of spoilage of bakery products. Sens. Actuators, B 106, 20–23.

Nishio, N, Tai, K, Nagai, S., 1979. Hydrolase production by *Aspergillus niger* in solid-state cultivation. Appl. Microbiol. Biotechnol. 8, 263–270.

Oda, Y., Park, B-S., Moon, K-H., Tonomura, K., 1997. Recycling of bakery wastes using an amylolytic lactic acid bacterium. Bioresour. Technol. 60, 101–106.

Oostra, J, Tramper, J, Rinzema, A., 2000. Model-based bioreactor selection for large-scale solid-state cultivation of *Coniothyrium minitans* spores on oats. Enzyme Microb. Technol. 27, 652–663.

Pandey, A., 1991. Effect of particle size of substrate of enzyme production in solid-state fermentation. Bioresour. Technol. 37, 169–172.

Parry, A., 2007. Food Waste Reduction: How Can Technology Help? Waste & Resources Action Programme (WRAP).

Pasha, C., Nagavalli, M., Rao, L.V., 2007. *Lantana camara* for fuel ethanol production using thermotolerant yeast. Lett. Appl. Microbiol. 44, 666–672.

Pestana, F., Castillo, F.J., 1985. Glucoamylase production by *Aspergillus awamori* on rice flour medium and partial characterization of the enzyme. World J. Microbiol. Biotechnol. 1, 225–237.

Ramesh, M.V., Lonsane, B.K., 1990. Critical importance of moisture content of the medium in alpha amylase production by *Bacillus licheniformis* M 27 in a solid state fermentation system. Appl. Microbiol. Biotechnol. 33, 501–505.

Ranken, M.D., Kill, R.C., Baker, C.G.J., 1997. Food Industries Manual, twenty fourth ed. Springer-Verlag.

Ribotta, P.D., Bail, A.L., 2007. Thermo-physical assessment of bread during staling. LWT 40, 879–884.

Rosales, E., Couto, S.R., Sanroman, M.A., 2007. Increased laccase production by *Trametes hirsuta* grown on ground orange peelings. Enzyme and Microb. Technol. 40, 1286–1290.

Sandhya, C., Sumantha, A., Szakacs, G., Pandey, A., 2005. Comparative evaluation of neutral protease production by *Aspergillus oryzae* in submerged and solid-state fermentation. Proc. Biochem. 40, 2689–2694.

Suhr, K.I., Nielsen, P.V., 2004. Effect of weak acid preservatives on growth of bakery product spoilage fungi at different water activities and pH values. Int. J. Food Microbiol. 95, 67–78.

Wang, R., Godoy, L.C., Shaarani, S.M., Melikoglu, M., Koutinas, A., Webb, C., 2009. Improving wheat flour hydrolysis by an enzyme mixture from solid state fungal fermentation. Enzyme. Microb. Technol. 44, 223–228.

Wang, R., Ji, Y., Melikoglu, M., Koutinas, A., Webb, C., 2007. Optimization of innovative ethanol production from wheat by response surface methodology. Process Saf. Environ. Protect. 85, 404–412.

Wang, R., Shaarani, S.M., Casas Godoy, L., Melikoglu, M., Sola Vergara, C., Koutinas, A., Webb, C., 2010. Bioconversion of rapeseed meal for the production of a generic microbial feedstock. Enzyme. Microb. Technol. 47, 77–83.

Webb, C., Wang, R., 1997. Development of a generic fermentation feedstock from whole wheat flour. In: Campbell, G.M., Webb, C., McKee, S.L. (Eds.), Cereals: Novel Uses and Processes. Plenum Press, New York, pp. 205–218.

Webb, C., Koutinas, A.A., Wang, R., 2004. Developing a sustainable bioprocessing strategy based on a generic feedstock. Adv. Biochem. Eng./Biotechnol. 86, 195–268.

Wiseman, A., 1985. Topics in Enzyme and Fermentation Biotechnology. Ellis Horwood Limited.

Wrigley C. and Walker C. E. (2004) Bakeries. In: Wrigley, C (Ed) Encyclopedia of Grain Science. Elsevier, Oxford, 21–27.

Yahagi, K., Yamaguchi, T., Konakawa, J., 2003. Manufacture of industrial starch from food wastes. Kawasaki Heavy Industries, Ltd., Japan, Jpn. Kokai Tokkyo Koh, 3 pp. CODEN: JKXXAF JP 2003201302 A 20030718.

Yamashita, M., Miwa, K., 2003. Method and Apparatus for treating food wastes by hydrothermal reaction. Ishikawajima–Harima Heavy Industries Co., Ltd., Japan, Jpn. Kokai Tokkyo Koho, 11 pp. CODEN: JKXXAF JP 2003251306 A 20030909.

Zadrazil, F., Brunert, H., 1981. Investigation of physical parameters important for solid-state fermentation of straw by white rot fungi. Eur. J. Appl. Microbiol. Biotechnol. 11, 183–188.

Zadrazil, F., Puniya, A.K., 1995. Studies on the effect of particle size on solid-state fermentation of sugarcane bagasse into animal feed using white-rot fungi. Bioresour. Technol. 54, 85–87.

5

Recovery of Commodities from Food Wastes Using Solid-State Fermentation

Maria R. Kosseva

1. INTRODUCTION

Aligned with the green production concept, any new chemical product today must be manufactured in a manner that minimizes the impact on the environment at every stage: (1) the use of raw materials and non-renewable forms of energy; (2) emissions and effluents; and (3) the life-cycle cost of the product. Biochemical processing of food waste materials satisfies the above constraints for the green production of a wide variety of chemical and biochemical products, because these processes are environmentally friendly, cost-effective, and carried out at ambient conditions. There are two types of bioconversion methods in operation: submerged fermentation (SmF), which is well established, and solid-state or solid-substrate fermentation (SSF). Although SSF has been used for centuries throughout the world, it is still in an evolutionary state and under intensive research (Prabhakar et al., 2005). Traditionally, SSF was mainly used for producing fermented foods such as dairy products, soy sauce, tempe (an Indonesian fermented food based on soybeans), and fermented sorghum (Campbell-Platt, 1994).

Solid-state fermentation is a cultivation technique in which microorganisms are grown under controlled conditions on moist solid particles, in beds within which there is a continuous gas phase between the particles, and sufficient moisture is present to maintain microbial growth and metabolism (Rahardjo et al., 2006; Mitchell et al., 2011).

This article focuses on these SSF processes, which use agricultural and food wastes as substrates to produce the following microbial products: aroma and flavor compounds, antibiotics, enzymes, ethanol, polysaccharides, protein-enriched fermented feeds, and organic acids. SSF technology provides many new opportunities as it allows the use of agricultural and food waste products as fermentation substrates without the need for extensive pretreatment of the substrate. Because SSF products are not highly diluted, they can be easier to recover than products produced in submerged culture. Moreover, SSF can improve economic feasibility of biotechnological processes offering waste reduction in design and operation.

1.1 Economically and Industrially Important Advantages of SSF

The main advantage of the SSF process is reduced cost because of the lower water content in upstream and consequently in downstream processing. SSF requires lower capital investments. Since the substrates used are usually agro-industrial residues, fermentation media are simpler, requiring no complex nutrients and vitamins. This leads to lower recurring expenditure. Another advantage is the application of fermented solids directly, without isolating the product. One example is production of glucoamylase enzyme using wheat bran mixed with corn flour in SSF by *Aspergillus niger*. After the fermentation, wet fermented matter containing glucoamylase can be directly used for the hydrolysis of cassava flour to produce fermentable sugars. Hydrolysis efficiencies are as good as obtained with the purified enzyme. Fermented matter can be dried and stored at room temperature without any significant loss in activity. Generally, SSF offers potential benefits in the following:

- Higher product titers
- Lower capital and recurring expenditure
- Lower wastewater output, reduced water requirement
- Reduced energy requirement

- Absence of foam formation
- Simplicity
- High reproducibility
- Simpler fermentation media
- Smaller fermentation space
- Easier aeration
- Economical to use even in smaller scale
- Easier control of contamination
- Applicability of using fermented solids directly
- Easy storage of dried fermentation matter
- Lower cost of downstream processing
- Saving of water.

1.2 Comparison of SSF and SmF

Table 5.1 emphasizes the major differences in characteristics of solid and submerged fermentations

TABLE 5.1 Basic Differences between Solid-State Fermentation (SSF) and Submerged Fermentation (SmF)

Characteristics	Solid-state fermentation	Submerged fermentation
Microorganism, substrate	Static	Agitated
Water usage	Limited	Unlimited
Oxygen supply via	Diffusion	Aeration
Distribution of parameters	Gradients of T, pH, Cs, Cn, P	Uniform
Parameters controlled	T, O_2, H_2O (H_2O control critical)	T, O_2
Inoculum ratio	Large	Low
Intra-particle resistances	Exists	No such resistance
Bacterial and yeast cells	Adhere to solid or grow inside the substrate matrix	Uniformly distributed
Product	Highly concentrated	Low concentration
Volume of fermentation mash	Smaller	Larger
Liquid waste produced	Negligible	Significant volume
Physical energy requirement	Low	High
Human energy requirement	High	Low
Capital investment	Low	High

Adapted from Prabhakar et al., 2005; Pandrey et al., 2008.

(Prabhakar et al., 2005; Pandrey et al., 2008). Because of the static nature of substrate and microorganisms, the system poses difficulties for heat and mass-transfer effects. Due to respiration, carbon dioxide evolves and, being heavier than oxygen, remains accumulated in the substrate bed. This, in turn, results in an increase of temperature in the fermented bed.

The process needs to be controlled (Pandrey et al., 2008). The low moisture content means that fermentation can only be carried out by a limited number of microorganisms, therefore spoilage or contamination by unwanted bacteria is reduced. The low volume of water present in the media per unit mass of substrate (typically 1 kg dry matter contains 1−5 L of water) can substantially reduce the space occupied by the bioreactor without severely sacrificing the yield of the product. For comparison, an SmF system requires between 10 and 20 L of water for 1 kg dry matter in order to prevent the culture medium from being too viscous (Mitchell et al., 2011). Aeration and mixing requirements may also be easily met. On the negative side, SSF processes are slower than liquid fermentations because of the additional barrier from the bulk solid. They also present heat dissipation problems, which can be limited by inter- as well as intra-particle resistances and are more difficult to control (Raghavarao et al., 2003).

SSF involves heterogeneous interactions of microbial biomass with moistened solid substrate. In SSF, the microorganisms can grow between the substrate fragments, i.e., inside the substrate matrix, or on the substrate surface. The microbial biomass inside the substrate matrix and on the substrate surface consumes substrates and secretes metabolites and enzymes. As there is no convective transport in the solid mass, concentration gradients are needed to supply the substrates and to remove the products. Gradients in the concentrations of substrates and products may cause local differences in metabolic activity. Similarly, gradients in the concentrations of inducers or repressors may affect enzyme production. These gradients are the most typical difference between SSF and SmF and can therefore be assumed to contribute to the observed differences in gene expression, metabolism, product spectrum, and process efficiency between SSF and equivalent SmF processes. Such gradients will occur regardless of the type of microorganisms cultivated, but filamentous fungi are frequently used in SSF because of their high potential to excrete hydrolytic enzymes, their relatively high tolerance to low water activities, and their morphology (Rahardjo et al., 2006).

One of the important features of SSF is the continuous gas phase existing between the particles, which has poor thermal conductivity compared with water. Another point is the wide variety of matrices used in SSF, which vary in terms of composition, size, mechanical resistance, porosity, and water-holding capacity. All these factors can affect the reactor design and the control strategy for the parameters. In SSF, besides the oxygen transfer which can be a limiting factor for some designs, the problems are more complex and affect the control of two important parameters: the temperature and the water content of the solid medium. Other factors also affect the bioreactor design: (1) the morphology of the fungus (presence or not of septum in the hyphae) and, related to this, its resistance to mechanical agitation; (2) the necessity or not to have a sterile process (Durand, 2003).

As shown in Box 5.1, a typical bioreactor for SSF will involve three phases: (1) the body of the bioreactor itself; (2) a bulk gas phase, which, if it is above the bed, is typically referred to as the headspace; and (3) the substrate bed. The substrate bed itself may be thought of as consisting of two subphases, namely the particles of solid material, to which the growing microorganism is attached, and the inter-particle gas phase.

2. SELECTION OF BIOREACTOR DESIGN FOR SSF

The bioreactor is the heart of a fermentation process, wherein the raw material, under suitable conditions, is converted to the desired product. Maximization of the rate of formation and yield of product within the bioreactor is a key part of the optimization of the production process. In contrast to SmF systems, SSF bioreactor systems have yet to reach a high degree of development, mainly

---

BOX 5.1

THE PHASES WITHIN A SOLID-STATE FERMENTATION SYSTEM

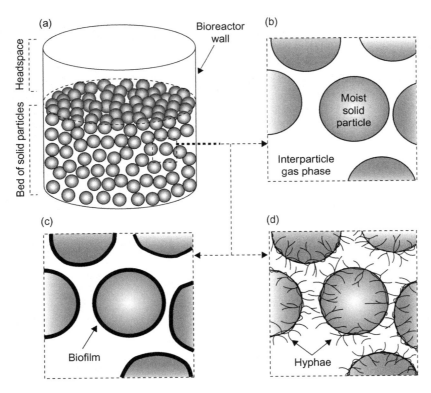

(a) Appearance of the system at the macroscale. At this scale it is possible to distinguish the substrate bed, the headspace above the bed, and the wall of the bioreactor. (b) Appearance of un-inoculated substrate at the microscale. At this scale it is possible to distinguish the individual particles and the gas spaces between them. (c) Growth of a biofilm of unicellular organisms (bacteria or yeasts) on the surfaces of the particles, as indicated by the thick black layer at the particle surface. (d) Growth of a network of fungal hyphae. This network grows across the surface while some hyphae penetrate into the substrate and others extend into the spaces between the particles.

because of the problems associated with solid beds, such as poor mixing and heat transfer characteristics and material handling (Raghavarao et al., 2003). Analogous with SmF, Lonsane et al. (1992) proposed a system of four stages for the development of SSF processes, with the selection of one size of container/bioreactor for each stage, as follows.

1. *Flask level:* ~5–1000 g working capacity for the selection of the culture, optimization of the process and experimental variables. Data collection is facilitated in a short time and at low cost.

2. *Laboratory bioreactor level:* ~5–20 kg working capacity for the selection of procedures for inoculum development, medium sterilization, aeration, agitation, and downstream processing; standardization of various parameters such as oxygen transfer rate, carbon dioxide evolution rate, biomass formation, product biosynthesis profiles; studies on the effect of pH, aeration-agitation rates, continuous or batch nutrient feeding policies; selection of control strategies and instruments; evaluation of economics of the process and its commercial feasibility.

3. *Pilot bioreactor level:* ~50–5000 kg mainly for confirmation of data obtained in the laboratory bioreactor, selection of best inoculum procedure, medium sterilization, and downstream processing strategies. This facilitates market trials of the product, its physico-chemical characterization or toxicity testing, and determination of the viability of the process.

4. *Production bioreactor level:* ~25–1000 tons for streamlining the process and achieving a financial return on the investments made so far on process development.

2.1 Classification of Bioreactors for SSF

SSF bioreactor designs can be divided into four groups on the basis of the operational characteristics of the bioreactor, specifically how it is mixed and aerated to minimize heat and mass transfer resistance, as follows?

- Group 1: With neither forced aeration nor forced agitation
- Group 2: Static bed with forced aeration
- Group 3: Continuous/intermittent agitation with forced aeration or air circulation
- Group 4: With continuous/ intermittent agitation as well as forced aeration.

The following strategies are applied: (a) the air circulates around the substrate layer or (b) it goes through it. Within the second strategy, three possibilities are available: static (unmixed), intermittently agitated, or continuously agitated beds.

2.2 Group 1: SSF Bioreactors without Forced Aeration (Tray Bioreactors)

Several types of containers are used for SSF at laboratory scale. Petri dishes, jars, wide mouth Erlenmeyer flasks, Roux bottles, and roller bottles offer the advantage of simplicity (Durand et al., 1995; Pandrey et al., 2001). Without forced aeration or agitation, only the temperature of the room where they are incubated is regulated. Easy to use in large numbers, they are particularly well adapted for the screening of substrates or microorganisms in the first steps of a research and development program. A typical representative of Group 1 vessels is a tray bioreactor. It consists of a chamber, which may be an incubator or a room, in which the temperature and humidity are controlled to some degree, and in which various recipients containing solid substrate are placed (Mitchell et al., 2011) as shown in Figure 5.1.

Various ancient civilizations have used this technology domestically for fermenting miscellaneous raw agricultural products in baskets. Applied on a commercial scale, this concept corresponds to the tray bioreactors as distinguished by the famous *Koji* process—an integral part of soya sauce production (Cannel, 1980; Christi, 1999). Made of wood, metal, or plastic, perforated or not, these trays, containing the solid medium at a maximum depth of 0.15 m, are placed in temperate rooms. The trays are stacked in tiers, one above the other with a gap

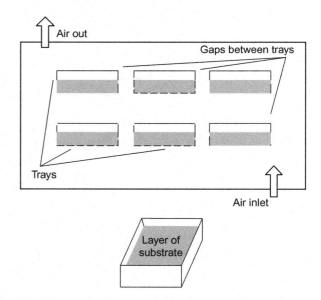

FIGURE 5.1 Schematic diagram of a tray bioreactor and individual tray. *Source: Ali and Zulkali (2011).*

of a few centimeters. This technology can be scaled-up easily because only the number of trays is increased. Although it has been extensively used in industry (mainly in Asian countries), this technology requires large areas (incubation rooms) and is labor intensive. It is difficult to apply this technology to sterile processes unless sterile rooms are built and procedures and equipment provided for the employees. An alternative could be to use polypropylene semi-permeable sterilizable bags to maintain sterility. Moreover, some bags have a microporous zone, which allows a passive airflow rate from 20 to 2000 $cm^3/cm^2/min$. The bags can be placed on a special rocking platform. The rocking motion provides some mixing of the substrate and oxygen transfer necessary for cell growth.

2.2.1 Current Challenges in Design, Operation and Scale-Up of Tray Bioreactors

So far, little attention has been given to important issues that will affect the success of tray bioreactors, such as how to optimize the spacing between trays, the positioning of the tray stacks, the geometry of the tray chamber, and the design of the air conditioning and circulation system. Attention also needs to be given to questions such as scheduling and automating of tray-handling steps such as loading, unloading, and cleaning. For covered trays, it is necessary to investigate which design of the tray cover will best prevent wetting of the bed while maximizing gas exchange between the headspace of the individual tray and the bulk gas phase of the tray chamber (Mitchell et al., 2011).

2.3 Group 2: Static Bed with Forced Aeration (Packed-Bed Bioreactors)

Traditionally, packed-bed bioreactors involve a static bed of substrate that sits on a perforated base, through which air is blown forcefully. The air must flow through the inter-particle spaces of the bed in order to leave a bioreactor. Various modifications to this basic design are possible, for example: (1) in addition to introducing air at the base, aeration can be applied via hollow perforated tubes inserted into the bed; (2) the air can be introduced at the top of the bed; (3) the bioreactor can be cylindrical with air being introduced through a perforated pipe at the central axis and removed through a perforated drum wall (a "radial packed-bed"); (4) the bed can be divided into compartments by heat transfer plates that are oriented parallel to the airflow; and (5) the bed can be broken up into relatively shallow layers with heat transfer plates, oriented perpendicularly to the airflow, inserted under each layer to cool the incoming air (Mitchell et al., 2011). Some examples are given below.

One of the interesting lab-scale units is the equipment developed and patented by an ORSTOM team in the late 1970s (Raimbault and Germon, 1976). It is composed of small columns (Ø 0.04 m, length 0.20 m) filled with a medium previously inoculated and placed in a temperate water bath. Water-saturated air passes through each column. This equipment is widely used by many researchers and offers the possibility to aerate the culture and also analyze the microorganism respiration by connecting the columns to a gas chromatograph with an automated sampler that routinely samples each column. This equipment is convenient for screening studies, optimization of the medium composition, and measurement of CO_2 produced. A small quantity of medium (a few grams) is used, and the geometry of the glass column is suitable for maintaining the temperature in the reactors. This equipment, with its advantages (forced aeration, cheap, relatively easy to use), can constitute a first step in research.

A new generation of small reactors was developed by an INRA-team in France in the late 1980s. A model built and tested during the year 2000 is shown in Figure 5.2. It has a working volume of about 1 L (Durand, 2003). Temperature and water amount in the medium can be monitored by means of the regulation of the temperature and relative humidity as well as flow rate of the air going through the substrate layer. Different profiles for the air-inlet temperature and flow rate can be elaborated and can generate useful information for scaling-up studies.

An AORSTOM team in France developed a packed bed reactor named Zymotis which is suitable for processes where the substrate bed must remain static (Mitchell et al., 2000). It consists of vertical internal heat transfer plates in which cold water circulates (Figure 5.3). Between each plate the previously inoculated solid medium is loaded. Temperate air is introduced through the bottom of each partition. This reactor, which looks like a tray reactor where the layers of substrate are set vertically, appears difficult to work within aseptic conditions. Very often in SSF, shrinkage of the volume of the medium occurs during mycelium growth. With this type of device, the risk is that the contacts with the vertical plates will decrease as the fermentation progresses, which will lead to poor heat transfer and air channeling. Finally, the scale-up of such a design appears very difficult.

2.3.1 Key Considerations in Designing Packed Beds

As a rule, in packed-bed bioreactors, the air flows unidirectionally from the air inlet to the air outlet, with a uniform velocity profile across the bed. This flow regime

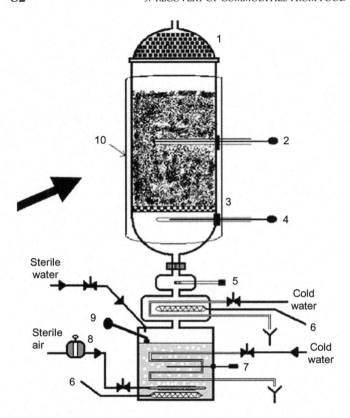

FIGURE 5.2 **Diagram of a lab-scale sterile column reactor.** (1) Heating cover, (2) medium temperature probe, (3) stainless steel sieve, (4) air-inlet temperature probe, (5) relative humidity probe, (6) resistive heater, (7) water temperature probe, (8) massic flow meter, (9) level probe, (10) insulating jacket. *Source: Durand (2003).*

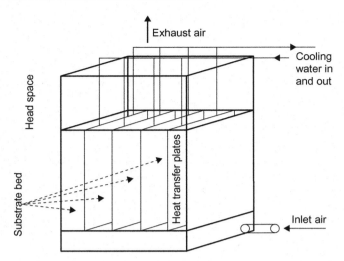

FIGURE 5.3 Zymotis packed-bed bioreactor with internal heat exchanger plates for circulation of cooling water. *Source: Ali and Zulkali (2011).*

is known as "plug flow". The unidirectional flow of air, combining with the release of metabolic waste heat, leads unavoidably to temperature profiles in the bioreactor, with the temperature increasing between the air inlet and air outlet. One of the major considerations in designing and operating large-scale packed-bed bioreactors is to prevent the temperature at the outlet air end of the bed reaching such high values that it will have an adverse effect on growth and product formation in the region. The use of internal heat transfer plates will minimize axial temperature gradients and therefore minimize evaporation. Other important considerations are the pressure drop through the bed and channeling. At the low superficial air velocities (of the order of 10 cm/s) that are most frequently used in packed-bed reactors, pressure drops are typically relatively small early in the process. As filamentous fungi grow, however, they form a network of hyphae, which fills the inter-particle spaces originally available between the particles, up to around 35%. This is sufficient for the pressure drop in the bed to become quite significant, in some cases reaching values around 0.1 atm per meter of bed. One strategy that can be used to prevent the pressure drop from reaching high

values is to mix the bed once, early during the process, before the hyphae have grown significantly into the inter-particle spaces. If channels appear within the bed, the only effective remedy is to agitate it, in the attempt to separate individual particles and let them settle again to form a bed of uniformly distributed particles (Mitchell et al., 2011).

2.4 Group 3: Continuously Agitated SSF Bioreactors with Air Circulation (Rotating and Stirred Drums)

This type of bioreactor can be a rotating-drum, a perforated-drum (Figure 5.4), or a horizontal paddle mixer. With or without a water jacket, this type of reactor is required to be continuously mixed to increase the contact between the reactor wall and the solid medium and also to provide oxygen to the microorganism. For rotating-drum bioreactors, as an horizontal cylinder, the mixing is provided by the tumbling motion of the solid medium, which may be aided by baffles on the inner wall of the rotating drum (perforated or not). However, in all these reactors, the mixing is less efficient than with a paddle mixer (Mitchell et al., 1992). Indeed, agglomeration of substrate particles during the growth of the mycelium can occur, which increases the difficulty of regulating the temperature of the solid medium. This bioreactor has a long history, having been used for the production of α-amylase in the early 1900s and penicillin after the Second World War.

A continuous-mixing horizontal paddle mixer was developed by a Dutch team at Wageningen University. This aseptic bioreactor was used for various purposes

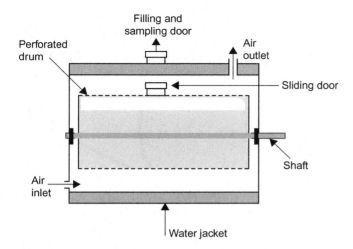

FIGURE 5.4 Perforated-drum bioreactor. *Source: Durand (2003).*

and to improve simultaneous control of temperature and moisture content. Although heat transport to the bioreactor wall was improved, this device becomes inefficient for larger volume (Nagel et al., 2001), because heat removal only through the wall becomes increasingly inefficient as the volume increases.

This category is essentially a rotating drum because continuous mixing is necessary to maximize the exposure of each substrate particle to the tempered air circulating in the headspace. The first main characteristic of this type of bioreactor is the bed of substrate contained in a horizontal or inclined drum, and the bed is mixed through the rotation of the drum. The second characteristic is air blown into the headspace above the solid bed. The bed is aerated by the exchange of gas between the bed and the headspace. Design variations include the use of an inclined central axis and the mounting of baffles on the inside of the drum wall. Rotating drums have been described in the past, the largest reactor recently cited in the literature was a 200 L stainless steel rotating drum (Ø 0.56 m and 0.90 m long) which used 10 kg of steamed wheat bran as substrate (Fung and Mitchell, 1995) for kinetic studies of *Rhizopus*. For this rotating drum, a strategy for regulating the medium temperature was described by De Reu et al. (1993). It consisted of activating the rotation of the drum in response to the temperature measured by a thermocouple in the medium. Efficient for a 4.7 L working volume of soy beans with *Rhizopus* for tempe production, no scale-up studies have been attempted.

2.4.1 Key Considerations in Designing and Operating Rotating and Stirred Drums

The most important considerations for rotating and stirred-drum bioreactors are related to the effectiveness of heat and mass transfer between the substrate bed and the headspace air, which depends mainly on the radial flow regime of the solids and the airflow pattern in the headspace. The radial flow regime of the solids determines how effectively the substrate particles are brought into contact with the headspace gases. The flow regime is affected by the velocity of rotation of the drum, the percentage filling of the drum, and the presence of baffles (or lifters). At low rotational speeds in rotating drums without baffles, the bed follows a slumping flow regime in which the solids tend to rise up as a single mass with the rotating wall of the drum and then slide back down. This regime therefore neither mixes the bed well nor promotes heat and mass transfer at the bed surface, leading to dead zones in the bed that are poorly agitated and inefficiently cooled. This problem can be overcome by installing baffles on the inside of the drum wall. In drums without lifters, slumping flow can be avoided by using high rotational rates. When the rotational rate is high enough, the bed enters a cataracting flow regime in which not only is the bed well mixed across its cross section, but also solid particles are thrown out into the airstream. In the case of stirred drums, the paddle design must ensure good radial mixing within the bed. Ideally, there should be good axial flow past the bed surface in order to maximize heat and mass transport between the bed and the headspace. The airflow patterns within the headspace will be affected by the designs and positioning of the air inlet and the air outlet, the aeration rate, the rotation rate, and the presence of baffles, which have not been particularly well studied (Mitchell et al., 2011).

2.4.2 Intermittently Mixed Bed Bioreactors with Forced Aeration (Mixed and Aerated)

In general, these bioreactors can be described as packed beds in which conditioned air passes through the bed. An agitation device is periodically used to mix the bed and, at the same time, water is sprayed if necessary. The design of these reactors, the capacity of which varies from a few kilograms to several tons, is influenced by the necessity or not to operate in sterile conditions. *For non-sterile processes*, a number of advances have been shown in the design and application of such bioreactors. One design is represented by the rotary type automatic Koji-making equipment marketed by Fujiwara in Japan (Figure 5.5). The treated substrate is heaped up on a rotary disk. Depending on the diameter of this disk, different working volumes are available but always with a layer of maximum thickness 0.50 m. This non-sterile reactor operates with a microcomputer which controls all the parameters (temperature of the air-inlet, air flow rate, and agitation periods). The main drawback of this equipment is the need to prepare and inoculate the substrate in

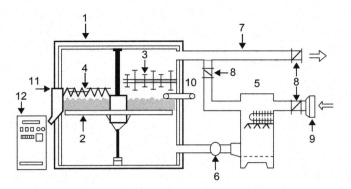

FIGURE 5.5 Scheme of Koji-making equipment. (1) Koji room, (2) rotating perforated table, (3) turning machine, (4, 11) screw and machine for unloading, (5) air conditioner, (6) fan, (7) air outlet, (8) dampers (9) air filter, (10) machine for filling, (12) control board. *Source: Durand (2003).*

other equipment before filling the reactor. Nevertheless, this type of design is widely used in Asian countries.

An INRA team in Dijon (France) has developed a non-sterile process strategy based on an intermittently mixed packed-bed reactor with forced aeration. This packed-bed bioreactor is mixed by a set of screws mounted on a carriage that sits on rails above the bed and moves continuously from one end of the bioreactor to the other. Although the screws operate continuously, any one part of the bed is agitated only intermittently. The temperature and the moisture of the medium are maintained by regulation of the temperature, relative humidity, and flow rate of the air input. It is also necessary to spray water and agitate periodically. The volume of water sprayed is calculated from online measurements of the total mass of the medium and by estimating the mass losses due to respiration (CO_2). On this basis, a reactor of 1.6 m^3 capacity using 1 tonne of sugar beet pulp (with 25% of dry matter) per batch was reported (Durand and Chereau, 1988). This reactor has been successfully used for various applications, including protein enrichment of agro-industrial by-products and production of enzymes (Durand et al., 1996). The system has since been continually improved.

2.5 Group 4: Bioreactors with Both Continuous Mixing and Forced Aeration (Mixed with Forced Aeration)

Three types of bioreactor belong to this group: the gas–solid fluidized bed, the continuously stirred aerated bed, and the rocking-drum bioreactor (Figure 5.6). With continuous agitation, it is practical to replenish water by spraying a fine mist onto the substrate or, in the case of the rocking-drum bioreactor, by dripping water through

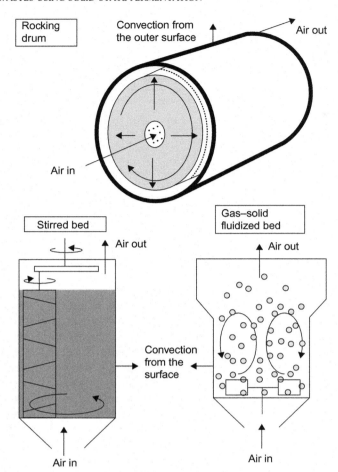

FIGURE 5.6 Bioreactors with mixing and forced aeration: the rocking drum, the stirred bed, and the gas–solid fluidized bed. Within the substrate bed the most important processes in the energy balance are waste heat generation, convective heat removal, and evaporation of water into the flowing air stream. Perfect mixing of the bed is usually assumed. Note that stirred beds are often operated with only intermittent mixing, and in this case they behave like packed beds during the static periods. *Source: Mitchell et al. (2000).*

the perforations in the inner drum (Lonsane et al., 1985). This permits the use of high evaporation rates as a cooling mechanism—a strategy which is not available for static beds, because of the problems in achieving an even distribution of added water in the bed. The design and velocity of the agitator are important in determining the effectiveness of mixing; however, this subject has received relatively little attention in SSF systems. In general, continuously mixed beds with forced aeration have the potential to perform better than other bioreactors because of the good heat and mass transfer characteristics. It is necessary to investigate the extent to which the continuous agitation is harmful to the microorganism, especially for processes involving fungi. In the final evaluation of performance, the costs of the extra energy consumed by the continuous aeration and agitation must be balanced against any increases in productivity achieved in comparison with other bioreactor types (Mitchell et al., 2011).

SSF processes can be operated in batch, fed-batch, or continuous modes, although batch processes are the most common. It should be noted that the performance of continuous SSF processes carried out in continuous stirred-tank reactor (CSTR) mode is different from that of CSTR processes in SmF. This is because each substrate particle acts as an independent micro-bioreactor. Based on the concentration profiles of the substrate and product achieved when continuous operation is used in SSF processes, the regime is most likely to be of the plug-flow type (Mitchell et al., 2011).

2.6 Examples of SSF Bioreactor Applications

A range of bioreactor applications for SSF of organic wastes to produce valuable commodities is shown in Table 5.2.

Many aspects of bioreactor design have yet to be studied in detail. To date, there has been little experimental effort to measure nutrient concentration gradients within particles. An effort in this direction will improve our understanding of the microscale phenomenon of intra-particle diffusion. Further, in the case of tray bioreactors, it is quite probable that natural convection will occur within the bed in response to an increase in temperature, simultaneously aiding transfer of heat and of CO_2, oxygen, and water vapor. This phenomenon has, however, attracted little attention. Similarly, in the case of the Zymotis design, attention is required to analyze the pressure drops, change in bed structure during the fermentation, and the flow patterns of air through the bed. In the case of gas—solid fluidized bed bioreactors, little information is available about design and operation, which needs detailed studies. Greater automation of the SSF process is needed for its increased industrial exploitation

TABLE 5.2 SSF Bioreactor Applications in the Last Decade to Produce Valuable Commodities

Bioreactor	Substrate	Capacity	Microorganism	Products	Yield	Reference
Tray (Erlenmeyer flask)	Coffee husk	15 g	*Ceratocystis fimbriata*	Fruity flavour	8.29 mmol/L per g	Soares et al. (2000)
Tray (tray bioreactor)	Wheat bran & bean cake powder	70 m³	*Bacillus thuringiensis* (Bt)	Cultivation of Bt	18,000 IU/mg	Hongzhang et al. (2002)
Tray (Erlenmeyer flask)	Tea waste	5 g	*Aspergillus niger*	Gluconic acid	82.2 g/L	Sharma et al. (2008)
Tray (Erlenmeyer flask)	Wheat bran & sesame oil cake	5 g	*Zygosaccharomyces rouxii*	L-glutaminase	11.61 U/g substrate	Kashyap et al. (2002)
Tray (Erlenmeyer flask)	Jackfruit seed powder	5 g	*Monascus purpureus*	Pigment	25 OD U/g dry substrate	Babitha et al. (2007)
Tray (Erlenmeyer flask)	Wheat bran	10 g	*Rhizopus oligosporous*	Lipase	48.0 U/g substrate	Haq et al. (2002)
Tray (Erlenmeyer flask)	Wheat bran & sesame oil cake	5 g	*Mucor racemosus*	Phytase	32.2 U/g dry substrate	Roopesh et al. (2006)
Tray (Erlenmeyer flask)	Soy bran	4 g	*Fomes sclerodermeus*	Laccase manganese peroxidase	520 U/g 14.5 U/g	Papinutti et al. (2007)
Packed-bed bioreactor	Whole rice	60 g	*Monascus* sp.	Biopigments	500 AU/g dry substrate	Carvalho et al. (2006)
Packed-bed bioreactor	Sugar cane bagasse	12 g	*Bacillus subtilis*	Penicillin	Respiration study	Dominguez et al. (2000)
Rotating drum bioreactor	Pineapple waste	600 g	*Aspergillus niger*	Citric acid	194 g/kg dry substrate	Tran et al. (1998)
Continuous mixing, forcefully aerated	Glucose	2.5 L	*Phanerochaete chrysosporium*	Ligninolytic enzymes	239 U/day	Couto et al. (2002)
Intermittent mixing, forcefully aerated	Corn	—	*Aspergillus niger*	Biomass	—	Meien et al. (2002)
Continuous mixing, forcefully aerated	Whole wheat grains	35.3 L	*Aspergillus oryzae*	Cultivation	—	Nagel et al. (2001)

Source: Ali and Zulkali (2011).

(Raghavarao et al., 2003). Despite challenges of process control of SSF, monitoring of these processes involves such key variables as the amount of microbial biomass, the bed temperature, the bed moisture content, the pressure drop across the bed, and off-gas composition (Mitchell et al., 2011).

3. MASS AND HEAT TRANSFER PHENOMENA IN SSF

The phenomena occurring within SSF bioreactors can conceptually be divided into microscale and macroscale.

3.1 Microscale Phenomena

After germination, filamentous fungi form tubular hyphae that elongate at the tips and at the same time form new branches along the hyphae. Their morphology allows filamentous fungi to colonize the surface of the substrate and penetrate into the substrate matrix in search for nutrients. The fungal hyphae form a porous three-dimensional net that is known as mycelium. Initially, sparse mycelia grow inside the substrate matrix (layer 3), on the surface of the substrate (layer 2), and into the air (layer 1). As the mycelia continue to grow, the following occurs: (a) layer 1 becomes so dense that its pores become filled with water and it transforms into layer 2; (b) the packing density and/or thickness of layer 2 increases to such an extent that its lower part becomes anaerobic; and (c) oxygen is depleted in the substrate matrix. Under anaerobic conditions, the mycelia in layers 2 and 3 stop growing or start fermenting. When the pores of the mycelial mat are filled with water, it can be regarded as a biofilm layer or a thin layer of water filled with growing biomass. As a consequence of the air-filled pores in the aerial mycelia layer 1, rapid diffusion of oxygen is expected there, but diffusion of all non-volatile compounds (e.g., the carbon source and enzymes) in the cytoplasm of the hyphae is likely to be comparatively slow. In contrast, as a consequence of the water-filled pores in the biofilm and substrate layers (i.e., layers 2 and 3), the supply of all nutrients (including oxygen) and removal of all products in these layers can be hampered by slow diffusion (Rahardjo et al., 2006). In brief, growth of the microorganisms is dependent on the inter- and intra-particle diffusion of gases like O_2 and CO_2 as well as enzymes, nutrients, and products of metabolism. Mitchell et al. (2004) have addressed the interaction of growth with intra-particle diffusion of enzymes, hydrolysis products, and O_2, proposing mechanistic equations for their description. The same authors have also given

insights into how these microscale processes can potentially limit the overall performance of a bioreactor.

3.2 Macroscale Phenomena
3.2.1 Mass Transfer Aspects
At the macroscale, mass transfer processes occurring include:

- The bulk flow of air into and out of the bioreactor and, as a consequence, changes in the sensible energy and concentrations of O_2, CO_2, and water;
- Natural convection, diffusion, and conduction taking place in a direction normal to the flow of air during unforced aeration;
- Conduction across the bioreactor wall and convective cooling to the surroundings;
- Shear effects caused by mixing within the bioreactor, including damage to either the microorganism or the integrity of the substrate particles.

3.2.2 Heat Transfer Aspects
In general, during SSF, a considerable amount of heat is evolved, which is a function of the metabolic activities of the microorganisms (Chahal, 1983). In the initial stages of fermentation, the temperature and oxygen concentrations are the same at all locations in the SSF bed. As the fermentation progresses, oxygen diffuses and undergoes bioreactions, liberating heat, which is not easily dissipated because of the poor thermal conductivity of the substrate. With the progress of the fermentation, shrinkage of the substrate bed occurs and the porosity also decreases, further hampering the heat transfer. Under these circumstances, temperature gradients develop in the SSF bed, which can sometimes be steep, giving rise to high temperatures. The transfer of heat into or out of the SSF system is closely associated not only with the metabolic activity of the microorganism, but also with the aeration of the fermenting system.

The temperature of the substrate is very critical in SSF. High temperatures affect spore germination, growth, product formation, and sporulation (Moreira et al., 1981), whereas low temperatures are not favorable for growth of the microorganisms and for the other biochemical reactions. The low moisture content and poor conductivity of the substrate make it difficult to achieve good heat transfer in SSF. Significant temperature gradients are reported to exist even when small depths of substrates are employed (Rathbun and Shuler, 1983); hence, it is very difficult to control the temperature of the bioreactors on a large scale. In fact, limiting the heat dissipation is one of the major drawbacks of SSF in comparison with conventional SmF, where good mixing provides for efficient dispersal of sparged oxygen and also serves to give better temperature control. Mixing not only aids homogeneity of the bed but also ensures an effective heat and mass transfer. Thus water addition coupled with continuous mixing is advantageous for simultaneous control of temperature and moisture control in large-scale SSF (Raghavarao et al., 2003).

Figure 5.7 summarizes the interaction between the various factors in a SSF bioreactor. The performance of the bioreactor is controlled by (1) the interactions between the microorganism and its local environment, and (2) how effectively the design and operating strategies influence the conditions in the local environment of the microorganism (Mitchell et al., 2011).

4. APPLICATIONS OF SSF

The solid-state fermentation of agro-industrial residues (such as rice bran, rice husk, potato wastes, cassava husk, wheat bran, sugar cane bagasse, sugar beet pulp, palm kernel cake, rice straw, cocoa pod,

fruit/vegetable wastes, etc.) into bulk chemicals and fine products (e.g., biofuel, enzymes, ethanol, organic acids, amino acids, antibiotics, and biologically active secondary metabolites) have been well documented (Couto, 2008; Laufenberg et al., 2003; Paganini et al., 2005; Pandey et al., 1999; Soccol and Vandenberghe, 2003; Villas-Boas et al., 2002). As described, SSF offers a feasible alternative for many operations, including waste disposal and the derivation of valuable products derived from it. Some of the applications which appear to have a promising future include the production of enzymes, organic acids, pigments, and aroma compounds (Raghavarao et al., 2003).

4.1 Bulk Chemicals and Products: Organic Acids, Ethanol, Enzymes, Polysaccharides, and Feed Protein

Recovery of commodities from fruit-and-vegetable processing wastes, with their further application in the

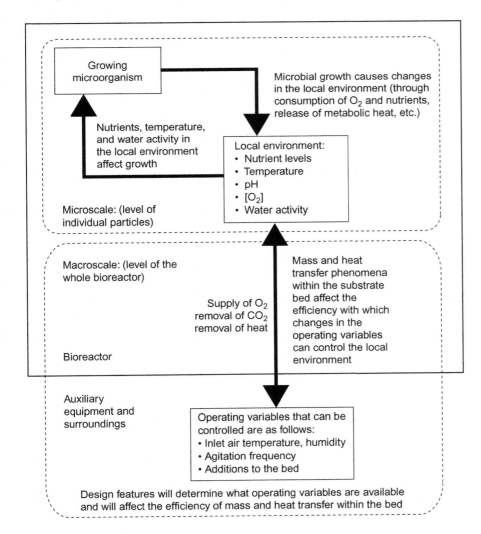

FIGURE 5.7 The interaction between the various factors that affect the performance of a SSF bioreactor. *Source: Mitchell et al. (2011).*

food industry, is of economic and environmental benefit. Vendruscolo et al. (2008) presented a review on the vast potential of apple pomace as a substrate for various biotechnological processes and possible strategies for its utilization by SSF. Apple pomace is the main solid waste generated in cider and apple juice making factories, accounting for approximately 25% to 35% of the weight of the processed raw material. It is a heterogeneous mixture consisting of peel, core, seed, calyx, stem, and soft tissue (Grigelmo-Miguel and Martin-Belloso, 1999). Bacteria, yeast, and fungi have been cultivated on apple pomace for different purposes.

Filamentous fungi, especially basidiomycetes, are the most suitable microorganisms for growing on fruit processing residues. Table 5.3 lists biotechnological applications of apple pomace, with the respective microorganism and the fermentation process applied.

Several studies and patents have been described regarding the utilization of this residue for the production of valuable compounds, such as organic acids, enzymes, single cell protein (SCP), biopolymers, fatty acids, and polysaccharides among others. Quite a few operational variables must be considered and optimized in order to cost-effectively use the apple pomace

TABLE 5.3 Bioprocesses Using Apple Pomace as Substrate

Application	Microorganism	Process	Reference
ENZYME PRODUCTION			
β-glucosidase	*Aspergillus foetidus*	SSF	Hang and Woodams (1994)
Lignocellulolytic enzymes	*Candida utilis*	SSF	Villas-Boas et al. (2002)
Pectin methylesterase	*Aspergillus niger*	SSF/SmF	Joshi et al. (2006)
Pectinases	*Polyporus squamosus*	SmF	Pericin et al. (1999)
Pectolytic enzymes	*A. niger*	SSF	Berovic and Ostroversnik (1997)
Polygalacturonase	*Lentinus edodes*	SSF	Zheng and Shetty (2000b)
AROMA COMPOUND PRODUCTION			
Aroma compounds	*Rhizopus* sp., *Rhizopus oryzae*	SSF	Christen et al. (2000)
Aroma compounds	*Kluyveromyces marxianus*	SSF	Medeiros et al. (2000)
Fruity aroma	*Ceratocystis fimbriata*	SSF	Bramorski et al. (1998)
Phenolic compounds	*Trichoderma viride, Trichoderma harzianum, Trichoderma pseudokoningii*	SSF	Zheng and Shetty (2000a)
NUTRITIONAL ENRICHMENT			
Single cell protein	*Candida utilis*	SmF	Albuquerque (2006)
Animal feed	*Gongronella butleri*	SSF	Vendruscolo (2005)
Nutritional enrichment	*Candida utilis, Kloeckera* sp.	SSF	Devrajan et al. (2004)
Protein enrichment	*Rhizopus oligosporus*	SSF	Albuquerque et al. (2006)
HETEROPOLYSACCHARIDE PRODUCTION			
Chitosan	*G. butleri*	SSF	Streit et al. (2004)
Chitosan	*G. butleri*	SmF	Streit et al. (2004); Vendruscolo (2005)
Heteropolysaccharide	*Beijerinckia indica*	SmF	Jin et al. (2002)
Xanthan	*Xanthomonas campestris*	SSF	Stredansky and Conti (1999)
OTHER PRODUCTS			
Citric acid	*A. niger*	SSF	Shojaosadati and Babaeipour (2002)
Ethanol	*S. cerevisiae*	SSF	Ngadi and Correa (1992)
γ-Linolenic acid	*Thamnidium elegans, Mortierella isabelina, Cunninghamella elegans*	SSF	Stredansky et al. (2000)

Source: Vendruscolo et al. (2008).

in bioprocesses: strain type, reactor design, aeration, pH, moisture, and nutrient supplementation are only a few examples (Vendruscolo et al., 2008).

4.1.1 Organic Acids from Fruit Pomace

4.1.1.1 LACTIC ACID PRODUCTION

Lactic acid has a number of applications in food technology (as acidulant, flavor, and preservative), pharmaceuticals and chemicals (Hofvendalh and Hahn-Hagerdal, 2000). Current production of lactic acid is about 68 million kg per year, and it is expected that the world market for this acid will grow by up to 10–15% per year (Wassewar, 2005). Apple pomace shows several advantages as a raw material for lactic acid manufacture, including: (i) high content of free glucose and fructose, which are excellent carbon sources for lactic acid production (Hofvendalh and Hahn-Hagerdal, 2000); (ii) high content of polysaccharides (cellulose, starch, and hemicelluloses) which can be enzymatically hydrolyzed to give monosaccharides; (iii) presence of other compounds (e.g., monosaccharides other than glucose and fructose, di- and oligosaccharides, citric acid, and malic acid) which can be metabolized by lactic bacteria (Carr et al., 2002); and (iv) presence of metal ions (Mg, Mn, Fe, etc.) which could limit the cost of nutrient supplementation for fermentation media.

In order to increase the lactic acid yield from apple pomace, polysaccharides have to be hydrolyzed, leading to solutions containing high concentrations of sugars and other fermentable compounds. Gullon et al. (2008) evaluated: (i) the analytical characterization of apple pomace, (ii) its enzymatic saccharification by cellulase—cellobiase mixtures, and (iii) lactic acid production by fermentation of enzymatic hydrolysates obtained under selected operational conditions. The effect of the cellulase-to-solid ratio (CSR) and the liquor-to-solid ratio (LSR) on the kinetics of glucose and total monosaccharide generation was studied. They also developed a set of statistical regression models to reproduce and predict the effect of the operational conditions on the hydrolysate composition, using the Response Surface Methodology.

When samples of apple pomace were subjected to enzymatic hydrolysis, the glucose and fructose present in the raw material as free monosaccharides were extracted at the beginning of the process. Using low cellulase (activity expressed in terms of filter paper units) and cellobiase (activity reported as international units) charges (8.5 FPU/g-solid and 8.5 IU/g-solid, respectively), 79% of total glucan was saccharified after 12 hours, leading to solutions containing up to 43.8 g monosaccharides per liter (glucose, 22.8 g/L; fructose, 14.8 g/L; xylose + mannose + galactose, 2.5 g/L; arabinose + rhamnose, 2.8 g/L). These results correspond to a monosaccharide/cellulase ratio of 0.06 g/FPU and to a volumetric productivity of 3.65 g of monosaccharides per liter hour. Liquors obtained under these conditions were used for fermentative lactic acid production with *Lactobacillus rhamnosus* CECT-288, leading to media containing up to 32.5 g/L of L-lactic acid after 6 hours (product yield $Y_{P/S} = 0.88$ g/g). Figure 5.8 shows the time course of the fermentation medium composition. Glucose and FXO were rapidly metabolized, and lactic acid was generated at high volumetric productivity ($Qp = 5.41$ g/(Lh)) after 6 hours). Box 5.2 illustrates the mass balance of the studied process.

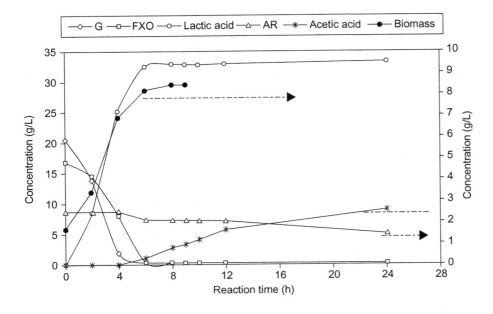

FIGURE 5.8 Evaluation of glucose, fructose + xylose + mannose + galactose + malic acid (FXO), arabinose + rhamnose (AR), lactic acid, acetic acid, and biomass concentrations obtained during the fermentation of enzymatic hydrolysates with *Lactobacillus rhamnosus. Source: Gullon et al. (2008).*

BOX 5.2

MATERIAL BALANCES OF THE LACTIC ACID PRODUCTION

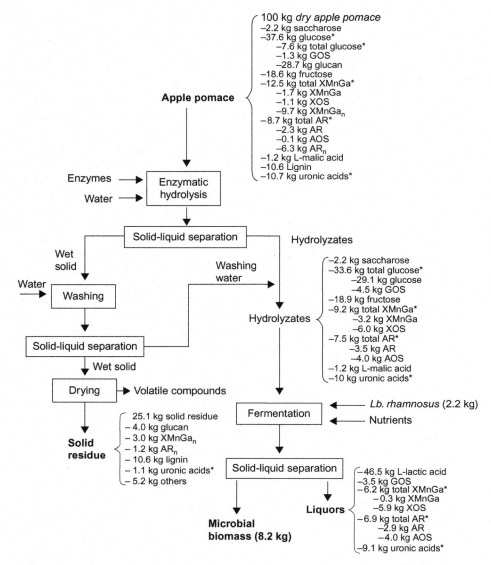

*Expressed as monomer equivalents but present as monomer, oligomers, and polymers.

Starting from 100 kg of apple pomace (containing 80.8 kg of potential sugars and malic acid), 72.6 kg of saccharose + monosaccharides + malic acid (suitable carbon sources for lactic acid production) can be obtained after enzymatic hydrolysis carried out under the conditions (CSR = 8.5 FP/g, LSR = 12 g/g, t = 12 h). Lignin was retained in the residual solid, and 93% of uronic acids were solubilized during the enzymatic hydrolysis. Fermentation of hydrolysates resulted in the production of 46.5 kg of lactic acid. Moreover, 13.4 kg of oligosaccharides (which can be used as ingredients for functional foods) and 8.2 kg of microbial biomass (which could be used as a probiotic) were simultaneously produced. Uronic acids suitable for food applications were not metabolized by these bacteria and remained in the fermentation medium. (Proposed by Gullon et al., 2008)

4.1.1.2 CITRIC ACID PRODUCTION

Citric acid is a commercially valuable product, widely used in the food, pharmaceutical, and beverage industries as an acidifying and flavor-enhancing agent. It is commercially produced mainly by SmF of sucrose or molasses-based media (Kubicek and Rhor, 1986).

Shojaosadati and Babaeipour (2002) used apple pomace as a substrate for the production of citric acid with *Aspergillus niger* via SSF in column reactors. They evaluated several cultivation parameters, such as aeration rate (0.8, 1.4, and 2.0 L/min), bed height (4, 7, and 10 cm), particle size (0.6—1.18, 1.18—1.70, and 1.70—2.36 mm), and moisture content (70%, 74%, and 78%). For citric acid yield, the aeration rate and particle size were the most important parameters. Neither the bed height nor the moisture content was found to significantly affect citric acid production. The operating conditions that maximized citric acid production consisted of low aeration rate (0.8 L/min), high bed height (0.10 m), large particle size (1.70—2.36 mm), and elevated moisture content (78%).

Kumar et al. (2003a) investigated the production of citric acid by *A. niger* under SSF conditions using pineapple, mixed fruit, and maosmi waste as substrates. They obtained a maximum citric acid yield of 51.4%, 46.5%, and 50% (based on sugar consumed), respectively. In another study (Kumar et al., 2003b) they tested wheat bran and sugar cane bagasse as substrates. They found that the latter gave the highest production (9.4 g/100 g of dried substrate (ds) after 9 days of fermentation).

Imandi et al. (2007) studied the production of citric acid by *Yarrowia lipolytica* under SSF conditions using pineapple waste as the sole substrate. After optimizing the composition of the culture medium, a citric acid production of 202.35 g/kg ds (g citric acid produced per kg of dried pineapple waste as a substrate) was obtained.

4.1.1.3 FATTY ACID PRODUCTION

Apple pomace has also been used for fatty acid production. Stredansky et al. (2000) evaluated γ-linolenic acid (GLA) production in *Thamnidium elegans* by SSF. Apple pomace and spent malt grain were used as the major substrate components for the production of high-value fungal oil containing up to 11.43% biologically active GLA.

4.1.2 Production of Ethanol

A solid-state fermentation process for the production of ethanol from apple pomace by *Saccharomyces cerevisiae* was described by Khosravi and Shojaosadati (2003). A moisture content of 75% (wt/wt), an initial sugar concentration of 26% (wt/wt), and a nitrogen content of 1% (wt/wt) were the conditions used to obtain 2.5% (wt/wt) ethanol without saccharification and 8% (wt/wt) with saccharification. The results indicate that alcohol fermentation from apple pomace is an efficient method to reduce waste disposal, with the concomitant production of ethanol (Paganini et al., 2005).

Nogueira et al. (2005) evaluated the alcoholic fermentation of the aqueous extract of apple pomace. Apple juice, pomace extract, and pomace extract with added sucrose provided after fermentation 6.90%, 4.30%, and 7.30% ethanol, respectively. A fermentation yield of 60% was obtained when pomace extract was used as a substrate.

For bioconversion of bean curd refuse, a processing by-product of bean curd, ethanol-producing anaerobic thermophiles were newly isolated. Both of them degraded hemicellulose, but not cellulose at all. Phylogenetically the strains belong to the *Clostridium* and *Thermoanaerobacterium* genus. Aerobic thermophiles degrading cellulose were also newly isolated. This strain belongs to the *Geobacillus* genus phylogenetically. The co-culture also significantly reduced CH_3SH production, leading to the prevention of offensive odor (Miyazaki et al., 2008).

4.1.3 Production of Enzymes

One important area of apple pomace utilization is the production of enzymes. Polygalacturonases or hydrolytic depolymerases are enzymes involved in the degradation of pectic substances. They have a wide range of applications in food and textile processing, degumming of plant rough fibers, and treatment of pectic wastewaters.

Villas-Boas et al. (2002) found a novel lignocellulolytic activity of *Candida utilis* during SSF on apple pomace. Hydrolytic and oxidative enzymes of *C. utilis*, excreted to the culture medium during solid-substrate cultivation, were identified, evaluated, and quantified. The soluble lignin fraction of the apple pomace was consumed at very significant levels (76%) compared with the non-fermented apple pomace. The enzyme produced by *C. utilis* with the highest activity was a pectinase (23 U/mL). The yeast showed a significant manganese-dependent peroxidase activity (19.1 U/mL) and low cellulase (3.0 U/mL) and xylanase (1.2 U/mL) activities, suggesting that *C. utilis* has the ability to use lignocellulose as a substrate.

Joshi et al. (2006) reported the production of pectin methylesterase by *Aspergillus niger* using apple pomace as culture medium, comparing SmF and SSF. The pectin methylesterase activity was 2.3 times higher when produced by SSF than by SmF.

Dry citrus peels are rich in pectin, cellulose, and hemicellulose and may be used as a fermentation substrate. Production of multi-enzyme preparations containing pectinolytic, cellulolytic, and xylanolytic enzymes by the mesophilic fungi *Aspergillus niger* BTL, *Fusarium oxysporum* F3, *Neurospora crassa* DSM 1129, and *Penicillium decumbens* under SSF on dry orange peels was enhanced by optimization of initial pH of the culture medium and initial moisture level (Mamma et al., 2007). Under optimal conditions *A. niger* BTL was by far the most potent strain in polygalacturonase and pectate lyase production, followed by *F. oxysporum* F3, *N. crassa* DSM 1129, and *P. decumbens*. *N. crassa* DSM 1129 produced the highest endoglucanase activity and *P. decumbens* the lowest. *N. crassa* DSM 1129 and *P. decumbens* did not produce any β-xylosidase activity, whereas *A. niger* BTL produced approximately 10 times more β-xylosidase than *F. oxysporum* F3. The highest invertase activity was produced by *A. niger* BTL and the lowest by *F. oxysporum* F3 and *P. decumbens*. After SSF of the four fungi, under optimal conditions, the fermented substrate was either directly exposed to autohydrolysis or new material was added, and the *in situ* produced multi-enzyme systems were successfully used for the partial degradation of orange peel polysaccharides and the liberation of fermentable sugars.

Feruloyl esterase (FAE) and xylanase activities were detected in culture supernatants from *Humicola grisea* var. *thermoidea* and *Talaromyces stipitatus* grown on brewers' spent grain (BSG) and wheat bran (WB), two agro-industrial by-products. Maximum activities were detected from cultures of *H. grisea* grown at 150 rpm, with 16.9 U/mL and 9.1 U/mL of xylanase activity on BSG and WB, respectively. Maximum FAE activity was 0.47 U/mL and 0.33 U/mL on BSG and WB, respectively. Analysis of residual cell wall material after microbial growth shows the preferential solubilization of arabinoxylan and cellulose, two main polysaccharides present in BSG and WB. The production of low-cost cell-wall-degrading enzymes on agro-industrial by-products could lead to the production of low-cost enzymes used for valorization of food processing wastes (Mandalari et al., 2008).

SSF is commercially used in Japan to produce industrial enzymes. This is not surprising since Japan is a very large producer of soya sauce, and SSF (Koji fermentation) is a part of this production process. The main enzymes produced during the growth phase of *Aspergillus oryzae* on roasted cracked soya beans and wheat are amylase and protease, which are then used to hydrolyze the substrates completely.

4.1.3.1 α-AMYLASE

α-amylases (endo-1,4-β-D-glucan glucohydrolases EC 3.2.1.1) are extracellular enzymes secreted as a primary metabolite that randomly cleave the 1,4-α-D glucosidic linkages between adjacent glucose units in the linear amylose chain. The extensive application of amylases in the food industry, such as baking, brewing, preparation of digestive aids, production of chocolate cakes, moist cakes, fruit juices, and starch syrups, has paved a way for their large-scale commercial production. Generally, SSF systems for production of α-amylase appear promising in view of the natural potential and advantages they offer (Pandey et al., 1999).

Mulimani and Ramalingam (2000) tested several substrates (rice straw, rice bran, red gram husk, jowar straw, jowar spathe, wheat bran) for the production of α-amylase by the fungus *Gibberella fujikuroi* in SSF. Among the tested substrates, wheat bran led to the maximum α-amylase production. Baysal et al. (2003) evaluated two agro-wastes—wheat bran and rice husk—as substrates for the production of α-amylase by *B. subtilis* under thermophilic SSF conditions. They found that α-amylase yield was 7.3-fold higher on wheat bran than on rice husk. Thus, under the optimized conditions (0.1 M phosphate buffer at pH 7, 30% initial moisture, 1000 μm particle size, 20% inoculum concentration), a maximum yield of 159,520 U/g was obtained after 48 hours.

Kunamneni et al. (2005) studied the production of extracellular amylase by the thermophilic fungus *Thermomyces lanuginosus* in SSF. Solid substrates such as wheat bran, molasses bran, rice bran, maize meal, millet cereal, wheat flakes, barley bran, crushed maize, corncobs, and crushed wheat were evaluated for enzyme production. Growth on wheat bran gave the highest amylase activity. The maximum enzyme activity obtained was 534 U/g of wheat bran under optimum conditions of an incubation period of 120 hours, an incubation temperature of 50°C, an initial moisture content of 90%, a pH of 6.0, an inoculum level of 10% (v/w), a salt solution concentration of 1.5:10 (v/w) and a ratio of substrate weight to flask volume of 1:100 with soluble starch (1% w/w) and peptone (1% w/w) as supplements.

Erdal and Taskin (2010) reported production of α-amylase by *Penicillium expansum* MT-1 in SSF using waste Loquat (*Eriobotrya japonica* Lindley) kernels as a possible substrate. Loquat, an Asian fruit tree, is a member of the Rosaceae family, which includes apples, pears, and quinces. It is cultivated mainly in China, Japan, India, Pakistan, Madagascar, Spain, Turkey, Italy, Greece, Israel, Brazil, Venezuela, and Australia (Morton 1987; Martinez-Calvo et al., 1999). According to the 2003 data, Turkey has some 288,000 loquat trees and these yield 13,000 tones of fruit (Sutcu and Demiral, 2009). In terms of weight, the seeds comprise about 20−30% of the weight of the whole fruit (Freihat et al., 2008). The kernels accounted for 22.5% of whole fruit (by wet weight). They were rich in protein (22.5%) and total carbohydrate (71.2%). The starch accounted for 25.6% of total carbohydrate and 36% of whole kernel, on a dry weight basis. The fungus was isolated from fermented loquat kernels. Loquat kernel flour (LKF) could serve as a sole source of nitrogen and carbon for the fungus to grow and synthesize α-amylase. However, additional carbon and nitrogen sources increased the enzyme production. Supplementation with each of three alcohols (methanol, ethanol, or propanol) had a positive effect on the enzyme production. Optimal conditions for the production of α-amylase by the fungus on LKF were determined as initial moisture content of 70%, particle size of 1 mm, pH 6.0, incubation temperature of 30°C, starch and peptone as supplements, 1 mL methanol as supplement alcohol, and an incubation period of 6 days. Under the optimized culture conditions, the maximum enzyme production was 1012 U/g of LKF.

Thermostable alkaline α-amylase-producing bacterium Bacillus firmus CAS 7 strain isolated from marine sediment of the Parangipettai coast grew maximally in both shake flasks. Elayaraja et al. (2011) found that potato peel is a superior substrate for the production of α-amylase at 35°C, pH 7.5 and 1.0% of substrate concentrations. Under optimal conditions, Bacillus firmus produced 676 U/mL of α-amylase, which was optimally active at 50°C and pH 9.0.

4.1.3.2 XYLANASE

From a commercial point of view, xylanases (EC 3.2.1.8) are a significant group of carbohydrolases and have a worldwide market of around 200 million dollars. Xylanases are used in a variety of applications: biobleaching of kraft pulp, clarification of juice and wine, starch separation and production of functional food ingredients, improving the quality of bakery products, and in animal feed biotechnology (Bhat, 2000). The most important end user of xylanase is the paper and pulp industry, where thermostable cellulase-free alkaline active xylanase is required for the biobleaching of kraft pulp. The addition of xylanase in the pre-bleaching process can replace up to 20–30% of chlorine and can reduce up to 50% organic halogens, which are known to form toxic dioxins.

Gawande and Kamat (1999) tested several lignocellulosic residues such as sugar cane bagasse, wheat bran, rice straw, and soya bean hulls for the production of xylanase by Aspergillus terreus and Aspergillus niger under SSF conditions. Wheat bran led to the best xylanase production. This was a very interesting result because economically, wheat bran is cheaper than pure xylan as a substrate for xylanase production. In addition, both Aspergillus strains produced xylanase with very low levels of cellulase activity, facilitating the subsequent enzyme purification processes.

Virupakshi et al. (2005) found that, among different easily available lignocellulosic substrates tested, rice bran was the best substrate for xylanase production by Bacillus sp. under SSF conditions.

Seyis and Aksoz (2005) investigated the use of apple pomace, orange pomace, orange peel, lemon pomace, lemon peel, pear peel, banana peel, melon peel, and hazelnut shell as substrates for xylanase production using Trichoderma harzianum. The maximum enzyme activity was observed when melon peel was used as the substrate for SSF, followed by apple pomace and hazelnut shell. Botella et al. (2007) showed the enormous potential of grape pomace to produce xylanase by Aspergillus awamori in SSF. They found that the particle size did not influence the enzyme production. However, moisture content and the addition of extra carbon sources had an acute effect on the yield of xylanase. Mamma et al. (2007) found that Aspergillus niger produced high xylanase activities when grown on orange peels under SSF. Comparison of xylanase production revealed that Aspergillus niger BTL produced the highest activity followed by Neurospora crassa DSM 1129, Penicillium decumbens, and Fusarium oxysporum F3.

4.1.3.3 PROTEASE

Proteases (EC 3.4.21.19) constitute one of the major collections of industrial enzymes and are used in a variety of manufacturing processes: in the detergent, food, pharmaceutical, leather and silk industries, and for recovery of silver from used X-ray films (Anwar and Saleemuddin, 1998) Proteases hydrolyze peptide bonds in aqueous environments and synthesize peptide bonds in micro-aqueous environments. Proteases (in detergent industries) account for 30% of the total world enzyme production. The detergent industry is one of the major consumers of hydrolytic enzymes working at alkaline pH and accounts for over a quarter of global enzyme production. Alkaline proteases are particularly important because they are both stable and active under harsh conditions such as temperatures of 50–60°C, high pH, and in the presence of surfactants or oxidizing agents.

Germano et al. (2003) used defatted soya bean cake as a support-substrate for the production of protease by a wild strain of Penicillium sp. in SSF. The crude enzyme obtained was compatible with commercial detergents, retaining 50–60% of its activity. Tunga et al. (2003) reported the production of an extracellular alkaline serine protease by Aspergillus parasiticus under SSF, using wheat bran as a support-substrate. The produced protease was stable in various detergents, retaining 85% of its activity.

Sandhya et al. (2005) performed a comparative study on the production of neutral protease by

A. oryzae using several agro-industrial residues (wheat bran, rice husk, rice bran, spent brewing grain, coconut oil cake, palm kernel cake, sesame oil cake, jackfruit seed powder, and olive oil cake) as substrates in SSF and SmF. They found that wheat bran was the best substrate in both systems. They also found that SSF produced 3.5-fold more enzyme than SmF.

Paranthaman et al. (2009) carried out a comparative study on the production of protease using different varieties of rice brokens (PONNI, IR-20, CR-1009, ADT-36 and ADT-66) from rice mill wastes as substrates in SSF with *Aspergillus niger*. Among the tested varieties of rice, broken PONNI produced the highest activity, at 67.7 U/g, while ADT-66 produced the lowest protease, at 44.7 U/g, under SSF conditions. The optimized conditions for producing the maximum yield of protease were incubation at 35°C and pH 7.0 for 96 hours. Protease production from waste treatment could be commercially used in the detergents and leather industries.

4.1.3.4 LACCASE

Laccases (benzenediol: oxygen oxidoreductase, EC 1.10.3.2) are multi-copper enzymes belonging to the group of blue oxidases. They are defined as oxidoreductases by the Enzyme Commission (EC), oxidizing diphenol and allied substances and using molecular oxygen as an electron acceptor. Fungal laccases are of particular interest with regard to potential industrial applications, because of their capability to oxidize a wide range of toxic and environmentally problematic substrates. Oxidation reactions are comprehensively used in industrial processes, for instance in the textile, food, wood processing, pharmaceutical, and chemical industries. Enzymatic oxidation is a potential substitute for chemical methods, since enzymes are very specific and efficient catalysts and are ecologically sustainable (Desai and Nityanand, 2011).

Studies on fungal enzyme production via SSF have shown that it provides higher volumetric productivities, is less prone to problems with substrate inhibition, and yields enzymes with a higher temperature or pH stability, compared with SmF. The fermentation time is shorter and the degradation of the produced enzymes by undesirable proteases is minimized. Production of laccase from different fungal sources employing SSF is illustrated in Table 5.4.

Ellen et al. (2008) exploited the potential of orange bagasse, major industrial food waste arising from processing orange for juice, as solid support for laccase production from *Botryosphaeria rodhina*. A good enzyme titer was obtained in SSF without added nutrients, indicating nutrient sufficiency of orange bagasse at a solids concentration of 16% (w/v) to sustain growth and high enzyme titers.

Kapoor et al. (2009) reported the potential of wheat straw as a natural support for the production of laccase from *Lentinus edodes*. The study also reported the efficiency of several organic compounds such as rice bran, corn steep meal, peanut meal, soya meal, and wheat bran as supplements to wheat straw for laccase production. When cultivated on both supplemented wheat straw combinations and un-supplemented wheat straw (control) with 70% humidity, solid support with optimal concentration of supplement yielded good growth and enzyme activity.

Rosales et al. (2005) aimed to investigate the feasibility of kiwi fruit wastes as a support-substrate for laccase production by the white-rot fungus *Trametes hirsute* under solid-state conditions. Experiments were conducted to select variables that maximized levels of laccase activity. In particular, the effect of the initial ammonium concentration, the amount of support employed, the need for pre-treating the support, and the part of kiwi fruit wastes utilized (peelings or peelings plus pulp) was evaluated. The highest laccase

TABLE 5.4 Laccase Production from a Variety of Fungal Sources Grown on Various Natural Supports via SSF

Support	Microorganism	Reference
Corn cob	*Phanerochaete chrisosporius*	Cabaleiro et al., 2002
Wheat bran, wheat straw	*Pluerotus pulmonarius*	De Souza et al., 2002
Neem hull, wheat bran, and sugar cane bagasse	*Pluerotus ostreatus, Phanerochaete chrisosporius*	Verma and Madamwar, 2002
Wheat bran	*Fomes sclerodermeus*	Papinutti et al., 2003
Chestnut shell, barley bran	*Coriolopsis rigida*	Gomez et al., 2005
Orange peeling	*Trametes hirsute*	Rosales et al., 2007
Orange bagasse	*Botryosphaeria rodhina*	Ellen et al., 2008
Wheat straw	*Lentinus edodes*	Kapoor et al., 2009

Source: Desai and Nityanand (2011).

value (around 90,000 nkat/L) was obtained at an initial ammonium concentration of 0.150 g/L and with 2.5 g of pre-treated peelings of kiwi fruit. This activity is higher than that reported to date with no added inducers. These promising results open the way for the valuable utilization of other similar food-processing wastes.

4.1.3.5 TANNASE

Tannin acylhydrolase, also known as tannase, is an enzyme (EC 3.1.1.20) that catalyzes the hydrolysis of ester bonds present in gallotannins, complex tannins, and gallic acid esters (Ramırez et al., 2008; Rodrıguez et al., 2009). Tannase has several important applications in food, feed, chemical, and pharmaceutical industries, but high-scale use of this enzyme is severely restricted by high production costs (Aguilar and Gutierrez-Sanchez, 2001; Aguilar et al., 2007). Tannase is an ecologically important biocatalyst. Several research groups have developed economic and ecofriendly processes for the production and purification of the enzyme at laboratory level. On the other hand, the use of modern techniques of molecular biology has allowed the development of highly efficient processes for production and recovery of the enzyme, and this trend appears likely to continue in the coming years. The results have been positive, but more research is needed on basic and applied aspects of tannase, such as regulation of the enzyme, metagenomics, new expression systems, design of new bioprocesses using emerging large-scale cultivation technologies, efficient and cost-effective downstream processing, and design of new applications for the enzyme such as the production of antioxidants from waste materials or the bioremediation of tannery effluents (Rodriguez-Duran et al., 2011).

4.1.4 Production of Polysaccharides

Jin et al. (2002) examined the potential of three agro-industrial by-products to be used as substrate for the production of heteropolysaccharide-7 (PS-7) by *Beijerinckia indica* in SmF under the same cultivation conditions. By-products from apple juice production, soy sauce production, and the manufacturing processes of Sikhye (fermented rice punch), a traditional Korean food, were tested. The apple pomace was found to be the best carbon source for PS-7 production, giving a production of 4.09 g/L after 48 hours of cultivation. When Sikhye by-product was used as substrate, 3.00 g/L of PS-7 was formed; and, using the soy sauce residue, 0.96 g/L of PS-7 was obtained.

Xanthan gum is the most important microbial polysaccharide from the commercial point of view, with a worldwide production of about 30,000 tons/year. It has widespread commercial applications as a viscosity enhancer and stabilizer in the food, pharmaceutical, and petrochemical industries (Papagianni et al., 2001). The rheological behavior of the fermentation broth causes serious problems of mixing, heat transfer, and oxygen supply, thus limiting the maximum gum concentration achievable as well as the product quality (Wecker and Onken, 1991). Several strategies have been developed to overcome these problems, including use of two-level impellers. The use of cheap substrates, instead of the commonly used glucose or sucrose, might result in a lower cost of the final product. Stredansky and Conti (1999) proposed the use of SSF as an alternative strategy for the production of xanthan by *Xanthomonas campestris*, since solid substrates reproduce the natural habitat of this phytopathogenic bacterium. This technique allows the resolution of problems connected with broth viscosity and, in addition, utilizes cheap substrates.

Streit et al. (2004) studied the production of fungal chitosan in SmF and SSF (column reactors) using the watery extract of apple pomace and pressed apple pomace as substrate, respectively. Among the microorganisms studied, the fungus *Gongronella butleri* yielded the best results for the production of chitosan in SmF and SSF. Grown on the watery extract of apple pomace, *G. butleri* presented the highest productivity, 0.091 g/(Lh), and chitosan content in the biomass, 0.1783 g/g of apple pomace, for a medium supplemented with 40 g/L of reducing sugars and 2.5 g/L of sodium nitrate.

Vendruscolo (2005) used an external loop airlift bioreactor for chitosan production by *G. butleri* CCT 4274 on the watery extract of apple pomace. The experiments using higher levels of aeration (0.6 volume of air per volume of liquid per minute) provided greater concentrations of biomass, attaining 8.06 g/L and 9.61 g/L, in the production of 873 mg/L and 1062 mg/L of chitosan, respectively. These findings demonstrated the adequacy of the airlift bioreactor for the cultivation of microorganisms, with emphasis on the production of chitosan.

4.1.5 Production of Baker's Yeast

Bhushan and Joshi (2006) used apple pomace extract as a carbon source in an aerobic-fed batch culture for the production of baker's yeast. The fermentable sugar concentration in the bioreactor was regulated at 1–2%, and a biomass yield of 0.48 g/g of sugar was obtained. Interestingly, the dough-raising capacity of the baker's yeast grown on the apple pomace extract was apparently the same as that of commercial yeast. The use of apple pomace extract as substrate is a useful alternative to molasses, traditionally used as a carbon source for baker's yeast production.

4.1.6 Feed Protein

Song et al. (2005a) developed a method for producing SCP from apple pomace by dual SSF. This method comprises four different steps: (1) preparation of an SSF medium with pulverized apple pomace, (2) inoculating a cultured mixed mature strain for SSF, (3) rapidly drying the resulting fermentation product at low temperature, and (4) subjecting the dried product to solid fermentation in another SSF medium to obtain the final product (patent No.CN1673343). The same group of researchers (Song et al., 2005b) developed another method for producing feed protein by liquid-solid fermentation of apple pomace (patent No. CN1663421). As compared with solid fermentation, this method has the advantage of reduced consumption of medium for seed culture, reduced cost, and applicability to large-scale production.

4.2 Production of Fine Chemicals: Aroma Compounds, Antibiotics and Pigments

4.2.1 Aroma Compounds

The European Community guidelines 88/388/EWG and 9/71/EWG subdivide aromas into six categories, the first of which describes regulations for food labeling as "natural flavor". Natural flavors are chemical substances with aroma properties that are produced from feedstock of plant or animal origin by means of physical, enzymatic, or microbiological processing. The microbial synthesis of these natural flavors is generally carried out by SmF. Due to the high costs of this currently used technology on an industrial scale, there is a need to develop low-cost processes even for cheaper molecules like benzaldehyde. This could be achieved by exploration of the metabolic pathways and by alternative technology such as SSF (Couto, 2008). Table 5.5 lists flavors and biofine chemicals produced by SSF of vegetable residues (Laufenberg et al., 2003). The use of biotechnology for the production of natural aroma compounds by fermentation or bioconversion using microorganisms is an economic alternative to the difficult and expensive extraction from raw materials such as plants (Daigle et al., 1999). Currently, it is estimated that around one hundred different aroma compounds are produced commercially by fermentation (Medeiros et al., 2006). Furthermore, the world market of aroma chemicals, fragrances and flavors has a growth rate of 4—5% per year. Because of higher consumer acceptance, there is increasing economic interest in natural aromas.

Bramorski et al. (1998) analyzed the production of aroma compounds by *Ceratocystis fimbriata* under seven different medium compositions (prepared by mixing cassava bagasse, apple pomace, amaranth, and soya bean). The aroma production was growth dependent, and the maximum aroma intensity was detected in a few hours around the maximum respirometric activity. The medium containing apple pomace produced a strong fruity aroma after 21 hours of cultivation. The same medium was used by Christen et al. (2000) for the production of volatile compounds by *Rhizopus* strains. Authors found that the production of volatile compounds was related mainly to the medium used, and no difference was observed among the strains studied. The odors detected had a slight alcoholic note, and the apple pomace produced intermediate results, compared with the amaranth grain supplied with mineral salt solution.

Another source of aroma compounds is grape pomace, the main by-product of wine production, which consists of skins, seeds, and stalks, reaching an estimated amount of 13% by weight of processed grape (Torres et al., 2002). The evaluation of the qualitative aspects of a grape pomace is carried out in view of the production of high quality grappa; otherwise the grape pomace is used for alcohol distillation, or thrown away. The best grape pomaces are highly rich in vinous liquid, with a moisture degree ranging from 55% to 70%, which allows better exploitation of the raw material and extraction of the organoleptic characteristics of the native vine.

4.2.2 Antibiotics

Antibiotics are required in large quantity for the persistent battle against bacterial diseases. Research has thus concentrated on producing them at the highest concentration with minimal energy input. Yang (1996) reported the production of oxytetracycline by *Streptomyces rimosus* in SSF using corncob, a cellulosic waste, as a substrate. This author found that the oxytetracycline produced by SSF was more stable than that produced by SmF and the energy input was also less (Yang and Ling, 1989). In addition, the product presented the advantage that it could be temporarily stored without losing activity significantly. Adinarayana et al. (2003a) tested several substrates (wheat bran, wheat rawa, bombay rawa, barley, and rice bran) to produce cephalosporin C by *Acremonium chrysogenum* under SSF. Physical and chemical parameters were also optimized. A maximum productivity of cephalosporin C (22,281 μg/g) was achieved using wheat rawa and soluble starch (1%) and yeast extract (1%) as additives, an incubation period of 5 days, an incubation temperature of 30°C, an inoculum level of 10%, a ratio of salt solution to weight of wheat bran of 1.5:10, a moisture content of solid substrate of 80% and pH 6.5. Adinarayana et al. (2003b) also reported the production of neomycin by *Streptomyces marinensis* under SSF using wheat rawa as a support substrate.

TABLE 5.5 Flavors and Fine Chemicals Produced by SSF of Vegetable Residues

Residual matter	Description/conversion principle	Product	Reference
Spent malt grains, apple pomace	*T. elegans* CCF 1456 degraded the substrate in a ratio of 3 to 1 (AP to SMG); precursor peanut oil increased the yield	γ-Linolenic acid was produced in a yield of 5.17 g/kg dry substrate; with peanut oil precursor 8.75 g/kg DM	Stredansky et al. (2000)
Cassava bagasse, apple pomace	Four strains of *Rhizopus*, two residues and two precursors, mixed substrate combinations	Volatile carbons as flavors; acetaldehyde, ethanol, propanol, esters	Christen et al. (2000)
Cassava bagasse, wheat bran and sugar cane bagasse	*C. fimbriata*, ability to generate fruity aromas dependent on the substrate used	Banana flavor and fruity complex flavors	Bramorski et al. (1998)
Citrus, apple, sugar beet pomace	Microbial conversion by enzymatic hydrolysis	Pectin, substrate, liquid biofuel	Grohmann and Bothast (1994)
Cranberry pomace (fish offal-peat compost)	*Trichoderma viride, Rhizopus* CaCO₃ was added as neutralizer, water for aw adjustment	Polymeric dye decolorizing isolate for wastewater treatment, extracellular enzymes	Zheng and Shetty (1998)
Linseed cake, castor oil cake, olive press cake, sunflower cake	*Moniliella suaveolens, Trichoderma harzianum, Pityrosporum ovale*, and *Ceratocytis moniliformis* form decalactones (problems with phenolic components)	Acceptable yields on olive press cake and castor oil cake, d- and c-decalactone are produced	Laufenberg et al. (2001)
Olive cake, sugar cane bagasse	Lipase degrading fat in olive cake	Enzyme product applied in bakery goods, confectionery, pharmaceuticals	Cordova et al. (1998)
Olive pomace	Four microorganisms, delignification, saccharification with *Trichoderma* sp., biomass formation with *Candida utilis* and *Saccharomyces cerevisiae*	Crude protein enriched from 5.9% to 40.3%. Source for animal fodder	Haddadin et al. (1999)
Pineapple waste	*A. foetidus* produces citric acid 16.1 g/100 g DM and 3% methanol	Pharmaceuticals, food industry, preserving agent	Tran and Mitchell (1995)
Potato waste	Amylases	Bakery goods, breweries, textile industry	Lucas et al. (1997)
Sugar beet pulp, cereal bran	Commensalism of two microorganisms degrading the substrate	Flavor vanillin	Asther et al. (1997)
Tomato pomace	Co-cultures of *Trichoderma reesei* and *Sporotrichum* sp. degrade cellulose and hemicellulose fraction	67% less cellulose, 73% less hemicellulose, enhanced lignin and protein content	Carvalheiro et al. (1994)

Adapted from Laufenberg et al. (2003).

Ellaiah et al. (2004) tested several support substrates (wheat bran, wheat rawa, rice bran, rice rawa, rice husk, rice straw, maize bran, ragi bran, green gram bran, black gram bran, red gram bran, corn flour, jowar flour, sago, and sugar cane bagasse) for neomycin production by a mutant strain of *Streptomyces marinensis*, under SSF. The accumulation of neomycin by SSF was 1.85 times higher than during the SmF.

Asagbra et al. (2005) assessed the ability of *Streptomyces* sp. OXCI, *S. rimosus* NRRL B2659, *S. rimosus* NRRL B2234, *S. alboflavus* NRRL B1273, *S. aureofaciens* NRRL B2183, and *S. vendagensis* ATCC 25507 to produce tetracycline under SSF conditions using peanut shells, corn cob, corn pomace, and cassava peels as substrates. They found that peanut shells were the most effective substrate. Mizumoto et al. (2006) reported the production of the lipopeptide antibiotic

iturin A by *Bacillus subtilis* using soya bean curd residue, okara, a by-product of tofu manufacture in SSF. After 4 days of incubation, iturin A production reached 3300 mg/kg wet solid material (14 g/kg dry solid content material), which was approximately tenfold higher than that in SmF.

4.2.3 Production of Pigments

Attri and Joshi (2005) used an apple pomace-based medium to examine the effect of carbon and nitrogen sources on carotenoid production by *Micrococcus* sp. Using 20 g/L of apple pomace in the basic medium provided the best growth conditions for the microorganism. Maximum biomass (4.13 g/L) and pigment (9.97 mg per 100 g of medium) yields were achieved when the medium was supplemented with 0.2%

fructose. Optimal conditions for carotenoid production were 35°C, pH 6.0, and a cultivation time of 96 h. The same authors (Attri and Joshi, 2006) studied carotenoid production by *Chromobacter* sp. Using the same basic medium (20 g/L of apple pomace), they found a high production of biomass (6.6 g/L) and carotenoids (46.6 mg per 100 g of medium) and with a shorter incubation period (48 hours). These differences showed that the production of carotenoids can be improved by an accurate choice of organism.

Extracting carotenoids from shrimp waste by using different proteolytic enzymes gave improved extraction of carotenoids and maximum yield over the traditional, solvent extraction process and SC−CO$_2$ extraction (Babu et al., 2008). Trypsin recovered the highest amount of carotenoids from all types of head wastes, but pepsin and papain also showed good recoveries of carotenoids. The percent of recovery varied with the raw materials and the trend was *Penaeus indicus* > *Penaeus monodon* (culture) > *Metapenaeus monocerous* > *Penaeus monodon* (wild). The loss of carotenoids during processing of frozen carotenoid-protein cake (CPC) to freeze-dried product was noticed in all trials. Astaxanthin was the main stable pigment and its proportion in total carotenoids increased in freeze-dried product with the loss of minor carotenoids such as β-carotene and their derivatives.

5. CONCLUSIONS

Recovery of commodities from agricultural and food waste products can be successfully accomplished using SSF technology. It provides many novel opportunities as it allows the use of the wastes without the need for extensive pretreatment. Since the substrates used are usually agro-industrial residues, fermentation media are simpler, requiring no complex nutrients or vitamins. In contrast to SmF, products of SSF can be easier to recover, as they are not highly diluted. SSF requires lower capital investments, and it leads to lower recurring expenditure. Moreover, SSF can improve economic feasibility of the biotechnological processes, offering waste reduction in design and operation.

The SSF cultivation technique will be increasingly important in the future, as population growth puts ever greater pressure on world resources, stimulating the development of biorefineries. The use of solid-state fermentation in many of the biological processing steps in the biorefinery will help to minimize water use. However, in order to fulfill this potential, it will be necessary to have reliable large-scale solid-state fermentation bioreactors and strategies for optimizing

their operation (Mitchell et al., 2011). This chapter provides an overview of the engineering aspects of SSF and their application to the design and operation of various bioreactor types, identifying where further work is necessary. It presents the range of bioreactors that have been used to date, classifying them on the basis of aeration and agitation strategies: trays, packed-beds, rotating/stirred drums, and forcefully aerated agitated bioreactors. Examples of products of SSF, illustrated here, include bulk chemicals like organic acids (lactic, citric and γ-linolenic acids), ethanol, industrial enzymes, polysaccharides, and nutrient enriched animal feeds, as well as fine chemicals: antibiotics from a cellulosic waste, pigments and flavor compounds from fruit/vegetable residues. Numerous studies and patents are described regarding the employment of apple pomace for the production of valuable compounds. Several operational variables must be considered and optimized: strain type, reactor design, aeration, pH, moisture, and nutrient supplementation. These are examples of the basic process variables that are crucial for the cost-effective use of apple pomace as a substrate for biotechnological applications (Vendruscolo et al., 2008).

Agro-industrial residues such as rice bran, rice husk, potato wastes, cassava husk, wheat bran, sugar cane bagasse, sugar beet pulp, wheat/rice straw, cocoa pod, etc. are used for the production of a range of industrial enzymes by SSF processes. We have presented an illustrative survey on various individual groups of enzymes such as amylolytic, cellulolytic, ligninolytic, pectinolytic, proteolitic, etc. The major factors that affect microbial synthesis of enzymes in a SSF system include: selection of a suitable substrate and microorganism; pretreatment of the substrate; particle size, inter-particle space and surface area of the substrate; water content and activity of the substrate; relative humidity; type and size of the inoculum; control of temperature of fermenting matter/removal of metabolic heat; period of cultivation; maintenance of uniformity in the environment of the SSF system, as well as oxygen consumption and carbon dioxide evolution rates (Pandrey et al., 2008). Filamentous fungi are metabolically versatile organisms that are exploited commercially as cell factories for the production of enzymes and a wide variety of metabolites. The fungi are particularly well adapted for SSF as they have a high capacity for the secretion of hydrolytic enzymes necessary for the degradation of plant cell walls and storage carbohydrates (Archer et al., 2008). It was possible to control simultaneous production of pectinolytic, cellulolytic, and xylanolytic enzymes by strains of the genera *Aspergillus*, *Fusarium*, *Neurospora*, and *Penicillium* and generate multi-enzyme activities using a simple growth medium consisting of citrus processing waste (orange peels) and a mineral medium (Mamma et al., 2007). Further application of contemporary biotechnical knowledge and process control technologies can lead to significant productivity increases from SSF processes.

References

Adinarayana, K., Prabhakar, T., Srinivasulu, V., Rao, A.M., et al., 2003a. Optimization of process parameters for cephalosporin C production under solid state fermentation from *Acremonium chrysogenum*. Proc. Biochem. 39, 171–177.

Adinarayana, K., Ellaiah, P., Srinivasulu, B., Bhavani Devi, R., Adinarayana, G., 2003b. Response surface methodological approach to optimize the nutritional parameters for neomycin production by *Streptomyces marinensis* under solid-state fermentation. Proc. Biochem. 38, 1565–1572.

Aguilar, C.N., Gutierrez-Sanchez, G., 2001. Review: sources, properties, applications and potential uses of tannin acyl hydrolase. Food Sci. Technol. Int. 7, 373–382.

Aguilar, C.N., Rodriguez, R., Gutierrez-Sanchez, G., 2007. Microbial tannases: advances and perspectives. Appl. Microbiol. Biotechnol. 76, 47–59.

Albuquerque, P.M., Koch, F., Trossini, T.G., Esposito, E., Ninow, J.L., 2006. Production of *Rhizopus oligosporus* protein by solid state fermentation of apple pomace. Braz. Arch. Biol. Technol. 49, 91–100.

Ali, H., Kh., Q., Zulkali, M.M.D., 2011. Design aspects of bioreactors for solid-state fermentation: a review. Chem. Biochem. Eng. Q. 25, 255–266.

Anwar, A., Saleemuddin, M., 1998. Alkaline proteases: a review. Bioresour. Technol. 64, 175–183.

Archer, D.B., Connerton, I.F., MacKenzie, D.A., 2008. Filamentous fungi for production of food additives and processing aids. Adv Biochem Engin/Biotechnol. 111, 99–147.

Asagbra, A.E., Sanni, A.I., Oyewole, O.B., 2005. Solid-state fermentation production of tetracycline by *Streptomyces* strains using some agricultural wastes as substrate. World J. Microb. Biotechnol. 21, 107–114.

Asther, M., et al. (1997). Fungal biotransformation of European agricultural by-products to natural vanillin: a two-step process. In: Food Ingredients Porte de Versailles, Paris, France, 12–14 November 1996, pp. 123–125.

Attri, D., Joshi, V.K., 2005. Optimization of apple pomace based medium and fermentation conditions for pigment production by *Micrococcus* species. J. Sci. Ind. 64, 598–601.

Attri, D., Joshi, V.K., 2006. Optimization of apple pomace based pigment production medium and fermentation conditions for by *Chromobacter* species. J. Food Sci. Technol. 43, 484–487.

Babitha, S., Soccol, C., Pandey, A., 2007. Solid-state fermentation for the production of *Monascus* pigments from jackfruit seed. Bioresource Technolog. 98, 1554–1560.

Babu, C.M., Chakrabarti, R., Sambasivarao, K.R.S., 2008. Enzymatic isolation of carotenoid-protein complex from shrimp head waste and its use as a source of carotenoids. LWT 41, 227–235.

Baysal, Z., Uyar, F., Aytekin, C., 2003. Solid state fermentation for production of α-amylase by a thermotolerant *Bacillus subtilis* from hot-spring water. Proc. Biochem. 38, 1665–1668.

Berovic, M., Ostroversnik, H., 1997. Production of *Aspergillus niger* pectolytic enzymes by solid state bioprocessing of apple pomace. J. Biotechnol. 53, 47–53.

Bhat, M.K., 2000. Cellulase and their related enzymes in biotechnology. Biotechnol. Adv. 18, 355–383.

Bhushan, S., Joshi, V.K., 2006. Baker's yeast production under fed batch culture from apple pomace. J. Sci. Ind. Res. 65, 72–76.

Botella, C., Diaz, A., de Ory, I., Webb, C., Blandino, A., 2007. Xylanase and pectinase production by *Aspergillus awamori* on grape pomace in solid state fermentation. Proc. Biochem. 42, 98–101.

Bramorski, A., Soccol, C.R., Christen, P., Revah, S., 1998. Fruit aroma production by *Ceratocystis fimbriata* in solid cultures from agroindustrial wastes. Rev. Microbiol., 29.

Cabaleiro, D.R., Rodriguez, S., Sanroman, A., Longo, M.A., 2002. Comparison between the protease production ability of lininolytic fungi cultivated in solid state media. Proc. Biochem. 37, 1017–1023.

Campbell-Platt, G., 1994. Fermented foods—a world perspective. Food Res. Int. 27, 253–257.

Cannel, E., Moo-Young, M., 1980. Solid state fermentation. Proc. Biochem. 15, 2–7.

Carr, F.J., Chill, D., Maida, N., 2002. The lactic acid bacteria: a literature survey. Crit. Rev. Microbiol. 28, 281–370.

Carvalheiro, F., Roseiro, J.C., Collaco, M.T.A., 1994. Biological conversion of tomato pomace by pure and mixed fungal cultures. Proc. Biochem. 29, 601–605.

Carvalho, J.C., Pandey, A., Oishi, B.O., Brand, D., Leon, J.A.R., Soccol, C., 2006. Relation between growth, respirometric analysis and biopigments production from *Monascus* by solid-state fermentation. Biochem. Eng. J. 29, 262–269.

Chahal, D.S., 1983. Foundation in biochemical engineering kinetics and thermodynamics in biological systems. In: American Chemical Society Symposium Series, vol. 207. American Chemical Society, Washington, DC, p. 421.

Christen, P., Bramorski, A., Revah, S., Soccol, C.R., 2000. Characterization of volatile compounds produced by *Rhizopus* strains grown on agro-industrial solid wastes. Bioresour. Technol. 71, 211–215.

Christi, Y., 1999. Solid substrate fermentations, enzyme production, food enrichment. In: Flickinger, M.C., Drew, S.W. (Eds.), Encyclopedia of Bioprocess Technology: Fermentation, Biocatalysis and Bioseparation, vol. 5. Wiley, New York, pp. 2446–2462.

Cordova, J., et al., 1998. Lipase production by solid state fermentation of olive cake and sugar cane bagasse. J. Mol. Catal. B: Enzymatic 5, 75–78.

Couto, S.R., 2008. Exploitation of biological wastes for the production of value-added products under solid-state fermentation conditions. Biotechnol. J. 3, 859–870.

Couto, S.R., Barreiro, M., Rivela, I., Longo, M.A., Sanroman, A., 2002. Performance of a solid-state immersion bioreactor for ligninolytic enzyme production: evaluation of different operational variables. Process Biochem. 38, 219–227.

Daigle, P., Gelinas, P., Leblanc, D., Morin, A., 1999. Production of aroma compounds by *Geotrichum candidum* on waste bread crumb. Food Microbiol. 16, 517–522.

De Reu, J.C., Zwietering, M.H., Rombouts, F.M., Nout, M.J.R., 1993. Temperature control in solid-substrate fermentation through discontinuous rotation. Appl. Microbiol. Biotechnol. 40, 261–265.

De Souza, C.G.M., Zilly, A., Peralta, R.M., 2002. Production of laccase as the sole phenoloxidase by a Brazilian strain of *Pluerotus pulmonarius* in SSF. J. Basic Microbiol. 42, 83–90.

Desai, S.S., Nityanand, C., 2011. Microbial laccases and their applications: a review. Asian J. Biotechnol. 3, 98–124.

Devrajan, A., Joshi, V.K., Gupta, K., Sheikher, C., Lal, B.B., 2004. Evaluation of apple pomace based reconstituted feed in rats after solid state fermentation and ethanol recovery. Braz. Arch. Biol. Technol. 47, 93–106.

Dominguez, M., Mejia, A., Gonzalez, J.B., 2000. J. Biosc. Bioeng. 89, 409.

Durand, A., 2003. Bioreactor design for solid-state fermentation. Biochem. Eng. J. 13, 113–125.

Durand, A., Chereau, D., 1988. A new pilot reactor for solid-state fermentation: application to the protein enrichment of sugar beet pulp. Biotechnol. Bioeng. 31, 476–486.

Durand, A., Renaud, R., Maratray, J., Almanza, S., 1995. The INRA-Dijon reactors: designs and applications. In: Roussos, S., Lonsane, B.K., Raimbault, M., Viniegra-Gonzalez, G. (Eds.), Advances in Solid State Fermentation. Kluwer Academic Publishers, Dordrecht, pp. 71–92. Proceedings of the 2nd international symposium on solid state fermentation FMS-95, Montpellier, France.

Durand, A., Renaud, R., Maratray, J., Almanza, S., Diez, M., 1996. INRA Dijon reactors for solid-state fermentation: design and applications. J. Sci. Ind. Res. 55, 317–332.

Elayaraja, S., Velvizhi, Maharani, T. V., Mayavu, P., Vijayalakshmi, S., Balasubramanian, T., 2011. Thermostable α-amylase production by Bacillus firmus CAS 7 using potato peel as a substrate. African Journal of Biotechnology, 10, 11235–11238.

Ellaiah, P., Shrinivasulu, B., Adinarayana, K., 2004. Optimization studies on neomycin production by a mutant strain of Streptomyces marinensis in solid-state fermentation. Proc. Biochem. 39, 529–534.

Ellen, C.G., Dekker, R.F.H., Barbosa, A.M., 2008. Orange baggase as a substrate for the production of pectinase and laccase by Botryosphaeria rodhina MAMB-05 in submerged and SSF. Bioresources 3, 335–345.

Erdal, S., Taskin, M., 2010. Production of α-amylase by Penicillium expansum MT-1 in solid-state fermentation using waste Loquat (Eriobotrya japonica Lindley) kernels as substrate. Rom. Biotechnol. Lett. 15, 5343–5350.

Freihat, N.M., Al-Ghzawi, A.A., Zaitoun, S., Alqudah, A., 2008. Fruit set and quality of loquats (Eriobotrya japonica) as effected by pollinations under sub-humid Mediterranean. Sci. Horticult. 117, 58–62.

Fung, C.J., Mitchell, D.A., 1995. Baffles increase performance of solid state fermentation in rotating drums. Biotechnol. Tech. 9, 295–298.

Gawande, P.V., Kamat, M.Y., 1999. Production of Aspergillus xylanase by lignocellulosic waste fermentation and its application. J. Appl. Microbiol. 87, 511–519.

Germano, S., Pandey, A., Osaku, C.A., Rocha, S.N., Soccol, C.R., 2003. Characterization and stability of proteases from Penicillium sp. produced by solid-state fermentation. Enzyme Microb. Tech. 32, 246–251.

Gomez, J.M., Pazos, M., Rodriguez-Couto, S., Sanroman, M.A., 2005. Chestnut shell and barley bran as potential substrates for laccase production by Coriolopsis rigida under SS conditions. J. Food Eng. 68, 315–319.

Grigelmo-Miguel, N., Martın-Belloso, O., 1999. Comparison of dietary fibre from by-products of processing fruits and greens and from cereals. LWT-Food Sci. Technol. 32, 503–508.

Grohmann, K., Bothast, R.J., 1994. Pectin-rich residues generated by processing of citrus fruits, apples, and sugar beets: enzymatic hydrolysis and biological conversion to value-added products, ACS-symp-ser., 566. American Chemical Society, Washington, pp. 372–390.

Gullon, B., Yanez, R., Alonso, J.L., Parajo, J.C., 2008. L-Lactic acid production from apple pomace by sequential hydrolysis and fermentation. Bioresour. Technol. 99, 308–319.

Haddadin, M.S., Abdulrahim, S.M., Al-Kawaldeh, G.Y., Robinson, R. K., 1999. Solid state fermentation of waste pomace from olive processing. J. Chem. Technol. Biotechnol. 74, 613–618.

Hang, Y.D., Woodams, E.E., 1994. Apple pomace: a potential substrate for production of β-glucosidase by Aspergillus foetidus. LWT-Food Sci. Technol. 27, 587–589.

Haq, I., Idrees, S., Rajoka, M.I., 2002. Proc. Biochem. 37, 637.

Hofvendahl, K., Hahn-Hagerdal, B., 2000. Factors affecting the fermentative lactic acid production from renewable resources. Enz. Microb. Technol. 26, 87–107.

Hongzhang, C., Fujian, X., Zhonghou, T., Zuohu, L., 2002. A novel industrial-level reactor with two dynamic changes of air for solid-state fermentation. J. Biosci. Bioeng. 93, 211–214.

Imandi, S.B., Bandaru, V.V.R., Somalanka, S.R., Bandaru, S.R., Garapati, H.R., 2007. Application of statistical experimental designs for the optimization of medium constituents for the production of citric acid from pineapple waste. Bioresour. Technol. 99, 4445–4450.

Jin, H., Kim, H.S., Kim, S.K., Shin, M.K., Kim, J.H., Lee, J.W., 2002. Production of heteropolysaccharide-7 by Beijerinckia indica from agroindustrial byproducts. Enzyme Microb. Technol. 30, 822–827.

Joshi, V.K., Parmar, M., Rana, N.S., 2006. Pectin esterase production from apple pomace in solid-state and submerged fermentations. Food Technol. Biotechnol. 44, 253–256.

Kapoor, S., Khanna, P.K., Katyal, P., 2009. Effect of supplement of wheat straw on growth and lignocellulolytic enzyme potential of Lentinus edodes. W.J. Agric. Sci. 5, 328–381.

Kashyap, P., Sabu, A., Pandey, A., Szakacs, G., Soccol, C., 2002. Extra-cellular L-glutaminase production by Zygosaccharomyces rouxii under solid-state fermentation. Process Biochemistry 38, 307–312.

Khosravi, K., Shojaosadati, S.A., 2003. A solid state fermentation system for production of ethanol from apple pomace. Fanni va Muhandisi-i Mudarris 10, 55–60.

Kubicek, C.P., Rhor, M., 1986. Citric acid fermentation. CRC Crit. Rev. Biotechnol. 33, 1500–1504.

Kumar, D., Jain, V.K., Shanker, G., Srivastava, A., 2003a. Utilisation of fruits waste for citric acid production by solid state fermentation. Proc. Biochem. 38, 1725–1729.

Kumar, D., Jain, V.K., Shanker, G., Srivastava, A., 2003b. Citric acid production by solid state fermentation using sugarcane bagasse. Proc. Biochem. 38, 1731–1738.

Kunamneni, A., Permaul, K., Singh, S., 2005. Amylase production in solid state fermentation by the thermophilic fungus Thermomyces lanuginosus. J. Biosci. Bioeng. 100, 168–171.

Laufenberg, G., Rosato, P., Kunz, B., 2001. Conversion of vegetable waste into value added products: oil press cake as an exclusive substrate for microbial d-decalactone production. Lecture at Lipids, Fats, and Oils: Reality and public perception. AOCS press, Berlin, 24th world congress and exhibition of the ISF, 16–20.09.01. p. 10.

Laufenberg, G., Kunz, B., Nystroem, M., 2003. Transformation of vegetable waste into value added products: (A) the upgrading concept; (B) practical implementations. Bioresour. Technol. 87, 167–198.

Lonsane, B.K., Ghildyal, N.P., Budiatman, S., Ramakrishna, S.V., 1985. Engineering aspects of solid state fermentation. Enzyme Microbiol. Technol. 7, 258–265.

Lonsane, B.K., Saucedo-Castaneda, G., Raimbault, M., Roussos, S., Viniegra-Gonzales, G., Ghildyal, N.P., et al., 1992. Scale-up strategies for solid state fermentation systems. Proc. Biochem. 27, 259–273.

Lucas. J., Filipini, M., Gruess, O., et al. (1997) Fermentative utilization of fruit and vegetable pomace (biowaste) for the production of novel types of products—results of an air project. In: Proceedings of the Eleventh Forum for Applied Biotechnology, vol. 62, pp. 1865–1867. Gent, Belgium, 25–26 September. Gent: Universiteit-Gent.

Mamma, D., Kourtoglou, E., Christakopoulos, P., 2007. Fungal multi-enzyme production on industrial by-products of the citrus-processing industry. Bioresour. Technol. 99, 2373–2383.

Mandalari, G., Bisignano, G., Lo Curto, R.B., Waldron, K.W., Faulds, C.B., 2008. Production of feruloyl esterases and xylanases by *Talaromyces stipitatus* and *Humicola grisea* var. thermoidea on industrial food processing by-products. Bioresour. Technol. 99, 5130–5133.

Martinez-Calvo, J., Badenes, M.L., Llacer, G., Bleiholder, H., Hack, H., Meier, U., 1999. Phenological growth stages of loquat tree (*Eriobotrya japonica* (Thunb.) Lindl.). Ann. Appl. Bid. 134, 353–357.

Medeiros, A.B.P., Pandrey, A., Freitas, R.J.S., Christen, P., Soccol, C. R., 2000. Optimization of the production of aroma compounds by *Kluyveromycesmarxianus* in solid-state fermentation using factorial design and response surface methodology. Biochem. Eng. J. 6, 33–39.

Medeiros, A.B.P., Pandey, A., Vandenberghe, L.P.S., Pastore, G.M., Soccol, C.R., 2006. Production and recovery of aroma compounds produced by solid-state fermentation using different adsorbents. Food Technol. Biotechnol. 44, 47–51.

Meien, O.F., Mitchell, D.A., 2002. A two-phase model for water and heat transfer within an intermittently-mixed solid-state fermentation bioreactor with forced aeration. Biotech. Bioeng. 79, 416–428.

Mitchell, D.A., Lonsane, B.K., Durand, A., Renaud, R., Almanza, S., Maratray, J., et al., 1992. General principles of reactor design and operation for solid substrate cultivation. In: Doelle, H.W., Mitchell, D.A., Rolz, C.E. (Eds.), Solid Substrate Cultivation. Elsevier Applied Science, Amsterdam, pp. 115–139.

Mitchell, D.A., Krieger, N., Stuart, D.M., Pandey, A., 2000. New developments in solid-state fermentation. Part II. Rational approaches to the design, operation and scale-up of bioreactors. Proc. Biochem. 35, 1211–1225.

Mitchell, D.A., Von Meien, O.F., Krieger, N., 2004. A review of recent developments in modeling microbial growth kinetics and intraparticle phenomena in solid-state fermentation. Biochem. Eng. J. 17, 15–26.

Mitchell, D.A., de Lima Luz, L.F., Krieger, N., 2011. Bioreactors for solid-state fermentation. In: Second ed. Murray Moo-Yong (Ed.), Comprehensive Biotechnology, vol.2. Elsevier, pp. 347–360.

Miyazaki, C.I., Takada, J., Matsuura, A., 2008. An ability of isolated strains to efficiently cooperate in ethanolic fermentation of agricultural plant refuse under initially aerobic thermophilic conditions. Bioresour. Technol. 99, 1768–1775.

Mizumoto, S., Hirai, M., Shoda, M., 2006. Production of lipopeptide antibiotic iturin A using soybean curd residue cultivated with *Bacillus subtilis* in solid-state fermentation. Appl. Microbiol. Biotechnol. 72, 869–875.

Moreira, A.R., Phillips, J.A., Humphrey, A.E., 1981. Biotechnol. Bioeng. 23, 1325.

Morton, J.F., 1987. Loquat. In: Fruits of Warm Climates. Creative Resource Systems, Winterville, Fl. 103–108.

Mulimani, V.H., Ramalingam, P.G.N., 2000. Amylase production by solid state fermentation: a new practical approach to biotechnology courses. Biochem. Educ. 28, 161–163.

Nagel, F.J., Tramper, J., Bakker, M., Rinzema, A., 2001. Temperature control in a continuously mixed bioreactor for solid-state fermentation. Biotechnol. Bioeng. 72, 219–230.

Ngadi, M.O., Correia, L.R., 1992. Kinetics of solid-state ethanol fermentation from apple pomace. J. Food Eng., 17, 97–116.

Nogueira, A., Santos, L.D., Paganini, C., Wosiacki, G., 2005. Evaluation of alcoholic fermentation of aqueous extract of the apple pomace. Semina: Ciencias Agrarias, Londrina 26, 179–193.

Paganini, C., Nogueira, A., Silva, N.C., Wosiacki, G., 2005. Utilization of apple pomace for ethanol production and food fiber obtainment. Ciencia Agrotecnica, Lavras 29, 1231–1238.

Pandey, A., Selvakumar, P., Soccol, C.R., Nigam, P., 1999. Solid state fermentation for the production of industrial enzymes. Curr. Sci. 79, 149–162.

Pandey, A., Soccol, C.R., Rodriguez-Leon, J.A., Nigam, P., 2001. Aspects of design of fermenter in solid stale fermentation. In: Solid State Fermentation in Biotechnology: Fundamentals and Applications. Asiatech Publishers Inc, New Delhi, pp. 73–77.

Pandey, A., Soccol, C.R., Larroch, C. (Eds.) 2008. Current development in solid-state fermentation. New York. Springer, Science + Business Media, LLC.

Papagianni, M., Psomas, S.K., Batsilas, L., Paras, S., Kyriakidis, D.A., Liakopoulou, K.M., 2001. Xanthan production by *Xanthomonas campestris* in batch cultures. Proc. Biochem. 37, 73–80.

Papinutti, V.L., Diorio, L.A., Forchiassin, F., 2003. Production of laccase and manganese peroxidase by *Fomes sclerodermeus* grown on wheat bran. J. Indus. Microbiol. Biotechnol. 30, 157–160.

Papinutti, V.L., Diorio, L.A., Forchiassin, F., 2007. Production of laccase and manganese peroxidase by *Fomes sclerodermeus* grown on wheat bran. J. Indus. Microbiol. Biotechnol. 30, 157–160.

Paranthaman, R., Alagusundaram, K., Indhumathi, J., 2009. Production of protease from rice mill wastes by *Aspergillus niger* in solid state fermentation. World J. Agric. Sci. 5, 308–312.

Pericin, D.M., Antov, M.G., Popov, S.D., 1999. Simultaneous production of biomass and pectinases by Polyporus squamosus. Acta Periodica Technol. 29, 183–189.

Prabhakar, A., Krishnaiah, K., Janaun, J., Bono, A., 2005. An overview of engineering aspects of solid state fermentation. Malays. J. Microbiol. 1, 10–16.

Raghavarao, K.S.M.S., Ranganathan, T.V., Karanth, N.G., 2003. Some engineering aspects of solid-state fermentation. Biochem. Eng. J. 13, 127–135.

Rahardjo, Y.S.P., Tramper, J., Rinzema, A., 2006. Modeling conversion and transport phenomena in solid-state fermentation: a review and perspectives. Biotechnol. Adv. 24, 161–179.

Raimbault, M. and Germon, J.C. (1976). Procédé d'enrichissement en protéines de produits comestibles solides, French Patent No. 76-06-677.

Ramırez, L., Arrizon, J., Sandoval, G., 2008. A new microplate screening method for the simultaneous activity quantification of feruloyl esterases, tannases, and chlorogenate esterases. Appl. Biochem. Biotechnol. 151, 711–723.

Rathbun, B.L., Shuler, M.L., 1983. Heat and mass transfer effects in static solid-substrate fermentations: design of fermentation chambers. Biotechnol. Bioeng. 25, 929–938.

Rodrıguez, H., Curiel, J.A., Landete, J.M., 2009. Food phenolics and lactic acid bacteria. Int. J. Food Microbiol. 132, 79–90.

Rodrıguez-Duran, L., Valdivia-Urdiales, B., Contreras-Esquivel, J.C., Rodrıguez-Herrera, R., Aguilar, C.N., 2011. Novel strategies for upstream and downstream processing of tannin acyl hydrolase. Enzyme Res.20.

Roopesh, K., Ramachandran, S., Nampoothiri, K.M., Szakacs, G., Pandey, A., 2006. Biores. Tech. 97, 506.

Rosales, E., Couto, S.R., Sanromán, M.A., 2005. Reutilisation of food processing wastes for production of relevant metabolites: application to laccase production by *Trametes hirsute*. J. Food Eng. 66, 419–423.

Rosales, E., Couto, S.R., Sanromán, M.A., 2007. Increased laccase production by *Trametes hirsute* growth on crushed orange peelings. Eng. Microb. Technol. 40, 1286–1290.

Ruberto, G., Renda, A., Amico, V, Tringali, C., 2008. Volatile components of grape pomaces from different cultivars of Sicilian *Vitis vinifera* L. Bioresour. Technol. 99, 260–268.

Sandhya, C., Sumantha, A., Szakacs, G., Pandey, A., 2005. Comparative evaluation of neutral protease production by *Aspergillus oryzae* in submerged and solid-state fermentation. Proc. Biochem. 40, 2689–2694.

Seyis, I., Aksoz, N., 2005. Xylanase production from *Trichoderma harzianum* 1073 D3 with alternative carbon source and nitrogen sources. Food Technol. Biotechnol. 43, 37–40.

Sharma, A., Vivekanand, V., Singh, R.P., 2008. Solid-state fermentation for gluconic acid production from sugarcane molasses by *Aspergillus niger* ARNU-4 employing tea waste as the novel solid support. Biores. Technol. 99, 3444–3450.

Sharoni, Y., Danilenko, M., Levy, J., 2000. Molecular mechanisms for the anticancer activity of the carotenoid lycopene. Drug. Dev. Res. 50, 448–456.

Shojaosadati, S.A., Babaeipour, V., 2002. Citric acid production from apple pomace in multi-layer packed bed solid-state bioreactor. Proc. Biochem. 37, 909–914.

Soares, M., Christen, P., Pandey, A., Soccol, C.R., 2000. Fruity flavour production by *Ceratocystis fimbriata* grown on coffee husk in solid-state fermentation. Proc. Biochem. 35, 857–861.

Song, J., Xu, K., Ma, H., Huang, J., 2005a. Method for producing single cell protein from apple pomace by dual solid state fermentation. Patent no. CN 1673343-A.

Song, J., Xu, K., Ma, H., Huang, J., 2005b. Method for producing feed protein by liquid-solid fermentation of apple pomace. Patent no. CN 1663421-A.

Stredansky, M., Conti, E., 1999. Xanthan production by solid state fermentation. Proc. Biochem. 34, 581–587.

Stredansky, M., Conti, E., Stredanska, S., Zanetti, F., 2000. γ-Linolenic acid production with *Thamnidium elegans* by solid state fermentation on apple pomace. Bioresour. Technol. 73, 41–45.

Streit, F., Koch, F., Trossini, T. G., Laranjeira, M. C. M., Ninow, J. L., 2004. An alternative process for the production of an additive for the food industry: chitosan. In: International Conference Engineering and Food—ICEF 9, Montpellier France.

Sutcu, H., Demiral, H., 2009. Production of granular activated carbons from loquat stones by chemical activation. J. Anal. Appl. Pyrolysis 84, 47–52.

Torres, J.B., Varela, M.T., Garcia, J., Carilla, C., Matito, J.J., Centelles, M., et al., 2002. Valorization of grape (*Vitis vinifera*) by products. Antioxidant and biological properties of polyphenolic fractions differing in procyanidin composition and flavonol content. J. Agr. Food. Chem. 50, 7548–7555.

Tran, C.T., Mitchell, D.A., 1995. Pineapple waste—a novel substrate for citric acid production by solid-state fermentation. Biotechnol. Lett. 17, 1107–1110.

Tran, C.T., Sly, L.I., Mitchell, D.A., 1998. Selection of a strain of *Aspergillus* for the production of citric acid from pineapple waste in solid-state fermentation. W. J. Micro. Biotech. 14, 399–404.

Tunga, R., Shrivastava, B., Banerjee, R., 2003. Purification and characterization of a protease from solid state cultures of *Aspergillus parasiticus*. Proc. Biochem. 38, 1553–1558.

Vendruscolo, F., 2005. Cultivo em meio solido e submerso do bagaco de maca por *Gongronella butleri* e avaliacao do seu potencial biotecnologico. Florianopolis: UFSC, 2005. Dissertation (Master's degree in Food Engineering), Departamento de Engenharia Química e Engenharia de Alimentos, Universidade Federal de Santa Catarina.

Vendruscolo, F., Albuquerque, P.M., Streit, F., Esposito, E., Ninow, J.L., 2008. Apple pomace: a versatile substrate for biotechnological applications. Crit. Rev. Biotechnol. 28, 1–12.

Verma, P, Madamwar, D., 2002. Production of ligninolytic enzymes for dye decolorization by cocultivation of white-rot fungi *Pluerotus ostreatus* and *Phanerochaete chrisosporius* under SSF. Appl. Biochem.Biotechnol. 102, 109–118.

Villas-Boas, S.G., Esposito, E., Mendonca, M.M., 2002. Novel lignocellulolytic ability of *Candida utilis* during solid state cultivation on apple pomace. World J. Microbiol. Biotechnol. 18, 541–545.

Virupakshi, S., Gireesh Babu, K., Gaikwad, S.R., Naik, G.R., 2005. Production of a xylanolytic enzyme by a thermoalkaliphilic *Bacillus* sp. JB-99 in solid state fermentation. Proc. Biochem. 40, 431–435.

Wassewar, K.L., 2005. Separation of lactic acid: recent advances. Chem. Biochem. Eng. Q. 19, 159–172.

Wecker, A., Onken, V., 1991. Influence of dissolved oxygen concentration and shear rate on the production of pullulan by *Aureobasidium pullulans*. Biotechnol. Lett. 13, 155–160.

Yang, S.S., 1996. Antibiotics production of cellulosic wastes with solid state fermentation by *Streptomyces*. Renew. Energy. 9, 876–979.

Yang, S.S., Ling, M.Y., 1989. Tetracycline production with sweet potato residue by solid state fermentation. Biotechnol. Bioeng. 33, 1021–1028.

Zheng, Z., Shetty, K., 1998. Cranberry processing waste for solid state fungal inoculant production. Proc. Biochem. 33, 323–329.

Zheng, Z., Shetty, K., 2000a. Enhancement of pea (*Pisum sativum*) seedling vigour and associated phenolic content by extracts of apple pomace fermented with *Trichoderma* spp. Proc. Biochem. 36, 79–84.

Zheng, Z., Shetty, K., 2000b. Solid state production of polygalacturonase by *Lentinus edodes* using fruit processing wastes. Proc. Biochem. 35, 825–830.

Functional Food and Nutraceuticals Derived from Food Industry Wastes

Maria R. Kosseva

1. INTRODUCTION

Vegetables and fruits are known to contain components, such as vitamins, essential minerals, antioxidants, and prebiotics (fibers), with several types of health-promoting actions, and most of these have been evaluated in intervention studies. Many epidemiological studies show negative correlations between the intake of vegetables and fruits and the incidence of several important diseases, including cancer and atherosclerosis (Kris-Etherton et al., 2002; Gundgaard et al., 2003; Maynard et al., 2003; Trichopoulou et al, 2003). Optimization of composition of plant-derived food would be a very cost-effective method for disease prevention, since diet-induced health improvements would not carry any added costs for the health sector (Gundgaard et al., 2003). If improvements can be obtained with existing or slightly adapted food technology, the production costs will be similar (Brandt et al., 2004).

The current global market size of functional foods has already reached $73.5 billion from a modest base just 10 years ago (www.just-food.com, 2006). The US market dominates (>30% of the total global market) and is showing a sustained growth of ~14% per year. Other significant markets include the EU and Japan. Growth in the functional foods market across the world is currently ~8% per year, and at this rate the market will be valued at >$100 billion by 2012. In this large and burgeoning marketplace, the food industry is demanding economical, high-quality, novel, and substantiated ingredients (Smithers, 2008).

The first part of this chapter is focused on functional foods/nutraceuticals derived from fruit-and-vegetable waste (FVW) and phytochemical extraction methods. Then a brief overview of bioactive peptides and their occurrence in whey and dairy by-products is provided. Finally, consumer acceptance of the concept of functional foods is recognized as a key success factor for new product development and market orientation.

1.1 Definition of Nutraceuticals and Functional Food

Products isolated or purified from food but that are generally sold in medicinal forms not usually associated with food, such as capsules, are referred to as *nutraceuticals*. Nutraceuticals are demonstrated to have a physiological benefit or provide protection against chronic disease. A working definition of nutraceutical from a *Science* forum states: "a diet supplement that delivers a concentrated form of a biologically active component of food in a non-food matrix to enhance health" (Zeisel, 1999). This distinguishes nutraceuticals from *functional foods*, which according to the forum: "are consumed as part of a normal diet and deliver one or more active ingredients (that have physiologic effects and may enhance health) within the food matrix".

The European Commission's Concerted Action on Functional Food Science in Europe (FuFoSE), coordinated by the International Life Science Institute (ILSI) Europe, defined functional food as follows: "a food product can only be considered functional, if together with the basic nutritional impact it has beneficial effects on one or more functions of the human organism thus either improving the general and physical conditions or/and decreasing the risk of the evolution of diseases. The amount of intake and form of the functional food should be as it is normally expected for dietary purposes. Therefore, it could not be in the form of pill or capsule just as normal food form" (Diplock et al., 1999). Contrary to this latter statement,

Food Industry Wastes.
DOI: http://dx.doi.org/10.1016/B978-0-12-391921-2.00006-8

since 2001 FOSHU (Food for Specified Health Uses) products in Japan can also take the form of capsules and tablets like nutraceuticals, although a great majority of products are still in more conventional forms (Ohama et al., 2006).

European legislation, however, does not consider functional foods as specific food categories but rather as a concept (Coppens et al., 2006). According to the European regulation on nutrition and health claims made on foods (EC No.1924/2006), a list of authorized claims has to be published for all member states, and nutrient profiles also have to be established for foods containing health claims (Siro et al., 2008). Bagchi (2006) has discussed relevant legislation in the USA.

2. PHENOLIC COMPOUNDS DERIVED FROM FRUIT-AND-VEGETABLE PROCESSING WASTES

The beneficial effects of plant-derived food are attributed mainly to high-molecular-weight dietary fiber on one hand, and low-molecular-weight secondary plant metabolites on the other. The latter components are chemically very heterogeneous and comprise carotenoids, polyphenols, glucosinolates, saponins, alkaloids, and so on. The polyphenols are broadly classified into phenolic acids (hydroxybenzoic and hydroxycinnamic acids), flavonoids, xanthones, and stilbenes and constitute an extremely diverse class of secondary metabolites. They have been associated with a number of health-promoting properties such as antioxidant, anticarcinogenic, anti-inflammatory, antidiabetic, antithrombotic, and vasoprotective activities.

Why are wastes from fruit-and-vegetable (FV) processing such a rich source of bioactive compounds? The trend to produce functional foods by adding bioactive compounds has entailed numerous investigations on the extraction of secondary plant metabolites, which in turn raises the question as to the sources of these components. In this context, wastes from FV processing such as peels, seeds, and stems have attracted intense interest during the past decade. Depending on the raw material and the technologies applied, they emerge in large quantities and are often a considerable disposal problem for the food industry. For example, during wine and apple juice production, approximately 20% and 35% respectively of the raw material remains as pomace. Even higher proportions of by-products emerge from processing of some exotic fruits such as mangoes, where the peels and seeds may amount to up to 60% of the total fruit weight. Secondary plant metabolites such as polyphenols play an important role in the defense system of the plant, protecting it from biotic and abiotic stress. For example, flavonoids act as UV absorbing compounds and signal molecules. Phenolic compounds also show antimicrobial activity against plant pathogens. Because of their biological role in plants, secondary metabolites are located primarily in the outer layers of fruits and vegetables and in the seeds. During processing, these plant parts are usually removed by peeling or are retained in the press residues (e.g. skins and seeds in grape pomace). For this reason, the by-products contain large amounts of secondary plant metabolites in concentrated form and represent promising sources of bioactive compounds, which may be included in functional foods. Because of their high water contents, the by-products are prone to microbial spoilage and need to be dried immediately after processing, which is an economically limiting factor (Schieber, 2009).

Selected sources of FV processing residues and the major fractions of phenolics are shown in Table 6.1.

2.1 Flavonoids

The flavonoids are a group of plant metabolites that are the most common group of polyphenolics in the human diet (Figure 6.1). They are subdivided into several

TABLE 6.1 Waste from Fruit-and-Vegetable Processing as a Source Of Phenolic Compounds

FV-derived By-products	Phenolic Compounds (Major Fractions)
FRUIT-DERIVED	
Apple pomace	Chlorogenic acid, quercetin glycosides, dihydrochalcones, flavanols
Black currant residues	Anthocyanins
Blueberry processing waste	Anthocyanins, hydroxycinnamates, flavonol glycosides
Cranberry pomace	Caffeic acid, ellagic acid
Grape pomace	Anthocyanins, flavonol glycosides, stilbenes, phenolic acids
Mango peels/ kernels	Flavonol glycosides, xanthone glycosides, hydrolyzable tannins, alk(en)ylresorcinols
Star fruit residues	Procyanidins
VEGETABLE-DERIVED	
Artichoke pomace	Hydroxycinnamates, flavonoids
Cauliflower by-products	Kaempferol glycosides, hydroxycinnamates
Olive mill waste	Oleuropein, hydroxytyrosol, verbascoside, dihydroxyphenylglycol
Onion peels	Quercetin glycosides
Potato peels	Phenolic acids

Adapted from Schieber (2009).

other groups including flavone, flavonol, flavanone, and isoflavones. Isoflavonoids differ from other flavonoids by having ring B attached to the C-3 position of ring C (e.g., puerarin, daidzein, and genistein). In plants, they are especially important in guarding against oxidant damage, and they provide to the plant the color that attracts pollinators and repels attacks by insects and microbes.

Recent research suggests that, in humans, these plant polyphenols provide important health benefits related to metabolic syndrome, cancer, brain health, and the immune system. The relatively low toxicity and potential efficacy of most of these agents make them attractive to a large sector of the population. While dietary flavonoids are primarily obtained from soy, many are found in fruits, nuts, and more exotic sources. Perhaps the strongest evidence for the benefits of flavonoids in diseases of aging relates to their effect on components of the metabolic syndrome. Flavonoids from grape seed, soy, kudzu (*Pueraria lobata*) roots, and other sources all lower arterial pressure in hypertensive animal models and in a limited number of tests in humans. They also decrease the plasma concentration of lipids and buffer plasma glucose. The underlying mechanisms appear to include

antioxidant actions, central nervous system effects, gut transport alterations, fatty acid sequestration and processing, peroxisome proliferators-activated receptor (PPAR) activation, and increases in insulin sensitivity. In animal models of disease, dietary flavonoids also demonstrate a protective effect against cognitive decline, cancer, and metabolic disease. However, research also indicates that the flavonoids can be detrimental in some settings and, therefore, are not universally safe. Thus, as the population ages, it is important to determine the impact of these agents on prevention of disease, including optimal exposure (intake, timing/duration) and potential contraindications (Prasain et al., 2010).

The citrus flavonoids, naringin, and naringenin, were found to significantly lower the expression levels of vascular cell adhesion molecule-1 (VCAM-1) and monocyte chemotactic protein-1 (MCP-1), with potential applications in the prevention of atherosclerosis (Lee et al., 2001). Bioflavonoids like hesperidin (from orange peel), naringin (grapefruit peel), or rutin can normalize capillary permeability and vascular brittleness, therefore they are frequently called vitamin P factors. Hesperidin is applied in vein medication, acts as an antiviral in flu

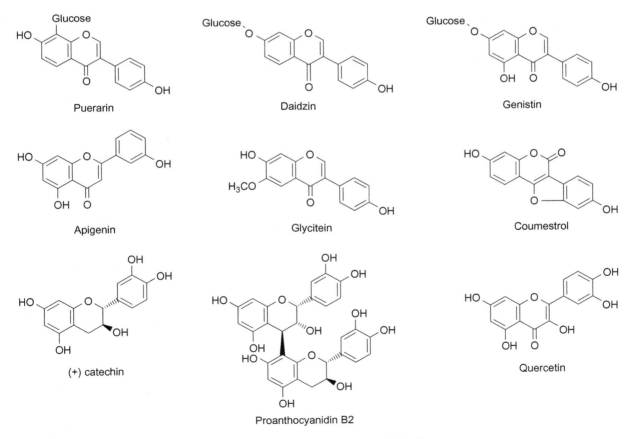

FIGURE 6.1 Chemical structures of some dietary flavonoids. *Source: Prasain et al. (2010).*

therapy, and possesses artificial sweetener properties; hydrated naringin is about 300 times sweeter than saccharose, neohesperidin almost 2000 times (Laufenberg et al., 2003).

2.2 Polyphenol Content of Grape Wine Wastes

The phenolic compounds of wine, particularly the flavanols (e.g., catechins, proanthocyanidins), have been the focus of a number of studies because of their relation to the beneficial effects attributed to a moderate consumption of wine (Shrikhande, 2000). These compounds have their origin in grape, and only a part of them is transferred to the must. Their extractability mainly depends on the technological conditions during vinification (Kammerer et al., 2004). For this reason, important quantities of phenolic compounds still remain in the wine by-products and there is great interest in the exploitation of this type of grape by-product to obtain potentially bioactive phenolic compounds (Santos-Buelga and Scalbert, 2000; Moure et al., 2001; Ray et al., 2001). Grape seeds are an abundant source of proanthocyanidins with varying degrees of polymerization, which find application as nutraceuticals in numerous products. In the USA, grape seed extracts have GRAS status, which means that they are generally recognized as safe. Pomace from red wine production has long been used for the extraction of anthocyanins, which are the red and blue pigments also found in cranberry, elderberry, blueberry, blackberry, black currant, strawberry, red cabbage, and purple carrots. The highest concentrations of grape polyphenols are found in the skin, stems, and seeds (Table 6.2).

2. 2. 1 Proanthocyanidins

Proanthocyanidins are a class of compounds that are found in many plants and that can be extracted from grape seed. Their basic structural unit is catechin (Figure 6.1). Proanthocyanidins contain catechin

TABLE 6.2 Quantity of Total Phenolic Substances, Total Flavanoids, and Proanthocyanidins Reported in Grape Extract and Grape Seeds

Compounds	Quantity in Grape Extract (g/L)	Quantity in Grape Seeds (g/100 g DM)
Total phenols (GAE)	2.86 ± 0.01	8.58 ± 0.03
Total flavanoids (CE)	2.79 ± 0.01	8.36 ± 0.04
Proanthocyanidins (CyE)	1.38 ± 0.06	5.95 ± 0.17

Source: Adapted from Negro et al. (2003)
CE, catechin equivalent; GAE, gallic acid equivalent; CyE, cyanidin equivalent.

monomer, dimer, and trimer, all of which are water-soluble molecules that contain a number of phenolic hydroxyls (Bagchi et al., 2002). Polyphenolic compounds have a very important antioxidant function; they can clean off free radicals in the body and reduce membrane lipid peroxidation, so they can reduce the occurrence of free-radical-related diseases and possibly delay aging (Morillas-Ruiz et al., 2006; Iacopini et al., 2008). Current studies have shown that grape seed proanthocyanidin extract can neutralize free radicals, protect the over-oxidative damage caused by free radicals (Feng et al., 2005; Spranger et al., 2008), and reduce the incidence of a range of diseases caused by free radicals, such as myocardial infarction, atherosclerosis, and drug-induced liver and kidney injury. Moreover, they have antithrombotic, antitumor, antimutagenic, anti-radiation-damage, and antifatigue effects (Yamakoshi et al., 1999; Sano et al., 2005; Engelbrecht et al., 2007).

Shan et al. (2010) carried out a preliminary study of the effect of grape seed proanthocyanidin extract (GSPE) on the free radical and energy metabolism indicators during movement. The extract rate of proanthocyanidins was 6.17%, according to Feng and Chen (2003). They explored the antifatigue mechanism of GSPE with regard to antioxidation system and energy metabolism, trying to provide theoretical guidance and experimental evidence for using GSPE in sport practice. Their results show that GSPE can significantly increase the activity of antioxidant enzymes in mice and clear free radicals in the body, so it can protect the body against free-radical damage. At the same time, GSPE can affect glucose metabolism in mice, increase the liver and muscle glycogen reserves, reduce the consumption of glucose, and keep the level of blood glucose stable. In addition, GSPE can affect fat metabolism in mice and promote the utilization of fat.

2. 2. 2 Resveratrol

Resveratrol is found abundantly in the skin of grapes; peanuts, itadori tea, and wine also contain resveratrol in appreciable amounts (Burns et al., 2002). Among the non-flavonoid polyphenolic compounds, trans-resveratrol (3, 4′, 5-trihydroxy-trans-stilbene), used for analgesic and therapeutic purposes in oriental folk medicine (Pace-Asciak et al., 1995), has been proposed as one of the components in red wine that might confer specific protection against coronary heart disease (CHD). Preliminary evidence from experimental animals (Kimura et al., 1983) and more recent *in vitro* studies on human plasma (Frankel et al., 1993) suggest that its antioxidant activity might also be relevant *in vivo*. Moreover, *in vitro* antiplatelet activity of trans-resveratrol has been observed and subsequently confirmed (Rotondo et al., 1998).

The additional parts of the grape such as the skin, the whole grape by itself, grape-derived raisins, and phytochemicals present within the grapes have also demonstrated potential anticancer efficacy in various preclinical and clinical studies. The underlying mechanisms of action of these grape-waste products are summarized by Kaur et al. (2009).

Experimental studies (Dohadwala and Vita, 2009) indicate that grape polyphenols could reduce atherosclerosis by a number of mechanisms, including inhibition of oxidation of low density lipoprotein (LDL) and other favorable effects on cellular redox state, improvement of endothelial function, lowering of blood pressure, inhibition of platelet aggregation, reducing inflammation, and activating novel proteins that prevent cell senescence (e.g., Sirtuin 1). Translational studies in humans support these beneficial effects. More clinical studies are needed to confirm these effects and formulate dietary guidelines. The available data, however, strongly support the recommendation that a diet rich in fruits and vegetables, including grapes, can decrease the risk of cardiovascular disease. *In vitro* studies have shown that grape-derived polyphenols inhibit platelet activity, and a number of potential mechanisms have been elucidated. Flavonoids inhibit cyclooxygenase and reduce production of thromboxane A2. Red wine polyphenols also decrease platelet production of hydrogen peroxide and inhibit activation of phospholipase C and protein kinase C (Pignatelli et al., 2000).

2. 2. 3 *Anthocyanins*

Anthocyanins are a group of phenolic compounds that belong to the flavanoid family. They are responsible for the coloration (orange, rose, red, violet, and blue) of the petals of flowers and fruit of a great variety of plants (Strack and Wray, 1989). There are numerous sources of anthocyanins, but the main raw material is the pomace from the red wine vinification process (Table 6.3). Currently, the European Union allows the use of anthocyanins as food dyes in drinks, marmalades, candies, ice creams, and pharmaceutical products (EU, 1994).

2.3 Polyphenols in Apple Pomace

Apple pomace contains various types of polyphenols, especially chlorogenic acid, dihydrochalcone derivatives, quercetin glycosides, and flavanols. A number of studies indicate that phenolic compounds from apples might reduce the risk of colon cancer because of their antioxidative and antiproliferative activities and by favourably modulating gene expression. Therefore, the extraction of polyphenols for use as functional food ingredients appears to be a promising approach (Schieber et al., 2001).

3. VEGETABLE FLAVONOIDS

3.1 Onion Flavonoids

A wide variety of flavonoids are distributed in vegetables. Onion bulbs (*Allium cepa* L.) are among the richest sources of dietary flavonoids and contribute to a large extent to the overall intake of flavonoids. Flavonoids continue to attract attention as potentially useful agents with implications for inflammation, cardiovascular diseases, and cancer (Middleton et al., 2000; Okamoto, 2005). In their review, Slimestad et al. (2007) report a compilation of more than 50 flavonoids identified in pigmented scales of onions. The majority of these structures have been confirmed through NMR investigations.

3.2 Flavonols of Onions

Flavonols are the predominant pigments of onions. From ancient times, dried pigmented scales of onions have been used to provide yellow coloration to textiles and Easter eggs. Flavonols are the main flavonoids of pigmented scales of onions. The main flavonols are based on quercetin (3, 5, 7, 3', 4'-pentahydroxyflavone) (Box 6.1). The structural diversity of the minor flavonols of onions is extensive and includes derivatives of kaempferol, isorhamnetin, and possibly myricetin. Altogether at least 25 different flavonols have been

TABLE 6.3 Anthocyanin Content of Grape Skins

Compound	Value (mg/kg DM)
Delphinidin 3-O-glucoside	68−5552
Cyanidin 3-O-glucoside	37−1903
Petunidin 3-O-glucoside	65−6680
Peonidin 3-O-glucoside	515−12,450
Malvidin 3-O-glucoside	1117−50,981
Delphinidin 3-O-acetglucoside	392−956
Petuidin 3-O-acetglucoside	545−1375
Peonidin 3-O-acetglucoside	1371−1484
Peonidin 3-O-acetglucoside	45−8688
Cyanidin 3-O-coumaroylglucoside	374−1071
Petunidin 3-O-coumaroylglucoside	974−2458
Peonidin 3-O-coumaroylglucoside	68−6828

Source: Adapted from Kammerer et al. (2004).

BOX 6.1

STRUCTURE OF QUERCETIN

o-dihydroxyl structure (catechol group)

2, 3-double bond with conjugation to 4-oxo group

hydroxyl group at the 3 and 5 position

Bors et al. (1990) were the first to claim three partial structures contributing to the radical-scavenging activity of flavonoids: (a) o-dihydroxyl structure in the B ring (catechol structure) as a radical target site; (b) 2, 3-double bond with conjugation to 4-oxo group, which is necessary for delocalization of an unpaired electron from the B ring, and (c) hydroxyl groups at the 3 and 5 position, which are necessary for enhancement of radical-scavenging activity. The circles indicate partial structures contributing to the free-radical-scavenging activity of flavonoids (Terao, 2009).

characterized from onion bulbs. The glycosyl unit(s) of these pigments has in most cases been identified as glucose. Most of the anthocyanins reported to occur in various cultivars of red onion are cyanidin derivatives. The main anthocyanins of all cultivars investigated are exclusively glycosylated at the anthocyanidin 3-position. Yellow onions contain 270−1187 mg of flavonols per kilogram of fresh weight, whereas red onions contain 415−1917 mg of flavonols per kilogram of fresh weight. The anthocyanins of red onions are mainly cyanidin glucosides acylated with malonic acid or non-acylated. Some of these pigments facilitate unique structural features like 4′-glycosylation and unusual substitution patterns of sugar moieties. Altogether at least 25 different anthocyanins have been reported from red onions, including two novel 5-carboxypyranocyanidin-derivatives. The quantitative content of anthocyanins in some red onion cultivars has been reported to be approximately 10% of the total flavonoid content or 39−240 mg/kg fresh weight.

At least 25 different flavonols have been characterized, and quercetin derivatives are the most important ones in all onion cultivars. Their glycosyl moieties are almost exclusively glucose, which is mainly attached to the 4′, 3, and/or 7-positions of the aglycones. Quercetin 4′-glucoside and quercetin 3, 4′-diglucoside are in most cases reported as the main flavonols in recent literature. Analogous derivatives of kaempferol and isorhamnetin have been identified as minor pigments. Recent reports

indicate that the outer dry layers of onion bulbs contain oligomeric structures of quercetin in addition to condensation products of quercetin and protocatechuic acid (Slimestad et al., 2007).

3.3 Functionality of Flavonoids

3. 3. 1 Prevention of Atherosclerosis and Cardiovascular Disease

Quercetin 4′-glusoside and quercetin 3, 4′-diglucoside are exclusively present in onion, whereas quercetin 3-glucoside (isoquercitrin) and quercetin-3-rutinoside (rutin) are predominant glycosides in common vegetables (Terao et al., 2008). Hertog et al. (1993) found that flavonoid intake was inversely correlated with CHD mortality in elderly men. Nowadays, epidemiological studies strongly suggest that the intake of flavonoids from diet is helpful in the prevention of atherosclerosis and its related events, including CHD. However, the molecular mechanism for their anti-atherosclerotic action and the absorption mechanism related to their bioavailability should be clarified for the practical use of dietary flavonoids in lowering the risk of atherosclerosis.

An ultimate target for dietary quercetin as anti-atherosclerotic agent is undoubtedly the blood aorta, where plaque formation happens in relation to atherosclerotic injury. However, work on the accumulation of dietary flavonoids in this target site is limited. Terao et al. (2008) designed an experiment using high cholesterol-fed rabbits to examine whether or not dietary quercetin actually accumulates in the aorta tissue and exerts antioxidant activity. By HPLC and HPLC-MS analyses, they detected quercetin metabolites in the aorta tissue of quercetin glucoside-fed rabbits. It is therefore likely that a high-cholesterol diet induces oxidative stress in the blood vessel and that quercetin metabolites contribute to an antioxidant network to counteract reactive oxygen species (ROS)-induced injury directly or indirectly. Recently the same researchers prepared a novel monoclonal antibody, which specifically recognizes quercetin glucuronide, and succeeded in detecting quercetin glucuronides in human atherosclerotic aorta by an immunohisto chemical approach. Their findings suggest that quercetin metabolites are incorporated into the atherosclerotic region and act as complementary antioxidants when oxidative stress is loaded on the vascular system. It is likely that plasma albumin is a carrier for translocation of quercetin metabolites to the vascular target.

3. 3. 2 Antioxidant Activity

Although flavonoids in the diet have a bitter or astringent taste and inhibit digestive enzymes, recent studies strongly suggest that dietary flavonoids may have a favorable role in human health through their antioxidant activity (Hooper et al., 2008). Oxidative stress is frequently referred to as an essential factor in the initiation and/or promotion of degenerative diseases such as atherosclerosis. Much attention has therefore been paid to the antioxidant activity of dietary flavonoids from the viewpoint of food factors that modulate oxidative stress (Rice-Evans et al., 1996). Dietary flavonoids seem to participate in the antioxidant network together with vitamin E, vitamin C, and other biological antioxidants in the human body (Terao, 2009). Recent studies also suggest that dietary flavonoids can exert various effects by a mechanism different from classical antioxidant activity: regulation of the activity and protein expression of specific enzymes (Virgili and Marino, 2008).

The mechanism of antioxidant activity of flavonoids can be characterized by direct scavenging or quenching of oxygen free radicals or excited oxygen species as well as inhibition of oxidative enzymes that generate these reactive oxygen species. The essential part of the free radical-scavenging activity of flavonoids is attributed to the o-dihydroxyl group in the B ring (catechol group) in their diphenyl propane structure. Catechol type flavonoids therefore possess powerful antioxidant activity (Terao, 2009).

Accumulation of flavonoid metabolites in the appropriate target site is probably required to exert their antioxidant activity. Conversion of inactive metabolites to active aglycones via a deconjugation reaction in the target site may be a key process for efficient exertion of their antioxidant activity. The significance of dietary flavonoids as antioxidants in vivo is much more complicated than that expected from in vitro assays. The specific target should be taken into account when evaluating the antioxidant activity of dietary flavonoids in vivo.

The levels of antioxidative quercetin derivatives are considerably higher in onion peels than in the flesh. Since onion flavonoids are readily absorbed, they could contribute significantly to antioxidant defense. Protocols for their recovery based on subcritical water extraction and water/ethanol/citric acid extraction have recently been developed (Schieber, 2009).

3. 3. 3 Metabolic Syndrome

The strongest evidence for the benefits of flavonoids in diseases of aging relates to their effect on components of the metabolic syndrome. The metabolic syndrome has three major contributors: hypertension, dyslipidemia/obesity, and hyperglycemia/hyperinsulinemia, all of which act synergistically to greatly increase morbidity and mortality (Prasain et al., 2010). While the incidence of all three contributors is increasing exponentially in adults, a parallel rise is also occurring in children (Nathan and Moran, 2008). It is estimated that clustering of these metabolic risk factors occurs in up to 50% of overweight adolescents, leading to an increased appearance in early onset type-2 diabetes and cardiovascular disease (Nelson and Bremer, 2009). The treatment of metabolic syndrome in both young and aging populations has greatly increased pharmaceutical expenditures, and antihyperglycemic drugs are projected to become the largest single component of all prescription drug spending in the near future (Hoerger and Ahmann, 2008), making the metabolic syndrome a very significant burden on individual health and the economy. Research is increasingly exploring the ability of botanical supplements to reduce metabolic syndrome risk factors, since these compounds could provide greater efficacy and tolerability at lower cost than current pharmaceutical options (Figure 6.2).

3. 3. 4 Hormonal Activity

Currently, a broad range of phyto-pharmaceuticals with a claimed hormonal activity, called phytoestrogens, is recommended for prevention of various diseases related to a disturbed hormonal balance (Dijsselbloem et al., 2004). In this respect, soy isoflavones (genistein, daidzein, biochanin), as potential superior alternatives to the synthetic selective estrogen receptor modulators (SERMs), are currently applied in hormone replacement therapy (HRT). As phytochemicals integrate hormonal ligand activities and interference with signaling cascades, therapeutic use may not be restricted to

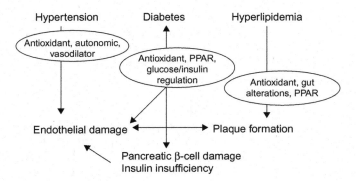

FIGURE 6.2   Mechanisms by which flavonoids are most likely to reduce the three main contributors to metabolic syndrome in aging adults. Peroxisome Proliferator-Activated Receptor (PPAR). *Source: Prasain et al. (2010).*

hormonal ailments only, but may have applications in cancer chemoprevention and/or NF-κB-related inflammatory disorders as well.

Accordingly, intense investigations point to the important role of consumption of dietary soy, as the major source of isoflavones, in low cancer incidence (Birt et al., 2001; Adlercreutz, 2002). Particular Japanese population groups are estimated to register the highest intake of soy products, with levels up to 200 mg/day. Generally, the Asian consumption of legumes is assumed to supply 20–50 mg of isoflavones in the daily diet, which sharply contrasts with the western negligible amount of less than 1 mg isoflavones/day (Nagata et al., 1998). Consistent with the epidemiological studies are the findings that soy phytoestrogens improve bone mass in peri- and postmenopausal women (Potter et al., 1998; Alekel et al., 2000). Furthermore, they prevent atherosclerosis of coronary arteries in monkeys (Anthony et al., 1996) and significantly reduce cholesterol levels in hypercholesterolemic subjects (Crouse et al., 1999).

4. COLORING AGENTS AND ANTIOXIDANTS

Anthocyanins, carotenoids, betalains, lycopenes, and leucoanthocyanidin represent the major groups of colored phenolic compounds in FV residues. They are powerful antioxidants and may possess pharmacological properties, which could make them desirable ingredients in the developing market of "functional foods" for health. For example, grape skin extract in powder form is commercially available as a natural food-coloring agent. Besides the blue-red color, the food will be enriched with "healthy" polyphenols (Laufenberg et al., 2003).

4.1 Betalains

More than 200,000 tonnes of red beet are produced in Western Europe annually, most of which (90%) is consumed as vegetable (Schieber et al., 2001). The remainder is processed into juice, coloring foodstuff, and food colorant, the latter commonly known as beetroot red. Though still rich in betalains, the pomace from the juice industry, accounting for 15–30% of the raw material, is disposed of as feed or manure. The colored portion of the beetroot ranges from 0.4% to 2.0% of the dry matter, depending on intraspecific species, edaphic factors, and postharvest treatments. Whereas the colored fraction consists of betacyanins and betaxanthins, the phenolic portion of the peel shows L-tryptophane, p-coumaric, and ferulic acids, as well as cyclodopa glucoside derivatives. Beets are ranked among the 10 most potent vegetables with respect to antioxidant capacity, ascribed to a total phenolic content of 50–60 mmol/g dry weight. Toxicological studies reveal that betanin, the major compound from red beet, does not exert allergic potential, nor mutagenic or hepatocarcinogenic effects. High content of folic acid, up to 15.8 mg/g dry matter, is another nutritional feature of beets (Schieber et al., 2001).

4.2 Lycopenes

Skin, rich in lycopene, is an important component of waste originating from tomato paste manufacturing plants. Lycopene is the principal carotenoid, causing the characteristic red hue of tomatoes. Several epidemiological studies have reported that lycopene-rich diets have beneficial effects on human health. A possible role has been suggested for tomatoes and tomato products in preventing cardiovascular disease and protecting against some types of cancer (based on lycopene content). Maximum lycopene (1.98 mg/100 g) was extracted when the solvent/meal ratio adjusted to 30:1 v/w, number of extractions 4, temperature 50°C, particle size 0.15 mm, and extraction time 8 min (Schieber et al., 2001).

5. DIETARY FIBERS

Numerous studies indicate that dietary fibers (DF) counteract obesity, cardiovascular disease, and type 2 diabetes. DF also play an important role in the prevention and treatment of gastrointestinal disorders and large-intestine cancer (Champ et al., 2003). DF comprise a combination of many compounds (pectin, lignin, cellulose) differing in physical and chemical properties. FV by-products such as apple, pear, orange, peach, black currant, cherry, artichoke, asparagus, onion, and carrot pomace are used as sources of DF supplements in refined food (Larrauri et al., 1999). DF concentrates from vegetables and fruits may contain important amounts of DF and have a better ratio between soluble (SDF) and insoluble dietary fiber (IDF) than cereal brans (Garau et al., 2007). SDF:IDF ratio is important for health properties and also for technological characteristics; 30–50% of SDF and 70–50% of IDF is considered a well-balanced proportion in order to obtain the physiological effects associated with both fractions (Grigelmo-Miguel and Martin-Belloso, 1999). Onions show an important quantity of DF and a good SDF:IDF ratio that will have various metabolic and physiological effects (Jaime et al., 2002). Ability of the fiber matrix to maintain its physical properties after being processed is quite essential (Femenia et al., 1997).

Sterilization produces a decrease of IDF and an increase of SDF, improving the SDF:IDF ratio. In general, uronic acids of IDF undergo solubilization after thermal treatment, with transition to SDF. Swelling capacity (SWC), which is related to fiber structure, suffers a decrease after sterilization as a result of IDF losses, since insoluble fibers can adsorb water like a sponge. However, water-holding capacity (WHC) does not undergo relevant changes with sterilization, since this treatment does not produce drastic changes in the free hydroxyl groups that adsorb water through hydrogen bonds. Cation exchange capacity (CEC) reduction could be related to the changes that pectic substances undergo after sterilization. Despite changes produced by thermal treatment, sterilization would be a good method to stabilize onion by-products in order to use them as a potential DF ingredient, since physico chemical properties of sterilized by-products are generally higher than those of cellulose. Therefore, these by-products might have potential applications as low-calorie bulk ingredients in DF enrichment, and would be interesting in food products requiring oil and moisture retention (Benítez et al., 2011).

Potato peel waste has been proposed as DF for baking products (Arora and Camire, 1994). Cauliflower has a very high waste index and is an excellent source of protein (16.1%), cellulose (16%), and hemicellulose (8%). It is considered a rich source of DF and it possesses both antioxidant and anticarcinogenic properties. Stojceska et al. (2008) studied the incorporation of cauliflower trimmings into ready-to-eat expanded products (snacks) and their effect on the textural and functional properties of extrudates. It was found that addition of cauliflower of up to 10% significantly increased the dietary fiber and levels of proteins. The high crude fiber content of the vegetable pomace (in total 20–65% DM) suggests its utilization as a crude

fiber *bread improver*. In bread and bakery goods, as well as in pastry, cereals, and dairy products, the investigated carrot pomace works as a stabilizer, acidifying agent, preservative, or antioxidant.

6. SULFUR-CONTAINING BIOACTIVE COMPOUNDS

6.1 Cabbage Glucosinolates

Glucosinolates are a group of sulfur-containing plant secondary metabolites that are widely distributed in brassica vegetables including broccoli, cauliflower, lettuce, and cabbage (Cartea and Velasco, 2008). Intact glucosinolates are generally located in vacuoles and are inactive until plant cells are damaged, resulting in the release of glucosinolates from the vacuoles. Sulforaphane is a hydrolysis product of glucosinolates, which is formed via the conversion of glucoraphanin (one of the glucosinolates) by myrosinase under neutral or close to neutral conditions (pH 5–8). Sulforaphane is heat sensitive and its thermal susceptibility is much dependent on an experimental system (Shen et al., 2010). Outer leaves of white cabbage (*Brassica oleracea* L. var. capitata), a typical by-product from a cabbage processing plant, have the potential of being transformed into dietary fiber (DF) powder with high levels of antioxidants and anticarcinogenic activities. However, losses of health-beneficial bioactive compounds in cabbage leaves may occur during processing. As the processing of DF powder involves mechanical tissue damage (e.g., slicing, chopping) and thermal treatment (e.g., drying), it is interesting to determine how the processing steps affect the sulforaphane content and degradation in the DF powder.

6.2 Methods of Processing

Tanongkankit et al. (2011) studied the effect of hot air drying at 40–70°C on the evolution of sulforaphane in cabbage. They found that the drying temperature had a significant effect on both the formation and degradation rates of sulforaphane. The results showed that the formation of sulforaphane occurred when the cabbage temperature during drying was in the range of 25–53.5°C, and thermal degradation took place once the cabbage temperature exceeded this range. A semi-empirical heat transfer and kinetic model was proposed to describe the change of sulforaphane throughout the drying process and it gives a very good fit to the experimental data. The results also showed that almost all sulforaphane formed during an early stage of drying degraded by the end of the drying process, and only a small fraction of sulforaphane was retained in the final DF powder. However, a previous study reported that sulforaphane content of only approximately 2.7 mg/L could inhibit the growth of HT29 human colon cancer cells (Gamet-Payrastre et al., 2000). This indicates that the present DF powder could still provide anticarcinogenic properties. Alternatively, the hot air drying process may be stopped when the sulforaphane content is at a maximum level; sulforaphane can then be extracted at that maximum point. Drying at 60°C is suggested as an optimum condition to obtain the highest retention of sulforaphane in cabbage DF powder.

7. EXTRACTION PROCESSES FROM FOOD-AND-VEGETABLE WASTE

7.1 Extraction of Phenolic Compounds from Olive Pomace

The high-pressure–high-temperature reactor has been shown to be efficient for extraction of bioactive compounds from olive pomace. Maximum total polyphenols yield, expressed as units of caffeic acid equivalents per gram of dried pomace (45.2 mg$_{CAE}$/g$_{DP}$), was achieved at 180°C and 90 minutes, while total flavonoids reached maximum value, expressed as units of catechin equivalents per gram of dried pomace (15.3 mg$_{CE}$/g$_{DP}$), at 150°C and 60 min. A shorter contact time (30 min) was needed to extract the maximum yield of o-diphenols (5.6 mg$_{CAE}$/g$_{DP}$) at 180°C. The preliminary results suggest that further complete optimization of the proposed treatment, taking into account the effects of extraction solvent and solid/liquid ratio, is necessary. Solid residue, derived from the olive milling process, can be successfully used as an inexpensive source of phenolic compounds so that food and pharmaceutical industries may benefit from this emerging technology (Aliakbarian et al., 2011).

7.2 Solvent and Enzyme-Aided Aqueous Extraction of Goldenberry

Goldenberry (Physalis peruviana L.) pomace (seeds and skins) represents the waste obtained during juice processing (around 27.4% of fruit weight). Ramadan et al. (2008) evaluated the potential of goldenberry pomace for use as a substrate for the production of edible oil. The results provide important data that may encourage development of goldenberry as a commercial crop and its industrial application. Three extraction methods were examined for the best oil

yield. The n-hexane-extractable oil content of the raw by-products were estimated to be 19.3%. Enzymatic treatment with pectinases and cellulases followed by centrifugation in aqueous system, or followed by solvent extraction, were also investigated for recovery of oil from pomace fruit. Enzymatic hydrolysis of pomace followed by extraction with n-hexane reduced the extraction time and enhanced oil extractability up to a maximum of around 7.60%. The latter processes increased the levels of protein, carbohydrates, fiber, and ash in the remaining meal. Regarding the oil composition, there were no substantial changes noted in the fatty acid pattern of the oils extracted with different techniques. Although goldenberry is a part of a supplemental diet in many parts of the world, information on the phytochemicals in this fruit is limited. Yet these phytochemicals may bring nutraceutical and functional benefits to food systems. A variety of health-promoting products improved from goldenberry pomace may include ground-dried skins and extracts obtained from skins and/or seeds. The levels of polar lipids, unsaponifiables, peroxides, and phenolics in various extracts are associated with oxidative stability and radical scavenging activity.

7.3 Extraction of Antioxidants from Potato Peels by Pressurized Liquids

Aqueous potato peel extracts were shown to be a source of phenolic acids, especially of chlorogenic, gallic, protocatechuic, and caffeic acids (Mäder et al., 2009). The extracts display species-dependent antibacterial but no mutagenic activity, and concentrations of the glycoalkaloids solanine and chaconine are below toxic threshold levels, if peel extracts are added at 200 ppm to a foodstuff (Sotillo et al., 1998). Wijngaard et al. (2011) studied the extraction of antioxidants from industrial potato peel waste and optimized it with regards to ethanol concentration, temperature, and time using solid–liquid extraction and pressurized liquid extraction of polyphenols. Both techniques were optimized by response surface methodology. Efficiency of extraction was optimized by measuring antioxidant activity, phenol content, and the level of caffeic acid. Conditions for optimal antioxidant activity as measured by the 2, 2-diphenyl-1-picrylhydrazyl assay were 75% ethanol, 80°C, and 22 minutes with solid–liquid extraction, resulting in an optimum activity of 352 mg Trolox Equivalents/100 g DW potato peel. Both ethanolic extractions resulted in higher antioxidant activities, polyphenol levels, and glycoalkaloids content than conventional extractions with 100% methanol and 5% acetic acid. Before the use as a food ingredient, glycoalkaloids should be removed; or the extracts could be used for other purposes, for example in the pharmaceutical industry. Glycoalkaloids have gained interest as possible precursors for the production of hormones (Schieber and Saldaña, 2009), antibiotics, and for use during certain skin diseases (Friedman, 2006).

7.4 Extraction of Phytochemicals from Common Vegetables

Chu et al. (2002) established a complete profile of total phenolic contents in vegetables by further digesting and extracting the bound phytochemicals. They developed the process of extraction of polyphenols from 10 common vegetables as shown in Figure 6.3. Phenolic compounds in the edible part of the vegetables are present in both free and bound forms. Bound phenolics, mainly in the form of β-glycosides, may survive human stomach and small intestine digestion and reach the colon intact, where they are released and exert bioactivity. Bound phenolic contents are composed of bound-E and bound-W, which have distinct extraction properties. Phenolic compounds extracted by ethyl acetate are styled "bound-E", and phenolic compounds recovered with water are styled "bound-W". Bound phenolics of vegetables, mostly in ester forms, are associated with cell wall components. Notably among them is ferulic acid (FA). FA has been found to be esterified to families, including pectic polysaccharides, and cross-linked as a result of peroxidative activity. Owing to this protective mechanism, it is possible that bound phenolics can survive upper gastrointestinal digestion and may ultimately be broken down in the colon by the microflora of the large intestine. On average, approximately a fourth of the fresh vegetable phenolic compounds may be released and absorbed in the colon to furnish additional health benefits locally, whereas potato and carrot could release approximately half of their phytochemical contents in the colon. Epidemiological studies have shown an inverse correlation between vegetable consumption and colon cancer occurrence (Voorrips et al., 2000).

Broccoli possessed the highest total phenolic content, followed by spinach, yellow onion, red pepper, carrot, cabbage, potato, lettuce, celery, and cucumber. Red pepper had the highest total antioxidant activity, followed by broccoli, carrot, spinach, cabbage, yellow onion, celery, potato, lettuce, and cucumber. The phenolics antioxidant index (PAI) was proposed to evaluate the quality/quantity of phenolic contents in these vegetables and was calculated from the corrected total antioxidant activities by eliminating vitamin C contributions. Antiproliferative activities were also studied *in vitro* using HepG2 human liver cancer cells (Table 6.4). Spinach showed the highest inhibitory effect, followed by cabbage, red pepper, onion, and broccoli. On the basis of these results, the bioactivity index (BI) for dietary cancer prevention was proposed to provide a simple reference for consumers to choose vegetables in accordance with their beneficial activities. The BI is a half of the sum of total antioxidant activity (AOA) score and antiproliferative activity (AA) score against liver cancer cells (Chu et al., 2002). The BI could be a new alternative biomarker for future epidemiological studies in dietary cancer prevention and health promotion.

Focusing on separation processes, cost-effective approaches proportional to the commercial value of the recovered molecule have been applied for recovery of specific fine chemicals, such as solvent extraction, ion-exchange chromatography, and super-critical CO_2 extraction. These processes often need specific

TABLE 6.4 Bioactivity Index (BI), Antiproliferative Activity (AA) Score, and Antioxidant Activity (AOA) Score of Ten Vegetables

Vegetable	BI	AA	AOA
Spinach	0.95	1.00	0.90
Red pepper	0.78	0.55	1.00
Broccoli	0.66	0.38	0.94
Cabbage	0.57	0.76	0.38
Carrot	0.45	0	0.91
Yellow onion	0.36	0.42	0.30
Celery	0.05	0	0.11
Potato	0.05	0	0.10
Lettuce	0.03	0	0.06
Cucumber	0.01	0	0.03

Source: Adapted from Chu et al. (2002).

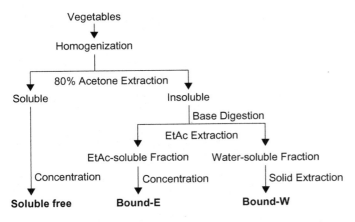

FIGURE 6.3 Flowchart of extraction of phytochemicals from vegetables. *Source: Adapted from Chu et al. (2002).*

pretreatments to preserve the raw materials, such as storage under modified atmosphere, microbial or chemical acidification, and pasteurization (Laufenberg et al., 2003). Milling, crushing, steam explosion, and/or use of depolymerizing enzymes (e.g., macerase, β-glucosidase, and feruloyl esterase) are essential to favor the extraction (Benoit et al., 2006).

8. WHEY AS A SOURCE OF BIOACTIVE PEPTIDES

8.1 Occurrence of Bioactive Peptides in Whey and Other Dairy By-Products

Whey contains a large array of biologically active proteins and peptides which, together with other components, have formed the basis for the use of whey in medicinal applications for centuries. Apart from the major whey proteins, β-lactoglobulin, α-lactalbumin, and glycomacropeptide, whey contains a number of other proteins with potent bioactivity (Table 6.5) (Smithers, 2008).

8.2 Functionality of Bioactive Peptides

A number of the above-mentioned bioactive whey peptides, together with the parent protein source and proposed biological functionality, are shown in Table 6.6 (Smithers, 2008). Commercially, the most promising of these peptides is the effective antimicrobial lactoferricin.

TABLE 6.5 Content of Minor Bioactive Proteins in Cheese Whey

Protein	Content (mg/L)
Lactoferrin	50−70
Lactoperoxidase	8−20
Immunoglobulins	300−600
Growth factors*	<0.06
IGF-I	<0.001
IGF-II	<0.001
PDGF	<0.0002
TGF-β	<0.01
FGF	<0.0001
Betacellulin	<0.002

Source: Adapted from Smithers (2008).
IGF, insulin-like growth factor; PDGF, platelet-derived growth factor; TGF, transforming growth factor; FGF, fibroblast growth factor.

TABLE 6.6 Bioactive Whey-Derived Peptides

Protein Source	Peptide	Bioactivity (Some Putative)
α-Lactalbumin	α-Lactophorin	ACE inhibitor
β-Lactoglobulin	β-Lactophorin	Ileum stimulation
	β-Lactotensin	Ileum contraction
Serum albumin	Albutensin	ACE inhibitor, ileum contraction
	Serophorin	Opioid activity
Glycomacropeptide	108−110; 106−116	Antithrombotic
Lactoferrin	Lactoferrin	Antimicrobial

Source: Adapted from Smithers (2008).
ACE, angiotensin-converting enzyme.

Milk-derived peptides have been shown *in vivo* to exert various activities affecting, for example, the digestive, cardiovascular, immune, and nervous systems. Dietary proteins are traditionally known to provide a source of energy and the amino acids essential for growth and maintenance of various body functions. In addition, they contribute to the physico chemical and sensory properties of protein-rich foods. In recent years, food proteins have gained increasing value due to the rapidly expanding knowledge about physiologically active peptides. Milk proteins provide a rich source of peptides which are latent until released and activated (e.g., during gastrointestinal digestion or milk fermentation). Once activated, these peptides are potential modulators of many regulatory processes in living systems. Upon oral administration, bioactive peptides may affect the major body systems, namely, the cardiovascular, digestive, immune, and nervous systems (Figure 6.4), depending on their amino acid sequence. In the future, milk-derived bioactive peptides may be important health-sustaining components in food and in the prevention of diseases and conditions such as cardiovascular diseases, obesity, osteoporosis, and stress (Korhonen and Pihlanto, 2006).

8.2.1 Regulation of the Gastrointestinal System

Food-derived proteins and peptides may play important functions in the intestinal tract before hydrolysis to amino acids and subsequent absorption. These include regulation of digestive enzymes and modulation of nutrient absorption in the intestinal tract (Shimizu, 2004).

8.2.2 Regulation of the Immune System

Milk protein hydrolysates and peptides derived from caseins and major whey proteins can enhance

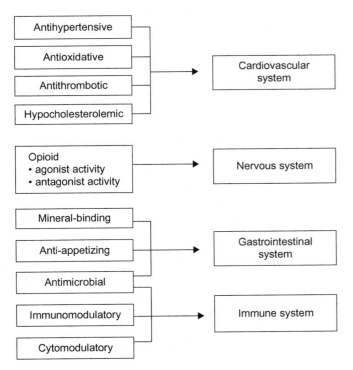

FIGURE 6.4 Physiological functionality of milk-derived bioactive peptides. *Source: Korhonen and Pihlanto (2006).*

immune cell functions, measured as lymphocyte proliferation, antibody synthesis, and cytokine regulation (Gill et al., 2000). Of special interest are peptides released during milk fermentation with lactic acid bacteria, as these peptides have been found to modulate the proliferation of human lymphocytes, to down-regulate the production of certain cytokines, and to stimulate the phagocytic activities of macrophages (Matar et al., 2003; Meisel and FitzGerald, 2003). It has been also suggested that immunomodulatory milk peptides may alleviate allergic reactions in atopic humans and enhance mucosal immunity in the gastrointestinal tract (Korhonen and Pihlanto, 2003). Furthermore, immunopeptides formed during milk fermentation have been shown to contribute to the antitumor effects observed in many studies with fermented milks (Matar et al., 2003). The fact that caseinophosphopeptides have been shown to exert cytomodulatory effects is of particular interest in this context. Cytomodulatory peptides derived from casein fractions inhibit cancer cell growth or stimulate the activity of immunocompetent cells and neonatal intestinal cells (Meisel and FitzGerald, 2003). Glycomacropeptides (GMP) derived from κ-casein may have a beneficial role in modulating the gut microflora, as this macropeptide is known to promote

the growth of bifidobacteria due to its carbohydrate content (Manso and Lopez-Fandino, 2004).

8.2.3 Regulation of the Cardiovascular System

Bioactive peptides from casein and whey may be deployed in disease-specific applications such as blood pressure reduction and stress-related areas (e.g., in the case of mood-influencing peptides rich in tryptophan). Peptides from casein may be used to enhance the solubility of minerals such as calcium and zinc, and hence increase the bioavailability of these minerals, or to inhibit angiotensin-converting enzyme, which causes vasoconstriction and hence raises blood pressure.

8.2.4 Regulation of the Nervous System

Peptides with opioid activity have been identified in various casein fractions hydrolyzed by digestive enzymes (Teschemacher, 2003). These peptides are opioid receptor ligands with agonistic or antagonistic activities. Opioid receptors are located in the nervous, endocrine, and immune systems as well as in the gastrointestinal tract of mammals. Thus, orally administered opioid peptides may modulate absorption processes in the gut and influence the gastrointestinal function in two ways: first, by affecting smooth muscles, which reduces the transit time, and second, by affecting the intestinal transport of electrolytes, which explains their antisecretory properties. The actual physiological effects of milk-derived opioid peptides remain, however, to be confirmed. β-Casein-derived opioid peptides (β-casomorphins) or their precursors have been detected in the human small intestine upon oral administration of casein or milk (FitzGerald and Meisel, 2003).

8.2.5 Antimicrobial Function

Immunoglobulins from vaccinated cows may be considered as natural antimicrobials with certain advantages over synthetic antibiotics. Lactoferrin is an example of a minor whey protein that has been studied in great detail; it is becoming increasingly clear that it is important for nonspecific defense against bacteria, fungi, and viruses. Oligosaccharides, glycolipids, and glycoproteins containing sialic acid residues may have a role as anti-infectives.

8.2.6 Growth Factor Activity

Pioneering work over the past approximately fifteen years has laid the foundation for exploitation of the remarkable cell growth promotional activity of an extract from cheese whey containing a plurality of growth factors (Table 6.5).

8.3 Commercial Dairy Products Containing Bioactive Peptides

An increasing number of ingredients containing specific bioactive peptides based on casein or whey protein hydrolysates have been launched on the market within the past few years or are currently under development by international food companies. Korhonen and Pihlanto (2006) reported several examples of these commercial products, their health function, and manufacturers.

8.4 Commercial-Scale Production

Industrial-scale technologies suitable for the commercial recovery of bioactive milk peptides have been developed and launched recently. These technologies are based on novel membrane separation and ion exchange chromatographic methods being employed by the emerging dairy ingredient industry. Emerging technologies for generation and concentration of bioactive peptides include high-pressure processing, high-power ultrasound, pulsed electric field, and microfluidization (Smithers, 2008).

Controlled fermentation of raw materials rich in proteins by known lactic acid bacteria strains may be developed on the commercial scale with respect to continuously operating bioreactors. Alternatively, commercial production of specific peptide sequences could be enabled through recombined enzyme technology utilizing certain production strains, or through the use of purified proteolytic enzymes isolated from suitable microorganisms (Korhonen and Pihlanto, 2003). Before commencing the development of nutraceuticals from whey and milk by-products, a number of technological and marketing issues should be addressed. On the technological side, the question of the cost of the process required to manufacture the desired protein, peptide, or special lipid should be dealt with at a very early stage, together with how to valorize the non-bioactive residual raw material. The selected technology should preferably be proprietary, and the application of the bioactive should be protected by patents. Food safety is also an important feature. On the marketing side, cost-effectiveness of the ingredient and ease of incorporation into a good-tasting end product are vital. Claiming a message that can be understood by consumers or is allowed by legal authorities is a further prerequisite for successful market introduction. Quick entry into the market is sometimes possible when a large body of circumstantial evidence from the scientific literature is available. Animal models are sometimes used for proof of concept. To be able to make firm health claims, human clinical data are needed, preferably with the end product containing the bioactive ingredient (Steijns, 2001).

9. PRODUCT DEVELOPMENT, MARKETING, AND CONSUMER ACCEPTANCE OF FUNCTIONAL FOODS

Consumer acceptance of the concept of functional foods has been widely recognized as a key success factor for market orientation, consumer-led product development, and successfully negotiating market opportunities. Acceptance, however, is determined by a host of factors such as primary health concerns, consumers' familiarity with *functional food* concepts and with the functional ingredients, the nature of the carrier product, the manner of health effect communication, etc. Consumers' knowledge and awareness of the health effects of newly developed functional ingredients seems to be rather limited; therefore, there is a strong need for specific communication activities to consumers in this respect. Different surveys have shown that consumer acceptance of functional foods is far from being unconditional, with one of the main conditions for acceptance pertaining to taste, besides product quality, price, convenience, and trustworthiness of health claims. As a rule, consumers seem to evaluate functional foods first and foremost as foods. Functional benefits may provide added value to consumers but cannot outweigh the sensory properties of foods (Siro et al., 2008).

On the other hand, Europeans in general are far more critical of new products and technologies (e.g., GMO food, irradiated food) than are American consumers (Lusk and Rozan, 2005). They are not only suspicious of the safety of novel foods, but they are critical of the whole process through which food production becomes more and more anonymous and distanced from everyday life (Poppe and Kjnes, 2003). Therefore, it can be hypothesized that Europeans' acceptance of functional foods is less unconditional, better thought out, and with more concerns and reservations than in the USA. This may also be influenced by the recent sequence of food safety scares (e.g., BSE, dioxins, foot-and-mouth disease, *E. coli*, aviatic pneumonia, acrylamid) (Verbeke, 2005).

10. CONCLUSIONS

Fruits and vegetables represent an untapped reservoir of various nutritive and non nutritive phytochemicals with potential cancer chemopreventive activity. The phytochemicals such as phenolic compounds, namely flavonoids, are the major contributors to the antioxidant capacity of the apple, grape, and olive pomace, onion/potato peels, and other common FVW. Recent research suggests that in humans, these plant polyphenols provide important health benefits related to metabolic syndrome, cancer, brain health, and the immune system. The relatively low toxicity and potential efficacy of most of these agents make them attractive to a large sector of the population.

Various grape-based products of different origin are currently available in the US market: 22 grape seed products, 5 grape extracts, and 7 red wine powders (Shrikhande, 2000). Because of their beneficial properties, winery by-products are now being sold to the rapidly growing dietary supplement industry (Arvanitoyannis et al., 2006). However, more studies are needed in high-risk populations for cancer of specific organs or sites with standardized grape seed extract (GSE) preparations to establish the dose regimen and to determine pharmacologically achievable levels of biologically active constituents in the plasma/target organ. These studies would also help establish any toxicity associated with long-term administration of GSE (Kaur et al., 2009). Anthocyanins, carotenoids, betalains, and lycopenes represent the major groups of colored phenolic compounds in FV residues, which also exert a strong antioxidant activity. The European Commission (EC) allows the use of anthocyanins as food dyes in drinks, marmalades, candies, ice creams, and pharmaceutical products. FV by-products such as apple, pear, orange, peach, black currant, cherry, artichoke, asparagus, onion, cauliflower, potato peels, and carrot pomace are also used as sources of dietary fiber supplements in refined food. From a scientific point of view, further studies are needed to demonstrate the efficacy and safety of nutraceuticals recovered from FV by-products. Efficient purification steps are necessary before bioactive compounds can be used as natural food ingredients. These processing steps must remove any natural and anthropogenic toxins that might be present in the raw material (Schieber, 2009).

Today's achievements in separation techniques in the dairy industry and enzyme technology offer opportunities to isolate, concentrate, or modify bioactive peptides from whey and other milk by-products, so that their application in functional foods, dietary supplements, nutraceuticals, and healing foods has become possible. So far, antihypertensive, mineral-binding, and anticarcinogenic peptides have been most studied for their physiological effects. A few commercial developments have been launched on the market, and this trend is likely to continue alongside increasing knowledge about the functionalities of the peptides. However, molecular studies are needed to assess the mechanisms by which the bioactive peptides exert their activities. This research area is currently considered to be the most challenging one, due to the understanding that most known bioactive peptides are not absorbed from the gastrointestinal tract into the blood circulation. Their effect is likely to be mediated directly in the gut lumen or through receptors on the intestinal cell wall. In this respect, the target function of the peptide concerned is of utmost importance. It is expected that in the near future such targets shall be the following lifestyle-related disease groups: (a) cardiovascular diseases, (b) cancers, (c) osteoporosis, (d) stress, and (e) obesity. Peptides derived from dietary proteins offer a promising approach to prevent, control, and even treat such disease conditions through a regulated diet (Korhonen and Pihlanto, 2006). Whey components, particularly the proteins and peptides, will increasingly be preferred as ingredients for functional foods and nutraceuticals, and as active medicinal agents, built upon the strong consumer trend for health and wellbeing and continuing discovery and substantiation of the biological functionality of whey constituents (Smithers, 2008).

Increased health awareness and consumer confidence coupled with an aging population have driven demand for products that improve the quality of life, have specific health benefits, and can be used by consumers to self-medicate. The global consumption of functional foods and drinks increased by a compound annual rate of over 6% between 2003 and 2008, and growth is predicted to continue over the next years (www.just-food.com). However, many processes described so far have been performed on a laboratory scale, or at best on a pilot plant scale. In order for these approaches to find their way into industrial reality, they must be economically feasible, which sometimes may pose a challenge because of the additional processing steps needed for valorization.

References

Adlercreutz, H., 2002. Phyto-oestrogens and cancer. Lancet. Oncol. 3, 364–373.

Alekel, D.L., Germain, A.S., Peterson, C.T., Hanson, K.B., Stewart, J.W., Toda, T., 2000. Isoflavone-rich soy protein isolate attenuates bone loss in the lumbar spine of perimenopausal women. Am. J. Clin. Nutr. 72, 844–852.

Aliakbarian, B., Casazza, A.A., Perego, P., 2011. Valorization of olive oil solid waste using high pressure–high temperature reactor. Food Chem. 128, 704–710.

Anthony, M.S., Clarkson, T.B., Hughes Jr., C.L., Morgan, T.M., Burke, G.L., 1996. Soybean isoflavones improve cardiovascular risk factors without affecting the reproductive system of peripubertal rhesus monkeys. J. Nutr. 126, 43–50.

Arora, A., Camire, M.E., 1994. Performance of potato peels in muffins and cookies. Food Res. Int. 27, 15–22.

Arvanitoyannis, I.S., Ladas, D., Mavromatis, A., 2006. Potential uses and applications of treated wine waste: a review. Int. J. Food Sci. Technol. 41, 475–487.

Bagchi, D., Bagchi, M., Stohs, S.J., Ray, S.D., Sen, C.K., Pruess, H.G., 2002. Cellular protection with proanthocyanidins derived from grape seeds. Ann. New York Acad. Sci. 957, 260–270.

Bagchi, D., 2006. Nutraceuticals and functional foods regulations in the United States and around the world. Toxicology 221, 1–3.

Benítez, V., Mollá, M., Martín-Cabrejas, M.A., Aguilera, Y., López-Andréu, F.J., Esteban, R.M., 2011. Effect of sterilisation on dietary fibre and physicochemical properties of onion by-products. Food Chem. 127, 501–507.

Benoit, I., Navarro, D., Marnet, N., et al., 2006. Feruloyl esterases as a tool for the release of phenolic compounds from agro-industrial by-products. Carbohydr. Res. 341, 1820–1827.

Birt, D.F., Hendrich, S., Wang, W., 2001. Dietary agents in cancer prevention: flavonoids and isoflavonoids. Pharmacol Ther. 90, 157–177.

Bors, W., Heller, W., Michel, C., Saran, M., 1990. Flavonoids as anti-oxidants: determination of radical-scavenging efficiencies. In: Packer, L., Glazer, A.N. (Eds.), Methods in Enzymology, 186. Academic Press, San Diego, pp. 343–355.

Brandt, K., Christensen, L.P., Hansen-Møller, J., Hansen, S.L., 2004. Health promoting compounds in vegetables and fruits: a systematic approach for identifying plant components with impact on human health. Trends. Food Sci. Technol. 15, 384–393.

Burns, J., Yokota, T., Ashihara, H., Lean, M.E., Crozier, A., 2002. Plant foods and herbal sources of resveratrol. J. Agric. Food Chem. 50, 3337–3340.

Cartea, M.E., Velasco, P., 2008. Glucosinolates in brassica foods: bioavailability in food and significance for human health. Phytochem. Rev. 7, 213–229.

Champ, M., Langkilde, A.M., Brouns, F., Kettlitz, B., Collet, Y.L.B., 2003. Advances in dietary fibre characterisation. 1. Definition of dietary fibre, physiological relevance, health benefits and analytical aspects. Nutr. Res. Rev. 16, 71–82.

Chu, Y.F., Sun, J., Wu, X., Liu, R.H., 2002. Antioxidant and antiproliferative activities of common vegetables. J. Agric. Food Chem. 50, 6910–6916.

Coppens, P., Fernandes Da Silva, M., Pettman, S., 2006. European regulations on nutraceuticals, dietary supplements and functional foods: a framework based on safety. Toxicology 221, 59–74.

Crouse III, J.R., Morgan, T., Terry, J.G., Ellis, J., Vitolins, M., Burke, G.L., 1999. A randomized trial comparing the effect of casein with that of soy protein containing varying amounts of isoflavones on plasma concentrations of lipids and lipoproteins. Arch. Intern. Med. 159, 2070–2076.

Dijsselbloem, N., Berghe, W.V., Naeyer, A.D., Haegeman, G., 2004. Soy isoflavone phyto-pharmaceuticals in interleukin-6 affections. Multi-purpose nutraceuticals at the crossroad of hormone replacement, anti-cancer and anti-inflammatory therapy. Biochem. Pharmacol. 68, 1171–1185.

Diplock, A.T., Aggett, P.J., Ashwell, M., Bornet, F., Fern, E.B., Roberfroid, M.B., 1999. Scientific concepts of functional foods in Europe: consensus document. British J. Nutrition. 81 (Suppl.1), S1–S27.

Dohadwala, M.M., Vita, J.A., 2009. Grapes and cardiovascular disease. The Journal of Nutrition Supplement: Grapes and Health. Am. Soc. Nutr. 139, 1788S–1793S.

Engelbrecht, A.M., Mattheyse, M., Ellis, B., Loos, B., Thomas, M., Smith, R., et al., 2007. Proanthocyanidin from grape seeds inactivates the PI3-kinase/PKB pathway and induces apoptosis in a colon cancer cell line. Cancer. Lett. 258, 144–153.

EU, 1994. EU 94/39 EC: Colours for Use in Foodstuffs. Brussels, EU.

Femenia, A., Lefebvre, A.-C., Thebaudin, J.Y., Robertson, J.A., Bourgeois, C.M., 1997. Physical and sensory properties of model foods supplemented with cauliflower fibre. J. Food Sci. 62, 635–639.

Feng, J.G., Chen, L.Q., 2003. Determination of procyanidin in grape seed extracts. China Food Addit. 6, 103–105.

Feng, Y.Z., Liu, Y.M., Fratkins, J.D., LeBlan, M.H., 2005. Grape seed extract suppresses lipid peroxidation and reduces hypoxic ischemic brain injury in neonatal rats. Brain Res. Bull. 66, 120–127.

FitzGerald, R.J., Meisel, H., 2003. Milk protein hydrolysates and bioactive peptides. In: Fox, P.F., McSweeney, P.L.H. (Eds.), Advanced Dairy Chemistry, vol. 1. Kluwer Academic/Plenum Publishers, New York, NY, pp. 675–698.

Frankel, E.N., Waterhouse, A.L., Kinsella, J.E., 1993. Inhibition of human LDL oxidation by resveratrol. Lancet 341, 1103–1104.

Friedman, M., 2006. Potato glycoalkaloids and metabolites: roles in the plant and in the diet. J. Agric. Food Chem. 54, 8655–8681.

Gamet-Payrastre, L., Li, P., Lumeau, S., Cassar, G., Dupont, M.-A., Chevolleau, S., et al. (2000). Sulforaphane, a naturally occurring isothiocyanate, induces cell cycle arrest and apoptosis in HT29 human colon cancer cells. Cancer Research, 60, 1426–1433.

Garau, M.C., Simal, S., Rosselló, C., Femenia, A., 2007. Effect of air-drying temperature on physico-chemical properties of dietary fibre and antioxidant capacity of orange (Citrus aurantium v. Canoneta) by-products. Food Chem. 104, 1014–1024.

Gill, H.S., Doull, F., Rutherfurd, K.J., Cross, M.L., 2000. Immunoregulatory peptides in bovine milk. Br. J. Nutr. 84 (Suppl. 1), S111–S117.

Grigelmo-Miguel, N., Martin-Belloso, O., 1999. Comparison of dietary fibre from by-products of processing fruits and greens and from cereals. LWT–Food Sci. Technol. 32, 503–508.

Gundgaard, J., Nielsen, J.N., Olsen, J., Sørensen, J., 2003. Increased intake of fruit and vegetables: estimation of impact in terms of life expectancy and healthcare costs. Public Health Nutr. 6, 25–30.

Hertog, MG, Feskens, EJ, Hollman, PC, Katan, MB, Kromhout, D. 1993. Dietary antioxidant flavonoids and risk of coronary heart disease: the Zutphen Elderly Study. Lancet 342, 1007–1011.

Hoerger, T.J., Ahmann, A.J., 2008. The impact of diabetes and associated cardiometabolic risk factors on members: strategies for optimizing outcomes 280. J Manag. Care Pharm. 14 (1 Suppl C), S2–14.

Hooper, L., Kroon, P.A., Rimm, E.B., Cohn, J.S., Harvey, I., Le Cornu, K.A., et al. 2008. Flavonoids, flavonoid-rich foods, and cardiovascular risk: a meta-analysis of randomized controlled trials. Am. J. Clin. Nutr. 88, 38–50.

Iacopini, P., Baldi, M., Storchi, P., Sebastiani, L., 2008. Catechin, epicatechin, quercetin, rutin and resveratrol in red grape: content, in vitro antioxidant activity and interactions. J. Food Comp. Anal. 21, 589–598.

Jacobs, M.N., Lewis, D.F., 2002. Steroid hormone receptors and dietary ligands: a selected review. Proc. Nutr. Soc. 61, 105–122.

Jaime, L., Mollá, E., Fernández, A., Martín-Cabrejas, M.A., López-Andréu, F.J., Esteban, R.M., 2002. Structural carbohydrate differences and potential source of dietary fibre of onion (Allium cepa L.). J. Agric. Food Chem. 50 (1), 122–128.

Kammerer, D., Claus, A., Carle, R., Schieber, A., 2004. Polyphenol screening of pomace from red and white grape varieties (Vitis vinifera L.) by HPLC-DAD-MS/MS. J. Agric. Food Chem. 52, 4360–4367.

Kaur, M., Agarwal, C., Agarwal, R., 2009. Anticancer and cancer chemopreventive potential of grape seed extract and other grape-based products. J. Nutr. 139, 1806S–1812S.

Kimura, Y., Ohminami, H., Okuda, H., Baba, K., Kozawa, M., Arichi, S., 1983. Effects of stilbene components of roots of Polygonum ssp. on liver injury in peroxidized oil-fed rats. Planta Med. 49, 51–54.

Korhonen, H., Pihlanto, A., 2003. Food-derived bioactive peptides—opportunities for designing future foods. Curr. Pharm. Des. 9, 1297–1308.

Korhonen, H., Pihlanto, A., 2006. Bioactive peptides: production and functionality. Int. Dairy J. 16, 945–960.

Kris-Etherton, P.M., Etherton, T.D., Carlson, J., Gardner, C., 2002. Recent discoveries in inclusive food-based approaches and dietary patterns for reduction in risk for cardiovascular disease. Curr. Opin. Lipidol. 13, 397–407.

Larrauri, J.A., 1999. New approaches in the preparation of high dietary fibre powders from fruit by-products. Trends. Food Sci. Technol. 10, 3–8.

Laufenberg, G., Kunz, B., Nystroem, M., 2003. Transformation of vegetable waste into value added products: (A) the upgrading concept; (B) practical implementations. Bioresour. Technol. 87, 167–198.

Lee, C.H., Jeong, T.S., Choi, Y.K., Hyun, B.H., Oh, G.T., Kim, E.H., Kim, J.R., Han, J.I., Bok, S.H., 2001. Anti-atherogenic effect of citrus flavonoids, naringin and naringenin, associated with hepatic ACAT and aortic VCAM-1 and MCP-1 in high cholesterol-fed rabbits. *Biochem Biophys Res Commun.* 284 (3), 681–688.

Lusk, J.L., Rozan, A., 2005. Consumer acceptance of biotechnology and the role of second generation technologies in the USA and Europe. Trends. Biotechnol. 23, 386–387.

Mäder, J., Rawel, H., Kroh, L.W., 2009. Composition of phenolic compounds and glycoalkaloids α-solanine and α-chaconine during commercial potato processing. J. Agric. Food Chem. 57, 6292–6297.

Manso, M.A., Lopez-Fandino, R., 2004. κ-Casein macropeptides from cheese whey: physicochemical, biological, nutritional, and technological features for possible uses. Food Rev. Int. 20, 329–355.

Matar, C., LeBlanc, J.G., Martin, L., Perdigon, G., 2003. Biologically active peptides released in fermented milk: role and functions. In: Farnworth, E.R. (Ed.), Handbook of Fermented Functional Foods. Functional Foods and Nutraceuticals Series. CRC Press, Florida, pp. 177–201.

Maynard, M., Gunnell, D., Emmett, P., Frankel, S., Smith, G.D., 2003. Fruit, vegetables, and antioxidants in childhood and risk of adult cancer. J. Epidemiol Community Health 57, 218–225.

Meisel, H., FitzGerald, R.J., 2003. Biofunctional peptides from milk proteins: mineral binding and cytomodulatory effects. Curr. Pharm. Des. 9, 1289–1295.

Middleton Jr., E., Kandaswami, C., Theoharides, T.C., 2000. The effects of plant flavonoids on mammalian cells: implications for inflammation, heart disease, and cancer. Pharmacol. ReV. 52, 673–751.

Morillas-Ruiz, J.M., Villegas Garcia, J.A., Lopez, F.J., Vidal-Guevara, M.L., Zafrilla, P., 2006. Effects of polyphenolic antioxidants on exercise induced oxidative stress. Clin. Nutr. 25, 444–453.

Moure, A., Cruz, J., Franco, D., et al., 2001. Natural antioxidant from residual sources (review). Food Chem. 72, 145–171.

Nagata, C., Takatsuka, N., Kurisu, Y., Shimizu, H., 1998. Decreased serum total cholesterol concentration is associated with high intake of soy products in Japanese men and women. J. Nutr. 128, 209–213.

Nathan, B.M., Moran, A., 2008. Metabolic complications of obesity in childhood and adolescence: more than just diabetes 278. Curr. Opin. Endocrinol Diabetes Obes. 15 (1), 21–29.

Negro, C., Tommasi, L., Miceli, A., 2003. Phenolic compounds and antioxidant activity from red grape marc extracts. Bioresour. Technol. 87, 41–44.

Nelson, R.A., Bremer, A.A., 2009. Insulin Resistance and Metabolic Syndrome in the Pediatric Population 279. Metab. Syndr. Relat. Disord.

Ohama, H., Ikeda, H., Moriyama, H., 2006. Health foods and foods with health claims in Japan. Toxicology 22, 95–111.

Okamoto, T., 2005. Safety of quercetin for clinical application (Review). Int. J. Mol. Med. 16, 275–278.

Pace-Asciak, C.R., Hahn, S., Diamandis, E.P., Soleas, G., Goldberg, D.M., 1995. The red wine phenolics trans-resveratrol and quercetin block human platelet aggregation and eicosanoid synthesis: implication for protection against coronary heart disease. Clin. Chim. Acta 235, 207–219.

Pignatelli, P., Pulcinelli, F.M., Celestini, A., Lenti, L., Ghiselli, A., Gazzaniga, P.P., et al., 2000. The flavonoids quercetin and catechin synergistically inhibit platelet function by antagonizing the intracellular production of hydrogen peroxide. Am. J. Clin. Nutr. 72, 1150–1155.

Poppe, C., Kjnes, U., 2003. Trust in Food in Europe. A Comparative Analysis. National Institute for Consumer Research, Oslo. Available at: <http://www.trustinfood.org/SEARCH/BASIS/tif0/all/publics/DDD/24.pdf>.

Potter, S.M., Baum, J.A., Teng, H., Stillman, R.J., Shay, N.F., Erdman Jr., J.W., 1998. Soy protein and isoflavones: their effects on blood lipids and bone density in postmenopausal women. Am. J. Clin. Nutr. 68, 1375S–1379SS.

Prasain, J.K., Carlson, S.H., Wyss, J.M., 2010. Flavonoids and age related disease: risk, benefits and critical windows. Maturitas 66 (2), 163–171.

Ramadan, M.F., Sitohy, M.Z., Moersel, J.-T., 2008. Solvent and enzyme-aided aqueous extraction of goldenberry (*Physalis peruviana* L.) pomace oil: impact of processing on composition and quality of oil and meal. Eur. Food Res. Technol. 226, 1445–1458.

Ray, S., Bagchi, D., Lim, P., et al., 2001. Acute and longterm safety evaluation of a novel IH636 grape seed proanthocyanidin extract. Res. Commun. Mol. Pathol. Pharmacol. 109, 165–197.

Rice-Evans, C.A., Miller, N.J., Paganga, G., 1996. Structure-antioxidant activity relationships of flavonoids and phenolic acids. Free Radic. Biol. Med. 20, 933–956.

Rotondo, S., Rajtar, G., Manarini, S., Celardo, A., Rotilio, D., de Gaetano, G., et al., 1998. Effect of trans-resveratrol, a natural polyphenolic compound, on human polymorphonuclear leukocyte function. Br. J. Pharmacol. 123, 1691–1699.

Sano, T., Oda, E., Yamashita, T., Naemura, A., Ijiri, Y., Yamakoshi, J., et al., 2005. Anti-thrombotic effect of proanthocyanidin, a purified ingredient of grape seed. Thromb. Res. 115, 115–121.

Santos-Buelga, C., Scalbert, A., 2000. Review: proanthocyanidins and tannin-like compounds—nature, occurrence, dietary intake, and effects on nutrition and health. J. Sci. Food Agric. 80, 1094–1117.

Schieber, A., 2009. Nutraceuticals from by-products of plant food processing. Can. Chem. News.

Schieber, A., Saldaña, M.D.A., 2009. Potato peels: a source of nutritionally and pharmacologically interesting compounds—a review. Food 3, 23–29.

Schieber, A., Stintzing, F.C., Carle, R., 2001. By-products of plant food processing as a source of functional compounds—recent developments. Trends. Food Sci. Technol. 12, 401–413.

Shan, Y., Ye, X., Xin, H., 2010. Effect of the grape seed proanthocyanidin extract on the free radical and energy metabolism indicators during the movement. Sci. Res. Essay 5 (2), 148–153. Available at: <http://www.academicjournals.org/SRE>.

Shen, L., Su, G., Wang, X., Du, Q., Wang, K., 2010. Endogenous and exogenous enzymolysis of vegetable-sourced glucosinolates and influencing factors. Food Chem. 119, 987–994.

Shimizu, M., 2004. Food-derived peptides and intestinal functions. BioFactors 21, 43–47.

Shrikhande, A., 2000. Wine by-products with health benefits. Food Res. Int. 33, 469–474.

Siro, I., Kapolna, E., Kapolna, B., Lugasi, A., 2008. Functional food. Product development, marketing and consumer acceptance—a review. Appetite 51, 456–467.

Slimestad, R., Fossen, T., Vågen, I.M., 2007. Onions: a source of unique dietary flavonoids. J. Agric. Food Chem. 55, 10067–10080.

Smithers, G.W., 2008. Whey and whey proteins—from 'gutter-to-gold'. Int. Dairy J. 18, 695–704.

Sotillo, R.D., Hadley, M., Wolf-Hall, C., 1998. Potato peel extract: a non-mutagenic antioxidant with potential antimicrobial activity. J. Food Sci. 63, 907–910.

Spranger, I., Sun, B., Mateus, A.M., Freitas, V., Ricardo-da-Silva, J.M., 2008. Chemical characterization and antioxidant activities of oligomeric and polymeric procyanidin fractions from grape seeds. Food Chem. 108, 519–532.

Steijns, J.M., 2001. Milk ingredients as nutraceuticals. Int. J. Dairy Technol. 54 (3), 81–88.

Stojceska, V., Ainsworth, P., Plunkett, A., Ibanoglu, E., Ibanoglu, S., 2008. Cauliflower by-products as a new source of dietary fibre, antioxidants and proteins in cereal based ready-to-eat expanded snacks. J. Food Eng. 87, 554–563.

Strack, D., Wray, V., 1989. Anthocyanins. In: Dey, P., Harborne, J. (Eds.), Methods in Plant Biochemistry, Plant phenolics. vol. 1. Academic Press, London, pp. 325–356.

Tanongkankit, Y., Chiewchan, N., Devahastin, S. 2011. Evolution of anticarcinogenic substance in dietary fibre powder from cabbage outer leaves during drying. Food Chemistry 127, 67–73.

Terao, J., 2009. Dietary flavonoids as antioxidants. In: Yoshikawa, T. (Ed.), Food Factors for Health Promotion. Forum Nutr, 61. Karger, Basel, pp. 87–94.

Terao, J., Kawai, Y., Murota, K., 2008. Vegetable flavonoids and cardiovascular disease. Asia Pac. J. Clin. Nutr. 17 (S1), 291–293.

Teschemacher, H., 2003. Opioid receptor ligands derived from food proteins. Curr. Pharm. Des. 9, 1331–1344.

Trichopoulou, A., Naska, A., Antoniou, A., Friel, S., Trygg, K., Turrini, A., 2003. Vegetable and fruit: the evidence in their favour and the public health perspective. Int. J. Vitam. Nutr. Res. 73, 63–69.

Verbeke, W., 2005. Consumer acceptance of functional foods: sociodemographic, cognitive and attitudinal determinants. Food Qual. Prefer. 16, 45–57.

Virgili, F., Marino, M., 2008. Regulation for cellular signals from nutritional molecules; a specific role for phytochemicals, beyond antioxidant activity. Free Radic. Bol. Med. 45, 1205–1216.

Voorrips, L.E., Goldbohm, R.A., van Poppel, G., Sturmans, F., Hermus, R.J.J., van den Brandt, P.A., 2000. Vegetable and fruit consumption and risks of colon and rectal cancer in a prospective cohort study—the Netherlands cohort study on diet and cancer. Am. J. Epidemiol. 152, 1081–1092.

Wijngaard, H.H., Ballay, M., Brunton, N., 2011. The optimisation of extraction of antioxidants from potato peel by pressurised liquids. Food Chem. (online doi:10.1016/j.foodchem.2011. 01.136).

www.just-food.com (2006). Global market review of functional foods—forecasts to 2012. Report #44028, August 2006.

Yamakoshi, J., Kataok, S., Koga, T., Ariga, T., 1999. Proanthocyanidin-rich extract from grape seeds attenuates the development of aortic atherosclerosis in cholesterol-fed rabbits. Atherosclerosis 142, 139–149.

Zeisel, S.H., 1999. Regulation of "nutraceuticals". Science 285, 1853–1855.

CHAPTER

7

Manufacture of Biogas and Fertilizer from Solid Food Wastes by Means of Anaerobic Digestion

Naomichi Nishio and Yutaka Nakashimada

1. INTRODUCTION

Anaerobic microbial degradation of biomass, which is usually called "anaerobic digestion", has been used for the treatment of organic wastes and wastewater. In the process, microorganisms degrade organic matter and produce the mixed gases of methane and carbon dioxide (CO_2) in the absence of oxygen. Since methane is a useful renewable energy source, anaerobic digestion gets much attention as the technology for not only treatment of organic wastes but also biomass conversion to energy. Although anaerobic digestion is technologically simple, it can convert organic materials (sugars, proteins, lipids, xenobiotics, etc.) from a wide range of organic wastes including wastewaters, solid wastes, and biomass into methane. This is a great advantage for conversion technology of biomass compared with production of ethanol and lactate from only sugars in biomass. Thus, it is feasible to apply anaerobic digestion as the technology for renewable energy production from organic substances.

Food waste possesses the highest potential in terms of economic exploitation of the different types of organic wastes available because it contains a high amount of biodegradable carbon and can be efficiently converted into methane and organic fertilizer. In this chapter, the basic principles and process designs of anaerobic digestion are introduced; then some applications for solid food waste treatment, investigated by the authors, are presented. After that, fertilization of the residue after the digestion is briefly explained.

2. BASIC PRINCIPLES OF ANAEROBIC DIGESTION

2.1 Conversion Flow of Organic Matter to Methane

Anaerobic digestion is a highly sophisticated process in which a variety of microorganisms play various roles in the decomposition of organic material and production of methane and CO_2. Anaerobic digestion can be subdivided into four phases, as illustrated in Figure 7.1 and described below.

2.1.1 Disintegration and Hydrolysis

Disintegration and hydrolysis are extracellular biological and nonbiological processes. This phase mediates the breakdown and solubilization of complex organic material (Batstone et al., 2002). Obtained organic compounds are usually composed of carbohydrates, protein, and lipids as the substrates for anaerobic digestion. Nondegradable matter such as lignin and inorganic substances are also present.

The complex organic matter is disintegrated to small particles because extracellular enzymes secreted by hydrolyzing bacteria can attack reaction sites on the particles. The disintegration is sometimes facilitated by nonbiological pretreatment (e.g., physical shearing, heat treatment, and chemical treatment (Mata-Alvarez, et al., 2000; Kim et al., 2003)). Even if there is no such pretreatment, during this stage, complex molecules are depolymerized into soluble substrates by extracellular hydrolytic enzymes released by bacteria (cellulases, hemicellulases, amylases, lipases, proteases, etc.). These

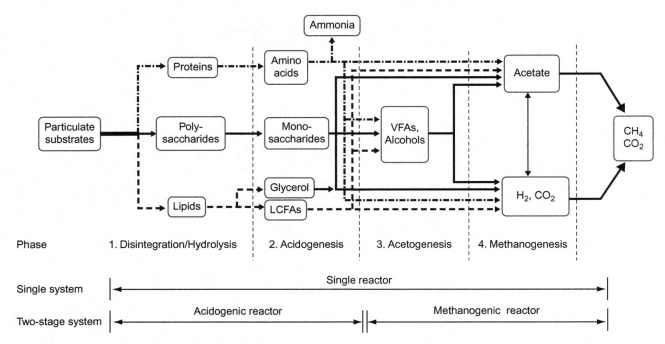

FIGURE 7.1 **Schematic representation of anaerobic digestion.** Particulate substrates in the wastes are disintegrated to polysaccharides, proteins, and lipids and simultaneously hydrolyzed to monosaccharides, amino acids, and long chain fatty acids (LCFAs)/glycerol. The hydrolyzed compounds are then fermented to acetate, other volatile fatty acids (VFAs), alcohols, hydrogen (H_2), and carbon dioxide (CO_2) by acidogenic microflora. VFAs and alcohols are further oxidized to acetate, H_2, and CO_2 by particular acetogenic bacteria. Acetate and H_2/CO_2 are finally converted to methane and CO_2 by acetoclastic and hydrogenotrophic methanogens (*adapted from de Mes et al., 2003*). Disintegration/ hydrolysis is widely regarded as the rate-limiting step, and microbial flora associated with this phase often overlap that of acidogenesis. Since acetogenesis needs to be carried out under a low H_2 partial pressure, H_2 scavenging by methanogens is essential for successful methane production. Thus, the two-stage system that consists of acidogenic and methanogenic reactors is one feasible solution for optimizing anaerobic digestion efficiency of organic solid substrates.

enzymes break down the carbohydrates, proteins, and lipids to produce soluble monosaccharides, amino acids, and long chain fatty acids (LCFAs) and glycerol, respectively. Several bacterial groups are involved in the hydrolysis process, such as phyla *Firmicutes* and *Bacteroidetes* (Leven et al., 2007; Goberna et al., 2009).

Disintegration/hydrolysis is widely regarded as the rate-limiting step of degradation of particulate organic matter like manure, sewage sludge, crop residues, and so on (Pavlostathis and Giraldo-Gomez, 1991). Therefore, the overall rate of the process is often determined by the disintegration/hydrolysis rate of this complex substrate. Despite being very complicated, this process is most commonly represented by first-order kinetics (Batstone et al., 2002; Sanders et al., 2002). Disintegration rate constants of solids are suggested to be 0.5 (mesophilic) or 1.0 (thermophilic) per day, and hydrolysis rate of coefficients are on the order of 0.1−0.25 (mesophilic) or 10 (thermophilic) per day (Batstone et al., 2002).

2.1.2 Acidogenesis

Soluble organic compounds are catabolically converted to a number of simpler molecules. Monosaccharides, such as glucose and fructose, are abundantly contained in organic wastes discharged from the food industry. They are usually catabolized to pyruvate via the Emden−Mayerhof−Parnas (EMP) or the Entner−Doudoroff (DE) pathway and converted to reduced products such as lactate and propionate, or further metabolized via acetyl-CoA to acetate and butyrate as the oxidized end products or ethanol as the reduced products of waste excess. H_2 and CO_2 are also produced from waste excess electrons and carbons, respectively, and used directly by the hydrogenotrophic methanogens. A wide variety of strict anaerobes such as *Clostridium*, *Bacteroides*, and *Butyrivibrio* spp. and facultative anaerobes such as *Bacillus*, *Enterobacter*, and *Lactobacillus* spp. participate in this acidogenesis. The reaction is usually called "fermentation", and it is a typical characteristic of anaerobic heterotrophic microorganisms to produce ATP, the energy molecule required for the growth and the maintenance of cells.

Regarding the acidogenesis of monosaccharides, their intermediates such as pyruvate and acetyl-CoA are used as electron acceptors to produce oxidized products and reduced products. On the other hand, degradation of mixed amino acids obtained by hydrolysis of proteins occurs via two main pathways: one is a coupled reaction of oxidation/reduction with pairs of amino acids, which is called the Stickland reaction (Winter et al., 1987). The other is an uncoupled reaction, in which a single amino acid is oxidized with protons or carbon dioxide as the external electron acceptor (Batstone et al., 2002).

In the Stickland reaction, one amino acid is oxidized to become one carbon shorter while the other is reduced. In the case of acidogenesis with alanine and glycine, alanine (C3) is oxidized to acetate (C2), CO_2, and ammonia, while glycine is reduced to acetate and ammonia.

Alanine (oxidized as electron donor):

$$CH_3CH(NH_2)COO^- + 2H_2O \rightarrow CH_3COO^- \\ + CO_2 + NH_3 + 4H^+ + 4e^- \tag{1}$$

Glycine (reduced as electron acceptor):

$$2CH_2(NH_2)COO^- + 4H^+ + 4e^- \rightarrow 2CH_3COO^- \\ + 2NH_3 \tag{2}$$

Different amino acids can act as electron donor, acceptor, or both in some cases (Ramsay and Pullammanappallil, 2001).

Although Stickland reactions occur more rapidly than uncoupled degradation (Baker, 1981), there is normally a 10% shortfall in acceptor amino acids in normal mixed-protein systems (Nagase and Matsuo, 1982). Thus, the remainder is oxidized by uncoupled degradation. When hydrogen concentrations are low and the energetics are suitable, uncoupled oxidation can also occur (Stams, 1994). In this case, from alanine, two molecules of hydrogen gas are produced, instead of electrons that are used to convert glycine to acetate.

2.1.3 Acetogenesis (H_2-producing)

The metabolites such as propionate, butyrate, and ethanol produced via the acidogenesis phase are further oxidized to acetate by H_2-producing acidogenic bacteria. Electrons produced from this reaction are transferred to protons to produce H_2 or bicarbonate to generate formate. For example, the stoichiometric equation, when propionate is oxidized, is represented as follows:

$$CH_3CH_2COO^- + 3H_2O \rightarrow CH_3COO^- + HCO_3^- \\ + H^+ + 3H_2 \tag{3}$$

$$(\Delta G^{0\prime} = +76 \text{ kJ/reaction})$$

In the case where butyrate is oxidized, $\Delta G^{0\prime}$ is +48 kJ/reaction. Since the oxidation reaction of these fatty acids is usually endergonic under the biological standard conditions (25°C, 1 atm, pH 7), bacteria can only derive energy for growth from these oxidation reactions if the concentration of the products is kept quite low and $\Delta G'$ becomes negative. This results in an obligate dependence of acetogenic bacteria on methanogens, or other hydrogen scavengers, such as hydrogenotrophic methanogens (see the next section) and sulfate reducers for product removal. Syntrophic bacteria that oxidize fatty acids cooperatively with

hydrogen scavengers, for example, *Syntrophomonas wolfei* (butyrate oxidation) and *Syntrophobacter wolinii* (propionate oxidation), have been isolated by Schink (1997). Because the accumulation of VFA causes a significant drop of pH, resulting in the inhibition of methanogenesis, acetogenesis plays an important role in anaerobic digestion.

2.1.4 Methanogenesis

Substrates that methanogens utilize are very limited. In the industrial process of anaerobic digestion, the usual substrates for methane formation are only H_2 and CO_2, or formate by hydrogenotrophic methanogen, or acetate by acetoclastic methanogen produced via acidogenesis and acetogenesis as follows, respectively:

$$4H_2 + HCO_3^- + H^+ \rightarrow CH_4 + 3H_2O \\ (\Delta G^{0\prime} = -136 \text{ kJ/reaction}) \tag{4}$$

$$CH_3COO^- + H_2O \rightarrow CH_4 + HCO_3^- \\ (\Delta G^{0\prime} = -31 \text{ kJ/reaction}) \tag{5}$$

All known methanogens belong to the archaea and are strict anaerobes. Several hydrogenotrophic methanogens have been isolated and characterized, such as *Methanobacterium*, *Methanobrevibacter*, *Methanosprillum*, *Methanococcus*, *Methanogenium*, and *Methanoculleus*.

The removal of H_2 by hydrogenotrophic methanogens is indispensable for syntrophic acetogenesis of higher organic acids to proceed because the total reaction becomes exergonic by coupling with acetogenesis and methanogenesis. For example, change of Gibbs free energy, when oxidization of propionate (Eq 3) is coupled with hydrogenotrophic methanogenesis (Eq 4), is negative under biological standard conditions:

$$4CH_3CH_2COO^- + 3H_2O \rightarrow 4CH_3COO^- + \\ HCO_3^- + H^+ + 3CH_4 \\ (\Delta G^{0\prime} = -104 \text{ kJ/reaction}) \tag{6}$$

The syntrophic relationship of hydrogen producers and hydrogen consumers (e.g., hydrogenotrophic methanogen) is called interspecies hydrogen transfer. Syntrophic communities of H_2-producing acetogenic bacteria and methanogens degrade fatty acids coupled to growth. However, neither the methanogens nor the acetogens alone are able to degrade these compounds. Inter-microbial distances between acetogens and methanogens can influence specific growth rates of these microorganisms (Batstone et al., 2006).

On the other hand, diversity of acetoclastic methanogens that are important for conversion of acetate to methane is very limited. The reported acetoclastic methanogens are only *Methanosarcina* and *Methanosaeta* spp. *Methanosarcina* can consume a relatively high concentration of acetate and have the ability to use methyl compounds such as methanol. Although *Methanosaeta* can consume a lower concentration of acetate and play a role in decreasing chemical oxygen demand (COD) in anaerobic digestion, since their growth rate is significantly low, the amount of *Methanosaeta* cells significantly affects the performance of anaerobic digestion.

2.2 Methane Production Potential of Organic Wastes

Since in organic matter various kinds of chemical compounds such as sugar, protein, and lipids are contained as mixtures at different compositions, COD is used to quantify the amount of organic matter in the wastes. COD is the amount of oxygen required to oxidize an organic compound to carbon dioxide, ammonia, and water. COD is usually measured using a strong chemical oxidizing agent (potassium dichromate is usually used in the case of application to anaerobic digestion) under an acidic condition and theoretically given by the following equation in the case that an organic compound $C_nH_aO_bN_c$ was tested:

$$C_nH_aO_bN_c + \left(n + \frac{a}{4} - \frac{b}{2} - \frac{3}{4}c\right)O_2 \rightarrow nCO_2$$
$$+ \left(\frac{a}{2} - \frac{3}{2}c\right)H_2O + cNH_3 \qquad (7)$$

Since COD is an index that does not depend on the kind of organic compound, it is useful to quantify organic wastes containing various kinds of compounds.

Furthermore, COD can be used to predict the potential for biogas production because COD also represents the amount of energy conserved in an organic matter as an alternative meaning. Therefore, during anaerobic digestion, COD in the wastes is conserved in the end products such as methane and the newly formed microbial cells. It can be easily understood using stoichiometric considerations. If an organic compound $C_nH_aO_bN_c$ is completely biodegradable and completely catabolized by the microorganisms without biomass formation into methane, carbon dioxide, and ammonia, the theoretical amount of the biogases can be calculated by using the following equation:

$$C_nH_aO_bN_c + \left(n - \frac{a}{4} - \frac{b}{2} + \frac{3}{4}c\right)H_2O$$
$$\rightarrow \left(\frac{n}{2} + \frac{a}{8} - \frac{b}{4} - \frac{3}{8}c\right)CH_4$$
$$+ \left(\frac{n}{2} - \frac{a}{8} + \frac{b}{4} + \frac{3}{8}c\right)CO_2 + cNH_3 \qquad (8)$$

For example, in the case of glucose ($C_6H_{12}O_6$), yield of methane can be calculated to be

$$6/2 + 12/8 - 6/4 = 3 + 1.5 - 1.5 = 3 \text{ mol/mol glucose.}$$

In Table 7.1, theoretical biogas yield from various types of organic matter is listed. The quantity of CO_2 is usually lower than the theoretical value. This is because solubility of CO_2 in water is significantly higher than that of methane. Since a part of CO_2 is solubilized in the form of carbonate ion (CO_3^{2-}) under weak alkali conditions feasible for anaerobic digestion, CO_3^{2-} can chemically bind with polyvalent metal ions to form insoluble salt and remain in water phase.

2.3 Environmental Factors Affecting Anaerobic Digestion

Anaerobic digestion is strongly affected by environmental factors, since it is a biological process. Although microorganisms are robust and can survive under certain a degree of severe conditions (psychrophilic, extreme thermophilic, high alkaline or acidic, etc.), for effective anaerobic digestion the process should be optimized so they can grow and function well. Environmental control factors important to the optimization of the process are primarily temperature, pH and alkalinity, and biological toxic compounds.

2.3.1 Temperature

Temperature can affect biochemical reactions (Batstone et al., 2002). Anaerobic digestion is usually classified into three major operating ranges: psychrophilic (4–15°C), mesophilic (20–40°C), and thermophilic (45–70°C). Anaerobic digestion can be performed effectively

TABLE 7.1 Theoretical Biogas Yield by Anaerobic Digestion of Various Types of Organic Matter

Substrate	Pseudo Formula	Biogas Yield* (m^3 N/kg-VS Degraded)	CH_4 Content (%)
Carbohydrate	$(C_6H_{10}O_6)_n$	0.83	50.0
Protein	$C_{16}H_{24}O_5N_4$	0.76	68.8
Lipid	$C_{50}H_{90}O_6$	1.42	69.5
Lignin	$(-CH_2-)_n$	1.60	75.0
Food waste	$C_{17}H_{29}O_{10}N$	0.88	57.8
Municipal waste	$C_{46}H_{73}O_{31}N$	0.89	53.3
Sewage sludge	$C_{10}H_{19}O_3N$	1.00	69.4
Waste sludge	$C_5H_7O_2N$	0.79	62.5
Night soil wastes	$C_7H_{12}O_4N$	0.77	60.4

*For calculation of the yield, cell biomass production from substrate is neglected.

between these ranges, while optimum temperatures for mesophilic and thermophilic conditions are approximately 35°C and 55°C, respectively.

2.3.2 pH and Alkalinity

Remembering that lactic acid bacteria grow at pH 4–5, since there are many kinds of hydrolyzing and acidogenetic microorganisms in nature, acidogenesis can occur at a wide range of pH values. On the other hand, the optimum pH for methanogenesis is narrow. Methanogenesis actively proceeds when the pH is around neutral (6.5–7.5), otherwise the rate of methane formation is lower (Lettinga and Haandel, 1993). In this context, a sufficient amount of buffer material in the solution is important to maintain the optimal pH range required for methanogenesis. The most abundant buffer material in anaerobic digestion is hydrogen carbonate, since it is continuously supplied as a by-product of methane fermentation. The amount of hydrogen carbonate is frequently denoted as bicarbonate alkalinity.

2.3.3 Biological Toxic Compounds

Several compounds have a toxic effect at excessive concentrations of ammonia, cations such as Na^+, K^+, and Ca^{2+}, sulfide, VFAs, and xenobiotics. Carbohydrates in food wastes are a favorable substrate for anaerobic digestion, but the accumulation of ammonia released during the degradation of proteins and amino acids is very toxic for methanogens. Free ammonia (NH_3) is more toxic than the ionized form (NH_4^+), since NH_3 can passively transport across the cell membrane and subsequently dissociate, resulting in changes of intracellular pH and disruption of cell homeostasis. Since the relative ratio of NH_3 to total ammonia ($= NH_3 + NH_4^-$) depends on pH, the inhibition is also pH-dependent. At neutral pH, the inhibition of methane production by ammonia is often a significant problem when the concentration of ammonia exceeds the critical level (usually 2,500–3,000 mg-N/L at thermophilic conditions and 4,000–5,000 mg-N/L at mesophilic conditions) (Koster and Lettinga, 1984; Hashimoto, 1986; Koster and Lettinga, 1988).

In this context, adjustment of the ratio of carbon to nitrogen (C/N) of feedstock should be used to prevent ammonia overload. Monitoring and prediction of ammonia level during the digestion process, as well as adjustment of feedstock C/N ratio, should be a part of daily operation of a high-solids anaerobic digester. Kayhanian suggests that a feedstock C/N ratio from 27 to 32, calculated using biodegradable carbon and total nitrogen values, will promote steady digester operation at optimum ammonia levels (Kayhanian, 1999). Adjustment of feedstock C/N ratio is also important since significant ammonia accumulation can increase the pH value in the case of very low C/N ratio of feedstock.

Since deviant accumulation of VFAs such as acetate, propionate, and butyrate, which are the main methanogenic precursors (with pK_a values from 4.7 to 4.9), causes significant decrease in pH, it is also inhibitory for methane production. The undissociated forms of VFAs are more toxic than the dissociated forms, applying a similar inhibition mechanism as in the case of ammonia, although VFAs cause inhibition at lower pH, whereas ammonia causes inhibition at higher pH.

Salt tolerance in methane fermentation is usually low (i.e., less than 3% NaCl in conventional systems (Sonoda and Seiko, 1977) and less than 1% for *Methanosaeta* spp. (Rinzema et al., 1988), which is the most dominant aceticlastic methanogen in a UASB system).

Sulfide is inhibitory at 3–6 mM of total S, of which the fully associated form (H_2S) is the inhibitory agent at levels of 2–3 mM H_2S (Speece, 1996). Hydrogenotrophic, acetogenic, and aceticlastic organisms are all affected, and other groups are inhibited by sulfide.

3. PROCESS DEVELOPMENT FOR ANAEROBIC DIGESTION OF ORGANIC WASTES

A variety of organic wastes are treated by anaerobic digestion. Anaerobic digestion systems are classified according to operation temperature, mixing method, and percentage of total solids (TS) in the waste (Table 7.2). Among them, the percentage of TS is an important factor to classify systems for anaerobic digestion because it affects configuration of reactor design and treatment rate.

3.1 Reactor Design for Anaerobic Digestion

3.1.1 Continuously Stirred Tank Reactor (CSTR)

For anaerobic digestion of low-solid wastes, the Continuously Stirred Tank Reactor (CSTR) is commonly used because it is simple and cost-effective. CSTR is generally used for digestion of wastes with a TS percentage of ca. 2–10% and applied to treat a variety of organic wastes such as kitchen wastes, food wastes, animal manure, and sewage sludge. Contents in the reactor are continuously stirred for complete mixing. Mixing is achieved by mechanical stirring (Figure 7.2A), biogas recirculation (Figure 7.2B), or their combination. Since contents in the reactor are completely mixed, sludge retention time (SRT) is equal to hydraulic retention time (HRT). Wastes are fed into the reactor continuously or intermittently, and the same amount of effluent is withdrawn from the reactor. Sludge retention time in the reactor, which

TABLE 7.2 Schematic Overview of Anaerobic Digestion Systems as Classified by the International Energy Agency (IEA)

TS Content	Mixing	Temperature	Commercial Plants
Low (~10%) Wet system	CSTR	Thermo	Herning, Vegger
		Meso	Bellaria, DSD-CTA
	Plugflow	Thermo/meso	S-Uhde
	UASB	Thermo/meso	BTA, Paques
Medium (10–25%) Semi-dry process	CSTR	Thermo/meso	WASSA
High (25–40%) Dry process	Batch	Thermo	ANM, BioFerm
		Meso	Biocel
	CSTR	Thermo	Snamprogetti
	Intermittent mixing	Meso	Valorga
	Plugflow	Thermo	Dranco, Kompogas
		Meso	Funnell

Adapted from International Energy Agency (1994).
CSTR, Continuously Stirred Tank Reactor.

is calculated by dividing content volume by the removal rate of effluent, is typically in the range of 20 to 30 days, although it can be varied according to the characteristics of the organic wastes and operating conditions such as temperature.

3.1.2 Repeated Batch System

In a repeated batch type system, to accelerate methane production the anaerobic digester and fresh substrates are mixed prior to fill into the reactor at the start of the process. After the digestion, the content is removed, leaving ca. 15–25% of the contents as the seed sludge for methane fermentation, and the fresh substrate is added and mixed. These steps are repeated. Since the operation is very simple and various types of mixing methods (stirring paddle, bulldozer, gas bubbling, etc.) can be used according to the nature of the wastes, a repeated batch system can be used for wet and dry digestion of wastes with a broad range of TS percentage. In the batch system, disintegration/hydrolysis of solid mater is usually the rate-limiting step, although it depends on the amount and figure of inoculum. In the case that the waste solid matter and liquid fraction can be easily separated, the liquid fraction is recirculated from the bottom of the

reactor to the top of the contents for effective contact of organic matter and bacterial biomass (Figure 7.2C). This type of repeated batch system has been developed as, for example, the BioFerm system (www.bioferm-energy.com/) and used for treatment of organic wastes with a high solid content, such as un- or partially crushed vegetable wastes and pruned branches and leaves. Since it is impossible to input fresh wastes once batch operation is started, multiple reactors are used for daily or weekly treatment of the wastes.

3.1.3 Plugflow Reactor System

A plugflow reactor uses slurries such as undiluted manure and crushed garbage with TS content of more than 10% and thus may be a semi-wet or a dry system. The digester design is a long horizontal trough, cylinder (Figure 7.2D) (e.g., KOMPOGAS system developed in Swizerland), or vertical vessel (Figure 7.2E) (e.g., DRANCO (Dry Anaerobic Composting) system). In this type of process, disintegration, hydrolysis, acidogenesis, and methanogenesis are separated over the length of the trough. Early in the process, solid matter is hydrolyzed and acidified mainly, whereas later in the process methanogenesis from fatty acids actively occurs. Since the rate-limiting step is disintegration/hydrolysis of organic solid matter in the wastes, treatment time is long (15–30 days) and organic loading rate is low.

3.2 High-Rate Methane Fermentation

Anaerobic digestion leads to the overall gasification of organic wastewaters and solid wastes into methane and carbon dioxide. Although the digestion processes have been practiced for decades, interest has increased recently in economical recovery of fuel gas from industrial and agricultural surpluses due to changing socioeconomical situations in the world. To achieve rapid and effective anaerobic digestion, some processes are being developed (i.e., the upflow anaerobic sludge blanket (UASB, Figure 7.3A), expanded granular sludge bed (EGSB, Figure 7.3B), and upflow anaerobic filter process (UAFP, Figure 7.3C)). The basic concept for improvement of methane fermentation is how to keep the methanogenic biomass concentration as high as possible in the reactor.

3.2.1 UASB System

The concept of a UASB system is based on the capability of forming a self-granulation (flocculation) of anaerobic microbes associated with methane formation (Figure 7.3A). The UASB system was first constructed by Lettinga and coworkers in The Netherlands (Lettinga et al., 1980). In the system, the wastewater

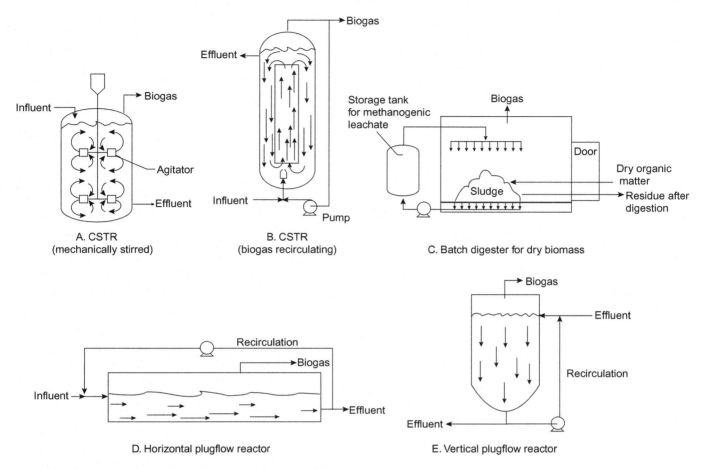

FIGURE 7.2 Schematic drawings of wet and dry anaerobic digesters.

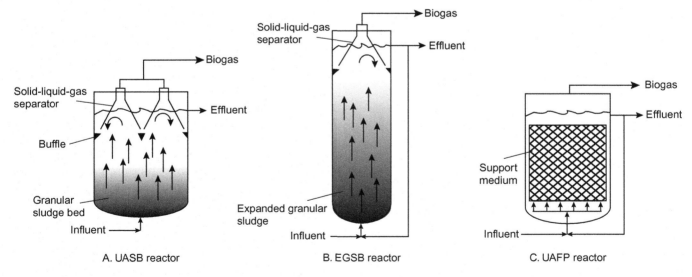

FIGURE 7.3 Schematic drawings of high-rate methane fermentation reactors. *Adapted from Bowker (1983).*

influent enters from the bottom of the reactor. The water flows upwards through the blanket of methanogenic sludge, where the organic matter is anaerobically decomposed and some of the methanogenic bacteria form granules by aggregation in the absence of any support matrix. The produced biogas is separated by a gas-solid-liquid separator and the clarified liquid is discharged over the weir. The dense compact granular sludge with high sedimentation property naturally settles to the bottom, being adverse to the water flow. Upward flow of gas-containing granules through the blanket, combined with return downward flow of degassed granules, creates continuous convection, enabling effective contact of granules and wastewater without the need for any energy-consuming mechanical or hydraulic agitation within the reactor. Although the figures of the granules vary according to environmental conditions in the reactor, the granule size is 0.5–2.5 mm, and the granule concentration is 50–100 kg-VS/m^3 at the bottom and 5–40 kg-VS/m^3 at the upper part of the reactor (Lettinga et al., 1980). A typical UASB reactor can operate at 10–20 kg-COD/m^3/day (Verstraete et al., 1996). UASB reactors are typically suited to TS < 3% of diluted wastewater stream, with particle size >0.75 mm.

3.2.2 EGSB System

The EGSB reactor is a variant of the UASB concept (Kato et al., 1994). Since internal mixing is suboptimal in a UASB reactor, the result is dead space in the reactor and reduced treatment efficiency. To solve the problem, a faster rate of upward flow velocity for the wastewater passing through the sludge bed is applied in the EGSB reactor. The increased superficial velocity is either accomplished by using tall reactors or by incorporating an effluent recycle, or both (Figure 7.3B). The increased superficial velocity (>4 m/h) causes partial expansion (fluidization) of the granular sludge bed. Partial expansion eliminates dead zones, resulting in better wastewater–sludge contact. Accumulation of flocculent excess sludge between the sludge granules is also prevented. Compared with UASB reactors, higher organic loading rates, up to 40 kg-COD/m^3/day (Seghezzo et al., 1998), can be accommodated in EGSB systems. The EGSB reactor is suitable for low-strength soluble wastewaters (<1–2 g-COD/L) or for wastewaters that contain inert or poorly biodegradable suspended particles, which should not be allowed to accumulate in the sludge bed.

3.2.3 UAFP System

A UAFP system contains a methanogenic microbial flora immobilized on the surfaces of support materials such as rocks and plastics (Young and McCarty, 1969). UAFP systems have been applied for domestic sewage and industrial wastewater containing relatively low organic materials. Methanogenic microbial flora not only attach on the surface of the media but also exist in the spaces of the media. Thus, a high-density microbial population can be retained in the reactor. Since support medium is usually fixed in the reactor, to avoid short-cut of the flow through the packed column, a distributer is set at the bottom to make a homogeneous up-flow of wastewater. Selection of the medium is a critical issue in successful development of a UAFP system, because microbial adhesion is greatly influenced by characteristics of the medium.

3.3 Multistage Systems

The anaerobic digestion of organic matter can be subdivided into the four phases: disintegration/hydrolysis, acidogenesis, acetogenesis, and methanogenesis. This means that optimum conditions at each phase will be different. Thus, two- and multistage reactor systems are proposed to optimize these reactions separately. A two-stage system is designed as a first reactor for disintegration/hydrolysis and acidogenesis, and a second for acetogenesis and methanogenesis (see Figure 7.1). Separate optimization of these reactions leads to an improvement of overall reaction rate and biogas yield. Especially, installation of a reactor for disintegration/hydrolysis and acidogenesis prior to high-rate methane fermentation using a UASB or EGSB reactor is effective in obtaining full performance of the high-rate reactor, since TS content acceptable for a UASB reactor is less than 2%.

3.3.1 Hydrogen–Methane Two-Stage Fermentation System (Hy–Met Process)

It is well known that many types of hydrogen-producing bacteria work in methane fermentation, especially at the acidogenesis stages. If high rate and high yield of hydrogen fermentation are achieved, the produced hydrogen gas might be connected directly to a "fuel cell" without reforming. Furthermore, methane can be produced from fatty acids such as acetate, propionate, and butyrate remaining in the liquid after hydrogen fermentation. Thus, a two-stage process is proposed (Figure 7.4) (Nishio and Nakashimada, 2004). In this process, hydrogen and methane are separately recovered from organic wastes.

There have been several reports on hydrogen production after the hydrolysis/acidogenesis stages. Ueno et al. (1995) reported that thermophilic anaerobic microflora enriched from sludge compost produced a significant amount of hydrogen (2.4 mol/mol-hexose). The microbial community in the microflora was investigated through isolation of the microorganisms by both plating and denaturing gradient gel electrophoresis (DGGE) of the PCR-amplified V3 region of 16S rDNA (Ueno et al., 2001). Most of the isolates belong to the cluster of the thermophilic Clostridium/Bacillus subphylum of low G + C gram-positive bacteria. Thermoanaerobacterium thermosaccharolyticum was isolated in the enrichment culture and was detected with strong intensity by PCR-DGGE. Two

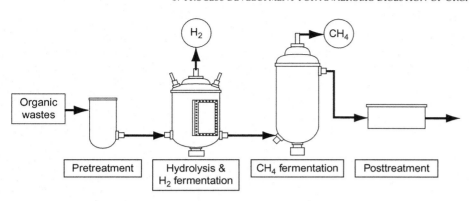

FIGURE 7.4 Schematic drawings of H_2-CH_4 two-stage fermentation process. *Reproduced, with permission, from Nishio and Nakashimada (2004).*

TABLE 7.3 Evaluation of Two-Stage Hydrogen and Methane Fermentation of Brewery Effluent

Process	Gas Production per Influent Volume	Energy Balance	
		Total	
TWO-STAGE PROCESS			
H_2 fermenter at first stage	2.2 L-H_2/L-supplied liquid	24 kJ/L-supplied liquid	
CH_4 fermenter at second stage	2.2 L-CH_4/L-supplied liquid	79 kJ/L-supplied liquid	103 kJ/L-supplied liquid
CH₄ SINGLE PROCESS			
	2.5 L-CH_4/L-supplied liquid	90 kJ/L-supplied liquid	90 kJ/L-supplied liquid

Reproduced, with permission, from Oki and Mitani (2008).

other thermophilic cellulolytic microorganisms, *Clostridium thermocellum* and *Clostridium cellulosi*, were also detected by PCR-DGGE, although they could not be isolated. To elucidate the potential of the Hy–Met process, H_2-CH_4 two-stage fermentation from brewery effluent (Nakashimada and Nishio, 2003; Mitani et al., 2005) and bread manufacturing wastes (Oki and Mitani, 2008) have been explored.

3.3.1.1 APPLICATION TO BREWERY EFFLUENT

The UASB methane fermentation process has been extensively used in a beer brewery. However, since the pressed filtrate from the spent malt remaining after the mashing and lautering process at the brewery contains a high density of suspended matter, such filtrate is difficult to process in a UASB reactor. Therefore, only the filtrate obtained from the pressed filtrate after SS removal has been treated using the UASB reactor. The H_2-CH_4 two-stage process is applied directly to this pressed filtrate. When continuous culture for hydrogen fermentation and subsequent UASB methane fermentation is carried out at 50°C and 37°C, respectively, COD removal is more or less the same compared with single UASB methane fermentation, which is currently in use in this brewery. However, the total amount of energy recovered, as the sum of hydrogen and methane, increases to 103 kJ/L from 90 kJ/L, which corresponds to the amount of the suspended matter solubilized

during the hydrogen fermentation (Table 7.3). These results also demonstrate that waste treatment could be carried out without the removal of suspended matter from the pressed filtrate by connecting the hydrogen fermenter prior to UASB methane fermentation.

3.3.1.2 APPLICATION TO BREAD MANUFACTURING WASTES

In Japan, a total of 100,000 tonne/year solid bread waste is discharged. For this waste, the H_2-CH_4 two-stage fermentation process is applied (Nakashimada and Nishio, 2003). In a batch culture, in which 100 g-wet-wt/L bread waste (43% water content, w/w) is treated at 55°C with 10% (w/v) of a thermophilic sludge, collected from an anaerobic digester of sewage sludge, the waste is fermented to hydrogen and volatile fatty acids. When culture pH is controlled at 7, 240 mM H_2 is produced with a 91% decrease in SS after 24 h. The culture broth contains 150 mM each of acetate and butyrate, and the TOC concentration is approximately 20,000 mg/L. Next, the culture broth of the hydrogen fermentation of the bread waste is used for methane production. The culture broth is diluted to yield a TOC concentration of 2000–5000 mg/L and supplied continuously to a UASB reactor, in which acclimatized methanogenic granules are inoculated. The optimum organic loading rate is 9.5 g-TOC/L/day, yielding 80% TOC removal, a methane production rate of 400 mmol/L/day, and a methane yield of approximately 0.6 on the carbon base. These results indicate that when reactor volumes for hydrogen and methane fermentations are set to a ratio of 1:2.1, SS level will be decreased by 91% at a loading rate of 29 g-wet-wt/L/day, and the hydrogen and methane yields will be 2.4 mol/kg wet weight and 8.6 mol/kg wet weight, respectively.

The amount of energy recovered from the process using bread waste is estimated on the basis of these results. To treat the waste discharged from one factory at 2.67 tonne/day,

- a 26.7 m^3 hydrogen fermentation reactor is used, in which 145 m^3 of hydrogen per day is produced, which corresponds to 214 kWh when the conversion efficiency of the fuel cell system is 50%, and
- a 56 m^3 methane fermentation reactor is used, in which 514 m^3 methane per day is produced, which corresponds to 530 L of oil/day.

3.3.2 Ammonia–Methane Two-Stage System

As already mentioned, the accumulation of ammonia, released during the degradation of proteins and amino acids, is very toxic for methanogens. At neutral pH, the inhibition of methane production by ammonia is often a significant problem when the concentration of total ammonia exceeds the critical level (>3000 mg-N/L). In the case of wet process, ammonia is diluted to the level at which inhibition does not occur. On the other hand, dry process is more sensitive to inhibitors, especially ammonia, because high organic material content causes high accumulation of ammonia (Mata-Alvarez et al., 2000). Therefore, several techniques have been developed to avoid ammonia inhibition. The acclimation of methanogenic bacteria to high ammonia concentration is effective for maintaining stable production of methane from an organic solution containing a high concentration of ammonia (van Velsen, 1979; Parkin et al., 1983; Koster and Lettinga, 1988; Robbins et al., 1989; Sung and Liu, 2003). Co-digestion of organic wastes that contain a high amount of nitrogen compounds, and other wastes such as garbage and paper that have a relatively low nitrogen content, has frequently been applied to decrease the ammonia concentration to less than the threshold level (Kayhanian, 1999; Sosnowski et al., 2003).

The stripping of ammonia prior to the dry anaerobic digestion of organic wastes that contain a high amount of nitrogen compounds is useful when it is difficult to collect a sufficient amount of low-nitrogen waste to maintain the ammonia at less than the threshold level. For this purpose, an ammonia–methane two-stage system is proposed as shown in Figure 7.5.

In this process, high-nitrogen organic wastes, such as dehydrated waste activated sludge (DWAS) (Nakashimada et al., 2008), model garbage (Yabu et al., 2011), and chicken manure (Abouelenien et al., 2009), are anaerobically digested in the first reactor to release ammonia from nitrogen-containing compounds such as protein and uric acid. After ammonia is removed from the digested sludge, dry methane fermentation is carried out with ammonia-removed sludge. Using this process, when the raw or ammonia-stripped DWAS is fermented to methane under a dry condition (water content = 80%) in repeated batch mode, ammonia-stripped DWAS is fermented to methane successfully, whereas methane production is completely inhibited after several batch operations if ammonia stripping is not carried out (Figure 7.6).

The system is also applied to model garbage (water content = 78%) containing ca. 6,000 mg-N/kg-wet-wt of total nitrogen (Yabu et al., 2011). When ammonia-stripped garbage is subjected to semicontinuous thermophilic dry anaerobic digestion over 180 days, the gas yield is in the range of 0.68 to 0.75 m^3-N/kg volatile solid, and the ammonia–nitrogen concentration in the sludge is successfully kept below 3,000 mg-N/kg total wet sludge.

4. FERTILIZATION OF RESIDUES AFTER ANAEROBIC DIGESTION

The residue slurry after anaerobic digestion contains not only solid organic compounds that are not degraded by the anaerobic microorganisms (e.g., lignin), but also ammonia, phosphate, and many sorts of minerals that are components of fertilizer. Examples of the components of residues after wet anaerobic digestion that can be used as fertilizer are shown in Table 7.4. The main component of the residue is total nitrogen, in which ammonium is a major fraction, while nitrite and nitrate are rarely detected. Since potassium and phosphate are also

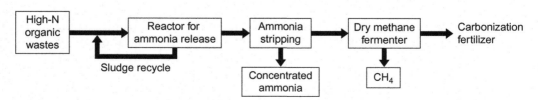

FIGURE 7.5 Schematic drawing of ammonia–methane two-stage process.

contained, although each concentration is lower than in chemical fertilizer, the residue can be used for fertilizer.

The residue slurry can be dewatered to produce a liquid stream and solid matter. The liquid fraction filtered out is recirculated to the digester (KOMPOGAS process) or can be used as mineral fertilizer. The solid fraction is treated aerobically to form compost. In the case of a dry process such as the DRANCO process, the residue can be directly composted. The compost is used as fertilizer as well as for amendment, conditioning, and improvement of organic soil. Another purpose of the composting of the solid fraction is to inactivate pathogenic microorganisms, which might be contained in the residue. However, pathogens can be

destroyed at thermophilic temperatures with a high SRT during the composting. Furthermore, Ottoson et al. (2008) reported that mesophilic digestion in high-ammonia reactors has a significant effect on the reduction of bacterial pathogens. Therefore, the residue is generally composted after the digestion in order to produce a high-quality end product.

The stripping of ammonia from nitrogen-rich wastes is useful for the production of nitrogen fertilizers such as ammonium sulfate and urea, which are produced from ammonia by Haber—Bosch synthesis from its constituent elements. The total annual production of ammonia was 121 million tonnes in 2005 (http://minerals.usgs.gov/minerals/pubs/commodity/-nitrogen/). Although chemical synthetic fertilizer might be currently cheaper than that recovered from organic wastes, ammonia stripping must be treated as something to be disposed of by means of nitrification and denitrification in wastewater treatment plants in the end. This treatment is an extremely energy-consuming process. If not undertaken, it would cause environmental pollution such as eutrophication of rivers and lakes. Moreover, recycling of ammonia from wastes would be helpful in protecting the environment through decrease in demand for chemical synthetic fertilizer.

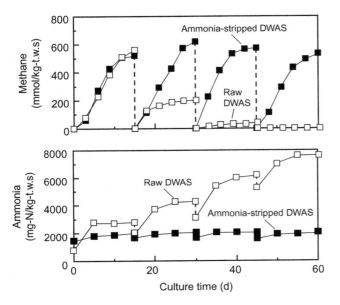

FIGURE 7.6 Methane fermentation of raw and ammonia-stripped dehydrated waste activated sludge (DWAS) with repeated batch culture. t.w.s., total wet sludge. *Adapted from results in Nakashimada et al. (2008).*

5. CONCLUSION

Since depletion of fossil resources such as petroleum and natural gas will become critical in the near future, the demand for saving and recovering energy is increasing. The tendency has been intensified by fear of global warming. In that context, organic substances discharged from the food industry should be used as energy and chemical resources, because they are originally produced by loading enormous energy from

TABLE 7.4 Components of Anaerobic Digestion Residues That Can Be Used for Fertilizer

Parameters	Garbage	Cow Manure + Food Processing Wastes	Cow, Pig, and Chicken Manure + Garbage	Pig Manure
Total nitrogen (mg-N/L)	2,067	3,336	2,520	2,370
Ammonia (mg-N/L)	1,600	1,686	2,194	1,454
Nitrite (mg-N/L)	<0.6	<0.6	—	—
Nitrate (mg-N/L)	<0.6	<0.6	2	—
Phosphate (mg/L)	125	1,028	41	1,196
Potassium (mg/L)	950	3,647	2,292	3,123
Magnesium (mg/L)	19	906	95	—
Calcium (mg/L)	80	2,043	139	—
Sodium (mg/L)	463	794	961	—

Adapted from Aizaki (2008).

farming and chemical fertilizers. Anaerobic digestion is one of the key technologies for recovery of cost-effective energy and chemicals from such organic substances. The number of industrial anaerobic digestion plants has recently increased all over the world. Thus, it might be considered that anaerobic digestion is an easy system for stable operation. However, since anaerobic digestion is a complicated system in which many kinds of microorganisms with different roles participate, operational stability is significantly affected by operation parameters such as pH, temperature, and concentration of toxic compounds. So, to continue successfully to operate anaerobic digestion systems and to improve process performance, it is necessary to understand not only technologies for plant development and operation, but also the microbial physiology and ecology of the process in more detail. We hope that this chapter can help our readers to a deeper understanding of the anaerobic digestion process.

References

Abouelenien, F., Kitamura, Y., Nishio, N., Nakashimada, Y., 2009. Dry anaerobic ammonia-methane production from chicken manure. Appl. Microbiol. Biotechnol. 82, 757–764.

Aizaki, M., 2008. Application to agriculture of the digestion residue discharged from biogas plant. In: Nakashimada, Y., Nishio, N. (Eds.), Recent Developments in Biogas Technology. CMC Publishing. Co., Ltd., Tokyo, Osaka, pp. 186–195. (in Japanese).

Baker, H.A., 1981. Amino acid degradation by anaerobic bacteria. Annu. Rev. Biochem. 50, 23–43.

Batstone, D.J., Keller, J., Angelidaki, I., Kalyuzhnyi, S.V., Pavlostathis, S.G., Rozzi, A., et al., 2002. Anaerobic Digestion Model No. 1 (ADM1). IWA Publishing, London.

Batstone, D.J., Picioreanu, C., van Loosdrecht, M.C.M., 2006. Multidimensional modelling to investigate interspecies hydrogen transfer in anaerobic biofilms. Water Res. 40, 3099–3108.

Bowker, R.P.G., 1983. New wastewater treatment for industrial applications. Environ. Prog. 2, 235–242.

de Mes, T.Z.D., Stams, A.J.M., Reith, J.H., Zeeman, G., 2003. Methane production by anaerobic digestion of wastewater and solid wastes. In: Reith, J.H., Wijffels, R.H., Barten, H. (Eds.), Biomethane & Bio-hydrogen. Dutch Biological Hydrogen Foundation, The Hague, pp. 58–102.

Goberna, M., Insam, H., Franke-Whittle, I.H., 2009. Effect of biowaste sludge maturation on the diversity of thermophilic bacteria and archaea in an anaerobic reactor. Appl. Environ. Microbiol. 75, 2566–2572.

Hashimoto, A.G., 1986. Ammonia inhibition of methanogenesis from cattle wastes. Agric. Wastes 17, 241–261.

International Energy Agency, 1994. Biogas from municipal solid waste: overview of systems and markets for anaerobic digestion of MSW. Report of International Energy Agency Task XI, Copenhagen.

Kato, M.T., Field, J.A., Versteeg, P., Lettinga, G., 1994. Feasibility of expanded granular sludge bed reactors for the anaerobic treatment of low-strength soluble wastewaters. Biotechnol. Bioeng. 44, 469–479.

Kayhanian, M., 1999. Ammonia inhibition in high-solids biogasification: an overview and practical solutions. Environ. Technol. 20, 355–365.

Koster, I.W., Lettinga, G., 1984. The influence of ammonium-nitrogen on the specific activity of pelletized methanogenic sludge. Agric. Wastes 9, 205–216.

Koster, I.W., Lettinga, G., 1988. Anaerobic digestion at extreme ammonia concentrations. Biol. Wastes 25, 51–59.

Lettinga, G., Haandel, A.C., 1993. Anaerobic digestion for energy production and environmental protection. In: Johansson, T.B.E.A. (Ed.), Renewable Energy; Sources for Fuels and Electricity. Island Press, California and London, pp. 817–839.

Lettinga, G., Vanvelsen, A.F.M., Hobma, S.W., Dezeeuw, W., Klapwijk, A., 1980. Use of the upflow sludge blanket (USB) reactor concept for biological wastewater-treatment, especially for anaerobic treatment. Biotechnol. Bioeng. 22, 699–734.

Leven, L., Eriksson, A.R.B., Schnurer, A., 2007. Effect of process temperature on bacterial and archaeal communities in two methanogenic bioreactors treating organic household waste. FEMS Microbiol. Ecol. 59, 683–693.

Mata-Alvarez, J., Mace, S., Llabres, P., 2000. Anaerobic digestion of organic solid wastes. An overview of research achievements and perspectives. Biores. Technol. 74, 3–16.

Mitani, Y., Takamoto, Y., Atsumi, R., Hiraga, T., Nishio, N., 2005. Hydrogen and methane two-stage production directly from brewery effluent by anaerobic fermentation. Master Brewers Assoc. Americas TQ 42, 283–289.

Nagase, M., Matsuo, T., 1982. Interactions between amino acid degrading bacteria and methanogenic bacteria in anaerobic digestion. Biotechnol. Bioeng. 24, 2227–2239.

Nakashimada, Y., Nishio, N., 2003. Hydrogen and methane fermentation of solid wastes from food industry. Food, Food Ingred. J. Japan 208, 703–708 (in Japanese).

Nakashimada, Y., Ohshima, Y., Minami, H., Yabu, H., Namba, Y., Nishio, N., 2008. Ammonia-methane two-stage anaerobic digestion of dehydrated waste-activated sludge. Appl. Microbiol. Biotechnol. 79, 1061–1069.

Nishio, N., Nakashimada, Y., 2004. High rate production of hydrogen/methane from various substrates and wastes. Adv. Biochem. Engin./Biotechnol. 90, 63–87.

Oki, Y., Mitani, Y., 2008. Hydrogen–methane two stage fermentation technology from food industry wastes. In: Nakashimada, Y., Nishio, N. (Eds.), Recent Developments in Biogas Technology. CMC Publishing. Co., Ltd., Tokyo, Osaka, pp. 139–146. (in Japanese).

Parkin, G.F., Speece, R.E., Yang, C.H.J., Kocher, W.M., 1983. Response of methane fermentation systems to industrial toxicants. J. Water. Poll. Con. Fed. 55, 44–53.

Pavlostathis, S.G., Giraldo-Gomez, E., 1991. Kinetics of anaerobic treatment. Wat. Sci. Technol. 24 (8), 35–59.

Ramsay, I.R., Pullammanappallil, P.C., 2001. Protein degradation during anaerobic wastewater treatment: derivation of stoichiometry. Biodegradation 12, 247–257.

Rinzema, A., van Lier, J.B., Lettinga, G., 1988. Sodium inhibition of acetoclastic methanogens in granular sludge from a UASB reactor. In: Lettinga, G., Zehnder, A., Grotenhuis, J., Hulschoff Pol, L. (Eds.), Granular Anaerobic Sludge; Microbiology and Technology. Puduc Wageningen, Wageningen, pp. 216–222.

Robbins, J.E., Gerhardt, S.A., Kappel, T.J., 1989. Effects of total ammonia on anaerobic digestion and an example of digestor performance from cattle manure-protein mixtures. Biol. Wastes 27, 1–14.

Sanders, W.T.M, Zeeman, G., Lettinga, G., 2002. Hydrolysis kinetics of dissolved polymers. Wat. Sci. Technol. 45 (10), 99–104.

Schink, B., 1997. Energetics of syntrophic cooperation in methanogenic degradation. Microbiol. Mol. Biol. Rev. 61, 262–280.

Seghezzo, L., Zeeman, G., van Lier, J.B., Hamelers, H.V.M., Lettinga, G., 1998. A review: the anaerobic treatment of sewage in UASB and EGSB reactors. Biores. Technol. 65, 175–190.

Sonoda, Y., Seiko, Y., 1977. Effects of heavy metal compounds, inorganic salts, hydrocarbon compounds and antibiotics in methane fermentation. Hakkoukougaku Kaishi 55, 22–29.

Sosnowski, P., Wieczorek, A., Ledakowicz, S., 2003. Anaerobic co-digestion of sewage sludge and organic fraction of municipal solid wastes. Adv. Environ. Res. 7, 609–616.

Speece, R.E., 1996. Anaerobic Biotechnology for Industrial Wastewaters. Archae Press, Nashville, TN.

Stams, A.J.M., 1994. Metabolic interactions between anaerobic bacteria in methanogenic environments. Antonie van Leeuwenhoek 66, 271–294.

Sung, S.W., Liu, T., 2003. Ammonia inhibition on thermophilic anaerobic digestion. Chemosphere 53, 43–52.

Ueno, Y., Haruta, S., Ishii, M., Igarashi, Y., 2001. Microbial community in anaerobic hydrogen-producing microflora enriched from sludge compost. Appl. Microbiol. Biotechnol. 57, 555–562.

Ueno, Y., Kawai, T., Sato, S., Otsuka, S., Morimoto, M., 1995. Biological production of hydrogen from cellulose by natural anaerobic microflora. J. Ferment. Bioeng. 79, 395–397.

van Velsen, A.F.M., 1979. Adaptation of methanogenic sludge to high ammonia-nitrogen concentrations. Water Res. 13, 995–999.

Verstraete, W., deBeer, D., Pena, M., Lettinga, G., Lens, P., 1996. Anaerobic bioprocessing of organic wastes. World J. Microbiol. Biotechnol. 12, 221–238.

Winter, J., Schindler, F., Wildenauer, F.X., 1987. Fermentation of alanine and glycine by pure and syntrophic cultures of Clostridium sporogenes. FEMS Microbiol. Ecol. 45, 153–161.

Yabu, H., Sakai, C., Fujiwara, T., Nishio, N., Nakashimada, Y., 2011. Thermophilic two-stage dry anaerobic digestion of model garbage with ammonia stripping. J. Biosci. Bioeng. 111, 312–319.

Young, J.C., McCarty, P.L., 1969. The anaerobic filter for waste treatment. J. Water. Poll. Con. Fed. 41, R160–R173.

PART III

IMPROVED BIOCATALYSTS AND INNOVATIVE BIOREACTORS FOR ENHANCED BIOPROCESSING OF LIQUID FOOD WASTES

CHAPTER

8

Use of Immobilized Biocatalyst for Valorization of Whey Lactose

Maria R. Kosseva

BOX 8.1

NOMENCLATURE

a	specific area of PUF particles (m^{-1})
A	area of the particles (m^2)
Bi	Biot number, $Bi = K_L h/D$
D	diffusivity ($m^2\,s^{-1}$)
G	mass concentration of immobilized biocatalyst ($kg\,m^{-3}$)
G_p	dimensionless inhibition parameter, $G_p = g\,h^2/D_s$
h	size of PUF particles (m)
k	rate constant of reaction ($kmol^{-2}\,m^6\,s^{-1}$)
K_L	mass transfer coefficient ($m\,s^{-1}$)
K_P	inhibition constant ($kmol\,m^3$)
K_S	constant in Monod equation ($kmol\,m^3$)
L	dimensionless number, $L = a\,h$
P	product concentration ($kg\,m^{-3}$)
Rd	ratio of product to substrate diffusivities, $Rd = Dp/Ds$
S	substrate concentration ($kg\,m^{-3}$)
t	time (h)
T	dimensionless time, $T = D_s.\,t/h^2$
V	volume of medium (m^3)
X	biomass concentration ($kg\,m^{-3}$) in the stationary phase of growth

y	spatial coordinate (m)
Y	dimensionless spatial coordinate, $Y = y/h$

Greek Symbols

β	productivity constant (h^{-1})
γ	inhibition constant (h^{-1})
η	degree of substrate to product conversion, $\eta = P/(S_0 + P_0)$
μ	specific growth rate of bacteria (h^{-1})
Φ^2	Thiele modulus, $\Phi^2 = \beta\,h^2\,X_0/D_S\,(S_0 + P_0)$

Subscripts

P	related to the product
S	related to the substrate or related to a quantity for the bulk liquid phase
max	denotes the maximum specific growth rate

Superscript

0	denotes an initial value

---

1. INTRODUCTION

Much of the material generated as waste by the dairy-processing industries contains components that could be utilized as substrates and nutrients in a variety of microbial/enzymatic processes, to give rise to added-value products. Added-value products actually produced from dairy industry wastes, or potentially so, include animal feed, single-cell protein and other fermented edible products, baker's yeast, organic acids, amino acids, enzymes, flavors and pigments, the biopreservative bacteriocin, microbial gums, and polysaccharides. Although these processes have been proven to be technically feasible, they are still far from being economical (Ghaly et al., 2007).

As broadly defined, whey is a milk serum, or the watery medium in which all milk phases are homogeneously dispersed. It is mainly formed during the coagulation of milk casein in cheese making or in casein manufacture. Cheese whey presents an important environmental problem because of the high volumes produced and its high organic matter content. Because of its high biochemical oxygen demand (BOD), whey disposal as waste poses serious pollution problems for the surrounding environment. Moreover, the pollutants in the cheese whey proved to be the most resistant to biodegradation as documented by Janczukowicz et al. (2008). Whey affects the physical and chemical structure of soil, resulting in a decrease of crop yield, and, when released in water bodies, it reduces aquatic life by depleting dissolved oxygen (Ghaly et al., 2007).

The World Market for Whey and Lactose Products 2006–2010—From commodities to value added ingredients (2007) clearly demonstrates how whey continues to show a significant growth rate both in volume terms and particularly in value terms. There has been a significant increase in consumer products launched containing Whey Protein Concentrate (WPC) from 2001–2003 to 2004–2006, of approximately 60%. Increased pharmaceutical and nutraceutical/functional food applications of protein fractions might further enlarge the market. All valuable enhancements by conversion of whey into whey protein concentrates create a larger stream of an aqueous lactose fraction, with the exception of lactoferrin extraction. This means that, besides the high price that can be obtained for whey protein isolate products, one has to take into account the large quantity of lactose permeate that will necessarily be created in parallel. The most beneficial step in increasing value for whey products would be to add more value to the lactose fraction. The production of galacto-oligosaccharides for the displacement of antibiotics in animal feeding is promising to influence the lactose market. It was calculated that the price of edible lactose has a greater influence on economics than the price of whey protein (Peters, 2005).

Whey constitutes about 85–95% of the milk volume and retains 55% of milk nutrients (Guimaraes et al., 2010). The type and composition of whey at dairy plants depends mainly upon the processing technique resulting in casein removal from fluid milk. The main components of both sweet (pH = 6.0–7.0) and acid (pH < 5) wheys after water are lactose (approximately 70–72% of the total solids (TS)), whey proteins (approximately 8–10% TS), and minerals (approximately 12–15% TS). The main differences between the two whey types are in mineral content, acidity, and the composition of the whey protein fraction (Jelen, 2003).

Lactose is the sugar present in the milk of most mammals. It is a disaccharide formed by galactose and glucose and is chemically defined as O-β-D-galactopyranosyl-(1-4)-β-D-glucose, $C_{12}H_{22}O_{11}$. The solubility and sweetness of lactose is low compared with other sugars, namely its hydrolysis products glucose and galactose, as well as fructose and sucrose. Most of the lactose produced is recovered from whey or whey permeate by a process involving crystallization (Gänzle et al., 2008). The major uses for lactose include as a food ingredient, ingredient in infant formula, filler or coating agent for tablets in the pharmaceutical industry, and raw material for the production of added-value lactose derivatives (such as lactulose, lactitol, lactobionic acid, lactosyl urea, galacto-oligosaccharides, and lactosucrose) (Guimaraes et al, 2010). The carbohydrate reservoir of lactose (4.5–5% w/v) in whey in addition to water-soluble vitamins, minerals, and proteins make it a good natural medium for the production of industrially important products, such as ethanol, lactic acid, galacto-oligosaccharides (GOS), citric acid, Single Cell Protein (SCP), biogas, vitamins, and fermented beverages. The use of immobilization technology can further improve the economics of the above processes because of various advantages over the free cell/enzyme systems (Kosseva et al., 2009).

The purpose of this chapter is to emphasize existing trends and recent advances in the application of immobilization technology for processing of cheese whey, with an emphasis on the processes of lactose hydrolysis, GOS production, and ethanol and lactic acid manufacture from whey lactose. Choice of support and method of enzyme and viable cell immobilization, as well as the type of bioreactor in which the immobilized biocatalyst is subsequently used, are presented. Discussion of the engineering aspects of immobilized cell techniques is focused on the mass-balance-based mathematical modeling of the system.

2. METHODS OF IMMOBILIZATION

2.1 Definition of Immobilized Biocatalyst

Immobilized enzymes/cells are defined as "the enzymes/microbial cells physically confined or localized in a certain defined region of space with retention of their catalytic activities, and which can be used repeatedly and continuously" (Chibata, 1978). Immobilization technology has several advantages: it permits higher cell densities in bioreactors, improves stability, makes reutilization and continuous operation possible, and advances operation control, separation processes, and bioreactor design.

2.2 Adsorption, Gel Entrapment, and Covalent-Binding

Adsorption, gel entrapment, and covalent-binding are the popular methods of immobilization used in

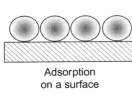
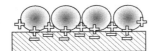
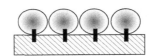

Adsorption
on a surface

Electrostatic binding
on a surface

Covalent binding
on a surface

FIGURE 8.1 **Basic methods of biocatalyst immobilization.** *Adapted from Kourkoutas et al. (2004).*

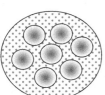
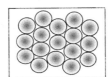
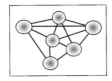

Entrapment within a
porous matrix

Natural flocculation
(aggregation)

Artificial flocculation
(cross-linking)

various bioprocesses. In adsorption, the biocatalysts are held to the surface of the carriers by physical (Van der Waals forces) or electrostatic forces (Figure 8.1). The advantages of adsorption are that it is simple to carry out and has little influence on conformation of the biocatalyst (Marwaha and Kennedy, 2006). However, a major disadvantage of this technique is the relative weakness of the adsorptive binding forces. The physical entrapment method is extremely popular for the immobilization of whole cells. The major advantage of the entrapment technique is the simplicity by which spherical particles can be obtained by dripping a polymer-cell suspension into a medium containing precipitate-forming counter ions or through thermal polymerization. The major limitation of this technique is the possible slow leakage of cells during continuous long-term operation. However, improvements can be made by using suitable cross-linking procedures. Mass transfer limitations are a significant drawback of most immobilized cell techniques (Kosseva et al., 2009).

The immobilization of enzymes/cells on solid supports by covalent coupling and metal coordination usually leads to very stable preparations with extended active life when compared with other methods of immobilization. Generally, it involves two steps: first, activation of the support, and second, coupling of enzyme to the activated support. The wide variety of binding reactions and of supports with functional groups capable of covalent coupling, or susceptible to being activated to give such groups, make it a generally applicable method of immobilization. The coupling of protein molecules to solid supports involves mild reactions between amino acid residues of the protein and several groups of functionalized carriers. The support materials most commonly used do not possess reactive groups but rather hydroxyl, amino, amide, and carboxy groups,

which have to be activated for immobilization of proteins. There are many reaction procedures for coupling enzymes/cells and a support within a covalent bond (Cabral and Kennedy, 1991). Some classes of coupling reaction used for immobilization of proteins are:

- Diazotization
- Amide (peptide) bond formation
- Alkylation and arylation
- Schiff's base formation
- Ugi reaction.

2.3 Microencapsulation

The microencapsulation method has been used since 1993 as an alternative technology to entrapment, over which it enjoys advantages of no leakage and higher cell loading. It has been applied to various bioprocesses such as whole cell enzymes, artificial cells, and biosorbents (Park and Chang, 2000).

2.3.1 Emulsion/Interfacial Polymerization

Interfacial polymerization occurs between monomers dissolved in the respective immiscible phases. Aqueous drops containing the water-soluble monomer are dispersed in the organic phase by stirring. The capsule membrane is then formed by addition of the other organic solvent-soluble monomer to the continuous organic phase (a procedure similar to that shown in Figure 8.2, from Park and Chang, 2000). The replacement of the monomer with nontoxic chitosan enables the encapsulated microbial cell to maintain the initial activity (Groboillot et al., 1993). For example, microaqueous droplets containing *Lactococcus lactis* and chitosan were formed in cyclohexane at a stirring speed of 200 rpm and became the liquid core of the capsule whose membrane was formed by cross-linking

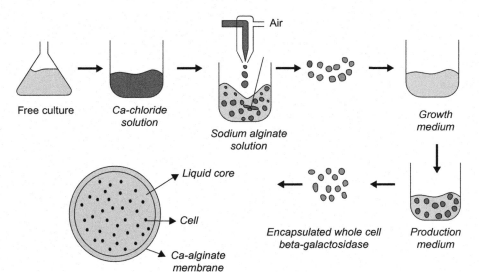

FIGURE 8.2 Schematic of capsule preparation method by liquid droplet forming—one-step method. *Adapted from Park and Chang (2000).*

between chitosan and terephthaloyl chloride, or glutaraldehyde.

2.3.2 *Liquid Droplet Forming—One-Step Method*

Microbial cells are mixed in a solution of calcium chloride and dropped into a swirling sodium alginate solution. A calcium alginate membrane is formed immediately on the surface of the microaqueous droplet by ionic interaction. Biomaterials can be immobilized inside the capsule and do not come into contact with the ionic prepolymer, which is converted to the capsule membrane. Wall thickness, pore size, surface charge, and mechanical strength of the capsules can easily be controlled by alteration of the concentrations of alginate, calcium, and gel-forming polymer (Figure 8.2). Industrial application of biotechnology requires immobilization in high density with low mass transfer resistance and stability, and the encapsulation method may satisfy these requirements better than other methods of immobilization (Park and Chang, 2000).

2.4 Stabilization of Enzymes via Immobilization

Technical applications of enzymes in industry are feasible only if the enzymes are stabilized against temperature, pH extremes, and in the presence of salts, alkalis, and surfactants. Immobilization to solid carriers is perhaps the most widely used strategy to improve operational stability of biocatalysts because of its reproducibility and recyclability, easier product recovery, and flexibility of reactor design. The use of an array of immobilization solutions that permits control of the support—enzyme interaction, to immobilize the enzyme via different orientations or in different conditions, is the key point that significantly increases enzyme stabilities (Mateo et al., 2007).

2.4.1 *Multipoint Covalent Attachment*

Imagine that a molecule of an enzyme is linked to a rigid support by several strong chemical bonds. Obviously the structure of the protein molecule will be much more rigid, and therefore unfolding as well as inactivation will be much more difficult to accomplish than in the case of the free enzyme (Klibanov, 1983). Multipoint covalent attachment of enzymes on highly activated preexisting supports via short spacer arms and involving many residues placed on the enzyme surface promotes rigidification of the structure of the immobilized enzyme (Figure 8.3). So the relative distances among all residues involved in the multipoint immobilization are kept unaltered during any conformational change induced by any distorting agent (heat, organic solvents, extreme pH values), which significantly increases the enzyme stability (Mateo et al., 2007).

2.4.2 *Multi-Subunit Immobilization*

A specific problem in the design of a biocatalyst occurs if the enzyme is a multimeric one. The inactivation of these enzymes starts, in many instances, by the dissociation of the enzyme in its individual subunits. In this case the immobilization—stabilization strategy should consider this multimeric nature of the enzyme, pursuing not only multipoint attachment of the protein, but also multi-subunit attachment. Thus, the multimeric structure of dimeric enzymes may be easily stabilized by immobilization on highly activated supports.

When the enzyme structure is complex, it is quite likely that not all the subunits of the enzyme may be attached to a plane surface. In these cases multi-subunit immobilization on preexisting solid supports may be complemented by chemical cross-linking with polyfunctional polymers of the already immobilized

Pure immobilization of enzyme   Multipoint immobilization

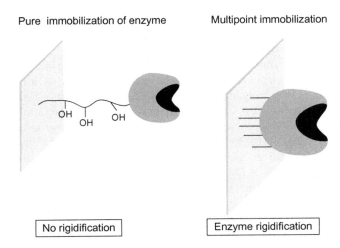

FIGURE 8.3 Effect of immobilization on enzyme stability. *From Mateo et al. (2007).*

Multi-subunit covalent immobilization

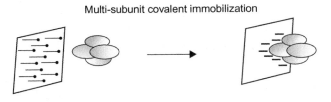

Subunit crosslinking with polyfunctional molecules

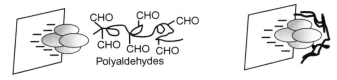

FIGURE 8.4 Stabilization of complex multimeric enzymes by multi-subunit stabilization plus cross-linking with polyfunctional polymers. *From Mateo et al. (2007).*

enzyme (Fernandez-Lafuente et al. 1999) (Figure 8.4). Following this two-step strategy, Pessela et al. (2004) have been able to stabilize several complex multimeric enzymes, including beta-galactosidase from *Thermus* sp.

Controlled immobilization of enzymes may be a very powerful and simple tool to modulate enzyme properties, with results such as those obtained using other techniques. Searching for good results, it is convenient to use an array of immobilization methods to immobilize the enzyme via different areas, with controlled intensity, and so on. That way, it may be possible to find a support and protocol that can produce the desired effect. As examples of the potential of these techniques, the hydrolysis of lactose by two different lactases is illustrated. Lactase from *Kluyveromyces lactis* presents competitive inhibition by galactose (χ was 45 mM) and noncompetitive inhibition by glucose (χ was 750 mM). Using soluble enzyme, these inhibition constants were enough to stop the reaction of

hydrolysis of 5% lactose (similar to the content in milk) after an 80% hydrolysis (Mateo et al. 2004). After immobilizing the enzyme on different supports, an immobilized preparation with a much higher inhibition constant by galactose (χ was over 40 M) was obtained and this preparation permitted the full hydrolysis of 5% lactose.

Lactase from *Thermus* sp. presented even stronger competitive inhibition by galactose (χ 3.1 mM) and noncompetitive inhibition by glucose (χ 50 mM); the reaction of hydrolyses of 5% lactose was apparently stopped at just 60% of conversion. Immobilization on different supports allowed the identification of one preparation where the χ by glucose was increased twofold, while the competitive χ by galactose increased fourfold. These changes in the kinetic constants permitted also in this case the full hydrolysis of 5% lactose (Pessela et al., 2003).

In all cases, enzyme engineering via immobilization techniques is perfectly compatible with other chemical or biological approaches to improve enzyme functions, and the final success depends on the availability of a wide array of immobilization protocols (Mateo et al., 2007).

2.4.3 Chemical Modifications

The stabilization of enzymes by chemical modification can usually be achieved by two major approaches:

- rigidification of the enzyme scaffold by the use of a bifunctional crosslinker; or
- engineering the microenvironment by introduction of new functional groups that favor (a) hydrophobic interaction (by hydrophobization of the enzyme surface), (b) hydrophilization of the enzyme surface (because of mitigation of unfavorable hydrophobic interaction), or (c) formation of new salt bridges or hydrogen bonds (because of the introduction of polar groups) (Mozheav et al, 1990).

Similarly, these two principles have been also increasingly applied to improve enzyme performance, for instance stability, selectivity, and activity (Cao, 2005). In general, the selectivity that can be influenced by the immobilization techniques can be classified into the following categories, according to the source of the effect:

1. Carrier-controlled selectivity
   a. pore-size-controlled selectivity,
   b. diffusion-controlled selectivity;
2. Conformation-controlled selectivity
   a. microenvironment-controlled selectivity,
   b. active-centre-controlled selectivity (Marwaha and Kenendy, 2006).

3. IMMOBILIZED ENZYMES

3.1 Lactose Hydrolysis

The soluble β-D-galactosidase (β-D-galactoside galactohydrolase, E.C. 3.2.1.23), most commonly known as lactase, is normally used for lactose hydrolysis to obtain glucose and galactose. This enzyme also catalyses the formation of galacto-oligosaccharides, which are prebiotic additives to "functional foods".

The hydrolysis of the lactose present in whey converts the whey into very useful sweet syrup, which can be used in the dairy, baking, and soft drink industries. Hydrolyzed lactose solutions possess greater sweetening power than lactose and have applications in the confectionery and ice cream industries, replacing saccharose or starch syrup. Sweetness can be further increased through bioconversion of glucose present in the lactose-hydrolyzed whey to fructose with immobilized glucose isomerase (Sienkiewicz and Riedel, 1990). Another advantage of enzymatic lactose hydrolysis is the simultaneous formation of GOS, used as prebiotic food ingredients. These compounds are indigestible, acting as dietary fiber. They promote the growth of intestinal bifidobacteria, with a subsequent healthy effect in the intestine and the liver. Nowadays, the demand for GOS production, as well as the development of effective and inexpensive GOS manufacture, has increased significantly (Tuohy et al., 2003).

β-D-Galactosidase is one of the most studied enzymes in terms of its immobilization. Currently, GRAS status is valid for *A. niger*, *A. oryzae*, *K. lactis*, and *K. fragilis*, which are the main producers of β-galactosidase used in the food industry. Fungal β-D-galactosidases are more suited for acidic whey hydrolysis than are yeast enzymes; fungal β-galactosidases are more thermostable, but they are more sensitive to product inhibition, mainly by galactose (Boon et al., 2000). β-D-Galactosidases have been immobilized to a variety of matrices by several methods, including entrapment, cross-linking, adsorption, covalent binding, or a combination of these methods (Table 8.1). Since each method has its own advantages and drawbacks, the selection of a suitable immobilization method depends on the enzyme (different properties of various β-D-galactosidases, such as molecule weight, protein chain length, and position of a the active site), matrix, reaction conditions, reactor, etc. (Tanaka and Kawamoto, 1999). Covalent binding of an enzyme to a support is the most interesting method of immobilization from an industrial point of view. Compared to other techniques this method has the following advantages: enzymes do not leak or detach from the carrier and the biocatalyst can easily interact with the substrate, since it is on the surface of the carrier. On the other hand, the major disadvantages are high costs and low activity yield owing to exposure of the biocatalyst to toxic reagents or severe reaction conditions (Tanaka and Kawamoto, 1999).

Several matrices have been used for β-D-galactosidase immobilization. Oxidized materials such as alumina, silica, and silicated alumina have been used for covalent binding of β-galactosidase from *K. marxianus* and applied in lactose hydrolysis processes. In spite of the good stability of the immobilizates shown, the immobilization yields were less than 5% (Di Serio et al., 2003). β-D-Galactosidase from *K. fragilis* was covalently linked to silanized porous glass beads via amino groups, using glutaraldehyde. The coupling efficiency was very high, since more than 90% of the enzyme was active and 87.5% of the protein was bound to the support (Szczodrak, 2000). This example of high lactose conversion (90%) of whey permeate was achieved in a recycle packed-bed reactor (PBR). Whey permeate (lactose 5%) was recycled through the column for 48 h at a flow rate of 0.3 mL/min and a residence time of 20.6 min.

Among different fibrous matrices tested (non-woven polyester fabric, cotton wool, terry cloth, rayon non-woven cloth, etc.), β-D-galactosidase (from *Aspergillus oryzae*) covalently bonded to cotton cloth activated with tosyl chloride showed the highest immobilized enzyme activity, with a coupling efficiency of 85% and an enzyme activity yield of 55%. Thermal stability of the enzyme was increased 25-fold upon immobilization and the immobilized enzyme had a half-life of 50 days at 50°C and more than one year at 40°C (Albayrak and Yang, 2002c). Glutaraldehyde, which interacts with the amino groups through a base reaction, has been the most extensively used cross-linking agent in view of its "generally regarded as safe" GRAS status, low cost, high efficiency, and stability (Kosseva et al., 2009).

In the case of β-D-galactosidase immobilization, cross-linking is often used in combination with other immobilization methods, mainly with adsorption and entrapment. β-D-Galactosidase isolated from *Aspergillus oryzae* was entrapped in lens-shaped polyvinyl alcohol capsules (with activity 25 U/g), giving 32% of its original activity (Grosová et al., 2008). No decrease of activity was observed after 35 repeated batch runs and during 530 h of continuous hydrolysis of lactose (10%, w/v) at 45°C. The immobilized enzyme was stable for 14 months without any change of activity during storage at 4°C and pH 4.5. For example, in β-D-galactosidase immobilization in fibers composed of alginate and gelatin, glutaraldehyde cross-links the enzyme and gelatin, forming an insoluble structure, and also stabilizes the alginate gel, helping to prevent leakage of the enzyme (Tanriseven and Dogan, 2002). The beneficial effect of glutaraldehyde

TABLE 8.1 Immobilized β-D-Galactosidase

Source of Enzyme	Method of Immobilization	Supports Used	Equipment / Bioreactor	Productivity / Conversion (%)	References
Aspergillus oryzae	Cross-linking with glutaraldehyde	Commercial chitosan beads (Chitopearl BCW-3007) from Fujibo	Plug reactor	∼15 % production of GOS	Sheu et al. (1998)
Kluyveromyces fragilis	Covalent coupling using glutaraldehyde	Silanized controlled porous glass (CPG)	Recycling packed-bed reactor, lab scale	86–90% saccharification of whey permeate	Szczodrak (2000)
Kluyveromyces fragilis	Silanized with a 10% g-APTES solution in water	Commercial silica-alumina (KA-3, from Südchemie)	Erlenmeyer flasks	ND	Ladero et al. (2000)
Kluyveromyces lactis	Entrapment	Calcium alginate beads	Laboratory scale bioreactor with re-circulation	99.5% of hydrolysis (30 h)	Becerra et al. (2001)
			Tubular plug flow bioreactor	50% (24 h)	
Aspergillus oryzae	Cross-linking	Poly(vinyl alcohol) chitosan	Fixed-bed reactor	∼95% hydrolysis of lactose	Rejikumar and Devi (2001)
Aspergillus oryzae	Enzyme coupling to cotton fibers activated with *p*-toluenesulfonyl chloride (tosyl chloride)	Cotton cloth	Plug flow reactor	>70% GOS production	Albayrak and Yang (2002c)
Kluyveromyces fragilis	Covalent coupling activated by epichlorohydrin	Cellulose beads	Fluidized bed reactor	>90% conversion in 5 h	Roy and Gupta (2003)
β-D-galacto-sidase	Core-shell microcapsulation	Alginate–chitosan. Alginate core cross-linked with Ca^{2+} and Ba^{2+} ions	ND	–	Taqieddin and Amiji (2004)
Kluyveromyces lactis	Covalent coupling using glutaraldehyde as activating agent	Polysiloxane–polyvinyl alcohol magnetic (mPOS–PVA) composite	Eppendorfs ($V = 1$ mL)	–	Neri et al. (2008)
β-D-galacto-sidase	Membrane retention	–	CSTMR	90% with 1 h residence time	Mehaia et al. (1993)
Aspergillus oryzae	Entrapment	Polyvinyl alcohol hydrogel capsules LentiKat®	Batch runs and 530 h of continuous hydrolysis	–	Grosová et al. (2008)

Adapted from Kosseva (2009).

as a cross-linker was also shown in immobilization of β-D-galactosidase from *A. oryzae* by entrapment in cobalt alginate beads.

Mineral supports can be added to the biopolymer materials, which, apart from improving their mechanical and barrier properties, have proved to be very efficient in enzyme binding. Silica has been widely used as an inert and stable matrix for enzyme immobilization owing to its high specific surface areas and controllable pore diameters, which can be tailored to the dimensions of a specific enzyme. Initial studies focused on the immobilization of enzymes within bio-silica nanoparticles that were formed by reaction of a silicate precursor with a silica-precipitating peptide (R5) (Luckarift, 2004). The R5 peptide is a synthetic derivative of a naturally occurring silaffin protein that is found in the silica skeleton of the marine diatom *Cylindrotheca fusiformis* (Poulsen et al., 2003; Kroger, 2007). The reaction rapidly forms a network of fused silica nanospheres with a diameter of ∼500 nm that entraps the scaffold peptide and any other material that is contained within the reaction mixture. β-D-Galactosidase, for example, was encapsulated directly onto a silicon wafer by entrapment in silica particles (Betancor et al., 2008). The silica was formed by R5 peptide, directly at the amino-activated surface of a silicon wafer. Amino groups are known to be critical in biological silicification reactions and, as such, interact with the silica particles as they form and covalently associate the resulting silica-immobilized enzymes directly at the silicon surface (Coffman et al., 2004). This silica encapsulation method provided a significant increase in enzyme-loading capacity relative to immobilization by alternative methods (Betancor et al., 2008). The morphology of the inorganic matrix can also be varied to create more functionalized and three-dimensional structures (Naik et al., 2003). Controlling specific morphology provides the opportunity to modify the surface area for catalysis and potentially control the mass transfer properties of the matrix for specific substrates and products.

3.2 Production of Galacto-Oligosaccharides

β-Galactosidase catalyses both hydrolysis and trans-galactosylation reactions. Compared with hydrolysis, requirements for GOS synthesis are altogether different. The reaction conditions should be those favoring trans-galactosylation, namely high lactose concentration, elevated temperature, and low water activity in the reaction medium (Boon et al., 2000). Hence, immobilized β-galactosidase should be stable at high temperature and low water content and give high trans-galactosylation activity. Many of the carriers used for immobilization of β-galactosidases applied in GOS production were several types of microbeads, such as chitosan (Sheu et al., 1998; Shin et al., 1998), cellulose (Grosová et al., 2008), and agarose beads (Berger et al., 1995). It was observed that the immobilized enzyme in these particle carriers often resulted in 20−30% reduction in GOS yield due to introduction of mass transfer resistance in the system (Sheu et al., 1998; Shin et al., 1998). An appropriate system of β-galactosidase immobilization leading to increase in its trans-galactosylation activity is still in development. Gaur et al. (2006) compared two different techniques for A. oryzae β-galactosidase immobilization—covalent coupling to chitosan beads and aggregation by cross-linking with glutaraldehyde—in terms of stability and efficiency in GOS synthesis. Using 20% (w/v) of lactose, the chitosan-immobilized β-galactosidase gave maximum trisaccharides yield (17.3% of the total sugar) within 2 hours as compared with 10% obtained with free enzyme and 4.6% obtained with cross-linked aggregates.

The main bioreactor systems for GOS production by immobilized β-galactosidase are packed-bed reactors (PBRs). Using a continuous PBR with β-galactosidase from *Bullera singularis* ATTC 24193 immobilized in chitosan beads, 55% (w/w) of GOS was produced continuously with a productivity of 4.4 g/(L h) for over 15 days. The substrate (100 g/L of lactose solution) was fed at flow rate 80 mL/h into a bioreactor (100 mL of bed volume), in which 970 GU/g (GU-galactosidase

unit defined as the amount of enzyme that liberated 109 mol of *o*-nitrophenol per min at 40°C) enzyme was immobilized (Shin et al., 1998). The PBR (60 mL) filled with 90 g of immobilized recombinant β-galactosidase from *Aspergillus candidus* CGMCC3.2919 (on adsorptive resin D113) was used for continuous production of GOS. The maximum productivity, 87 g/(L h), was reached when 400 g/L lactose was fed at a dilution rate of 0.8/h. The maximum GOS yield was 37% at a dilution rate of 0.5/h (Zheng et al., 2006). Stable continuous production of GOS was also demonstrated in a fibrous bed reactor (bed volume 37 mL) with β-galactosidase from *A. oryzae* immobilized on cotton cloth. Pieces of cotton cloth (total mass 20 g) were tightly rolled into a cylinder and then packed in the reactor. The high porosity, low pressure drop, and high mechanical strength of cotton cloth allowed the enzyme reactor to operate with a concentrated lactose feed (400 g/L) at a flow rate of 37 mL/h. At these conditions the maximum GOS production was 26% (w/w) of total sugars, and corresponding volumetric productivity was 106 g/(L h) (Albayrak and Yang, 2002a). Using polyethyleneimine (PEI) multilayered β-galactosidase immobilization on cotton cloth, several-hundred-fold higher productivity, 6 kg/(L h), was obtained in the same reaction conditions. PEI was used in such a way that the exterior surface of the cotton fibrils in the knitted form was coated with large PEI-enzyme aggregates of high activity. With the enzyme loading of 250 mg/g cotton cloth and 95% immobilization yield, the multilayered polyethyleneimine method is among the most successful ever reported in the literature (Albayrak and Yang, 2002b). A comparison of the above-mentioned continuous GOS production processes using PBRs is given in Table 8.2.

A continuous UF-hollow fiber membrane bioreactor (area 0.5 m²) was also applied for GOS production from whey as a substrate. The enzyme (*K. lactis* β-galactosidase) was kept in the ultrafiltration unit while the sugars (including GOS) permeated the membrane and were collected outside the vessel. The highest production of GOS obtained was 31% for whey UF permeate with initial 20% (w/v) lactose and 0.5% (v/v) initial enzyme concentration (flow rate 2.75 L/h). Corresponding productivity was 13.7 g/(L h) (Foda and Lopez-Leiva, 2000).

4. IMMOBILIZED CELL SYSTEMS

4.1 Ethanol Production

Ethyl alcohol is a versatile chemical because of its unique combination of properties as solvent, germicide, beverage, antifreeze, fuel, depressant, and

TABLE 8.2 Continuous GOS Production in Packed-Bed Bioreactors

Source of Enzyme	Immobilization Method	Reaction Conditions			Max GOS (wt%)	Productivity (g/(L h))	Operation Period (h)	References
		Lactose Conc. (g/L)	T (°C)	pH				
B. singularis	Immobilized in chitosan beads	100	45	4.8	55.0	4.4	360	Shin et al. (1998)
A. Oryzae	Immobilized on cotton cloth	400	40	4.5	26.6	106	400	Albayrak and Yang (2002a)
A. Oryzae	Polyethyleneimine multilayered immobilization on cotton cloth	400	40	4.5	26.0	6000	400	Albayrak and Yang (2002b)
A. Candidus	Immobilized on resin D113	400	40	6.5	37.1	87.1	>480	Zheng et al. (2006)

From Groshova et al. (2008).
GOS, content also includes disaccharides; Max GOS, weight percent of GOS based on the total sugars in the reaction mixture.

especially as an intermediate for other organic synthesis. Among the several microorganisms (e.g., *Kluyveromyces lactis* or *Kluyveromyces marxianus*, *Candida pseudotropicalis*) evaluated for direct production of ethanol, *Kluyveromyces fragilis* is the yeast of choice for most commercial plants. In batch fermentation *K. fragilis* utilizes more than 95% of the lactose of unconcentrated whey with a conversion efficiency of 80–85% of the theoretical value (Mawson, 1994). Lactose-fermenting yeast strains are more sensitive to high ethanol concentrations. Fewer than 10 commercial dairies worldwide ferment lactose in whey permeate directly into ethanol (Mawson, 1994; Murtagh, 1995). Various strains of *K. marxianus* have been used for alcoholic fermentation from deproteinized whey (Grba et al., 2002; Longhi et al., 2004).

Cell immobilization technology, applied to ethanol fermentation, has been shown to offer many advantages for biomass and metabolite production, such as high cell density and very high volumetric productivity, reuse of biocatalysts, high process stability (physical and biological) over long fermentation periods, retention of plasmid-bearing cells, improved resistance to contamination, uncoupling of biomass and metabolite productions, stimulation of production and secretion of secondary metabolites, and physical and chemical protection of the cells (Lacroix et al., 2005). Immobilization of microbial cells for fermentation has been developed to eliminate inhibition caused by high concentration of substrate and product and to enhance productivity and yield of ethanol production.

Kefir granular biomass used in the fermentation of sweet whey proved to be more effective than single-cell biomass of kefir yeast, and ethanol productivity levels reached 2.57 g/(L h), with a yield of 0.45 g/g (Athanasiadis et al., 2002). However, it is preferable to ferment mixtures of whey-molasses by adding molasses in whey after the completion of whey fermentation. The delignified cellulosic-supported biocatalyst, prepared by immobilization of kefir yeast on delignified cellulosic material, was suitable for continuous, modified whey (whey containing 1% raisin extract and molasses) fermentation (Kourkoutas et al., 2002). Ethanol productivities ranged from 3.6 to 8.3 g/(L day), and there is a possibility of using such a process for the production of potable alcohol or a novel, low-alcohol-content drink.

Alcoholic fermentation of cheese whey permeate using a recombinant flocculating *S. cerevisiae*, expressing the LAC4 (coding for β-D-galactosidase) and LAC12 (coding for lactose permease) genes of *K. marxianus* using a continuously operating bioreactor resulted in ethanol productivity near 10 g/(L h) (corresponding to 0.45/h dilution rate), which raises new perspectives for the economic feasibility of whey alcoholic fermentation (Domingues et al., 2001).

Another practical and economical approach for ethanol production from whey is co-immobilization of enzyme and yeast. A study (Lewandowska and Kujawski, 2007) was carried out to improve the effectiveness of a semicontinuous ethanol fermentation of lactose mash combined with a pervaporation module. The fermentation was conducted with a biocatalyst immobilized in calcium alginate and consisted of the yeast *S. cerevisiae* co-immobilized with β-D-galactosidase cross-linked with glutaraldehyde. A 5 liter bioreactor with a water jacket operated in circulation with mash feeding through the biocatalyst layer packed in a perforated cylinder was used in this study. The productivity of ethanol calculated for 24 hours of processing was in the range of 1.58–2.38 g/(L h), and the mean ethanol concentration in the received permeate was 44.0% w/w.

Cheese whey powder (CWP) solution containing 50 g/L total sugar was fermented to ethanol in a continuously operated packed-column bioreactor using olive pits as support particles for adsorption of the cells of *K. marxianus* (Ozmihci and Kargi, 2008). Sugar utilization and ethanol formation were investigated as a function of the hydraulic residence time (HRT) between 17.6 and 64.4 hours. The ethanol yield coefficient increased with increasing HRT and peaked at 0.54 g-E/g-S at an HRT of 50 hours.

Lately, improvements in lactose fermentation by using *Kluyveromyces* yeasts in coculture with lactose-negative microorganism *S. cerevisiae* have been reported by Guo et al. (2010).

During the last 30 years, many investigators have addressed the production of ethanol from lactose. Continuous-operation systems have been widely exploited using different bioreactor designs and, in most cases, using yeast immobilization strategies for obtaining high cell densities (Linko et al., 1981; Hahn-Hägerdal, 1985; Gianetto et al., 1986; Teixeira et al., 1990; and others).

Table 8.3 provides a comparison of immobilized systems proposed for ethanol production.

4.2 Lactic Acid Production

Lactic acid (LA) has found widespread applications as an acidulant, flavor, and preservative in the food, pharmaceutical, leather, and textile industries. It can also be used for the production of basic chemicals and for polymerization to biodegradable poly-lactides. The use of whey as a cheap substrate for LA bioproduction has proved to be more attractive economically than its organic synthesis. The genera that produce lactic acid from lactose are at their core lactic acid bacteria (LAB): *Lactobacillus* (*L.*), *Lactococcus* (*Lc.*), *Leuconostoc* (*Ln.*), *Pediococcus* (*P.*), and *Streptococcus* (*S.*) as well as the more peripheral *Aerococcus*, *Carnobacterium*, *Enterococcus*, *Oenococcus*, *Teragenococcus*, *Vagococcus*, and *Weissella*. They are recognized as GRAS bacteria. *Lactobacillus* is by far the largest genus included in LAB, and more than 125 species and subspecies names are currently recognized (Limsowtin et al., 2003; Axelsson, 2004). Some of the useful applications of LAB are in the biopreservation and aroma development of food. Exopolysaccharide (EPS)-producing lactic cultures have been used to modify the textural and functional properties of fermented dairy products, mainly cheese (Hassan, 2008). Understanding of the structure—function relationship of EPS would allow chemical, enzymatic, or genetic modification of the polysaccharide to obtain tailored characteristics in fermented dairy products (De Vuyst, 2001).

LAB have been immobilized by several methods on different supports (Table 8.4), and the immobilized systems have been investigated for LA production from whey. In search of economical immobilization supports, wood chips, brick particle, and porous glass and egg shells have been tested for immobilization of *L. casei* (Senthuran et al., 1999). Out of these, wood chips showed the highest adsorption capacity (Kazemi and Baniardalan, 2001; Nabi et al., 2004). This immobilized preparation displayed a high rate of production of LA (16 g/L) from whey in a batch system, and an LA production rate of 14.8 g/L with a dilution rate 0.2/h was observed in a continuous packed-bed bioreactor after 5 days.

We examined porous polyurethane foam (PUF) as a support for immobilization of *Lactobacillus casei* cells via adsorption (Kosseva et al., 1995). The rate of reaction obtained with immobilized cells was higher than that obtained with free cells. The microbial biocatalyst was stable and used repeatedly in a batch stirred

TABLE 8.3   Comparison of Immobilized Systems Proposed for Ethanol Production

Microorganism	Method of Immobilization	Matrix for Immobilization	Bioreactor	Conversion and Productivity	References
Kefir yeast	Adsorption	Delignified cellulosic material	Static fermentation	~90%; 5.9% v/v ethanol	Athanasiadis et al. (2002)
Kluyveromyces marxianus	Adsorption	Delignified cellulosic material	0.500 L shaking flask (150 rpm)	9.3 g/L	Kourkoutas et al. (2002)
Recombinant *Saccharomyces cerevisiae*	Aggregation (natural flocculation)	Yeast flocs —flocculent strain	0.600 L bubble column	53%; 7% (v/v) ethanol	Domingues et al. (2001)
Saccharomyces cerevisiae	Co-immobilized yeast cells with enzyme β-D-galactosidase	Ca-alginate beads cross-linked with GA	5 L PBR with circulation	67.7%; 44% w/w ethanol; 1.58—2.48 g/(L h)	Lewandowska and Kujawski (2007)
Kluyveromyces marxianus	Adsorption	Olive pits	Continuous packed-column bioreactor	~95%	Ozmihci and Kargi (2008)

Adapted from (Kosseva et al., 2009).

TABLE 8.4 Comparison of Immobilized Systems Proposed for Lactic Acid Production

Microorganism	Method of Immobilization	Matrix for Immobilization	Bioreactor	Conversion and Productivity	References
L. casei and *L. lactis* cells	Entrapment	Ca-alginate beads	6 L stirred tank reactor batch-fed	85.5%	Roukas and Kotzekidou (1991)
Lactobacillus casei subsp. *casei*	Adsorption	Porous sintered glass beads	Continuously stirred tank and fluidized bed reactors	100%; 93%	Krischke et al. (1991)
Lactobacillus casei	Adsorption	PUF cubes	0.5 L STR with circulation	99%	Kosseva et al. (1995)
Lactobacillus casei	Adsorption	Poraver® beads, Postbauer-Heng, Germany	Recycle packed-bed column ($V_m = 0.7$ L)	90–95%; 93 g/L	Senthuran et al. (1999)
L. helveticus	Adsorption	Activated alumina sphinx adsorbents	1 L packed-bed reactor + STR (PBR/CSTR)	96%	Tango and Ghaly (2002)
Lactobacillus brevis	Adsorption	Delignified cellulosics support	Batch stationary condition	80–100%	Elezi et al. (2003)
Lactobacillus casei	Adsorption	Fruit (apple and quince) pieces	Ferm. under stationary conditions	~40%	Kourkoutas et al. (2005)
Bifidobacterium longum L. helveticus	Entrapment	Sodium alginate beads	Spiral sheet bioreactor	69%; 79%	Li et al. (2005)
Lactobacillus casei	Entrapment	Ca-alginate–chitosan beads	0.1 L shaking flasks	90%	Göksungur et al. (2005)
L. helveticus	Entrapment	κ-carrageenan/locust bean gum	Two-stage batch and continuous	19–22 g/(L h)	Schepers et al. (2006)
Lactobacillus casei	Entrapment	Pectate beads	0.1 L shaking flasks	94.4%	Panesar et al. (2007a)

From (Kosseva et al., 2009).

bioreactor with circulation for a month. A mathematical model, which took into account product inhibition, internal diffusion in the biocatalyst particles, external mass transfer, and biomass growth rate, was developed (as explained in Case Study 1, below). It was found that the bacterial growth was not substrate-inhibited, particularly at higher substrate concentrations (>33 g/L). The maximal specific growth rate (μ_{max}) of the immobilized biomass was calculated as 0.33/h.

L. brevis cells immobilized by adsorption on delignified cellulosic (DC) material resulted in 70% yield of LA, while the remaining lactose in whey was converted to alcohol by-product, leading to 90% lactose exploitation (Elezi et al., 2003). The system showed high operational stability with 10 repeated batch fermentations without any loss in cell activity. *L. casei* cells immobilized by adsorption on fruit (apple and quince) pieces have been used for 15 successive fermentation batches of whey and milk (Kourkoutas et al., 2005). These immobilized biocatalysts proved to be very effective and suitable for food grade lactic acid production.

Fluidized-bed bioreactors with *L. casei* subsp. *casei* immobilized by adsorption on sintered glass beads showed higher productivities of LA than conventional stirred tank reactors in a continuous LA production (Krischke et al., 1991). A fibrous-bed bioreactor has also been tested for continuous LA production from unsupplemented acid whey using adsorption-biofilm immobilized cells of *L. helveticus* (Silva and Yang, 1995). Reactor performance was stable for continuous, long-term operation for both sterile and non-sterile whey feeds for a 6-month period. The chemostat system in salt whey permeate fermentation with *Lactobacillus* cells immobilized in agarose beads displayed a steady LA concentration of 33.4 g/L (Zayed and Winter, 1995). In a packed-bed bioreactor, a high lactic acid production rate of 3.90 g/(L h) was obtained with an initial lactose concentration of 100 g/L and a hydraulic retention time of 18 hours (Tango and Ghaly, 2002).

Among different matrices (calcium alginate, κ-carrageenan, agar, and polyacrylamide gels) tested for co-immobilization of *L. casei* and *Lc. lactis* cells, alginate proved to be a better matrix for the production of lactic acid from deproteinized whey (Roukas and

Kotzekidou, 1991). The polyacrylamide was polymerized *in situ*, and this could cause significant cell death due to toxicity of the monomer and activator present. The immobilization process protected the cells from adverse conditions and improved the yields of lactic acid.

Recently, a two-stage process has been used for continuous fermentation of whey permeate medium with *L. helveticus* immobilized by entrapment in κ-carrageenan/locust bean gum, which resulted in high LA productivity, 19–22 g/(L h), and low residual sugar (Schepers et al., 2006). However, after continuous culture operation with very low or no residual sugar for several days, loss of productivity was observed in the second reactor due to loss of biomass activity and cell death by starvation.

Immobilized *Bifidobacterium longum* in sodium alginate beads and on a spiral-sheet bioreactor have also been evaluated for production of LA from cheese whey (Li et al., 2005). *B. longum* immobilized in sodium alginate beads showed better performance in lactose utilization and LA yield than *L. helveticus*. In producing lactic acid, *L. helveticus* performed better when using the spiral-sheet bioreactor, and *B. longum* showed better performance with gel bead immobilization. Response surface methodology was used to investigate the effects of initial sugar, yeast extract, and calcium carbonate concentrations on the LA production from whey by immobilized *L. casei* NRRL B-441 (Göksungur et al., 2005). Higher LA production and lower cell leakage was observed with *L. casei* cells immobilized in alginate-chitosan beads compared with Ca-alginate beads, and these gel beads were used for five consecutive batch fermentations without any marked activity loss or deformation.

We optimized process conditions for the immobilization of *L. casei* using Ca-pectate gel, and the developed cell system was highly stable during whey fermentation to LA (Panesar et al., 2007a). A high lactose conversion (94.37%) to lactic acid (32.95 g/L) was also achieved. The long-term viability of the pectate-entrapped bacterial cells was tested by reusing the immobilized bacterial biomass, and the entrapped bacterial cells showed no decrease in lactose conversion to lactic acid for up to 16 batches, which demonstrated its high stability and potential for commercial application. Application of pectate gel as a support for LAB entrapment is very promising in LA production due to its good stability at low pH values and also biocompatibility and acceptability in the food industry. In lactic acid production, entrapment is the most common technique used by various scientists. The performance of an immobilized-cell biocatalyst, in which viable cells are entrapped in a gel matrix, depends on the coupled phenomena of cellular reaction kinetics, external and internal mass transfer of solutes, and cell release from the surface of the bead. LAB are sensitive to their microenvironment; they undergo substrate and product inhibitions. Usually substrate and product concentrations as well as pH profiles play a major role in immobilized cell productivity and growth. This competitive diffusion-reaction phenomenon explains the nonuniform cell growth in the colonized gel that results in the formation of high-density cell regions near the bead surface. The high entrapped biomass concentration or cell density at steady state corresponds to a high cell density reactor, where the maximum cell densities reported in batch fermentation of LAB were in the order of 10^{11} CFU/mL (Kosseva et al., 2009). Owing to use of chemicals that may be harmful to the cells, covalent binding is not the preferred method for LAB immobilization.

In most cases, LA productivity was limited by factors such as nonuniform pH control and clogging of the column reactors, destabilization of the alginate gel used for immobilization/entrapment by calcium-chelating lactates, and loss of biocatalyst activity. Mechanical stability of the beads and diffusion limitations of substrate and product within the gel bead matrix appeared to be the main problems encountered by previous researchers, particularly during continuous fermentation. Thus, the success of these processes could rely on the optimization of all fermentation parameters in order to achieve high stability, along with high productivity and low operating and capital costs. A clear understanding of the effects of immobilization on LAB kinetics is required to achieve this objective. Moreover, suitable bioreactor design is also very important for the success of LA production (Panesar et al., 2007b).

5. BIOREACTOR SYSTEMS WITH IMMOBILIZED BIOCATALYST

As a general rule, the choice of a suitable bioreactor system with immobilized biocatalyst would depend on the process requirements and conditions, type of support matrix, kinetics of reactions involved, hydraulic considerations, nature of substrate, enzyme activity, cost, etc. The main factors that significantly influence the choice of bioreactor type have been summarized by Kosseva (2011).

5.1 Packed-Bed Reactors (PBRs)

When the biocatalyst is in the form of spheres, chips, disks, sheets, beads, or pellets, it can be packed readily into a column. In a PBR, there is a steady movement of the substrate across a bed of immobilized biocatalyst. If the axial fluid velocity is perfectly uniform over the column cross section, the bioreactor is operated in a plug-flow regime.

PBRs have the advantage of simplicity of operation, high mass transfer rates, and high reaction rates (for non-substrate-inhibited kinetics). For immobilized cells, oxygen transfer can pose a serious problem in the scale-up of these reactors, unless they are staged or segmented. Another problem in such systems is the periodic fluctuation in the viable cell population, caused by nutrient depletion along the reactor length. Examples of PBRs applied in GOS, ethanol, and lactic acid production are given in Tables 8.2, 8.3, and 8.4.

PBR is a preferred bioreactor configuration for enzymatic lactose hydrolysis (Szczodrak, 2000; Becerra et al., 2001; Rejikumar and Devi, 2001) as well as for trans-galactosylation reactions. Generally, an enzyme is immobilized to a fairly rigid matrix with pellets with a diameter of about 1–3 mm. The PBR permits the use of the biocatalysts at high density, resulting in high volumetric productivities. These reactors are preferred in processes involving product inhibition, which occurs in enzymatic hydrolysis of lactose (especially for fungal β-D-galactosidases). The main disadvantage of PBR is that temperature and pH cannot be regulated easily (Grosová et al., 2008).

5.2 Continuous-Flow Stirred-Tank Reactors (CSTR)

All elements of the ideal STR have an essentially similar concentration, which is the same as the concentration of the outflow, or the CSTR bioreactor is operated in a perfectly mixed regime. This reactor configuration may be more suitable for substrate-inhibited reaction kinetics. The open construction of the CSTR permits ready replacement of the immobilized biocatalyst. It also facilitates easy control of temperature and pH. Thus the system may be suitable where substrate costs are not very important and where stable productivity is essential. CSTRs are often used to study reaction kinetics and measure the kinetic constants of the bioreaction. We applied this type of bioreactor combined with mild recirculation for the production of lactic acid from whey (Kosseva et al., 1995).

5.3 Fluidized-Bed Reactors (FBRs)

The individual catalyst particles are kept in motion by a continuous flow of the substrate in an FBR. The pressure drop of the fluid flow effectively supports the weight of the bed. The reactor thus provides for free movement of the catalyst particles throughout the bed. The energy input into the system or fluidization may be carried out either by liquid or by gas (e.g., air is used for aerobic immobilized cells). FBRs offer the advantage of good solid–fluid mixing and minimal pressure drops.

An FBR with recirculation of the substrate was used to hydrolyze lactose present in milk whey by β-D-galactosidase immobilized on epichlorohydrin-activated cellulose beads. Milk whey (90 mL) was loaded on a fluidized column of cellulose beads (bed volume 5 mL) with the immobilized enzyme, with an effectiveness factor of 0.5 at a flow rate of 2 mL/min. The effluent was recirculated through the column. About 94% conversion of the lactose in milk whey could be achieved by about 30 hours (Roy and Gupta, 2003). The effectiveness factor is defined as the ratio between the activity of immobilized enzyme and the amount of enzyme bound to the matrix.

The use of fluidized beds, as opposed to a packed-bed format, allows the use of feed without pretreatment. The main disadvantage of FBRs is that they are difficult to scale-up, and their use is generally restricted to small-scale high-priced products (Poletto et al., 2005).

5.4 Membrane Reactors (MRs)

A membrane bioreactor has a membrane immersed in an STR (e.g., a dialysis membrane that contains the enzyme (usually in free form) in a chamber where the substrate moves in and the product moves out). The main advantages of this process are the continuous operation of the bioreactor at low pressure and high enzyme concentration. On the other hand, compared with PBRs, the enzyme has less stability because of washout effects. Further disadvantages of MRs are the need for regular replacement of membranes and diffusion limitation through the membrane (Rios et al., 2004). The hollow fiber membrane has also been applied in continuous stirred tank MR. This reactor was the combination of a membrane (polysulfone hollow fiber of 30,000 molecular weight cutoff) and a reaction vessel to provide a continuous reaction and simultaneous separation of the product from the reaction mixture. The enzyme in the system was recycled and reused. The reactor offers effective lactose hydrolysis (>90% conversion) in cheese whey permeates, with a residence time of about 1 hour, at a flow rate of 5 mL/min, and at a substrate/enzyme ratio ≤2.5. The productivity of the continuous stirred tank MR (CSTMR) was 6 times higher than a comparable batch process, even after just 10 hours of operation (Mehaia et al., 1993)

6. KINETIC PERFORMANCE OF THE IMMOBILIZED CELLS (IMCS)

Although many advantages of IMCs are recognized and accepted, immobilization causes changes in the kinetics and properties of whole cells, with decrease of free-cell specific activity. The decrease of catalytic activity may be attributed to several factors, such as toxicity of the materials used in a specific immobilization method and others that have been implicated in the modification of immobilized-cell kinetics:

- A partitioning effect, related to the chemical nature of the support material, may arise from electrostatic or hydrophobic interactions between the matrix and low-molecular-weight species present in the solution, leading to a modified microenvironment.
- Mass transfer diffusional effects, owing to diffusional resistances to the translocation of substrates from the bulk solution to the catalytic sites and the diffusion of products of the reaction back to bulk solution, may operate. These diffusional resistances can be classified as (a) internal or intra-particular mass transfer effects, when the biocatalyst is located within a porous medium; and (b) external or inter-particular mass-tranfer effects between the bulk solution and the outer surface of the biocatalyst particles.

Consequently, a substrate-concentration gradient is established within the pores, resulting in concentration decreasing with increased distance from the surface of the IMC. A corresponding product-concentration gradient is obtained in the opposite direction. As is well established in the field of heterogeneous catalysis, external and internal mass transfer can play an important role in the overall reaction behavior, so transport phenomena have to be considered in each practical catalyst development. A size-dependent or cell-capacity-dependent reaction rate is a clear but indirect indication that transport steps are becoming rate controlling. In this context, it is desirable to obtain direct information on the yield of immobilized activity and on the transport coefficients—the effective diffusivity of the rate-controlling substances.

6.1 Kinetics of Free Cells

To understand the kinetics of immobilized cells, it seems appropriate to start with the kinetics of free cells exposed to similar solution conditions. The rate of a cellular reaction typically is first-order at vanishingly small substrate concentrations and zero-order at higher concentrations. The empirical Monod equation, which is often used to describe cell growth, is identical in form to both the first-order Langmuir–Hinshelwood and Michaelis–Menten equations. Many other rate expressions nevertheless have the same asymptotic behavior and give equally good results when applied to cell systems

(Roels, 1983). Because these rate laws are bounded by the zero- and first-order cases, the solutions of problems combining diffusion with these two simple rate laws are valuable in that they can be applied as lower or upper bounds to the general problem without requiring detailed knowledge of the rate expression. In some cases, more complex rate laws must be used. Very high substrate concentrations can inhibit reactions. Product inhibition is even more likely to be observed in a typical immobilized cell process. The wide variety of reactions catalyzed by a single cell means that side products cannot generally be ignored. The product selectivity in immobilized cell processes has been investigated in a number of studies (Kosseva, 2011).

6.2 Mass Transfer Considerations and the Observed Reaction Rate in an IMC System

The most important question that arises in the use of immobilized cells is that of mass transfer resistance either external to or within the immobilized cell aggregate. The reduced surface area of cell aggregates and the presence of the support constitute additional barriers to mass transfer relative to free cells in a well-mixed solution. This tends to lower the overall rate of reaction, as well as creating an environment within the aggregate different from that of the bulk solution. Living cells are quite sensitive to such changes in their environment. If the cells are growing within the aggregate, then the existence of internal mass transfer limitations may create a spatial distribution of the growth rate and thus redistribute the catalytic activity. The behavior of living cells is therefore a complex process, which only in certain simple cases is amenable to the type of theoretical treatment that has been successful in describing immobilized enzyme systems (Karel et al., 1985).

Mass transfer limitations are most striking in immobilized cell systems when the supply of oxygen to cells and the removal of carbon dioxide are required. The transfer of oxygen from the gas phase to the liquid phase has long been recognized as the major rate-limiting step in the aerobic growth of cells in suspension. The high volumetric reaction rates achieved with immobilized cells shift the relative importance of the resistance terms so that mass transfer within the aggregate may be rate-limiting. High concentrations of waste products also limit the reaction rate. Mass transfer of products away from the aggregate is therefore important. A common product is carbon dioxide, which can inhibit growth either directly or through its effect on the pH in the microenvironment. At sufficiently high concentrations, nucleation and bubble formation can occur. Strictly anaerobic organisms may be protected from trace quantities of oxygen by immobilization, as a result of the lowering of the diffusion rate of oxygen to the cells while the biochemical activity of the cells is maintained.

7. MATHEMATICAL MODELING OF IMMOBILIZED CELL SYSTEM

As a general rule, the immobilization method should be selected with great care, and the rate of its efficiency must be probed by employing conceptual models and mathematical methods. Furthermore, it is possible to improve the immobilization system characteristics by utilizing these models to materialize the intended process outcome (Biria et al., 2008). Mathematical models are also essential tools for the optimization and industrial implementation of fermentation systems.

Case Study 1: Lactic Acid Production from Lactose by Immobilized *Lactobacillus Casei* Cells

Applying the mathematical theory of reaction and diffusion, we have developed a mathematical model of the process of LA production from lactose with *Lactobacillus casei* cells attached to polyurethane foam (PUF) (Kosseva et al., 1995).

According to previous experimental studies, the production of LA follows non-growth-associated, product inhibition kinetics, and can be described by the following equation:

$$\frac{dP}{dt} = \beta X - \gamma P, \text{where } X = X_0 \exp(-\mu t) \qquad (1)$$

We found that specific growth rate of the suspended culture of *Lactobacillus casei* (NBIMCC1013), with lactose as a substrate, can be described by the equation of Monod–Yerousalimskii:

$$\mu = \mu_{\max} \frac{C_S}{K_S + C_S} \frac{K_P}{K_P + C_P} \qquad (2)$$

Both Eqs (1) and (2) are used in our mathematical model, which is based on the following assumptions:

1. The reaction is accomplished with the cells attached to the support and with free cells also existing in the fermentation medium (escaped cells).
2. The transfer of substrate and reaction product in the support particles is controlled by molecular diffusion only, with the corresponding diffusivities (D_s and D_p).

3. The equations expressing the kinetics of biomass growth and LA production are the same as those of a similar free culture, in which the product inhibition effects were taken into account.

We combined mass balance and kinetic Eqs (1) and (2) for a single particle with rectangular shape and finite thickness (*2h*) and obtained the following system of differential equations:

$$\frac{\partial S}{\partial t} = D_S \frac{\partial^2 S}{\partial y^2} - \beta X + \gamma P$$

$$\frac{\partial P}{\partial t} = D_P \frac{\partial^2 P}{\partial y^2} + \beta X - \gamma P \qquad (3)$$

In the considered case the following initial and boundary conditions are valid. The initial condition, Eq (3a), expresses the assumption that at the very beginning of the process the concentrations of substrate/product in the particles and in the solution are equal. The boundary condition Eq (3b) expresses the symmetry of mass transfer around a particular plane in the rectangular support particle or the impermeability of a solid support onto which a film, containing the biocatalyst, is attached. The boundary condition Eq (3c) expresses the quality of mass transfer rates on both sides of the particle surface: the molecular diffusion flux on the inner side and the convective flux on the side of the stirred solution, characterized by a mass transfer coefficient K_L.

$$t = 0, S = S_0, P = P_0, \qquad (3a)$$

$$t > 0, y = 0, \frac{\partial S}{\partial y} = \frac{\partial P}{\partial y} = 0, \qquad (3b)$$

$$y = h, D_S \frac{\partial S}{\partial y} = K_{Ls}(S_\infty - S) \text{ and } D_P \frac{\partial P}{\partial y} = K_{Lp}(P - P_\infty) \quad (3c)$$

Consistent with Fick's law and the mass balance between the support particles and the substrate solution, the time variation of substrate and product concentrations in a perfectly mixed batch bioreactor can be determined by the following equations:

$$\frac{d(VS_\infty)}{dt} = -D_S A \frac{\partial S}{\partial y}\Big|_{y=h} - \beta X_\infty + \gamma P_\infty$$

$$\frac{d(VP_\infty)}{dt} = -D_P A \frac{\partial P}{\partial y}\Big|_{y=h} + \beta X_\infty + \gamma P_\infty \qquad (4)$$

Introducing the dimensionless parameters R_d, Φ^2, Bi, G_p, spatial coordinate Y, time T, and concentration of the product (LA) $\overline{P}$, the system can be solved by implicit finite difference method combined with the

Thomas algorithm, using the estimated values for $R_d = 0.75$, $Bi = 1000$, $\mu_{max} = 0.33/h$, $K_S = 1.45\,kg/m^3$, $K_p = 1.84\,kg/m^3$, and $L = 0.3$. These data are taken from our own experiments and measurements.

$$\overline{P} = \frac{P}{S_0 + P_0} \qquad (5)$$

Equations (4) are very convenient for practical purposes, since the concentrations in the bulk, S_∞ and P_∞, were easily determined experimentally by sample analysis. The time variation of P concentration in the bulk is given by the sum of the P mass flux from the particles and the rate of S conversion due to the free suspended cells leaking from the particles. Solving the mathematical model, we proved that the rate constant for free escaped cells (β_∞) was an order of magnitude lower than that for the immobilized cells, but the net reaction rates were comparable (when agitation was used for mixing). Another kinetic equation, assuming non-substrate-inhibited bacterial growth, was also used for numerical simulation:

$$\mu = \mu_{max}\frac{K_P}{P + K_P} \qquad (6)$$

The statistical analysis via Fisher's test showed that there was no significant difference between the results obtained by the two kinetic equations (2) and (6) for the specific growth rate; or, in our case study, bacterial growth was not substrate inhibited, particularly at higher concentrations (e.g., initial lactose concentration $S \geq 33\,kg/m^3$). The LA production rate achieved with cells immobilized in PUF is larger than that obtained with the free cells at moderate lactose concentrations. This can be explained by the higher biomass concentrations attained in the reactor and the higher rate constant obtained with immobilized cells. The immobilized biocatalyst is stable for repeatable use but sensitive to agitation, so mild circulation was sufficient for good mixing. In summary, both our mathematical models—based on the intrinsic kinetics, the molecular diffusion of S and P in the pores of the support matrix, and on the external mass transfer resistance existing in the bulk—can predict the product formation and describe the overall performance of the process conducted with the IMCs well.

Comparison of the values of the degree of conversion, obtained by experiment and computed by the model, is shown in Figures 8.5 and 8.6. The experimental results obtained with the free cells show much lower degrees of conversion at initial lactose concentrations $33\,kg/m^3$ and $44\,kg/m^3$ than those obtained with the immobilized cells. However, it has to be noted that, at higher initial lactose concentrations ($55\,kg/m^3$), this difference disappeared and both sets of experimental data fit the model prediction (Figure 8.7). In conclusion, our numerical analysis shows that the growth of the

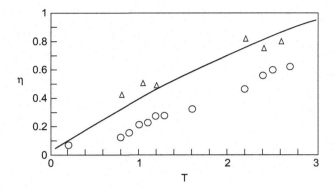

FIGURE 8.5 Experimental results achieved with free (o) and immobilized (△) cells at initial lactose concentration 33 kg/m³. Solid curve represents modeling results.

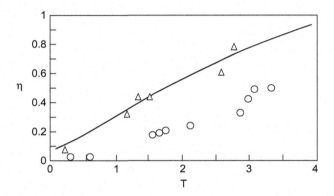

FIGURE 8.6 Experimental results achieved with free (o) and immobilized (△) cells at initial lactose concentration 44 kg/m³. Solid curve represents modeling results.

bacteria *L. casei* is not substrate inhibited in the range of the lactose concentration studied, but the process is considerably inhibited by the product (LA) accumulated in the fermentation medium. Kinetic constants are also estimated from the model parameters.

Our model was also successfully applied to predict the lactic acid fermentation with cells *L. rhamnosus* immobilized in polyacrylamide gel (Petrov et al., 2006).

8. INDUSTRIAL APPLICATIONS

8.1 Lactose Hydrolysis with Immobilized β-Galactosidase

Various immobilization systems for lactose hydrolysis have been investigated, but only a few of them were scaled up with success, and even fewer applied at an industrial or pilot scale. It is mainly because the materials and methods used for enzyme/cell immobilization are either too expensive or difficult to use at an industrial scale (Grosová et al., 2008). Specialist Dairy

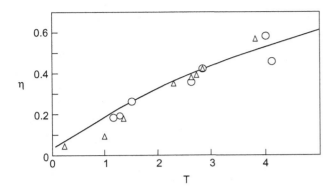

FIGURE 8.7 Experimental results achieved with free (o) and immobilized (Δ) cells at initial lactose concentration 55 kg/m³. Solid curve represents modeling results.

Ingredients, a joint venture between the Milk Marketing Board of England and Wales and Corning Glass Works, have set up an immobilized β-D-galactosidase plant in North Wales for the production of lactose-hydrolyzed whey with enzyme covalently bound to silica beads.

Rohm GmbH (Germany) developed a pilot plant for processing of whole milk using a PBR with fungal β-galactosidase covalently bound to macroporous beads made of plexiglass-like material. In several cases, low-lactose milk, obtained by β-galactosidase covalently bound to silica beads, has been used in order to accelerate the ripening of cheddar cheese (Milk Marketing Board, UK; Union Laitière Normande, France).

Sumitomo Chemical has developed high-purity immobilized β-galactosidase from *A. oryzae* covalently bonded to a macroporous amphoteric ion exchange resin of phenol formaldehyde polymer. This technology was used for producing market milk and hydrolyzed whey (Katchalski, 1993).

Since 1977, Valio laboratory in Finland has used fungal β-D-galactosidase adsorbed to phenol-formaldehyde resin Duolite ES-762 for whey processing. In this process, whey and whey permeate are hydrolyzed continuously by pumping through the column (Schmid et al., 2001).

8.2 Ethanol Production from Whey with Flocculated Yeasts

There are a few established industrial processes to produce ethanol from whey, which have been used in some countries, namely Ireland, New Zealand, USA, and Denmark (Pesta et al., 2007; Guimaraes et al., 2010). Carbery Group (formerly Carbery Milk Products) from Cork, Ireland, commenced operation of an industrial-scale whey-to-ethanol plant in 1978. Up to 2005 the main markets were beverages, pharmaceutical, and industrial (printing inks, etc.). Commencing with the production of potable ethanol, since 2005 the company has also been supplying fuel ethanol to a petrol company in Ireland for E85 and E5 blends (Doyle 2005; Ling 2008). The pioneer Carbery process was later adopted by plants in New Zealand and the USA (Ling, 2008). Currently, the Carbery plant operates with eleven cylindro-conical fermentation vessels, using compressed air for agitation and aeration. The whey permeate is fermented in batch mode for 12 to 20 hours, depending on the initial concentration and yeast activity. The yeast is recovered at the end of fermentation and reused a number of times before it is discarded. Ethanol titers at the end of fermentation are typically in the range 2.5–4.2% (v/v). Following fermentation, a continuous distillation process is used. Carbery produces about 11 thousand tonnes of ethanol per year (Doyle, 2005).

9. CONCLUSIONS

Immobilized biocatalyst has found important applications in the processing of milk by-products and whey lactose. Numerous benefits, intended for food fermentation processes, have been demonstrated, mainly because industrial processes using immobilized biosystems are usually characterized by lower capital/energy costs and better logistics. Immobilized biocatalyst permits higher cell densities in bioreactors, resulting in higher bioreactor volumetric productivity and improved stability. The microbial cells immobilized in a hydrogel matrix can be protected from harsh environmental conditions such as pH, temperature, chemical compounds acting as inhibitors, etc. Biocatalyst immobilization allows for easier separation/purification of the product and reutilization of the biocatalyst. Many advantages have been demonstrated for IMC systems that may be applied to LAB and probiotic bacteria in the dairy and starter industries, which are used to produce high-value products with positive effects on consumers' health. However, no industrial application has yet emerged (Kosseva, 2011).

The most common method used for enzyme immobilization is covalent-bonding and cross-linking with glutaraldehyde, which can also be used as a surface-activating agent. As enzyme immobilization supports, chitin and chitosan-based materials are used in the form of powders, flakes, and gels of various geometrical configurations. Chitin/chitosan-based powders and flakes are available as commercial products from Sigma-Aldrich, among others, and chitosan gel beads (Chitopearls) from Fuji Spinning Co. Ltd. (Japan). As functional materials, chitin and chitosan offer a unique set of characteristics: biocompatibility, biodegradability to harmless products, non-toxicity, physiological inertness, antibacterial properties, heavy metal ions chelation, gel-forming properties and hydrophilicity, and remarkable affinity to proteins. Owing to these characteristics, chitin- and chitosan-based materials are predicted to be widely exploited in the near future, especially in the category of biological systems for environmental applications (Krajewska, 2004). Recent developments in stabilizing enzymes within bioinspired inorganic matrices have substantially extended the range of operational stabilities of the immobilized enzymes. The resulting nano-sized materials offer several intrinsic advantages, such as larger surface areas that allow higher loading capacities (Betancor, 2008).

However, the major problem in enzyme immobilization is not how to immobilize enzymes but how to design the performance of the immobilized enzyme at will. Marwaha and Kennedy (2006) proposed to use a "Lego" or integration approach. In other words, if the enzyme-immobilization method (or approach) can be generally divided into several essential steps (or components), individual optimization of these by use of a rational design might lead to the more rational creation of a robust immobilized enzyme. Analysis of all the methods of immobilization currently available has led to the proposal of a rational general approach to enzyme immobilization based on three stages: selection of enzymes (activity, stability, selectivity), selection of carriers (properties, shape, pore and particle sizes), and selection of conditions (pH, ionic strength, temperature, loading and binding chemistry) as well as posttreatments (pH ripping, consecutive modification, washing).

Among the supports and methods for cell immobilization, the most widely used is entrapment into natural polymers like alginate, agarose, carrageenan, chitosan, and pectate. Natural gelling polysaccharides represent an emerging group due to their advantage of being nontoxic, biocompatible, and cheap, offering a versatile technique for biocatalyst preparation. Some of the whey-based products recovered have a very wide range of uses, such as food ingredients with unique functional and nutritional properties. Lactose in whey is also a renewable source of bioenergy: processing of cheese whey residues leads to the production of ethanol, SCP, and valuable organic acids, namely, acetic, citric, gluconic lactic, propionic, itaconic, and gibberellic (Kosseva et al., 2009). Among essential bio-products also derived from lactose are biogas (methane), amino acids (glutamic, lysine, threonine), vitamins (B_{12} and B_2, or cobalamins and riboflavin), polysaccharides (xanthan gum, dextran, phosphomannan, pullulan, gellan), oils (lipids), enzymes (β-galactosidase, polygalactorunase), and other compounds (fructosediphosphate, 2,3-butanediol, calcium magnesium acetate, ammonium lactate, butanol, glycerol) (Guimaraes et al., 2010). Exploitation of immobilized cells/enzymes and their

advantages to catalyze those fermentation processes, using whey lactose as a valuable feedstock, will lead to further benefits and improve the process economics.

References

Albayrak, N., Yang, S.T., 2002a. Production of galactooligosaccharides from lactose by Aspergillus oryzae β-galactosidase immobilised on cotton cloth. Biotechnol. Bioeng. 77, 8–19.
Albayrak, N., Yang, S.T., 2002b. Immobilization of betagalactosidase on fibrous matrix by polyethyleneimine for production of galacto-oligosaccharides from lactose. Biotechnol. Progr. 18, 240–251.
Albayrak, N., Yang, S.T., 2002c. Immobilization of Aspergillus oryzae β-D-galactosidase on tosylated cotton cloth. Enzym. Microb. Tech. 31, 371–383.
Athanasiadis, I., Boskou, D., Kanellaki, M., Kiosseoglou, V., Koutinas, A.A., 2002. Whey liquid waste of the dairy industry as raw material for potable alcohol production by kefir granules. J. Agric. Food Chem. 50, 7231–7234.
Axelsson, L., 2004. In: Salminen, S., von Wright, A., Ouwehand, A. (Eds.), Lactic Acid Bacteria: Microbiological and Functional Aspects. Marcel Dekker, Inc., New York, p. 1.
Becerra, M., Baroli, B., Fadda, A.M., Méndez, J.B., Siso, M.I.G., 2001. Lactose bioconversion by calcium-alginate immobilization of Kluyveromyces lactis cells. Enzym. Microb. Tech. 29, 506–512.
Berger, J.L., Lee, B.H., Lacroix, C., 1995. Immobilization of β-galactosidase from thermus aquaticus YT-1 for oligosaccharides synthesis. Biotechnol. Techniq. 9, 601–606.
Betancor, L., et al., 2008. Three dimensional immobilization of β-D-galactosidase on a silicon surface. Biotechnol. Bioeng. 99, 261–267.
Biria, D., Zarrabi, A., Khosravi, A., 2008. The application of corrugated parallel bundle model to immobilized cells in porous microcapsule membranes. J. Membr. Sci. 311, 159–164.
Boon, M.A., Janssen, A. E.M., van't Riet, K., 2000. Effect of temperature and enzyme origin on the enzymatic synthesis of oligosaccharides. Enzym. Microb. Tech. 26, 271–281.
Cabral, J.M.S., Kennedy, J.F., 1991. Covalent and coordination immobilization. In: Taylor, R. (Ed.), Protein Immobilization-Fundamentals and Applications, vol. 14. Marcel Dekker, New York, pp. 73–138.
Cao, L., 2005. Immobilised enzymes: science or art? Curr. Opin. Chem. Biol. 9, 219–226.
Chibata, I., 1978. Immobilised Enzymes-Research and Development. John Wiley and Sons, Inc., New York.
Coffman, E.A., et al., 2004. Surface patterning of silica nanostuctures using bio-inspired templates and directed synthesis. Langmuir 20, 8431–8436.
Di Serio, M., Maturo, C., De Alteriis, E., Parascandola, P., Tesser, R., Santacesaria, E., 2003. Lactose hydrolysis by immobilized β-D-galactosidase: the effect of the supports and the kinetics. Catalysis Today 79–80, 333–339.
Domingues, L., Lima, N., Teixeira, J.A., 2001. Alcohol production from cheese whey permeate using genetically modified flocculent yeast cells. Biotechnol. Bioeng. 72, 507–514.
Doyle, A., 2005. Another step in biofuel supply. Irish Farmers J. Interact. (available online <http://www.farmersjournal.ie/2005/1008/farmmanagement/crops/index.shtml>)
Elezi, O., Kourkoutas, Y., Koutinas, A.A., Kanellaki, M., Bezirtzoglou, E., Barnett, Y.A., et al., 2003. Food additive lactic acid production by immobilised cells of Lactobacillus brevis on delignified cellulosic material. J. Agric. Food Chem. 51, 5285–5289.
Fernandez-Lafuente, R., Rodríguez, V., Mateo, C., Penzol, G., Hernandez-Justiz, O., Irazoqui, G., et al., 1999. Stabilization of multimeric enzymes via immobilization and post-immobilization techniques. J. Mol. Catal. B: Enzym. 7, 181–189.

Foda, M.I., Lopez-Leiva, M., 2000. Continuous production of oligo-saccharides from whey using a membrane reactor. Process Biochem. 35, 581–587.

Gänzle, M.G., Haase, G., Jelen, P., 2008. Lactose: crystallization, hydrolysis and value-added derivatives. Int. Dairy J. 18, 685–694.

Gaur, R., Pant, H., Jain, R., Khare, S.K., 2006. Galacto-oligosaccharide synthesis by immobilized Aspergillus oryzae β-galactosidase. Food Chem. 97, 426–430.

Ghaly, A.E., Mahmoud, N.S., Rushton, D.G., Arab, F., 2007. Potential environmental and health impacts of high land application of cheese whey. Am. J. Agricul. Biol. Sci. 2, 106–117.

Gianetto, A., Berruti, F., Glick, B.R., Kempton, A.G., 1986. The production of ethanol from lactose in a tubular reactor by immobilized cells of Kluyveromyces fragilis. Appl. Microbiol. Biotechnol. 24, 277–281.

Göksungur, Y., Gunduz, M., Harsa, S., 2005. Optimization of lactic acid production from whey by L casei NRRL B-441 immobilized in chitosan stabilised Ca-alginate beads. J. Chem. Technol. Biotechnol. 80, 1282–1290.

Grba, S., Tomas, V.S., Stanzer, D., Vahcic, N., Skrlin, A., 2002. Selection of yeast strain Kluyveromyces marxianus for alcohol and biomass production on whey. Chem. Biochem. Eng. Q. 16, 13–16.

Groboillot, A.F., Champagne, C.P., Darling, G.D., Poncelet, D., 1993. Membrane formation by interfacial cross-linking of chitosan for encapsulation of Lactococcus lactis. Biotechnol. Bioeng. 42, 1157–1163.

Grosová, Z., Rosenberg, M., Rebroš, M., 2008. Perspectives and applications of immobilized β-D-Galactosidase in food industry – a review. Czech. J. Food Sci. 26, 1–14.

Guimaraes, P.M.R., Teixeira, J.A., Domingues, L., 2010. Fermentation of lactose to bio-ethanol by yeasts as part of integrated solutions for the valorisation of cheese whey. Biotechnol. Adv. doi:10.1016/j.biotechadv.2010.02.002.

Guo, X., Zhou, J., Xiao, D., 2010. Improved ethanol production by mixed immobilized cells of Kluyveromyces marxianus and Saccharomyces cerevisiae from cheese whey powder solution fermentation. Appl. Biochem. Biotechnol. 160, 532–538.

Hahn-Hägerdal, B., 1985. Comparison between immobilized Kluyveromyces fragilis and Saccharomyces cerevisiae coimmobilized with β-galactosidase, with respect to continuous ethanol production from concentrated whey permeate. Biotechnol. Bioeng. 27, 914–916.

Hassan, A.N., 2008. Possibilities and challenges of exopolysaccharide-producing lactic cultures in dairy foods. J. Dairy Sci. 91, 1282–1298.

Janczukowicz, W., Zielinski, M., Debowski, M., 2008. Biodegradability evaluation of dairy effluents originated in selected sections of dairy production. Bioresour. Technol. 99, 4199–4205.

Jelen, P., 2003. In: Roginski, H., Fuquay, J.W., Fox, P.F. (Eds.), Encyclopedia of Dairy Sciences, vol. 4. Academic press, London, p. 2739.

Karel, S., Libicki, S., Robertson, C., 1985. The immobilization of whole cells: engineering principles. Chem. Eng. Sci. 40, 1321–1354.

Katchalski, E., 1993. Immobilized enzymes—learning from past successes and failures. Trends Biotechnol. 11, 471–478.

Kazemi, A., Baniardalan, P., 2001. Production of lactic acid from whey by immobilised cells. Scient. Iranica 8, 218–222.

Klibanov, AM., 1983. Approaches to enzyme stabilization. Biochem. Soc. Trans. 11, 19–20.

Kosseva, M.R., 2011. Immobilization of microbial cells in food fermentation processes. Food Bioprocess. Technol. 4, 1089–1118.

Kosseva, M.R., Beschkov, V.N., Pilafova, E.I., 1995. Lactic acid production from lactose by immobilised Lactobacillus casei cells. Bulg. Chem. Comm. 28, 690–702.

Kosseva, M.R., Panesar, P.S., Kaur, G., Kennedy, J.F., 2009. Use of immobilized biocatalysts in the processing of cheese whey. Int. J. Biol. Macromol. 45, 437–447.

Kourkoutas, Y., Dimitropoulou, S., Kanellaki, M., Marchant, R., Nigam, P., Banat, I.M., 2002. High-temperature alcoholic fermentation of whey using Kluyveromyces marxianus IMB3 yeast immobilized on delignified cellulosic material. Biores. Technol. 82, 177–181.

Kourkoutas, Y., Bekatorou, A., Banat, I.M., Marchant, R., Koutinas, A.A., 2004. Immobilization technologies and support materials suitable in alcohol beverages production: a review. Food Microbiol. 21, 377–397.

Kourkoutas, Y., Xolias, V., Kallis, M., Bezirtzoglou, E., Kanellaki, M., 2005. Lactobacillus casei cell immobilization on fruit pieces for probiotic additive, fermented milk and lactic acid production. Process Biochem. 40, 411–416.

Krajewska, B., 2004. Application of chitin- and chitosan-based materials for enzyme immobilizations: a review. Enzym. Microb. Tech. 35, 126–139.

Krischke, W., Schroder, M., Trosch, W., 1991. Continuous production of L-lactic acid from whey permeate by immobilised Lactobacillus casei subsp. casei. Appl. Microbiol. Biotechnol. 34, 573–578.

Kroger, N., 2007. Prescribing diatom morphology: toward genetic engineering of biological nanomaterials. Curr. Opin. Chem. Biol. 11, 662–669.

Lacroix, C., Grattepanche, F., Doleyres, Y., Bergmaier, D., 2005. In: Nedovie, V., Willaert, R. (Eds.), Applications of Cell Immobilization Biotechnology. Springer Verlag, New York, p. 295.

Ladero, M., Santos, A., Garcia, J.L., Garcia-Ochoa, F., 2000. Kinetic modeling of lactose hydrolysis with an immobilised β-D-galactosidase from Kluyveromyces fragilis. Enzym. Microb. Tech. 27, 583–629.

Lewandowska, M., Kujawski, M., 2007. Ethanol production from whey in bioreactor with co-immobilised enzyme and yeast cells followed by pervaporative recovery of product. J. Food Eng. 79, 430–437.

Li, Y., Shahbazi, A., Coulibaly, S., 2005. Separation of lactic acid from cheese whey fermentation broth using cross-flow ultrafiltration and nanofiltration membrane system. AIChE Annual Conference, 2207–2216.

Limsowtin, G.K.Y., Broome, M.C., Powell, I.B., 2003. In: Roginski, H., Fuquay, J.W., Fox, P.F. (Eds.), Encyclopedia of Dairy Sciences, vol. 3. Academic Press, London, p. 2739.

Ling, K.C., 2008. Whey to Ethanol: A Biofuel Role for Dairy Cooperatives? USDA Rural Development, Washington DC (Available at: <http://www.rurdev.usda.gov/RBS/pub/RR214.pdf>).

Linko, Y.Y., Jalanka, H., Linko, P., 1981. Ethanol production from whey with immobilized living yeast. Biotechnol. Lett. 3, 263–268.

Longhi, L.G.S., Luvizetto, D.J., Ferreira, L.S., Rech, R., Ayub, M.A.Z., Secchi, A.R., 2004. A growth kinetic model of Kluyveromyces marxianus cultures on cheese whey as substrate. J. Ind. Microbiol. Biotechnol. 31, 35–40.

Luckarift, H.R., et al., 2004. Enzyme immobilization in a biomimetic silica support. Nat. Biotechnol. 22, 211–213.

Marwaha, S.S., Kennedy, J.F., 2006. Introduction: Immobilized Enzymes: Past, Present and Prospects, pp. 1–52.

Mateo, C., Monti, R., Pessela, B.C.C., Fuentes, M., Torres, R., Guisan, J.M., et al., 2004. Immobilization of lactase from Kluyveromyces lactis greatly reduces the inhibition promoted by glucose. Full hydrolysis of lactose in milk. Biotechnol. Prog. 20, 1259–1262.

Mateo, C., Palomo, J.M., Fernandez-Lorente, G., Guisan, J.M., Fernandez-Lafuente, R., 2007. Improvement of enzyme activity, stability and selectivity via immobilization techniques. Enzym. Microb. Tech. 40, 1451–1463.

Mawson, A.J., 1994. Bioconversion for whey utilisation and waste abatement. Biores. Technol. 47, 195–203.

Mehaia, M.A., Alvarez, J., Cheryan, M., 1993. Hydrolysis of whey permeate lactose in a continuous stirred tank membrane reactor. Int. Dairy J. 3, 179–192.

Mozheav, V.V., Melik-Nubarov, N.S., Sksnis, V., Martinek, K., 1990. Strategy for stabilizing enzymes. Part two: increasing enzyme stability by selective chemical modification. Biocatalysis 3, 181–196.

Murtagh, J.E., 1995. In: Lyons, T.P., Kelsall, D.R., Murtagh, J.E. (Eds.), The Alcohol Textbook. Nottingham University Press, Nottingham, UK.

Nabi, B., Gh, R., Baniardalan, P., 2004. Batch and continuous production of lactic acid from whey by immobilised Lactobacillus. J. Environ. Studies 30, 47–60.

Naik, R.R., et al., 2003. Controlled formation of biosilica structures in vitro. Chem. Comm., 238–239.

Neri, D.F.M., Balcão, V.M., Carneiro-da-Cunha, M.G., Carvalho Jr., L.B., Teixeira, J.A., 2008. Immobilization of β-D-galactosidase from Kluyveromyces lactis onto a polysiloxane-polyvinyl alcohol magnetic (mPOS-PVA) composite for lactose hydrolysis. Catal. Comm. 9, 2334–2339.

New Report, 2007. The World Market for Whey and Lactose Products 2006–2010—From Commodities to Value Added Ingredients. Available at: < http://www.3abc.dk/Report%20information%202007.pdf>.

Ozmihci, S., Kargi, F., 2008. Ethanol production from cheese whey powder solution in a packed column bioreactor at different hydraulic residence times. Biochem. Eng. J. 42, 180–185.

Panesar, P.S., Kennedy, J.F., Knill, C.J., Kosseva, M., 2007a. Applicability of pectate entrapped Lactobacillus casei cells for L(+) lactic acid production from whey. Appl. Microbiol. Biotechnol. 74, 35.

Panesar, P.S., Kennedy, J.F., Gandhi, D.N., Bunko, K., 2007b. Bioutilisation of whey for lactic acid production. Food Chem. 105, 1–14.

Park, J.K., Chang, H.N., 2000. Microencapsulation of microbial cells. Biotechnol. Adv. 18, 303–319.

Pessela, B.C.C., Fuentes, M., Vian, A., Garcia, J.L., Carrascosa, A.V., Guisan, J.M., et al., 2003. The immobilization of a thermophilic beta-galactosidase on Sepabeads supports promotes the reduction of product inhibition. Full enzymatic hydrolysis of lactose in dairy products. Enzym. Microb. Tech. 33, 199–205.

Pessela, B.C.C., Mateo, C., Fuentes, M., Vian, A., García, J.L., Carrascosa, A.V., et al., 2004. Stabilization of a multimeric β-galactosidase from Thermus sp. strain t2 by immobilization on novel hetero-functional epoxy supports plus aldehyde-dextran crosslinking. Biotechnol. Prog. 20, 388–392.

Pesta, G., Meyer-Pittroff, R., Russ, W., 2007. Utilization of whey. In: Oreopoulou, V, Russ, W (Eds.), Utilization of By-Products and Treatment of Waste in the Food Industry. Springer.

Peters, R.H., 2005. Economic aspects of cheese making as influenced by whey processing options. Int. Dairy J. 15, 537–545.

Petrov, K.K., Yankov, D.S., Beschkov, V.N., 2006. Lactic acid fermentation by cells of Lactobacillus rhamnosus immobilized in polyacrylamide gel. World J. Microb. Biotechnol. 22, 337–345.

Poletto, M., Parascandola, P., Saracino, I., Cifarelli, G., 2005. Hydrolysis of lactose in a fluidised bed of zeolite pellets supporting adsorbed β-D-galactosidase. Int. J. Chem. React. Eng. 3, A43.

Poulsen, N., et al., 2003. Biosilica formation in diatoms: characterisation of native silaffin-2 and its role in silica morphogenesis. Proc. Natl. Acad. Sci. U.S.A. 100, 12075–12080.

Rejikumar, S., Devi, S., 2001. Hydrolysis of lactose and milk whey using a fixed-bed reactor containing β-D-galactosidase covalently bound onto chitosan and cross-linked poly(vinyl alcohol). Int. J. Food Sci. Technol. 36, 91–98.

Rios, G.M., Beelleville, M.P., Paolucci, D., Sanchez, J., 2004. Progress in enzymatic membrane reactors—a review. J. Membr. Sci. 242, 186–196.

Roels, J.A., 1983. Energetics and Kinetics in Biotechnology. Elsevier, Amsterdam.

Roukas, T., Kotzekidou, P., 1991. Lactic acid production from deproteinized whey by coimmobilised Lactobacillus casei and Lactococcus lactis in packed bed reactor. Enzym. Microb. Tech. 13, 33–38.

Roy, I., Gupta, M.N., 2003. Lactose hydrolysis by Lactozym™ immobilised on cellulose beads in batch and fluidized bed modes. Process Biochem. 39, 325–332.

Schepers, A.W., Thibault, J., Lacroix, C., 2006. Continuous lactic acid production in whey permeate/yeast extract medium with immobilised Lactobacillus helveticus in a two-stage process: model and experiments. Enzym. Microb. Tech. 38, 324–337.

Schmid, A., Dordick, J.S., Hauer, B., Kiener, A., Wubbolts, M., Witholt, B., 2001. Industrial biocatalysis today and tomorrow. Nature 409, 258–268.

Senthuran, A., Senthuran, V., Mattiasson, B., Kaul, R., 1999. Lactic acid production by immobilised Lactobacillus casei in recycle batch reactor: a step towards optimisation. J. Biotechn. 73, 61–70.

Sheu, D.C., Li, S.Y., Duan, K.J., Chen, C.W., 1998. Production of galacto-oligosaccharides by β-D-galactosidase immobilized on glutaraldehyde-treated chitosan beads. Biotechnol. Techniq. 12, 273–276.

Shin, H.J., Park, J.M., Yang, J.W., 1998. Continuous production of galacto-oligosaccharides from lactose by Bullera singularis β-galactosidase immobilized in chitosan beads. Process Biochem. 33, 787–792.

Sienkiewicz, T., Riedel, C-L., 1990. Whey and Whey Utilisation. The Mann, Germany.

Silva, E.M., Yang, S.-T., 1995. Kinetics and stability of a fibrous-bed bioreactor for continuous production of lactic acid from unsupplemented acid whey. J. Biotechnol. 41, 59–70.

Szczodrak, J., 2000. Hydrolysis of lactose in whey permeate by immobilised β-D-galactosidase from Kluyveromyces fragilis. J. Mol. Catal. B: Enzymat. 10, 631–637.

Tanaka, A., Kawamoto, T., 1999. Cell and Enzyme Immobilization. American Society for Microbiology, Washington.

Tango, M.S.A., Ghaly, A.E., 2002. A continuous lactic acid production system using an immobilised packed bed of Lactobacillus helveticus. Appl. Microbiol. Biotechnol. 58, 712–720.

Tanriseven, A., Dogan, S., 2002. A novel method for the immobilization of β-D-galactosidase. Process Biochem. 38, 27–30.

Taqieddin, E., Amiji, M., 2004. Enzyme immobilization in novel alginate-chitosan core-shell microcapsules. Biomaterials 25, 1937–1945.

Teixeira, J.A., Mota, M., Goma, G., 1990. Continuous ethanol production by a flocculating strain of Kluyveromyces marxianus: bioreactor performance. Bioprocess Eng. 5, 123–127.

Tuohy, K.M., Probert, H.M., Smejkal, C.W., Gibson, G.R., 2003. Using probiotics and prebiotics to improve gut health. Drug Discovery Today 8, 692–700.

Zayed, G., Winter, J., 1995. Batch and continuous production of lactic acid from salt whey using free and immobilised cultures of Lactobacilli. Appl. Microbiol. Biotechnol. 44, 362–366.

Zheng, P., Hongfeng, Y., Sun, Z., Ni, Y., Zhang, W., Fan, Y., et al., 2006. Production of galacto-oligosaccharides by immobilized recombinant β-galactosidase from Aspergillus candidus. Biotech. J. 1, 1464–1470.

CHAPTER

9

Hydrogen Generation from Food Industry and Biodiesel Wastes

Naomichi Nishio and Yutaka Nakashimada

1. INTRODUCTION

Hydrogen gas is considered to be a clean energy gas because it is converted to water when it burns. Hydrogen can be produced from organic substrates via either photo-fermentation by photosynthetic microorganisms or dark fermentation by strict and facultative anaerobes (Table 9.1). Dark fermentation is capable of constantly producing hydrogen from organic compounds, whereas photo-fermentation only proceeds in the presence of light. Fermentative hydrogen production from food industry wastes is feasible because the wastes abundantly contain carbohydrates, which constitute the best substrate for such production. On the other hand, short-chain fatty acids, such as lactate or acetate, are also feasible for photobiological hydrogen fermentation. Furthermore, fermentative hydrogen production has some advantages over photosynthetic hydrogen production, such as (1) availability of conventional bioreactors, (2) compactness of process, (3) availability of various sugars, and (4) productivity at high rate. This chapter is focused on the advancement of fermentative hydrogen production with strict or facultative anaerobes, and some challenges experienced, when treating food industry wastes. Photo-fermentation of hydrogen is also briefly reviewed as one of the technologies for hydrogen production from by-products of dark fermentation.

2. BASIC PRINCIPLE OF DARK HYDROGEN FERMENTATION

Microbial hydrogen production is a simple reaction. The final step for hydrogen formation is denoted with the following equation:

$$X_{red} + H^+ \rightarrow X_{ox} + \tfrac{1}{2}H_2 \qquad (1)$$

where X represents electron carriers such as formate, ferredoxin, nicotine adenine dinuclotide (NAD), or chytocrome; red and ox in suffix represent reduced and oxidized form of the electron carrier. This means that an electron in the reduced carrier binds a proton to produce hydrogen and reoxidize the carrier. Since the proton is the one of the most abundant ions in water, any microorganism can always use it as an electron acceptor in the absence of any other electron acceptors such as oxygen, sulfate, or nitrate. Thus, hydrogen production is ubiquitous in the environment. Metabolic pathways for hydrogen production from organic matter are classified into two types: dark hydrogen production and photobiological production.

Dark hydrogen production occurs under anoxic or anaerobic conditions, because there is no oxygen present as an electron acceptor. A wide variety of microorganisms can reduce protons to hydrogen and re-oxidize the reduced form of the electron carrier that results from primary metabolism. For example, when microorganisms grow on glucose, glucose is mainly catabolized to pyruvate via a glycolytic pathway, in which excess electrons are generated and used for reduction of a native electron carrier (usually NAD^+ to $NADH_2$):

$$C_6H_{12}O_6 + 2NAD^+ + 2H^+ \rightarrow 2CH_3COCOOH + 2NADH_2 \qquad (2)$$

Reduced electron carriers must be reoxidized to maintain the metabolic activity. In aerobic conditions, oxygen is used for reoxidation of the electron carrier, with water and ATP formation (respiration). In anaerobic conditions, however, other compounds must

Food Industry Wastes.
DOI: http://dx.doi.org/10.1016/B978-0-12-391921-2.00009-3

157

act as electron acceptors, which include metabolic intermediates such as pyruvate, acetyl-CoA, end-products such as fatty acids (acetate, butyrate), sulfate, nitrate, and protons. The electron acceptor used for re-oxidation of a native electron carrier depends on the nature of the microorganism. For example, lactic acid bacteria use pyruvate as the electron acceptor to reoxidize NADH formed in the glycolytic pathway:

$$2CH_3COCOOH + 2NADH_2 \rightarrow 2CH_3CH(OH)COOH \\ + 2NAD^+ \tag{3}$$

A beer yeast, *Saccharomyces cerevisiae*, uses acetalde-hyde formed via decarboxylation as the electron acceptor:

$$2CH_3COCOOH \rightarrow 2CH_3CHO + 2CO_2 \tag{4}$$

$$2CH_3CHO + 2NADH_2 \rightarrow 2CH_3CH_2OH \tag{5}$$

In this context, hydrogen-producing microorgan-isms use a proton as the electron acceptor. Dark hydrogen-producing microorganisms can be further classified into two metabolic types: strict anaerobes or facultative anaerobes.

TABLE 9.1　Examples of Hydrogen-Producing Microorganisms

Microorganism (genera)	Substrates for H₂ Production
DARK H₂ FERMENTATION	
Strict anaerobes	
Clostridium	Various carbohydrates
Thermotoga	
Caldicellulosiruptor	
Archaea	
Pyrococcus	Starch, pyruvate
Thermococcus	
Facultative anaerobes	
Escherichia	Various carbohydrates
Enterobacter	
PHOTOBIOLOGICAL H₂ FERMENTATION	
Purple non-sulfur bacteria	
Rhodospirillium	Organic acids, amino acids
Rhodopseudomonas	
Rhodobacter	
Cyanobacteria	
Anabaena	Water, starch, glycogen
Synechococcus	
Spirulina	
Green algae	
Chlamydomonas	Water, starch, glycogen
Chlorella	

2.1 Hydrogen Production by Strict Anaerobes

Strict anaerobes such as *Clostridium* and *Ruminococcus* spp. reduce a proton to hydrogen with reduced electron carrier $NADH_2$ or reduced ferredoxin (Figure 9.1). If $NADH_2$ is directly used as the electron donor, 2 mol of hydrogen is produced using excess electrons generated at the glycolytic pathway from 1 mol of glucose:

$$2NADH_2 \rightarrow 2NAD^+ + 2H_2 \tag{6}$$

Furthermore, strict anaerobes generate excess elec-trons during decomposition of pyruvate to acetyl-CoA with pyruvate dehydrogenase and reduce ferredoxin:

$$2CH_3COCOOH + 2HS\text{-}CoA + 2Fd_{ox} \rightarrow 2CH_3 \\ - CO - S - CoA + 2CO_2 + 2Fd_{red} + 2H^+ \tag{7}$$

where Fd represents ferredoxin; CoA, coenzyme A. Hydrogen can be formed from reduced ferredoxin with hydrogenase:

$$2Fd_{red} + 2H^+ \rightarrow 2Fd_{ox} + 2H_2 \tag{8}$$

The formed acetyl-CoA can be further catabolized to acetate:

$$2CH_3 - CO - S - CoA + H_2O \rightarrow CH_3COOH \\ + HS - CoA \tag{9}$$

As the overall reaction, 4 mol of hydrogen are produced from 1 mol glucose:

$$C_6H_{12}O_6 \rightarrow 4H_2 + 2CH_3COOH + 2CO_2 \tag{10}$$

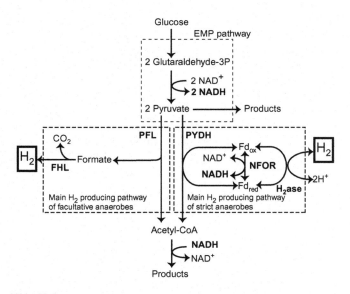

FIGURE 9.1　**Simplified metabolic pathways for H₂ production by strict and facultative anaerobes from various kinds of carbohy-drates.** FHL, formate hydrogen lyase; PFL, pyruvate formate lyase; PYDH, pyruvate dehydrogenase; NFOR, NADH-ferredoxin oxidore-ductase; H₂ase, hydrogenase.

Although this is the theoretical maximum of hydrogen yield from glucose by strict anaerobes, actual hydrogen yield is usually much lower than the theoretical value because of the formation of more reduced by-products besides acetate and the use of reducing power for anabolism (cell growth).

Strict anaerobes isolated and characterized for hydrogen production were mainly *Clostridium* spp. For example, *Clostridium beijerinckii* strain AM21B was isolated from termites (Taguchi et al., 1993). The strain AM21B produced 2.0 mol H_2/mol glucose at uncontrolled pH and 37°C. Since strain AM21B produced amylase, it can produce hydrogen from starch at the yield of 1.7 mol H_2/mol glucose at pH 6.0 (Taguchi et al., 1994). Strain AM21B also produced hydrogen from arabinose, cellobiose, fructose, galactose, lactose, sucrose, and xylose, which are abundantly contained in food wastes, with conversion efficiencies ranging from 15.7 to 19.0 mmol/g substrate for 24 h. Continuous fermentation of xylose and glucose to hydrogen using the other clostridia, *Clostridium* sp. strain No. 2, isolated from termites, was also carried out in a culture containing 0.3% substrate with the pH controlled at 6.0 (Taguchi et al., 1995). The maximal hydrogen production rates of 21.0 and 20.4 mmol/L/h were obtained from xylose and glucose with dilution rates of 0.96/h and 1.16/h, respectively.

C. paraputrificum M-21, isolated from a soil sample collected from Mie University campus, utilized chitin and N-acetyl-D-glucosamine (GlcNAc), a constituent monosaccharide of chitin, to produce a large amount of gas along with acetic acid and propionic acid as major fermentation products (Evvyernie et al., 2000). The bacterium grew rapidly on GlcNAc, with a doubling time of around 30 min, and produced hydrogen yielding 1.9 mol H_2/mol GlcNAc at initial medium pH 6.5 and 45°C. Strain M-21 produced 1.5 mol H_2 from ball-milled chitin equivalent to 1 mol of GlcNAc at pH 6.0 (Evvyernie et al., 2001). In addition, strain M-21 efficiently degraded and fermented ball-milled raw shrimp and lobster shells to produce hydrogen: 11.4 mmol H_2 from 2.6 g of the former and 7.8 mmol H_2 from 1.5 g of the latter.

Characteristics of continuous hydrogen production and fatty acid formation by *Clostridium butyricum* strain SC-El were examined under vacuum and non-vacuum culture systems (Kataoka et al., 1997). The cultures grown without vacuum showed 2.0 to 2.3 mol H_2/mol glucose and 1.4 to 2.0 mol H_2/mol glucose at 0.5% and 1.0% substrate concentration, respectively. The cultures conducted at 0.28 atm under vacuum gave 1.8 to 2.3 mol H_2/mol glucose and 1.3 to 2.2 mol H_2/mol glucose with the same substrate concentration, respectively. In addition, the total hydrogen production rate by a two-stage bioreactor consisting of a 1 L anaerobic fermenter (HRT 10 h) and a 4 L photobioreactor (HRT 36 h), feeding at 2.4 L of 1.0% glucose per day, was estimated at 1.4 to 5.6 mol H_2/mol glucose, which is 12–47% of theoretical values.

Fermentation processes with strict anaerobes grown under thermophilic (45–60°C) and extreme thermophilic (>65°C) conditions can be also investigated for H_2 production, because they possibly result in higher hydrogen yields due to favorable thermodynamics and lower variety of soluble by-products (van Groenestijn et al., 2002; Abreu et al., 2012). Actually, *Thermotoga maritima* grown at 80°C (Schröder et al., 1994) and *Caldicellulosirupter saccharolyticus* grown at 70°C (van Niel et al., 2002) produce H_2 at 4 and 3.3 mol/mol from glucose, respectively. Also, it is attractive that higher hydrolysis rates of cellulosic material have been observed in studies performed under thermophilic conditions, with the concurrent formation of higher amounts of fermentable sugars (Lu et al., 2008).

2.2 Hydrogen Production by Facultative Anaerobes

Facultative anaerobes such as *Escherichia coli* and *Bacillus* spp. possess another type of metabolic pathway for hydrogen production. Although such microbes also catabolize glucose to pyruvate using the glycolytic pathway, resulting in formation of $NADH_2$, they cannot reoxidize $NADH_2$ using a proton as the electron acceptor, perhaps because of lack or poor activity of NADH-dependent hydrogenase. Instead, $NADH_2$ is mainly reoxidized via formation of lactate or 2,3-butanediol from pyruvate, or ethanol from acetyl-CoA according to the nature of the microorganism. In the case of facultative anaerobes, hydrogen is produced by formate hydrogen lyase from formate that is formed in the process of conversion of pyruvate to acetyl-CoA with pyruvate formate lyase (Figure 9.1):

$$2CH_3COCOOH + 2HS-CoA \rightarrow 2CH_3CO-S-CoA + 2HCOOH \quad (11)$$

$$2HCOOH \rightarrow 2H_2 + 2CO_2 \quad (12)$$

Since $NADH_2$ has to be reoxidized by reduction of intermediary metabolites, the theoretical maximum yield of hydrogen from glucose is 2 mol/mol, when 1 mol ethanol and 1 mol acetate are produced from 1 mol glucose:

$$C_6H_{12}O_6 + H_2O \rightarrow 2H_2 + CH_3CH_2OH + CH_3COOH + 2CO_2 \quad (13)$$

There is an interest in hydrogen production by facultative anaerobes, especially *Enterobacter* spp., because they have a high growth rate, rapidly consuming oxygen and thereby restoring anaerobic conditions immediately

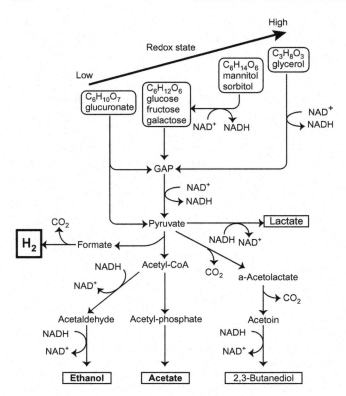

FIGURE 9.2 Simplified anaerobic catabolic pathway of various kinds of carbohydrates with different redox states by *Enterobacter aerogenes*. GAP, Glutaraldehyde 3P.

in the reactor, ability to utilize a wide range of carbon sources, and lack of an inhibitory effect on hydrogen generation under high hydrogen pressure. Strict anaerobes are very sensitive to oxygen and do not survive low oxygen concentration. Furthermore, strict anaerobes are also sensitive to high hydrogen pressure, and their growth and hydrogen production are severely inhibited.

Tanisho et al. first reported hydrogen production by *E. aerogenes* E.82005 isolated from leaves of a plant (Tanisho et al., 1983). It is known that *E. aerogenes* produces mainly ethanol, 2,3-butanediol (BD), lactate, acetate, and formate besides hydrogen (Figure 9.2) (Johansen et al., 1975; Magee and Kosaric, 1987). They investigated the effects of pH and temperature (Tanisho et al., 1987), the usability of various substrates (Tanisho et al., 1989b), and the effect of CO_2 removal (Tanisho et al., 1998) on hydrogen production. Continuous hydrogen production from molasses was also tested and yielded 20 mmol/L/h of hydrogen production rate at dilution rate 0.6/h and pH 6 (Tanisho and Ishiwata, 1994, 1995). However, the hydrogen yield from glucose was ca. 0.7−1.0 mol/mol, which was lower than the normal yield of *Clostridia*. This is due to the formation of by-products such as lactate and BD that are produced via pathways without participation in hydrogen formation.

In *E. aerogenes*, hydrogen is usually produced from formate, which is formed by the way of metabolism to acetyl-CoA from pyruvate via pyruvate formate-lyase. However, Tanisho et al. proposed that *E. aerogenes* possessed a hydrogen-producing pathway via NADH as the electron donor (Tanisho et al., 1989a). Provided that the hydrogen generation is derived from the pyruvate metabolism to ethanol and acetate formations in glucose fermentation, the maximum molar hydrogen yield should not exceed the sum of the ethanol and acetate produced. In the experiments by Rachman et al., however, even if acetate and ethanol production was reduced by mutation, hydrogen production was greatly enhanced (Rachman et al., 1997). This suggested that hydrogen should be generated from excess NADH. Furthermore, when *in vitro* enzymatic evolution of hydrogen from NADH was tested, both NADH and NADPH supported hydrogen formation on extract of *E. aerogenes* after a reaction time of 48 h. NAD(P)H-dependent hydrogenase was localized in the cell membrane (Nakashimada et al., 2002), like a membrane-bound hydrogenase previously reported for *Klebsiella pneumoniae*, a close relative of *E. aerogenes* (Steuber et al., 1999). If all NADH is converted to hydrogen, the theoretical maximum hydrogen yield is 4 mol/mol glucose, similar to that from clostridia. Breeding of such a mutant will make hydrogen production by *E. aerogenes* even more attractive.

To improve the hydrogen yield of facultative anaerobes, we have tested a mutation of *E. aerogenes* HU-101, isolated from anaerobic methanogenic sludge, as a hydrogen producer with high growth rate and hydrogen production rate (Rachman et al., 1997). Since hydrogen is produced from formate, which is formed with formation of acetyl-CoA from pyruvate (Figure 9.2), it would be possible to produce mostly 2 mol H_2/mol glucose by decreasing BD and lactate production that hampers hydrogen production. Such mutants can be screened by using the allyl alcohol (AA) method (Rachman et al., 1997). In this method, since AA is oxidized by ADH and/or BDDH to a toxic aldehyde (acrolein), mutants deficient in these enzymes can survive (Dürre et al., 1986). The hydrogen yield of AA-resistant mutant A-1 increased to 0.84 mol/mol glucose with less production of alcoholic metabolites but more production of acids, compared with the wild strain HU-101 (Table 9.2).

To isolate a non- or low-lactic-acid-producing mutant, a proton suicide method can be used (Rachman et al., 1997). This method is based on lethal effects of bromine and bromite produced from a mixture of NaBr and NaBrO$_3$ during production of acids such as lactate and acetate (Pablo and Mendez, 1990). The mutant HZ-3 screened by this method yielded hydrogen at 0.83 mol/mol glucose with less production of acids, but more alcoholic metabolite production compared with the wild strain HU-101 (Table 9.2). Double mutation by the AA

TABLE 9.2 Yield of End-Products of Fermentation by *E. aerogenes* HU-101 and Mutants (mol/mol glucose)

	Strain				
	HU-101*	A-1*	HZ-3*	AY-2*	VP-1[†]
H_2	0.56	0.84	0.83	1.17	1.76
Ethanol	0.49	0.32	0.54	0.34	0.70
BD	0.37	0	0.46	0.04	0.11
Formate	0.16	0.11	0.14	0.18	ND
Lactate	0.29	0.55	0.16	0.31	0.07
Acetate	0.17	0.48	0.13	0.14	1.02
Pyruvate	0.02	0.08	0.09	0.14	ND
CO_2	1.08	0.88	1.07	1.22	1.07

*Data from Rachman et al. (1997).
[†]Data from Ito et al. (2004).
BD, 2,3-butanediol; ND, not determined.

TABLE 9.3 Yields of Products from Various Carbon Sources by *E. aerogenes*

Substrate	Formula	C_{ave}	Yield (mmol/g-substrate)				
			H_2	Ethanol	Acetate	BD	Lactate
Gluconate	$C_6H_{12}O_7$	3.67	1.44	0.86	2.69	1.59	1.35
Glucose	$C_6H_{12}O_6$	4.00	1.97	2.59	0.81	2.66	1.99
Fructose	$C_6H_{12}O_6$	4.00	2.17	2.73	1.32	2.46	1.53
Galactose	$C_6H_{12}O_6$	4.00	1.90	2.65	1.02	2.61	1.28
Sorbitol	$C_6H_{14}O_6$	4.33	4.96	5.80	0.74	1.27	1.08
Mannitol	$C_6H_{14}O_6$	4.33	5.20	5.30	0.37	1.43	2.15
Glycerol	$C_3H_8O_3$	4.67	6.69	7.05	0.17	0.15	1.95

Adapted from Nakashimada et al. (2002).
BD, 2,3-butanediol; C_{ave} = (Available electrons in 1 mol compound)/(Number of carbon atoms in 1 mol compound).
Culture conditions: substrate, 10 g/L; culture time, 14 h.

and proton suicide methods successfully blocked production of both alcoholic and acidic metabolites and increased hydrogen yield compared with single mutation (Rachman et al., 1997). The hydrogen yield of AY-2 obtained by double mutation reached 1.17 mol mol/glucose, which is 2.1-fold higher than that of HU-101 (Table 9.2).

As the latest test, we screened a mutant strain VP-1 with decreased α-acetolactate synthase activity, which was isolated using the Voges-Proskauer (VP) test (Ito et al., 2004). Since the VP test detects acetoin that is the precursor of BD, a negative mutant for the VP test is expected to be a BD-deficient strain. In pH-uncontrolled batch culture with a complex medium, a screened mutant VP-1 showed the highest hydrogen yield of 1.76 mol/mol glucose with a decrease not only of 2,3-butanediol but also of lactate. Since the theoretical maximum yield of hydrogen is 2 mol/mol glucose, this result demonstrates that hydrogen yield can be improved to near optimum by using conventional breeding methods with mutagenesis and adequate screening strategy. Although improvement of hydrogen yield by means of genetic engineering has been intensively investigated, the availability of a conventional method to enhance hydrogen production is useful, because the use of a genetically engineered microorganism is usually very restricted and it would be difficult to gain approval to use one to treat organic wastes discharged from the food industry.

3. EFFECT OF INTRACELLULAR AND EXTRACELLULAR REDOX STATES ON HYDROGEN PRODUCTION

One of the important factors in determining the diversity of fermentation end products is the intracellular redox state. Intracellular electron carriers such as NADH

and NADPH especially play an important role in this state. Their actions in numerous anabolic and catabolic reactions range widely throughout the biological system (Foster and Moat, 1980). In the case of hydrogen production in clostridia under acidogenic conditions, NADH is reoxidized with H_2 formation via NADH-ferredoxin oxidoreductase and hydrogenase (see Figure 9.1) (Girbal et al., 1995). The activity of the former enzyme is regulated by the ratios of NADH/NAD and acetyl-CoA/CoA, resulting in a disposal of excess reducing power as hydrogen (Adams et al., 1980).

Regarding the facultative anaerobes such as the genera *Enterobacter*, *Klebsiella*, and *Bacillus*, NADH is usually used as the reductant for the production of 2,3-butanediol, ethanol, and lactate from pyruvate, but it is not used for hydrogen production. For these bacteria, hydrogen is produced from formate generated by splitting pyruvate. The intracellular redox state is affected by various environmental factors such as substrates (Tanisho et al., 1989b), culture pH (Jones and Woods, 1986), and the nature of the electron acceptor (de Graef et al., 1999). To optimize microbial hydrogen production, it is important to know the cellular responses to such conditions.

Although carbohydrates (sugars) are the preferred substrate for hydrogen production, it is difficult to compare hydrogen yield between different carbohydrates. As one example, end-product yields from various carbon sources by *E. aerogenes* are shown in Table 9.3. Considering the hydrogen yield based on weight, the yields on fructose and galactose are similar to that on glucose, while the H_2 yields on mannitol and sorbitol were 2.5-fold higher than that on glucose. The H_2 yield on glycerol was even 3.3-fold higher than that on glucose. On the other hand, H_2 yield on gluconate was much lower than that on glucose. Ethanol yield gives similar trends to those of H_2 yield, while that of acetate

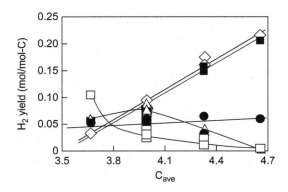

FIGURE 9.3 Dependence of redox state of carbohydrates on yields of metabolites per mol carbon in substrate for *E. aerogenes* **HU-101.** Symbols: yield of (■), H$_2$; (●), lactate; (□), acetate; (△) 2,3-butanediol; (◇), ethanol. *Adapted from Nakashimada et al. (2002).*

yield contradicts the trends. Lactate yield is not affected by carbon sources, but butanediol yield decreases drastically on glycerol. This result can be explained by introducing a concept of redox state of carbohydrate. To clarify the relationship between H$_2$ production and carbon source, the available electrons per one carbon atom for each carbon source (C$_{ave}$) can be defined as follows:

$$C_{ave} = \frac{\text{Available electrons in one mole of compound}}{\text{Number of carbon atoms in one mole of compound}}$$

where available electrons in one mole of compound are calculated as C = 4, O = −2, and H = 1. Using this equation, for example, C$_{ave}$ of glucose = 24/6 = 4 and C$_{ave}$ of mannitol = 26/6 = 4.33. Since C$_{ave}$ implies the redox state of one carbon in each compound, it can be used as the parameter to compare each redox state of compounds even if conformation and numbers of carbon of carbohydrate were different. In Figure 9.3, the relationship between C$_{ave}$ and product yields per mol carbon is illustrated. It is denoted that C$_{ave}$ of glucose, fructose and galactose, or mannitol and sorbitol is the same value because the chemical formula is the same. The figure clearly demonstrates that H$_2$ and ethanol yields increased linearly with C$_{ave}$, while the acetate yield decreased with C$_{ave}$. This indicates that the redox state of carbohydrate mainly affects hydrogen productivity by *E. aerogenes.*

4. BIOREACTOR SYSTEM FOR HIGH-RATE HYDROGEN PRODUCTION

For hydrogen production from organic wastes, although the Continuously Stirred Tank Reactor (CSTR) can be used, the hydrogen production rate is as important as its overall yield for industrial use. This means that, for the practical operation of a bioreactor, a high cell density

is required. To achieve this end, a variety of reactor systems with immobilized cells using several microbial supports have been investigated. Several examples of such attempts are listed in Table 9.4. More detailed reviews about bioreactor design for dark H$_2$ production are reported by Jung et al. (2011) and Ren et al. (2011). In an early study, the use of polyurethane foam for *E. aerogenes* E 82005 gave 13 mmol/L/h of hydrogen production rate using molasses as a substrate (Tanisho and Ishiwata, 1995). Agar gel or porous glass beads can be used for cell immobilization. When these supports were applied to aciduric *E. aerogenes* HO-39, the hydrogen production rate was 850 mL-H$_2$/L/h from glucose (Yokoi et al., 1997b).

It is also effective to use self-immobilized cells like "granules" in high-rate methane fermentation. Flocculation of *E. aerogenes* was reported by Tanisho and coworkers (Tanisho and Ishiwata, 1994). Additionally, Yokoi and coworkers reported that *Enterobacter* sp. BY-29 produced a new biopolymer flocculant consisting of polysaccharides, which caused flocculation of a suspension of kaolin, active carbon, cellulose, and yeast in the presence of Al^{3+}, Fe^{3+}, or Fe^{2+} (Yokoi et al., 1997c). We also observed strong flocculation of *E. aerogenes* HU-101 in a cylindrical column reactor during continuous culture (Rachman et al., 1998). An upflow anaerobic packed-bed (UAPB) reactor with immobilized cells has several advantages over a stirred tank reactor (Figure 9.4) in terms of the lower energy demands, the high cell density per reactor volume, and the ease of scale-up due to the simple construction of the reactor. The use of flocculated cells is advantageous compared with the use of support material because the space utility in the reactor is maximized.

In our study with a continuous culture, using glucose as the substrate in a UAPB reactor, cells from both strains of HU-101 and higher hydrogen-producing mutant AY-2 successfully flocculated and settled into the bottom of the reactor. Thereafter, the hydrogen production rate increased together with increase in the dilution rate in both strains. For the HU-101 strain, the hydrogen evolution rate was 30 mmol/L/h at a dilution rate 0.67/h, while for the AY-2 the hydrogen evolution rate reached 58 mmol/L/h at 0.67/h (Figure 9.5), giving a hydrogen yield of more than 1.1 at dilution rates from 0.13/h to 0.55/h, when the pH in the effluent was kept above 6.0.

To increase biomass in the reactor, the concept of the upflow anaerobic sludge blanket (UASB) reactor, developed originally for high-rate methane fermentation (see Chapter 7), has also been investigated for high-rate H$_2$ production. For example, Hu and Chen (2007) investigated H$_2$ production with methanogenic granules as a seeding source after heat, acid, or chemical shock and observed that only chemical shock with chloroform caused irreversible damage to methanogens, while H$_2$ productivity reached 0.48 L-H$_2$/L/h without granular structure breakage.

TABLE 9.4 Hydrogen Production with Cell Immobilization

Support Material	Bacteria	Substrate	Substrate Conc. (g/L)	H₂ Production Rate (mmol/L/h)	H₂ Yield (mol/mol-substrate)	Reference
Porous glass beads	*C. butyricum* IFO13949	Glu	5	51	1.9	Yokoi et al. (1997a)
Porous glass beads	*E. aerogenes* HO-39	Glu	10	38	0.73	Yokoi et al. (1997b)
Urethane foam	*E. aerogenes* E.82005	Mol	20	13	1.8	Tanisho and Ishiwata (1995)
Self-flocculated cells	*E. aerogenes* AY-2	Glu	15	58	1.1	Rachman et al. (1998)
Self-flocculated cells	*E. aerogenes* HU-101	Gly	10	80	0.95	Ito et al. (2005)
Lignocellulose carrier	*E. clocae* IIT-BT 08	Glu	10	76	1.8	Kumar and Das (2001)
Activated carbon	Microflora from anaerobic sludge	Suc	18	59	1.3	Chang et al. (2002)
Pretreated methanogenic granule	Microflora from granules	Glu	19	21	1.3	Hu and Chen (2007)

Glu, glucose; Gly, glycerol; Mol, Molasses; Suc, sucrose.

FIGURE 9.4 Photographs of (A) Upflow anaerobic packed-bed reactor for high-rate hydrogen production by *E. aerogenes* HU-101 and (B) self-flocculated cells in the bottom of the reactor.

5. HYDROGEN PRODUCTION FROM INDUSTRIAL ORGANIC WASTES

5.1 Carbohydrates

As mentioned earlier, there are two types of hydrogen production: by strict anaerobes or facultative anaerobes. The advantage of strict anaerobes is higher hydrogen yield than that of facultative anaerobes, but they are very sensitive to oxygen and do not survive a

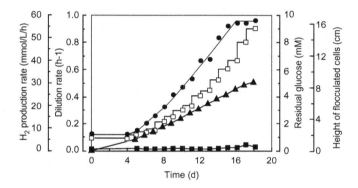

FIGURE 9.5 Hydrogen production of *E. aerogenes* mutant AY-2 in an upflow anaerobic packed-bed reactor. (●), H₂ production rate; (■), residual glucose; (▲) height of bed of flocculated cells from the bottom in the reactor; (□), dilution rate. *Adapted from Rachman et al. (1998).*

low oxygen concentration. On the other hand, facultative anaerobes rapidly consume oxygen, thereby restoring anaerobic conditions immediately in the reactor, although hydrogen yield is low. Thus, hydrogen production with a coculture of strict and facultative anaerobes is an attractive strategy to compensate for each disadvantage and treat actual industrial food wastes. Yokoi et al. (1998) reported that a continuous mixed culture of *C. butyricum* and *E. aerogenes* removed oxygen in a reactor and produced hydrogen from starch with a yield of more than 2 mol H₂/mol glucose without any reducing agents in the medium. The repeated batch culture using the same coculture produced hydrogen with a yield of 2.4 mol H₂/mol glucose under a controlled culture at pH 5.25 in a medium consisting of sweet potato starch residue and 0.1% polypepton without addition of any reducing agents (Yokoi et al., 2001).

The use of mixed culture acclimated for hydrogen production has been investigated by several researchers. Ueno et al. (1995) found that anaerobic microflora acclimated from anaerobic sludge compost produced hydrogen from cellulose with a high efficiency of 2.4 mol/mol hexose in batch experiments at 60°C. Furthermore, Ueno et al. (1996) reported stable hydrogen production for 190 days from industrial water from a sugar factory by the same microbial flora at 0.5–3 days of HRT in a chemostat culture, yielding 8.8 mmol/L/h hydrogen production rate and 2.5 mol H_2/mol hexose at 3 days of HRT. For continuous hydrogen production from a high-strength rice winery wastewater, Yu et al. (2002) applied microbial flora acclimated with glucose from the sludge derived from the secondary settling tank in a local municipal wastewater treatment plant. When the experiment was conducted in a 3.0 L UAPB reactor, an optimum specific hydrogen production rate of 9.33 L H_2/g-VSS/day and 7.1 mmol/L/h of volumetric rate were achieved at an HRT of 2 h, COD of 34 g/L, and pH 5.5 and 55°C, giving a hydrogen yield of 1.37–2.14 mol/mol hexose.

We have also investigated hydrogen production from industrial wastewater from a brewery (Mitani et al., 2005) and solid bread wastes (Oki and Mitani, 2008), using the same mixed microbial flora acclimated from anaerobic sludge mainly composed of *Thermoanaerobacterium thermosaccharolyticum* PHE9. Regarding the hydrogen production from solid bread wastes, stable hydrogen formation for 300 days was possible in the 900 L reactor at 60°C, giving a hydrogen yield of ca. 3.2 mol H_2/mol hexose and production rate of 1,600 m^3 H_2/day at the loading rate of 12.5 g wet weight per day of bread wastes (Box 9.1). In the effluent after hydrogen fermentation, although 5.7 g/L of hexose, 4 g/L of acetate, and 8 g/L of butyrate remained, they can be used as the influent for methane production (see Chapter 7 on methane fermentation) or photobiological hydrogen fermentation.

BOX 9.1

MATERIAL FLOW OF PILOT SCALE HYDROGEN FERMENTATION OF BREAD WASTES

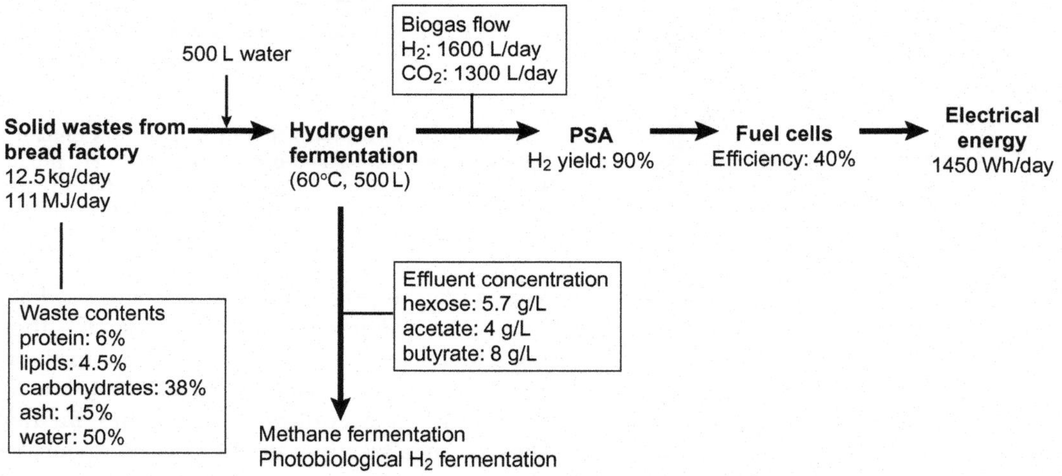

This is an example of material flow of hydrogen fermentation of bread wastes in a pilot scale plant carried out by Oki and Mitani (2008). Solid wastes from the bread factory are diluted with water and treated with a hydrogen fermenter, resulting in average hydrogen production of 1600 L (max. 2,500 L) from 12.5 kg wet weight of wastes per day. If hydrogen is used for electricity generation with a fuel cell system (assuming 40% of generating efficiency) after concentration of hydrogen with a pressure swing adsorption (PSA) system, 1450 Wh/day of electricity can be generated. In the effluent from the hydrogen fermenter, carbohydrates and fatty acids still remain. Thus, these can be used for further recovery of energy or material, such as methane production or hydrogen production.

Source: Reprinted, with permission, from Oki and Mitani (2008).

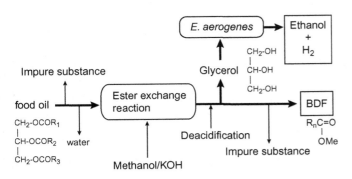

FIGURE 9.6 Schematic representation of production of biodiesel, via ester exchange reaction by alkali catalyst and H_2/ethanol with facultative anaerobe, from food oil.

5.2 Food Oil (Glycerol-Rich Residue Discharged after Biodiesel Manufacturing)

Biodiesel fuel (BDF) is defined as long chain fatty acid methyl or ethyl esters produced by the transesterification of vegetable oils or animal fats used as food oils (Figure 9.6). BDF has various advantages as an alternative to petroleum-based fuel: renewable, lower harmful emissions, and nontoxic. BDF can be used as fuel in diesel engines or heating systems (Eggersdorfer et al., 1992; Chowdhury and Fouhy, 1993). Although BDF is produced chemically (alkali catalyst) or enzymatically (lipase), glycerol is essentially generated as a by-product (Du et al., 2003; Vicente et al., 2004). The glycerol generated is presently applied, for example, as a resource for cosmetics, but a further increase in the production of BDF will raise the problem of how to treat efficiently the wastes containing glycerol.

Microbial conversion of glycerol to more useful compounds has been investigated recently, with particular focus on the production of 1,3-propanediol, which can be applied as a basic ingredient of polyesters (Gunzel et al., 1991; Biebl et al., 1992; Petitdemange et al., 1995). As mentioned, *E. aerogenes* HU-101 can convert crude glycerol to H_2 and ethanol with a slight production of other by-products (Table 9.5). Thus, biological production of H_2 and ethanol from crude glycerol is also an attractive technology for complete energy recovery from food oil wastes (Figure 9.7). Ito et al. (2005) reported efficient production of H_2 and ethanol by *E. aerogenes* HU-101 from crude glycerol discharged after a BDF manufacturing process. They reported that yields of H_2 and ethanol decreased with the increase of concentrations of biodiesel wastes, and the production rates of H_2 and ethanol in biodiesel wastes were lower than those for the same concentration of pure glycerol, partially because of high salt content in the biodiesel wastes. In continuous culture with a packed-bed reactor using self-immobilized cells, the maximum production rate of H_2 from pure glycerol was 80 mmol/L/h,

TABLE 9.5 Microbial Production of Useful Chemicals from Glycerol

Products	Microorganism	Reference
MATERIAL RESOURCES		
1,3-propanediol	*Klebsiella pneumoniae* (NG)	Moon et al. (2010)
	Clostridium butyricum (NG)	Hirschmann et al. (2005)
		Gonzalez-Pajuelo et al. (2004)
Succinate	*Anaerobiospirillum succiniciproducens* (NG)	Lee et al. (2001)
Lactate	*Escherichia coli* (NG)	Hong et al. (2009)
Poly-β-hydroxyalkanoates	*Burkholderia cepacia* (NG)	Zhu et al. (2010)
	Cupriavidus necator (NG)	Cavalheiro et al. (2009)
ENERGY RESOURCES		
H_2/Ethanol	*Escherichia coli* (GE)	Hu and Wood (2010)
	Enterobacter aerogenes (NG)	Ito et al. (2005)
Butanol	*Clostridium pasteurianum* (NG)	Biebl (2001)
Methane	Methanogenic sludge	Lopez et al. (2009)
OTHERS		
Propionic acid	*Propionibacterium acidipropionici* (GE)	Zhang and Yang (2009)
Succinate	*Yarrowia lipolytica* (NG)	Imandi et al. (2007)

GE, genetically engineered strain; NG, non-genetically engineered strain.

yielding 0.8 mol/mol of ethanol from 10 g/L glycerol, while hydrogen production rate from biodiesel wastes was only 30 mmol/L/h, due to low biomass retention in the reactor. The use of a porous ceramic as a support material was effective for fixing cells in the reactor, resulting in an increase in the maximum H_2 production rate from crude glycerol to 63 mmol/L/h, giving an ethanol yield of 0.85 mol/mol glycerol (Figure 9.7). This means 930 L of BDF, 145 m^3 of hydrogen, and 50 L ethanol can be produced from 1000 L of food oil.

6. TREATMENT OF EFFLUENT AFTER DARK HYDROGEN FERMENTATION

Dark hydrogen fermentation is an incomplete oxidation process. This means that organic matter is not completely oxidized to CO_2 but to intermediate

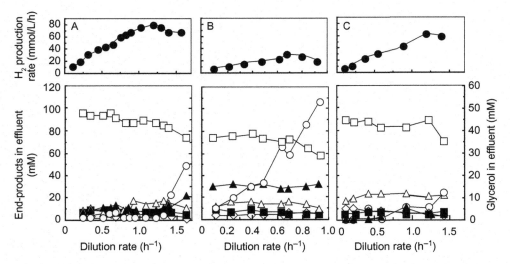

FIGURE 9.7 Continuous production of hydrogen and ethanol from pure glycerol with self-flocculated cells of *E. aerogenes* HU-101 in an upflow anaerobic packed-bed reactor, (A); crude glycerol discharged after BDF manufacturing process with self-flocculated cells (B); cells supported by a porous ceramic (C) (●), hydrogen; (□), ethanol; (▲), formate; (△), 1,3-propanediol; (■), lactate; (◇), acetate. *Adapted from Ito et al. (2005).*

compounds, like short-chain fatty acids—such as acetate, propionate, butyrate, and lactate—and alcohols such as ethanol and butanol. Even if a strict anaerobe ideally catabolizes glucose for hydrogen formation, acetate is produced as the by-product (see Eq (10)). Furthermore, since the actual industrial organic wastes are a mixture of several compounds with different characteristics, such as carbohydrates, lipids, and proteins, the composition of by-products is complicated. Although it is ideal to oxidize these intermediate compounds to H_2 and CO_2, further oxidation of these compounds is very unfavorable in the dark condition, since these reactions are endergonic under ordinary temperatures and pressures. However, since these compounds still retain chemical energy, the recovery of the energy will contribute to effective energy or material production from organic matter.

6.1 Methane Fermentation

As described in Chapter 7, the above-mentioned intermediate compounds can be further metabolized to methane, using a hydrogen and methane two-stage process. Briefly, fatty acids and alcohols can be converted to methane even if reactions are endergonic under ordinary temperatures and pressures—in methane fermentation the significant decrease of H_2 partial pressure by coupling with hydrogenotrophic methane formation enables further oxidation of these compounds to hydrogen and CO_2. Since methane fermentation is well established for practical use, the combination of hydrogen fermentation and methane fermentation is a realistic choice as an energy recovery process from organic wastes.

6.2 Photobiological Hydrogen Fermentation

In methane fermentation, the significant decrease of H_2 partial pressure by coupling with hydrogenotrophic methane formation enables further oxidation of these compounds to hydrogen and CO_2. The other way to shift the endergonic reaction to an exergonic reaction is to provide energy from outside. Photoheterotrophic bacteria such as purple non-sulfur bacteria or cyanobacteria can oxidize the intermediate products in hydrogen fermentation to hydrogen and CO_2 in the presence of light (Dasgupta et al., 2010). For purple non-sulfur bacteria, since fatty acids are preferred substrates rather than carbohydrates such as sugar, a two-stage process consisting of dark fermentation followed by photobiological hydrogen production can be used.

These bacteria fix N_2 with nitrogenase. The nitrogenase also catalyses the production of hydrogen, particularly in the absence of N_2. The overall reaction is:

$$N_2 + 8H^+ + 8e^- + 16ATP \rightarrow 2NH_3 + H_2 + 16ADP + 16P_i$$
(14)

In the case that the organic substrate is converted to hydrogen, energy obtained from light is needed. When acetate is theoretically converted to hydrogen,

$$CH_3COOH + 2H_2O + \text{''light energy''} \rightarrow 2CO_2 + 4H_2 \quad (15)$$

Theoretical hydrogen yield from acetate, lactate, and butyrate, which are major fatty acids in the effluent from dark hydrogen fermentation, can be calculated to be 4, 6, and 10. A study with immobilized cells of *Rhodopseudomonas* sp. RV demonstrated that hydrogen yields for acetate, lactate, and butyrate in the effluent from dark hydrogen fermentation were 1.6, 2.7, and 7.5 mol/mol substrate (Miyake et al., 1984). Maximum hydrogen production rate was reported to be 8 mmol/L/h (Dasgupta et al., 2010), which is much lower than that in dark fermentation (>60 mmol/L/h).

Similar to hydrogenase, nitrogenase is also highly sensitive to oxygen and inhibited by ammonia ions. Thus, for photoheterotrophic hydrogen production from organic substrates, the bioreactor must usually operate under anaerobic and light conditions with illumination and limiting concentrations of nitrogen sources. Additionally, since photoheterotrophic hydrogen production essentially needs light energy, the reactor must be designed to allow penetration of light sufficient for efficient production, resulting in a two-dimensional reactor such as a flat or tubular type. This requires a huge place for extremely large-scale production, compared with dark fermentation. Thus, although a combination of dark and photobiological fermentation for sole hydrogen production from organic wastes is very attractive for practical use, dramatic enhancement of hydrogen productivity would be needed by means of breeding of photoheterotrophic bacteria and optimization of reactor design and culture condition.

Several photobioreactors have been proposed for hydrogen production and recently reviewed by Dasgupta et al. (2010). Although the photobioreactors can be broadly classified into open system (i.e., raceway pond, lakes, etc.) and closed systems, the closed systems are used for hydrogen production because open systems cannot provide the anaerobic conditions needed and offer poor possibility for the collection of produced gas. Since light energy is essentially needed for H₂ production in a closed photobioreactor, the fundamental design factor of photobioreactors is transparency, which allows maximum penetration of light and a large surface area. Thus, the thicknesses of the reactors are usually small to avoid a shading effect by the growing cells, and a major issue limiting the large-scale production of hydrogen is the restricted light penetration into the deeper regions of the reactor.

Several types of bioreactor have been designed to improve hydrogen production (Figure 9.8). Among them, vertical-column (Figure 9.8A), tubular (Figure 9.8B), and flat-panel (Figure 9.8C) types of photobioreactor are widely used for hydrogen production.

Vertical-column photobioreactors may be airlift or bubble column reactors. They consist of vertical transparent tubes in which agitation is achieved with the help of bubbling at the bottom. The illumination area is small, but still they are widely used for photosynthetic bacteria, owing to the compactness, low cost, ease of operation, and low shear stress in airlift and bubble columns (Asada and Miyake, 1999).

Tubular photobioreactors consist of long transparent tubes with diameters ranging from 3 to 6 cm, and lengths ranging from 10 to 100 m. The culture liquid is pumped through these tubes by means of a mechanical or airlift pump. The tube can be positioned in many different ways such as in horizonal, vertical, or helical planes, although the shape of the light gradient in the tubes is similar in most designs. Compared with the vertical-column type, a large illumination area can be achieved in the tubular type (Molina et al., 2001). The length of the tubes is limited because accumulation of produced hydrogen gas inhibits hydrogenase activity, although this might not be so important for a nitrogenase-based process.

Flat-panel reactors consist of a rectangular transparent box with a depth of only 1 to 5 cm. The reactors are

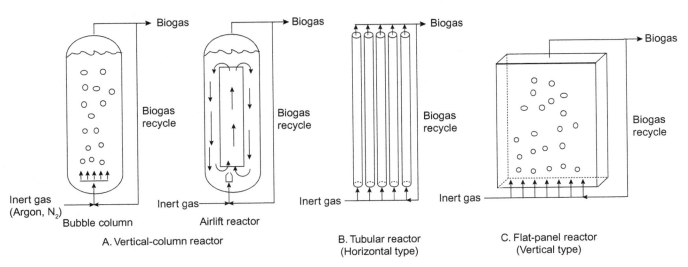

FIGURE 9.8 Schematic drawings of various kinds of photobioreactor for H₂ production.

mixed with gas introduced via a perforated tube at the bottom of the reactor. Usually the panels are illuminated from one side by direct sunlight, and the panels are placed vertically or inclined towards the sun. High photosynthetic efficiencies and effective control of gas pressure can be achieved in flat-panel photobioreactors, and they have been found to be more economical than other bioreactors (Lehr and Posten, 2009).

7. CONCLUDING REMARKS

Hydrogen has become a focus of attention as the energy of the future. Therefore, dark hydrogen fermentation from biomass wastes has been investigated using various types of microorganisms and bioreactors. The technology will make a large contribution to the improvement of the global environment, because biologically produced hydrogen from biomass is renewable and clean. Indeed, the development, construction, operation, and regulation of proper processes is still challenging, and further research is needed before practical application of biological hydrogen production is possible at a large scale. Especially, efficient use of soluble by-products of hydrogen fermentation, such as alcohols or fatty acids, is important. In this context, crude glycerol discharged after biodiesel production from vegetable oils is an ideal feedstock for hydrogen production, because the main by-product is ethanol that can be directly used as a liquid fuel. However, the usual by-products from other organic wastes are a combined mixture of alcohols and acids. For efficient use of such a mixture, a hydrogen production process combined with methane production might be a candidate because of improvement of the overall energy yield from feedstocks. Combination of dark fermentation with photofermentation is another candidate, because it will achieve maximum overall hydrogen yield, although drastic improvement of the hydrogen productivity of photofermentation is needed. Overall, dark hydrogen fermentation production is a promising process for sustainable supply of renewable hydrogen for economic and commercial applications in the future.

References

Abreu, A.A., Karakashev, D., Angelidaki, I., Sousa, D.Z., Alves, M.M., 2012. Biohydrogen production from arabinose and glucose using extreme thermophilic anaerobic mixed cultures. Biotechnol. Biofuels 5, 6. Available at: <http://www.biotechnologyforbiofuels.com/content/5/1/6>.

Adams, M.W.W., Mortenson, L.E., Chen, J.S., 1980. Hydrogenase. Biochim. Biophys. Acta 594, 105–176.

Asada, Y., Miyake, J., 1999. Photobiological hydrogen production. J. Biosci. Bioeng. 88, 1–6.

Biebl, H., 2001. Fermentation of glycerol by Clostridium pasteurianum—batch and continuous culture studies. J. Ind. Microbiol. Biotechnol. 27, 18–26.

Biebl, H., Marten, S., Hippe, H., Deckwer, W.D., 1992. Glycerol conversion to 1,3-propanediol by newly isolated clostridia. Appl. Microbiol. Biotechnol. 36, 592–597.

Cavalheiro, J.M.B.T., de Almeida, M.C.M.D., Grandfils, C., da Fonseca, M.M.R., 2009. Poly(3-hydroxybutyrate) production by Cupriavidus necator using waste glycerol. Process Biochem. 44, 509–515.

Chang, J.-S., Lee, K.-S., Lin, P.-J., 2002. Biohydrogen production with fixed-bed bioreactors. Int. J. Hyd. Energy 27, 1167–1174.

Chowdhury, J., Fouhy, K., 1993. Vegetable oils: from table to gas tank. Chem. Eng. 100, 35–39.

Dasgupta, C.N., Gilbert, J.J., Lindblad, P., Heidorn, T., Borgvang, S.A., Skjanes, K., et al., 2010. Recent trends on the development of photobiological processes and photobioreactors for the improvement of hydrogen production. Int. J. Hyd. Energy 35, 10218–10238.

de Graef, M.R., Alexeeva, S., Snoep, J.L., de Mattos, M.J.T., 1999. The steady-state internal redox state (NADH/NAD) reflects the external redox state and is correlated with catabolic adaptation in Escherichia coli. J. Bacteriol. 181, 2351–2357.

Du, W., Xu, Y.Y., Liu, D.H., 2003. Lipase-catalysed transesterification of soya bean oil for biodiesel production during continuous batch operation. Biotechnol. Appl. Biochem. 38, 103–106.

Dürre, P., Kuhn, A., Gottschalk, G., 1986. Treatment with allyl alcohol selects specially for mutants of Clostridium acetobutylicum defective in butanol synthesis. FEMS Microbiol. Lett. 36, 77–81.

Eggersdorfer, M., Meyer, J., Eckes, P., 1992. Use of renewable resources for nonfood materials. FEMS Microbiol. Rev. 9, 355–364.

Evvyernie, D., Yamazaki, S., Morimoto, K., Karita, S., Kimura, T., Sakka, K., et al., 2000. Identification and characterization of Clostridium paraputrificum M-21, a chitinolytic, mesophilic and hydrogen-producing bacterium. J. Biosci. Bioeng. 89, 596–601.

Evvyernie, D., Morimoto, K., Karita, S., Kimura, T., Sakka, K., Ohmiya, K., 2001. Conversion of chitinous wastes to hydrogen gas by Clostridium paraputrificum M-21. J. Biosci. Bioeng. 91, 339–343.

Foster, J.F., Moat, A.G., 1980. Nicotinamide adenine dinucleotide biosynthesis and pyridine nucleotide cycle metabolism in microbial systems. Microbiol. Rev. 44, 83–105.

Girbal, L., Croux, C., Vasconcelos, I., Soucaille, P., 1995. Regulation of metabolic shifts in Clostridium acetobutylicum ATCC824. FEMS. Microbiol. Rev. 17, 287–297.

Gonzalez-Pajuelo, M., Andrade, J.C., Vasconcelos, I., 2004. Production of 1,3-propanediol by Clostridium butyricum VPI 3266 using a synthetic medium and raw glycerol. J. Ind. Microbiol. Biotechnol. 31, 442–446.

Gunzel, B., Yonsel, S., Deckwer, W.D., 1991. Fermentative production of 1,3-propanediol from glycerol by Clostridium butyricum up to a scale of 2 m³. Appl. Microbiol. Biotechnol. 36, 289–294.

Hirschmann, S., Baganz, K., Koschik, I., Vorlop, K.D., 2005. Development of an integrated bioconversion process for the production of 1,3-propanediol from raw glycerol waters. Landbauforschung Volkenrode 55, 261–267.

Hong, A.A., Cheng, K.K., Peng, F., Zhou, S., Sun, Y., Liu, C.M., et al., 2009. Strain isolation and optimization of process parameters for bioconversion of glycerol to lactic acid. J. Chem. Technol. Biotechnol. 84, 1576–1581.

Hu, B., Chen, S.L., 2007. Pretreatment of methanogenic granules for immobilized hydrogen fermentation. Int. J. Hyd. Energy 32, 3266–3273.

Hu, H.B., Wood, T.K., 2010. An evolved Escherichia coli strain for producing hydrogen and ethanol from glycerol. Biochem. Biophys. Res. Commun. 391, 1033–1038.

Imandi, S.B., Bandaru, V.R., Somalanka, S.R., Garapati, H.R., 2007. Optimization of medium constituents for the production of citric acid from byproduct glycerol using Doehlert experimental design. Enzyme Microb. Technol. 40, 1367–1372.

Ito, T., Nakashimada, Y., Kakizono, T., Nishio, N., 2004. High yield production of hydrogen by Enterobacter aerogenes mutants with decreased α-acetolactate synthase activity. J. Biosci. Bioeng. 97, 227–232.

Ito, T., Nakashimada, Y., Senba, K., Matsui, T., Nishio, N., 2005. Hydrogen and ethanol production from glycerol-containing wastes discharged from biodiesel manufacturing process. J. Biosci. Bioeng. 100, 260–265.

Johansen, L., Bryn, K., Stormer, F.C., 1975. Physiological and biochemical role of butanediol pathway in Enterobacter aerogenes. J. Bacteriol. 123, 1124–1130.

Jones, D.T., Woods, D.R., 1986. Acetone-butanol fermentation revisited. Microbiol. Rev. 50, 484–524.

Jung, K.W., Kim, D.H., Kim, S.H., Shin, H.S., 2011. Bioreactor design for continuous dark fermentative hydrogen production. Biores. Technol. 102, 8612–8620.

Kataoka, N., Miya, A., Kiriyama, K., 1997. Studies on hydrogen production by continuous culture system of hydrogen-producing anaerobic bacteria. Water Sci. Technol. 36, 41–47.

Kumar, N., Das, D., 2001. Continuous hydrogen production by immobilized Enterobacter cloacae IIT-BT 08 using lignocellulosic materials as solid matrices. Enzyme Microb. Technol 29, 280–287.

Lee, P.C., Lee, W.G., Lee, S.Y., Chang, H.N., 2001. Succinic acid production with reduced by-product formation in the fermentation of Anaerobiospirillum succiniciproducens using glycerol as a carbon source. Biotechnol. Bioeng. 72, 41–48.

Lehr, F., Posten, C., 2009. Closed photo-bioreactors as tools for biofuel production. Curr. Opin. Biotechnol. 20, 280–285.

Lopez, J.A.S., Santos, M.D.M., Perez, A.F.C., Martin, A.M., 2009. Anaerobic digestion of glycerol derived from biodiesel manufacturing. Biores. Technol. 100, 5609–5615.

Lu, J.Q., Gavala, H.N., Skiadas, I.V., Mladenovska, Z., Ahring, B.K., 2008. Improving anaerobic sewage sludge digestion by implementation of a hyper-thermophilic prehydrolysis step. J. Environ. Manage. 88, 881–889.

Magee, R.J., Kosaric, N., 1987. The microbial production of 2,3-butanediol. Adv. Appl. Microbiol. 32, 89–159.

Mitani, Y., Takamoto, Y., Atsumi, R., Hiraga, T., Nishio, N., 2005. Hydrogen and methane two-stage production directly from brewery effluent by anaerobic fermentation. Master Brewers Assoc. Am. TQ 42, 283–289.

Miyake, Y., Mao, X.Y., Kawamura, S., 1984. Photoproduction of hydrogen from glucose by a co-culture of a photosynthetic bacterium and Clostridium butyricum. J. Ferment. Technol. 62, 531–535.

Molina, E., Fernandez, J., Acien, F.G., Chisti, Y., 2001. Tubular photobioreactor design for algal cultures. J. Biotechnol. 92, 113–131.

Moon, C., Ahn, J.H., Kim, S.W., Sang, B.I., Um, Y., 2010. Effect of biodiesel-derived raw glycerol on 1,3-propanediol production by different microorganisms. Appl. Biochem. Biotechnol. 161, 502–510.

Nakashimada, Y., Rachman, M.A., Kakizono, T., Nishio, N., 2002. H_2 production of Enterobacter aerogenes altered by extracellular and intracellular redox states. Int. J. Hyd. Energy 27, 1399–1405.

Oki, Y., Mitani, Y., 2008. Hydrogen-methane two stage fermentation technology from food industry wastes. In: Nakashimada, Y., Nishio, N. (Eds.), Recent Developments in Biogas Technology. CMC Publishing. Co. Ltd., Tokyo, Osaka, pp. 139–146.

Pablo, H.C., Mendez, B.S., 1990. Direct selection of Clostridium acetobutylicum fermentation mutants by a proton suicide method. Appl. Environ. Microbiol. 56, 578–580.

Petitdemange, E., Durr, C., Andaloussi, S.A., Raval, G., 1995. Fermentation of raw glycerol to 1,3-propanediol by new strains of Clostridium butyricum. J. Ind. Microbiol. 15, 498–502.

Rachman, M.A., Furutani, Y., Nakashimada, Y., Kakizono, T., Nishio, N., 1997. Enhanced hydrogen production in altered mixed acid fermentation of glucose by Enterobacter aerogenes. J. Ferment. Bioeng. 83, 358–363.

Rachman, M.A., Furutani, Y., Nakashimada, Y., Kakizono, T., Nishio, N., 1998. Hydrogen production with high yield and high evolution rate by self-flocculated cells of Enterobacter aerogenes in a packed-bed reactor. Appl. Microbiol. Biotechnol. 49, 450–454.

Ren, N.Q., Guo, W.Q., Liu, B.F., Cao, G.L., Ding, J., 2011. Biological hydrogen production by dark fermentation: challenges and prospects towards scaled-up production. Curr. Opin. Biotechnol. 22, 365–370.

Schröder, C., Selig, M., Schönheit, P., 1994. Glucose fermentation to acetate, CO_2 and H_2 in the anaerobic hyperthermophilic eubacterium Thermotoga maritima: involvment of the Embden-Meyerhof pathway. Arch. Microbiol. 161, 460–470.

Steuber, J., Krebs, S., Bott, M., Dimroth, P., 1999. A membrane-bound NAD(P)(+)-reducing hydrogenase provides reduced pyridine nucleotides during citrate fermentation by Klebsiella pneumoniae. J. Bacteriol. 181, 241–245.

Taguchi, F., Chang, J.D., Mizukami, N., Saitotaki, T., Hasegawa, K., Morimoto, M., 1993. Isolation of a hydrogen producing bacterium, Clostridium beijerinckii strain AM21b, from termites. Can. J. Microbiol. 39, 726–730.

Taguchi, F., Mizukami, N., Hasegawa, K., Saitotaki, T., Morimoto, M., 1994. Effect of amylase accumulation on hydrogen production by Clostridium beijerinckii, Strain AM21b. J. Ferment. Bioeng. 77, 565–567.

Taguchi, F., Mizukami, N., Taki, T.S., Hasegawa, K., 1995. Hydrogen production from continuous fermentation of xylose during growth of Clostridium sp. strain No. 2. Can. J. Microbiol. 41, 536–540.

Tanisho, S., Ishiwata, Y., 1994. Continuous hydrogen production from molasses by the bacterium Enterobacter aerogenes. Int. J. Hyd. Energy 19, 807–812.

Tanisho, S., Ishiwata, Y., 1995. Continuous hydrogen production from molasses by fermentation using urethane foams as a support of flocks. Int. J. Hyd. Energy 20, 541–545.

Tanisho, S., Wakao, N., Kosako, Y., 1983. Biological hydrogen production by Enterobacter aerogenes. J. Chem. Eng. Jpn. 16, 529–530.

Tanisho, S., Suzuki, Y., Wakao, N., 1987. Fermentative hydrogen evolution by Enterobacter aerogenes strain E 82005. Int. J. Hyd. Energy 12, 623–627.

Tanisho, S., Kamiya, N., Wakao, N., 1989a. Hydrogen evolution of Enterobacter aerogenes depending on culture pH: mechanism of hydrogen evolution from NADH by means of membrane-bound hydrogenase. Biochim. Biophys. Acta 973, 1–6.

Tanisho, S., Tu, H., Wakao, N., 1989b. Fermentative hydrogen evolution from various substrates by Enterobacter aerogenes. Hakko Kougaku Kaishi 67, 29–34.

Tanisho, S., Kuromoto, M., Kadokura, N., 1998. Effect of CO_2 removal on hydrogen production by fermentation. Int. J. Hyd. Energy 23, 559–563.

Ueno, Y., Kawai, T., Sato, S., Otsuka, S., Morimoto, M., 1995. Biological production of hydrogen from cellulose by natural anaerobic microflora. J. Ferment. Bioeng. 79, 395–397.

Ueno, Y., Otsuka, S., Morimoto, M., 1996. Hydrogen production from industrial waste water by anaerobic microflora in chemostat culture. J. Ferment. Bioeng. 82, 194–197.

van Groenestijn, J.W., Hazewinkel, J.H.O., Nienoord, M., Bussmann, P.J.T., 2002. Energy aspects of biological hydrogen production in high rate bioreactors operated in the thermophilic temperature range. Int. J. Hyd. Energy 27, 1141–1147.

van Niel, E.W.J., Budde, M.A.W., de Haas, G.G., van der Wal, F.J., Claassen, P.A.M., Stams, A.J.M., 2002. Distinctive properties of high hydrogen producing extreme thermophiles, *Caldicellulosirupter saccharolyticus* and *Thermotoga elfii*. Int. J. Hyd. Energy 27, 1391–1398.

Vicente, G., Martinez, M., Aracil, J., 2004. Integrated biodiesel production: a comparison of different homogeneous catalysts systems. Biores. Technol. 92, 297–305.

Yokoi, H., Maeda, Y., Hirose, J., Hayashi, S., Takasaki, Y., 1997a. H_2 production by immobilized cells of *Clostridium butyricum* on porous glass beads. Biotechnol. Tech. 11, 431–433.

Yokoi, H., Tokushige, T., Hirose, J., Hayashi, S., Takasaki, Y., 1997b. Hydrogen production by immobilized cells of aciduric *Enterobacter aerogenes* strain HO-39. J. Ferment. Bioeng. 83, 481–484.

Yokoi, H., Yoshida, T., Mori, S., Hirose, J., Hayashi, S., Takasaki, Y., 1997c. Biopolymer flocculant produced by an *Enterobacter* sp. Biotechnol. Lett. 19, 569–573.

Yokoi, H., Tokushige, T., Hirose, J., Hayashi, S., Takasaki, Y., 1998. H_2 production from starch by a mixed culture of *Clostridium butyricum* and *Enterobacter aerogenes*. Biotechnol. Lett. 20, 143–147.

Yokoi, H., Saitsu, A., Uchida, H., Hirose, J., Hayashi, S., Takasaki, Y., 2001. Microbial hydrogen production from sweet potato starch residue. J. Biosci. Bioeng. 91, 58–63.

Yu, H., Zhu, Z., Hu, W., Zhang, H., 2002. Hydrogen production from rice winery wastewater in an upflow anaerobic reactor by using mixed anaerobic cultures. Int. J. Hyd. Energy 27, 1359–1365.

Zhang, A., Yang, S.T., 2009. Propionic acid production from glycerol by metabolically engineered *Propionibacterium acidipropionici*. Process Biochem. 44, 1346–1351.

Zhu, C.J., Nomura, C.T., Perrotta, J.A., Stipanovic, A.J., Nakas, J.P., 2010. Production and characterization of poly-3-hydroxybutyrate from biodiesel-glycerol by *Burkholderia cepacia* ATCC 17759. Biotechnol. Progr. 26, 424–430.

CHAPTER

10

Thermophilic Aerobic Bioprocessing Technologies for Food Industry Wastes and Wastewater

Maria R. Kosseva and C.A. Kent

1. INTRODUCTION

Within many European regions, nitrogen imbalances—surplus versus shortage or demand—are unsustainable, owing to pollution, and inefficient in terms of output/input ratios of agricultural products, with regard to energy and nitrogen (N). Current trends in agricultural production will lead to more than a two-fold increase in global consumption of N fertilizers (Tilman et al., 2001). This is incompatible with the fact that, even today, the planetary limit for human N fixation is exceeded by a factor of four. The scarcity of N refers to the global constraints on the use of N that are created by the need to move to sustainable food production systems, in which reactive N is recycled rather than leaked to other ecosystems. N was classified as one of seven planetary boundaries (Rockstrom et al., 2009). The addition of reactive N to the environment acts primarily as a slow variable with time lag, eroding the resilience of ecosystems via acidification of terrestrial ecosystems and eutrophication of coastal and freshwater systems. Nitrous oxide emissions from agriculture also contribute to the climate change boundary. At field, farm, regional, national, and higher levels, agriculture must move towards N budgeting and recycling in agro-food and waste systems. There is a need for radically improved management and innovation within European farming and agro-food systems to improve the efficiency with which N is used and recycled and to reduce losses to the wider environment. N in waste must be seen as a resource rather than a problem (EC SCAR report, 2011). Legislative regulations for the dumping of sludge are forcing industries to come up with alternatives to make this process of elimination environmentally safer (Bruce et al., 1990; EPA, 1992). One potentially attractive option involves the use of thermophilic microorganisms to produce a pasteurized, easily dewatered sludge at temperatures that facilitate enhanced levels of energy recovery (Chiang et al., 2001). Processing options include the associated production of low-COD treated wastewater (Kosseva et al., 2001), or of added-value products such as xanthan gum (Papagianni et al., 2001) and polyhydroxyalkanoates (Pantazaki et al., 2003). The sludge, stabilized by thermophilic aerobic digestion (TAD) and rendered hygienic, can be exploited as a fertilizer. For example, the end product of thermophilic biodegradation of olive-mill wastewater provides a series of beneficial effects on the land, including increased microporosity, hence improved oxygenation of the surface profile of the soil; increased stability of aggregates; better hydrodynamic retention of the land; and greater bioavailability of microelements for vegetable nutrition (Niaounakis and Halvadakis, 2006).

Of particular acclaim is the potential for the use of TAD in upgrading by protein enrichment a variety of food and agricultural wastes for reuse in animal feed (Ugwuanyi et al., 2006). This application has implications for global food security, especially in the tropics, where, because of inadequate food supply, animals and humans often compete directly for the same sources of calories. Ugwuanyi et al. (2005) reported on the efficiency of TAD in the treatment of potato peel waste.

In this chapter we present results from investigations into TAD, which we carried out within an

Food Industry Wastes.
DOI: http://dx.doi.org/10.1016/B978-0-12-391921-2.00010-X

EC-funded project. Its overall aim was to accomplish effective *in situ* treatment of food industry wastes (FIW) under thermophilic aerobic conditions, achieving safer, cleaner, more energy-efficient bioremediation of the vast volumes of food wastes produced in Europe. One of the main objectives of our project was to reduce COD of cheese whey, distillery wastes, and potato processing wastewater, while complying with European safety and disposal regulations and simultaneously producing animal feed or fertilizer. Another objective was to generate added-value products from the wastes using the unique properties and outstanding stability of thermophilic microorganisms.

2. THERMOPHILIC AEROBIC DIGESTION

Over the years, the use of aerobic bioprocesses, in particular in the presence of thermophilic or mesophilic bacteria for the treatment of high-strength wastewater, has become increasingly frequent (Visvanathan and Nhien, 1995; LaPara and Alleman, 1999; Skjelhaugen, 1999; LaPara et al., 2000). Aerobic treatment involving populations of thermophilic bacteria offers a wide spectrum of benefits. One of these is the potential for the biodegradation of organics in high-temperature wastewaters (Sürücü et al., 1976), which eliminates the need for cooling them *prior* to treatment. Operation under thermophilic conditions gives a high rate of biodegradation—2 to 10 times higher than with a mesophilic process—and lends itself to high process stability. High temperatures also support the inactivation of the pathogens present in the wastewater (Becker et al., 1999; Skjelhaugen, 1999; Cheunbarn and Pagilla, 2000; Nakano and Matsamura, 2001), which is one of the main aims of the treatment process. That makes aerobic thermophilic processing suitable for stabilization of the sludge and for rendering it hygienic, so that it can be exploited as a fertilizer. For this application, it is not desirable to reduce COD too much, since a fertilizer should contain optimal concentrations of nutrients. This is different from a process for total degradation of organic matter (Skjelhaugen, 1999). In cases where there is a potential danger of the presence of pathogenic microorganisms (e.g., in municipal wastes), and therefore their inactivation is highly important, process effectiveness may be increased by introducing aerobic thermophilic degradation *prior* to anaerobic degradation by mesophilic microorganisms (Cibis et al., 2002). Currently "auto-thermal aerobic stabilization" is one of the most suitable methods for stabilization of municipal sludge with a high hygienic quality (Liu et al., 2012).

Various names have been used to designate the aerobic thermophilic treatment of organic wastes. This includes autothermal thermophilic aerobic digestion (ATAD) or treatment (ATAT), thermophilic aerobic digestion (TAD), aerobic thermophilic stabilization (ATS), aerobic thermophilic treatment (A-T treatment), etc. (Juteau, 2006). The exothermic reactions that occur in the course of this process can enable self-stabilizing temperature maintenance (Sürücü et al., 1975, 1976; Skjelhaugen 1999; Chiang et al., 2001). A minimal quantity of organic material is needed to sustain self-heating. Excluding any heat loss, for a 40°C increase (heating from 20°C to 60°C), a theoretical minimum of 24 g COD/L can be calculated using a specific heat of 4.2 kJ/(kg°C), a heat production of 20,000 kJ/kg of volatile solids destroyed (Metcalf & Eddy Inc. et al., 2003), an oxygen consumption of 1.42 kg O_2/kg organic matter oxidized (Henze et al., 2002), a COD/BOD ratio of 2, and a specific weight of 1 kg/L of slurry (Juteau, 2006). For self-heating to occur it is necessary to reach a COD reduction of about 20−40 g/L (LaPara and Alleman, 1999). Thus, aerobic thermophilic processing is particularly suitable for the degradation of wastes with high solids content. The ATAD process must have a feed control mechanism that ensures that the organic feed (volatile solids or soluble COD) is above a minimum concentration for adequate heat production.

A major benefit resulting from the use of thermophilic bacteria is the enhanced utilization of the nutrients present for bacterial respiration activity, which brings about lower biomass (sludge) yields than when the biodegradation process is carried out at lower temperatures (Visvanathan and Nhien, 1995) or in an anaerobic process. Any decrease in the sludge volume produced is of importance, since the problem of biomass separation from the wastewater under treatment, as well as that of biomass processing, is a difficult one to solve (LaPara and Alleman, 1999; Chiang et al., 2001).

Studies have been reported of the successful application of thermo- or mesophilic aerobic bacteria to the treatment of high-strength waste streams coming from food industry or agricultural production: effluents from breweries (Zvauya et al., 1994), from yeast factories (Loll, 1976), slaughterhouses (Couillard and Zhu, 1993), piggery wastes from fattening houses (Beaudet et al., 1990), etc., where the extent of COD reduction in some instances exceeded 90% (Couillard and Zhu, 1993).

Relatively little information is available concerning the diversity of the microflora composition over the course of a thermophilic aerobic process for wastewater treatment. Only a few researchers appear to have attempted to isolate cultures from bioreactors in which thermophilic aerobic wastewater and activated sludge treatment was taking place (Sonnleitner and Fiechter, 1983; LaPara et al., 2000; Liu et al., 2010). Most commonly, the isolated organisms belonged to bacteria from the genus *Bacillus* (Beaudet et al., 1990; Lim et al., 2001; Ugwuanyi et al., 2008).

3. THERMOPHILIC MICROORGANISMS

One of the main ways of classifying living organisms is in terms of their optimal growth temperature, or growth temperature range: psychrophiles (growing at low temperatures), mesophiles (growing at moderate temperatures), and thermophiles (growing at high temperatures). A practical approach for classification of thermophilic microorganisms is proposed by Wiegel and Ljungdahl (1986). They organized thermophiles as thermo-tolerant, moderate, and extreme thermophilic microorganisms, as shown in Table 10.1. Here, minimal temperature (T_{min}) is the lowest temperature at which growth and multiplication of the microorganisms occur at a reasonable rate; optimal temperature (T_{opt}) is the temperature at which the shortest doubling time of biomass occurs; and maximal temperature (T_{max}) is the highest temperature at which growth and multiplication of the microorganisms are observed.

Among bacteria, *Archaebacteria* represent the group of obligate thermophiles, with some species growing at temperatures up to 110°C. However the majority of thermophilic bacteria have optimal growth temperatures below 75°C. So far, 26 bacterial genera (over 50 species), including 22 genera of *Archaebacteria* (41 species), have been classified as thermophiles (Kristjansson, 1992). Therefore, it could be said that *Eubacteria* are not as well adapted to life at such high temperatures as *Archaebacteria*, although they are more numerous and are characterized by a more diverse metabolism. Most thermophilic bacteria belong to the *Bacillus* and *Thermus* genera, but some species can also be found in *Clostridium*, *Staphylococcus*, *Sarcina*, *Streptococcus*, and other genera, such as *Lactobacillus*, *Thermoactinomyces*, *Thermomonospora*, and also in cyanobacteria. Few thermophiles have been identified among Gram-negative bacteria. In nature they usually occur in soil, compost, and stored moist material (hay, corn), and extreme thermophiles are usually found in hot springs.

Thermophilic organisms have developed a number of mechanisms allowing growth at higher temperatures. It appears that thermal stabilization is accompanied by small structural modifications of their proteins: an increase in H-bonds and salt bridges. *Bacilli* are especially well adapted to higher temperatures thanks to the development of spores. These spores contain several thermophilic enzymes, which can sustain heating up to 100°C or above without losing their properties even after endospore sprouting. The thermophilic properties of endospores may be attributed to the presence of diaminopicolinic acid (DPA). This acid does not occur in vegetative cells, but considerable quantities of it are found in spores (5–15% of dry matter). DPA forms a complex with Ca^{2+} and, in combination with proteins, produces changes that lead to thermophilic properties.

In general, thermophilic organisms have a different cell structure compared with that of mesophiles, with increased quantities of lipid substances with higher melting points and containing higher amounts of saturated, branched-chain fatty acids. Structural differences between enzymes derived from mesophiles and thermophiles are slight but significant. It has been observed that thermophilic organisms contain enzymes and proteins that are resistant to temperature owing to the content of metal ions and hydrophobic amino acids in their molecules. These properties may well contribute to a rapid resynthesis of damaged cellular components. Remedial systems include heat shock proteins (HSP), the concentrations of which in the cells increase rapidly once the temperature exceeds 40°C (Suutari and Laakso, 1994; Tolner et al., 1997; Fields, 2001).

Pedro et al. (1999) characterized the bacteria isolated from a medium-scale EPGD treating 10–20 kg/day of restaurant refuse for over 2 years. The inside temperature was maintained at around 50°C to 60°C with a heating module. All isolates belonged to the genus *Bacillus*, and their 16S rDNA sequences were related to those of *B. licheniformis*, *B. subtilis*, or *B. thermoamylovorans*. These are general microorganisms in the mesophilic and thermophilic phases in field-scale composters (Ryckeboer et al., 2003). The same species were obtained in similar proportions within samples taken during a 6-month operation, indicating the stability of the bacterial community in the composter (Haruta et al., 2005).

The genus *Bacillus* comprises many species, but only some of them are thermophilic. Bacteria of this genus are Gram-positive and usually form rod-shaped cells. They can consume a variety of substrates owing to their formation of a wide spectrum of amylolytic, pectolytic, and proteolytic enzymes. The genus *Bacillus* is also well known for its spore production, which, as mentioned above, makes it well adapted for higher temperatures. Although *Bacillus* species may be either aerobic or facultatively anaerobic, thermophilic strains are usually aerobic. Many strains of *Bacillus* have shown potential for adapting to environmental conditions without apparent changes in their metabolism. The capabilities of mesophilic species (*B. licheniformis*,

TABLE 10.1 Classification of Thermophilic Microorganisms

Type of Microorganism	Temperature Minimal	Temperature Optimal	Temperature Maximal
Temperature-tolerant thermophiles	≤ 25°C	≥ 45°C	≥ 50°C
Moderate thermophiles	> 25°C	> 45°C	> 50°C
Extreme thermophiles	> 50°C	≥ 65°C	≥ 70°C

Adapted from Wiegel et al. (1985).

B. subtilis, B. coagulans) to adapt to growth at high temperatures, even at 60–70°C, deserve special attention. Such species are often described as thermotolerant. *Bacillus* rods can grow in environments characterized by considerable differences in temperature, pH, and salinity (Harwood, 1989).

As discussed, the usefulness of thermophilic bacteria in aerobic sludge treatment systems employing high temperatures is connected with the fact that such systems produce small quantities of biomass and at the same time break down organic substances rapidly and consequently reduce the costs of the entire process (Sonnleitner and Fiechter, 1983; LaPara and Alleman, 1999). Other benefits of TAD are the possibility of heat recovery, the efficient destruction of pathogens, the simplicity of the process, its robustness, a higher reaction rate and consequently smaller bioreactors, and the conservation of nitrogen. Nevertheless, there are very few examples of implementation of aerobic thermophilic technologies for treatment of FIW.

4. BIOREMEDIATION AND BIO-AUGMENTATION STRATEGIES

Wastes from the dairy and food industries are highly polluting: dairy industry wastes, for example, contain proteins, salts, fatty substances, lactose, and various kinds of cleaning chemicals (Kosseva et al., 2003). The use of bio-augmentation strategies, such as the addition of external microorganisms with a high capacity for degradation of specific compounds, can improve the overall performance of biological treatment systems. The relationship of the inoculated microorganism with its new biotic and abiotic environments, in terms of survival, activity, and migration, can be decisive in the outcome of any bio-augmentation strategy. There is increasing evidence from the literature that the best way to overcome the above ecological barriers is to look for microorganisms from the same ecological niche as the polluted area (El Fantroussi and Agathos, 2005; Thompson et al., 2005). The design of an inoculum for bio-augmentation strategies involves the isolation of individual strains from controlled testing to determine which ones show particular abilities to degrade different pollutant compounds present in the local wastewater. Thus, Loperena et al. (2009) isolated milk fat/protein-degrading microorganisms from different locations of a dairy wastewater treatment system with the goal of developing an inoculum for bio-augmentation strategies. Eight isolates, identified by 16S rRNA gene sequence analysis as belonging to the genera *Bacillus*, *Pseudomonas*, and *Acinetobacter*, were tested for their ability to remove COD and protein from a milk-based medium (3000 mg/L COD) and compared with a commercial bio-augmentation inoculum. Based on individual degradation capacity and growth behavior of the isolates, three microorganisms were further selected and tested together. This consortium exhibited a COD removal similar to that of a commercial inoculum (57% and 63%, respectively), but higher removals of protein (consortium 93%; commercial inoculum 54%) and fat (consortium 75%; commercial inoculum 38%).

Using this approach, we have isolated *Lactococcus* sp. from whey using general purpose medium supplemented with skimmed milk powder, as well as the thermotolerant yeast *Kluyveromyces* sp. using yeast and mold broth/agar. These had been employed as starting cultures in cheese making (Ercolini et al., 2003) and we used them to carry out biodegradation of lactose in whey at elevated temperatures (Kosseva et al., 2001, 2003, 2007). We also employed thermophilic digesters to isolate *Bacillus* sp. from a fruit-and-vegetable waste, using nutrient agar. The *Bacillus* sp. were identified using genotypic and phenotypic techniques by Silva et al. (2006). Three main species were found: *B. licheniformis*, *B. subtilis*, and *B. pumilus*. The *Bacillus* sp. organisms were grown in tryptone solution (1%, w/w) with addition of (g/L): lactate (13.8), citrate (5.0), and ethanol (8.0) in shake flasks at 150 rpm at 45°C for 24 h.

A key part of the project carried out by our research consortium sought to develop bioremediation technologies for the reduction of COD of selected food wastes at elevated temperatures. This novel approach should potentially comply with the standards for food industry environmental management systems, notably ISO 14000 (Boudouropoulos and Arvanitoyannis, 2000), and is described below.

4.1 Target Wastes

After considering a range of aqueous waste streams produced by European food processors, we selected a number of target wastes on which to focus our work. These streams contain a range of materials of greater or lesser intransigence to breakdown by biological processes. Each is produced in considerable volumes and presents a significant disposal problem within most countries of the European Union. Three main types of target waste were selected, as follows.

1. *Cheese whey*—containing lactose, with some organic acids and proteins as carbon sources. Stilton "acid" whey (pH ≤ 5) (Glanbia Foods, Tuxford and Tebbutt, UK) was used as the main material for our development of bioremediation technology (carried out at the University of Birmingham) (Table 10.2).
2. *Stillage* (distillers' slops) from the production of bioethanol—representing material from a starchy origin but containing significant amounts of organic

acids and potentially inhibitory components, including ethanol.

   a. Grain stillage—derived from mainly rye and wheat (Table 10.3) (carried out at the Institute of Chemical Technology Prague and at Wrocław University of Economics, respectively).

   b. Potato slops/stillage (Table 10.4) (carried out at Wrocław University of Economics).

3. *Potato processing waters*, from the production of potato starch. Two streams were involved: a hot *post* protein recovery wastewater (HPW) and a cold potato wash water (PSW) (carried out at the University of Agriculture in Poznań) (Tables 10.5 and 10.6)

When bioethanol is produced from starch-based substrates (approximately 39% of its total production), stillage represents the most severe hazard to the environment, owing to a very high COD (Cibis et al., 2006; Krzywonos et al., 2008). The quantities produced are too high to allow complete utilization for conventional agricultural purposes as a fodder or fertilizing agent (see Chapter 2). In search of alternative methods, we applied aerobic thermophilic treatment and studied the influence of aerobic conditions, pH, temperature, and organic load on the biodegradation rates.

4.1.1 Bioconversion of Cheese Whey

Our studies on bioconversion patterns using unpasteurized Stilton cheese whey were divided into three phases: (a) preliminary, shake-flask investigations, (b) development and characterization of staged processing, and (c) investigations into a one-stage process.

In phase (b), we investigated the following two processing strategies:

TABLE 10.2 Typical Composition of Blue Stilton Cheese Whey

Component	Sample 1 Concentration (g/L)	Sample 2 Concentration (g/L)	Sample 3 Concentration (g/L)
Lactose	49.7	45.4	14.4
Lactic acid	6.9	5.2	—
Ethanol	0	0	—
Citric acid	3.1	2.4	—
Acetic acid	2.4	0	—
Sum of organic acids + ethanol	12.4	7.6	10.6
COD (mg-O_2/L)	66,000	72,500	75,000
COD (mg-O_2/L) (filtered whey)	>25,000	~30,000	~30,000
Total protein (filtered whey)	2.8	2.6	5
pH	3.85–3.90	5.30–5.35	3.85–3.90

TABLE 10.3 Typical Analysis of the "Thin" Phase of Rye Stillage, or "Liquid" Phase after Solids Separation Using a Decanter Centrifuge

Component	Units	Value
pH	–	4.2
COD	mg-O_2/L^{-1}	31,500
Total organic carbon (TOC)	mg-C/L^{-1}	10,600
Acetic acid	g/L^{-1}	0.22
Lactic acid	g/L^{-1}	2.81
Succinic acid	g/L^{-1}	1.00
Glycerol	g/L^{-1}	3.69
Reducing substances before inversion (as glucose)	g/L^{-1}	7.71
Glucose	g/L^{-1}	1.43
N total (as N Kjeldahl)	g-N/L^{-1}	0.81
N (ammonia)	g-NH_4^+/L^{-1}	0.14
P total	g-P/L^{-1}	0.22
Total dry matter (after evaporation)	g/L^{-1}	58.5
Total dissolved matter	g/L^{-1}	33.1
Total suspended solids	g/L^{-1}	25.38

TABLE 10.4 Chemical Composition of the Liquid Phase of Potato Slops

Component	Stillage I	Stillage II
pH	3.88	3.88
Density (° Blg)	7.9	12.2
COD (mg-O_2/L^{-1})	103,760	116,330
TOC (mg-C/L^{-1})	35.15	45.60
Lactic acid (g/L^{-1})	17.531	61.138
Propionic acid (g/L^{-1})	2.635	2.772
Acetic acid (g/L^{-1})	2.101	4.135
Formic acid (g/L^{-1})	0.270	1.731
Butyric acid (g/L^{-1})	0.811	3.311
Isobutyric acid (g/L^{-1})	0.101	0.1598
Valeric acid (g/L^{-1})	0.359	1.315
Succinic acid (g/L^{-1})	0.430	0.2275
Malic acid (g/L^{-1})	0.154	0.1722
Citric acid (g/L^{-1})	0.070	0.1453
Reducing substances (g/L^{-1})	37.44	37.06
Glycerol (g/L^{-1})	5.96	3.81
Total nitrogen (g/L^{-1})	1.05	2.57
Ammonia nitrogen (g/L^{-1})	0.308	0.361
Total phosphorus (g/L^{-1})	0.277	0.816
Phosphate phosphorus (g/L^{-1})	0.165	0.588

- Strategy 1—An anaerobic, mesophilic first stage, followed by inoculation with a culture of thermophilic *Bacilli* and an aerobic, mesophilic second stage, both at 45°C;
- Strategy 2—An anaerobic, mesophilic first stage, at 45°C, followed by inoculation with a culture of thermophilic *Bacilli* and an aerobic, thermophilic second stage. We chose for our experimental work second stage temperatures of 55, 60, and 65°C.

Fermentations were carried out in batch mode in a Prelude™ bioreactor system (Biolafitte & Moritz, France), equipped with pH, temperature, and dissolved oxygen tension control. The working volume of the system was 1.3 L:1 L of cheese whey. Analysis of the whey is given in Table 10.2, samples 1 and 2.

4.1.1.1 STRATEGY 1 EXPERIMENT

The duration of the first, anaerobic stage was 48 h, after which we started aeration at 0.5 vvm (volume of

TABLE 10.5 Chemical Composition of Hot Waste after Protein Recovery (HPW)

Component	Units	Range
BOD	mg-O_2/L^{-1}	11,300–26,800
COD	mg-O_2 L^{-1}	15,410–41,940
TOC	mg-C/L^{-1}	6580–13,200
Ammonia nitrogen	mg-N-NH_4/L^{-1}	220–505
Organic nitrogen	mg-N/L^{-1}	438–1,050
Chlorides	mg/L^{-1}	1350–2550
Sulphates	mg/L^{-1}	520–680
Phosphates	mg/L^{-1}	34–164
Potassium	mg/L^{-1}	180–3860

TABLE 10.6 Chemical Composition of Potato Starch Production Waste (PSW)

Component	Units	Range
BOD	mg-O_2/L^{-1}	8,210–14,300
COD	mg-O_2/L^{-1}	12,200–27,080
TOC	mg-C/L^{-1}	4860–5430
Ammonia nitrogen	mg-N-NH_4/L^{-1}	420–557
Organic nitrogen	mg-N/L^{-1}	423–1036
Chlorides	mg/L^{-1}	250–1160
Sulphates	mg/L^{-1}	480
Phosphates	mg/L^{-1}	150
Potassium	mg/L^{-1}	115–3880

air per volume of whey per minute) and more active agitation (500–1000 rpm). The dissolved oxygen tension (DO) was maintained at or above 80% during aerobic processing. The DO variation during the two-stage process is shown in Figure 10.1A. The variations in carbon source concentration during both stages of the cultivation are given in Figure 10.1B. Figure 10.1C shows the corresponding profile of the off-gas carbon dioxide concentration.

From these results, we observed a slow but steady consumption of lactose during both stages, until it was totally depleted by 80 h. Of the main organic acids analyzed, only acetate appeared to rise as soon as lactose began to reduce, but acetate levels remained low, eventually becoming exhausted. Lactate rose sharply after a lag of approximately 20 h, and was the

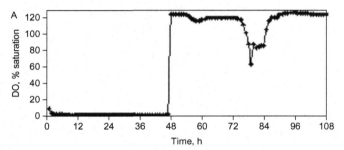

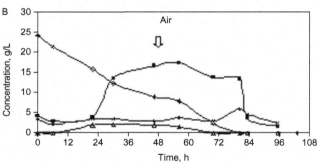

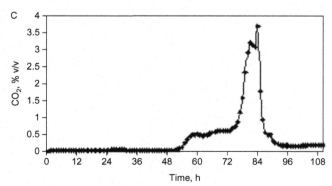

FIGURE 10.1 Data for a Strategy 1 fermentation of Stilton cheese whey at 45°C: **(A)** dissolved oxygen tension profile; **(B)** carbon source profiles; **(C)** off-gas CO_2 concentration profile. In Figure 1B: ◇, lactose; ■, lactate; △, acetate; **X**, citrate.

most important acid produced, accumulating until its consumption during the aerobic stage. Once lactose had been exhausted at approximately 80 h, lactate was consumed rapidly. Citrate accumulated slowly, in low concentrations, until its reduction after lactose exhaustion. An interpretation of this pattern is that initial lactose depletion was associated with biomass growth, until the organisms began to produce excess lactate too, along with smaller amounts of citrate. On aeration, lactose continued to be the main substrate for the population present—most likely, the whey organisms—until lactose exhaustion, when lactate would have taken over as main substrate. Its rapid removal after 80 h was probably due to the *Bacilli* inoculated at the start of aeration. Thus, the results from this experiment were consistent with the following reaction scheme:

1. *Activity of LAB* (starters in cheesemaking (e.g., *Lactococcus* sp.))
   a. Lactose hydrolysis to glucose and galactose: lactase
   $$C_{12}H_{22}O_{11} + H_2O \rightarrow C_6H_{12}O_6 + C_6H_{12}O_6$$
   b. Pyruvate formation: glycolytic system
   $$2C_6H_{12}O_6 \rightarrow 4C_3H_4O_3 + 8H^+$$
   c. Lactate formation from pyruvate: lactate dehydrogenase
   $$4C_3H_4O_3 + 8H^+ \rightarrow 4C_3H_6O_3 + 4ATP$$

2. *Heterofermentative LAB* (lactate and ethanol production)
   $$C_{12}H_{22}O_{11} + H_2O \rightarrow 2C_3H_6O_3 + 2C_2H_5OH + 2CO_2$$

3. *Citrate metabolism (LAB)* (to acetate and oxaloacetate)
   $$C_6H_8O_7 \rightarrow C_2H_4O_2 + C_4H_4O_5 \text{ (oxaloacetate)}$$

4. *Thermotolerant yeasts* (ethanol formation from lactose, glucose, galactose, and lactate) *Kluyveromyces* sp.
   $$C_6H_{12}O_6 \rightarrow 2C_2H_5OH + 2CO_2 \quad C_{12}H_{22}O_{11} + H_2O \rightarrow 4C_2H_5OH + 4CO_2 \quad C_3H_6O_3 \rightarrow C_2H_5OH + CO_2$$

5. *Acetic acid bacteria*
   a. Ethanol oxidation to acetate:
   $$C_2H_5OH + O_2 \rightarrow C_2H_4O_2 + H_2O$$
   b. Lactate oxidation to pyruvate:
   $$4C_3H_6O_3 \rightarrow 4C_3H_4O_3 + 8H^+$$

This reaction scheme features a thermotolerant/thermophilic population, made up from predominantly thermophilic bacteria (notably *Bacillus* sp.), but also containing thermotolerant yeasts (*Kluyveromyces* sp.). The *Bacilli* favor aerobic conditions, and yeasts can function under aerobic and anaerobic conditions, shifting their metabolism to acetic acid production.

Thermotolerant yeasts, under aerobic conditions, shift their metabolism to acetate and biomass production:

$$\text{Yeast } C_{12}H_{22}O_{11} + H_2O \rightarrow 6C_2H_4O_2$$

Thermophilic Bacilli oxidize ethanol, lactate, acetate, and citrate with carbon dioxide and biomass as main products:

$$C_2H_5OH + O_2 \rightarrow C_2H_4O_2 + H_2O$$
$$C_3H_6O_3 + 3O_2 \rightarrow 3CO_2 + 3H_2O$$
$$C_2H_4O_2 + 2O_2 \rightarrow 2CO_2 + 2H_2O$$
$$C_6H_8O_7 + 4.5O_2 \rightarrow 6CO_2 + 4H_2O$$

Over the reaction system as a whole, biomass formation occurs simultaneously:

$$aCH_xO_y + bO_2 + cH_1O_mN_n \rightarrow CH_\delta O_\varepsilon N_\phi + dH_2O + eCO_2$$

where CH_xO_y is a carbon source, $H_1O_mN_n$ is a nitrogen source, and $CH_\delta O_\varepsilon N_\phi$ is biomass formed.

The maximum off-gas carbon dioxide levels coincided with consumption of the main carbon source, lactate (Figures 10.1B and C). The curve showing oxygen depletion (Figure 10.1A) very largely complemented that of off-gas carbon dioxide enhancement, so that the respiratory quotient remained near a value of 1 during consumption of lactate. The total decrease of soluble COD of whey during the combination of anaerobic and aerobic stages of the scheme was approximately 68%. Soluble protein decreased by approximately 59%.

4.1.1.2 STRATEGY 2 EXPERIMENTS

The first, anaerobic mesophilic stage (at 45°C) was carried out in essentially the same way as with the Strategy 1 experiment, except that its duration was extended to approximately 70 h to reduce the lactose level to that in the Strategy 1 run. For the second, aerobic thermophilic stage, a vigorous aeration/agitation regime was employed for each run: 1 to 2 vvm with agitation at 1000 to 1500 rpm. In this way, the dissolved oxygen tension was maintained above 65% saturation during the aerobic stage. The pH was maintained around 7 for this stage too, in the experiments at 60°C and 65°C. The profiles of the carbon sources (i.e., ethanol, citrate, lactate, and acetate) for the operation of stage 2 at 65°C (i.e., pasteurization conditions), after the onset of aeration, are shown in Figure 10.2A. The profile for the off-gas carbon dioxide concentration after the start of aeration, for the same run, is shown in Figure 10.2B.

Comparing Figures 10.1B and 10.2A, we noted that the carbon source profiles for stage 2 of both experiments were similar, suggesting that similar reaction systems were operating under both mesophilic and thermophilic conditions. However, at 65°C, consumption of organic acids was very much faster than at 45°C: starting from the onset of aeration, it took some 40 h to reduce lactate

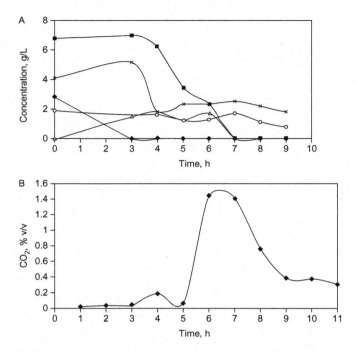

FIGURE 10.2 **Data for the second, aerated stage at 65°C of a Strategy 2 fermentation of Stilton cheese whey: (A) carbon source profiles; (B) off-gas CO₂ concentration profile.** In Figure 2A: ○, lactose; ■, lactate; △, acetate; X, citrate; ◇, ethanol.

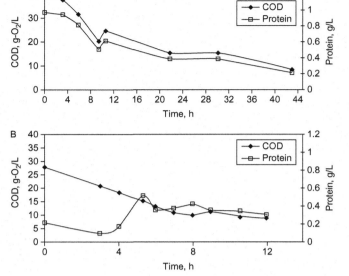

FIGURE 10.3 **Variations of COD and protein: (A) in the 2 L aerated stirred-tank reactor at 0.5 vvm air, 500 rpm, and 45°C; (B) in the 1.3 L aerated stirred-tank reactor at 1–2 vvm, 1000 rpm, and 65°C.**

from its peak value to its final value. However, at 65°C, this took only 7 h, by which time lactate had been totally removed. This speeding up of reaction rates is another advantage of thermophilic processing.

EFFECT OF PH AND TEMPERATURE ON THE EFFICIENCY OF THE BIOREMEDIATION OF STILTON WHEY We found that pH has a significant influence on the efficiency of the thermophilic process: at 65°C, we were unable to carry out successful experiments at pH values lower than 6.7. At 45°C and without pH control, however, the process functioned over a range of pH values: the process illustrated by Figure 10.2 showed a pH fall from an initially adjusted pH ~6 to ~3.5 in the first 15 h, followed by a constant pH region (~3.5) until 80 h, then a rise to ~8 after 100 h of operation. Therefore, we concluded that, where Strategy 2 was employed, pH control around 7 should be activated, at least from the onset of aeration and inoculation by the thermophilic *Bacilli*, which seemed to be the organisms most sensitive to pH at higher temperatures.

The characteristics of the decreases in COD and protein concentration at 45°C and 65°C are shown in Figure 10.3A (for 45°C) and Figure 10.3B (for 65°C). At 45°C, we observed an initial lag in the degradation of most of the carbon sources (citrate and lactate) (Figure 10.1B), while COD and soluble protein utilization patterns were very similar to each other, reflecting

falls and rises at the same time, after initial slow reductions (Figure 10.3A). At 65°C, we observed a similar but much shorter initial lag in the degradation of most of the carbon sources (citrate and lactate) (Figure 10.2A). However, utilization of soluble protein appeared to occur from the start of inoculation with *Bacillus* and aeration, as indicated in Figure 10.3B by the substantially reduced protein levels just after inoculation compared with those at 45°C (Figure 10.3A), while COD was reduced progressively. This was followed by a sharp rise in soluble protein after 4 to 5 hours, not matched by the COD plot, after which the protein level continued to fall gradually (Figure 10.3B). These differences suggest that the microbial consortium might exhibit a slightly different metabolic activity under thermophilic conditions than under mesophilic conditions.

The average rate of lactate biodegradation over the period of the two-stage digestion was approximately 0.5 g/(L h) at 45°C, whereas, at 65°C, it was twice as high: approximately 0.96 g/(L h). This is consistent with other reports in the literature (LaPara and Alleman, 1999; Abeynayaka and Visvanathan, 2011).

The off-gas oxygen concentration profile for stage 2 under both mesophilic and thermophilic conditions showed very similar patterns. Furthermore, curves of off-gas oxygen depletion were very similar to those of carbon dioxide enhancement (Figures 10.1C and 10.2B), so that, as with the totally mesophilic Strategy 1, the respiratory quotient remained near a value of 1 during consumption of lactate. Calculations for the 65°C aerobic thermophilic stage 2 showed that the average decrease of soluble COD was 62.5% based on the COD at the onset of aeration;

soluble protein decreased by approximately 47.5% during this stage. Taken over the whole of this two-stage batch run, approximately 100% reduction of soluble COD and lactose, as well as a 90% decrease in soluble protein, was obtained after 80 hours of cultivation.

4.1.1.3 INVESTIGATIONS INTO REDUCTION OF CHEMICAL OXYGEN DEMAND DURING A ONE-STAGE PROCESS

In an attempt to simplify the process, we decided to determine to what extent we could combine the activities of the two "clusters" of organisms into a single staged process at one temperature, and whether that could be made to function under thermophilic conditions. We started with Stilton whey (collected in summer) that had been allowed to partially degrade, and thus contained active LAB and yeast populations. Its analysis is given in Table 10.2, sample 3.

To 1 L of filtered whey in a 2 L stirred tank bioreactor (Biolafitte "Prelude"), 0.3 L of an inoculum of digester *Bacilli* was added, the pH was adjusted to 7, and maintained there throughout the run, and the system was aerated, with dissolved oxygen tension (DO) being controlled around a set point. We carried out several runs, with incubations at 55, 60, and 65°C and DO controlled at set points between 20% and 80% saturation. Each run was monitored for 24 h. Dry weight of biomass (DWB) was determined gravimetrically because of the turbidity of the whey. Profiles of

selected runs, choosing DO set points of 80% for each temperature for ease of comparison, are presented in Figure 10.4A to C.

Table 10.7 summarizes overall characteristics of the experimental profiles obtained during the one-stage process. These results suggest that, although thermophilic conditions might not have been at their most favorable for the whey organisms, a one-stage process operating at a single thermophilic temperature might well be a feasible option. Though lower than at 55°C, substantial levels of consumption of lactose and organic acids were obtained, along with significant reductions in COD. The overall biomass population grew more prolifically and rapidly at 55°C than at the higher two temperatures, but high biomass production need not necessarily be an advantage for a process designed primarily to reduce COD or to convert substrates to products. Reduced biomass yields in thermophilic aerobic processes have also been reported by others (Visvanathan and Nhien, 1995; Tripathi and Allen, 1999). Another point worth noting is the influence of dissolved oxygen tension. Previously, concern was expressed about the possible challenges posed by the reduced oxygen solubility in the aqueous phase at the elevated temperatures of thermophilic operation. In our experiments using the one-stage strategy, we observed no harmful effects on the process of reduced DO. Indeed, at 55°C and 60°C, we achieved the highest levels of COD reduction operating at a DO level of only 40% saturation, and the highest level of (lactose + acids)

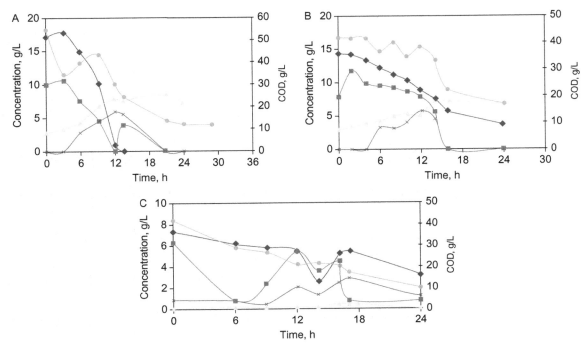

FIGURE 10.4 Profiles for single-stage whey fermentation at 80% dissolved oxygen tension set point: (A) at 55°C; (B) at 60°C; (C) at 65°C. ○, lactose (g/L); ■, LA (g/L); △, DWB (g/L); X, AA (g/L); ◇, COD (g/L).

conversion at 65°C was observed at a DO level of 40% saturation. Even at a DO level of 20% saturation, we did not see any drastic reduction in conversion. Oxygen limitation is more likely to be a problem for the thermophilic *Bacilli* than for the whey organisms (which are microaerophilic), but it would appear that the former are robust enough to operate over a wide range of oxygen availability. It would also appear that the whey organisms, or some of the species, are able to go on functioning even into the thermophilic operating region, to a limited extent: microscopic examination showed the presence of budding (i.e., active) yeasts in the system at 60°C but not at 65°C. Figure 10.5 illustrates this.

The results obtained suggest that temperature may have exerted a larger influence on the biodegradation process than dissolved oxygen, as the composition of the microbial community changed with temperature over the range 55–65°C.

4.1.2 Bioconversion of Grain Stillage/Distiller's Slops

The program of work was influenced by the challenging nature of this waste. Alcoholic fermented grain mash contains substances largely difficult to assimilate (see Table 10.3). This is an important factor to consider in any design of an autothermal process (ATAD). Therefore, among our studies on both "liquid phase" and untreated grain stillage were investigations into the mixing in of other, complementary wastes to facilitate the utilization of components of the stillage and contribute to more generation of heat. All mixed-species populations tested were able to reduce COD levels in distiller's slops (DS) by approximately 45–70%. Addition of mineral nutrients did not appear to improve on this. However, sparging with air enriched by pure oxygen could significantly speed up degradation and also increase COD reduction to approximately 90%.

TABLE 10.7   Summary of Profiles of the One-Stage Process

T (°C)	DO (%)	% COD Removal	% Removal of Lactose + Acids	DWB Initial (g/L)	DWB Final (g/L)	DWB Max (g/L)	Exponential Phase (h)	CO_2 Max (%)
55	20	92.8	96.1	1.48	4.12	4.63	9	3.5
55	40	93.8	96.4	2.5	7.72	8.07	8–10	3.5
55	80	78.6	~100	2.90	7.67	8.50	12	1.0
60	40	65.7	65.3	1.10	2.61	4.14	10	5.5
60	80	60.2	74.1	3.00	6.80	7.11	14	1.6
65	20	59.4	91.4	2.00	4.81	5.72	12	3.3
65	40	68.4	92.1	0.34	5.08	5.08	24	2.5
65	60	67.3	87.5	2.33	4.32	5.00	26	2.0
65	80	77.1	87.6	0.31	1.13	2.07	16	2.0

COD, chemical oxygen demand; DO, dissolved oxygen tension; DWB, dry weight of biomass.

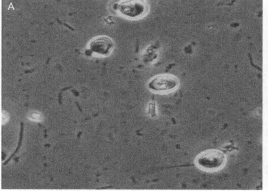 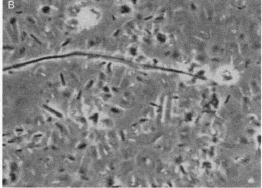

FIGURE 10.5   **Micrographs (× 1000) of single-stage whey fermentation at DO 40% saturation, after 6 hours: (A) bacteria and budding yeasts at 60°C; (B) bacteria at 65°C.**

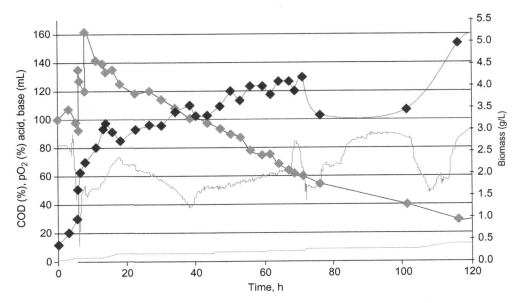

FIGURE 10.6 Fed-batch cultivation of mixed population on modified liquid phase of distiller's grain stillage supplemented with nutrients, at 55°C, waste glycerol feed. ○, pO$_2$ (%); ■, COD (%); Δ, acid (mL); **X**, base (mL); ◇, biomass (g/L).

It was observed that addition of carbon sources could improve the utilization of raw materials and increase the biomass concentration, indicating that the C:N ratio of the DS was too low to constitute a balanced medium for the organisms. Waste glycerol was considered to be a particularly useful source of additional carbon. Figure 10.6 shows the progress of a fed-batch biodegradation of DS at 65°C, with a waste glycerol feed; as a result, 83% reduction of COD was achieved.

From these observations, we would recommend a two-stage process for the bioconversion of this waste stream: an anaerobic digestion first stage, with biogas production, followed by a thermophilic aerobic second stage, in which the first stage effluent liquor (or even fresh stillage) together with excess sludge from the anaerobic digester would be degraded. This should result in a low COD and hygienic product.

4.1.3 Bioconversion of Potato Stillage/Distiller's Slops

The bulk of this work involved mixed-species investigations using digester populations similar to those involved in the whey biodegradation work, but adapted for use in potato slops. The potato stillage (Elipsa Ltd, Katy Wrocławskie, Poland) was filtered through filter paper. The liquid-phase composition of the medium (Stillage I) was as described in Table 10.3. Batch biodegradation processes were carried out in a 5 L working volume stirred-tank reactor (B. Braun) at 45°C and pH 7, which was kept constant throughout, via adjustment with 2M H$_2$SO$_4$ and 2M NaOH. Each aerobic treatment involved agitation at 550 rpm for 93 h. Nutrient addition—(NH$_4$)$_2$SO$_4$ at 2 g/L and (NH$_4$)$_2$HPO$_4$ at 1 g/L—

produced considerable improvements in reduction of COD, BOD$_5$, organic acids, and reducing substances. Thus, the process operated in the bioreactor yielded a COD reduction of 84.2%.

Continuous experiments were also run in the same aerated STR at various temperatures and pH 7.0. Of the main carbon sources (reducing substances, organic acids, glycerol), organic acids were utilized to the highest extent. Analyzing the data from the continuous biodegradation of Stillage II (Table 10.3), our colleagues found that the best results were those obtained with a dilution rate of 0.0121/h, an agitation rate of 900 rpm, and a pH of 6.5. Under such conditions, COD reduction was 92%, but at a comparatively low COD reduction rate of 1.25 g-O$_2$/(L h). Also, the large quantity of suspended solids produced might be considered a drawback. The recommended dilution rate, $D_2 = 0.0121$/h, corresponds to a retention time of 3.4 days. Were feasible processing carried out under thermophilic conditions, such a retention time should guarantee pasteurization of the product.

More recently, Krzywonos et al. (2008) looked at the effect of temperature on the efficiency of the thermo- and mesophilic aerobic batch biodegradation of potato stillage. The experiments were performed in a 5 L stirred-tank reactor at 20, 30, 35, 40, 45, 50, 55, 60, 63, and 65°C with pH 7. Over the temperature range of 20–65°C, the removal efficiency was very high (with a COD reduction following solids separation that varied between 77.57% and 89.14% after 125 h). The highest reaction rate was when the temperature ranged from 30 to 45°C; after at most 43 h, COD removal was 90% of the final removal value obtained for the process. At 20, 55,

60, and 63°C, a 90% removal was achieved after 80 h. Later the same authors (Krzywonos et al., 2009) investigated the extent to which temperature influences the utilization of main carbon sources (reducing substances determined before and after hydrolysis, glycerol, and organic acids) by the same mixed culture in the course of aerobic batch biodegradation of potato stillage, a high-strength distillery effluent (COD = 51.88 g-O_2/L). The experiments were performed at 20, 30, 35, 40, 45, 50, 55, 60, and 63°C, at pH 7, in a 5 L working volume STR with agitation at 550 rpm and aeration at 1.6 vvm. Particular consideration was given to the following issues: (1) the sequence in which the main carbon sources in the stillage were assimilated and (2) the extent of their assimilation achieved under these conditions. The extent of reduction in the content of reducing substances determined before and after hydrolysis ranged from 90.01% to 94.34% and from 84.32% to 95.98%, respectively. The utilization of glycerol varied between 90.23% and 95.94%, while that of total organic acids ranged from 91.70% to 99.63%. Apart from a few exceptions, all the processes led to a complete utilization of those organic acids with an initial content in the potato stillage higher than 1 g/L. 100% use was achieved of lactic acid (except for the processes conducted at 35, 45, and 55°C, where consumption exceeded 98%), propionic acid (except at 55°C, where consumption was 87.69%), and acetic acid (except at 35°C with an extent of utilization amounting to 99.18%). The content of isobutyric acid at 30−63°C, citric acid at 45−63°C, and valeric acid at 45°C was higher at the end of the process than at its start.

4.1.4 Bioconversion of Potato Starch Production Wastes

In this area, two process waters from the production of potato starch were used: a hot *post* protein recovery wastewater (HPW: Table 10.5) and a cold potato wash

water (PSW: Table 10.6). Batch biodegradation profiles of these wastes are depicted in Figure 10.7A and B, respectively.

To investigate the feasibility of an alternative operating mode, and to compare steady-state process behavior with that under unsteady state batch operation, several continuous-feed experiments in a 2 L stirred-tank bioreactor (B. Braun) were carried out. This mode of operation was effective: COD reductions in the range of 60−70%, and approximately 90% reduction in reducing substances, were achieved for dilution rates between 0.02/h and 0.1/h, at 60°C. While consumption of reducing substances seemed to reach a steady level, following start-up, after approximately 2 days' operation (i.e., after passage of approximately 1 reactor volume of feed), it took about 10 days for COD reduction and cell numbers to stabilize. Once this had occurred, performance remained constant for the remaining 10 days of operation.

In starch-producing plants such as the one from which our colleagues obtained the HPW and PSW wastewaters, both streams are produced simultaneously during the production campaign. Therefore, in treating these, it makes sense to consider mixing the two streams to treat them together. Doing this would both decrease the temperature of the hot stream, bringing it into a more suitable operating range for the process, and also reduce the initial load on the process. They therefore investigated this possibility by running a continuous experiment on a 1:1 v/v mixture of HPW:PSW at two dilution rates, 0.02/h and 0.05/h, at 60°C. The results are shown in Figure 10.8.

Figure 10.8 shows similar biodegradation profiles for COD estimated both with and without biomass and at both dilution rates. There was a constant difference between them, presumably representing the COD of the biomass. This is consistent with the approximately constant cell numbers detected after start-up.

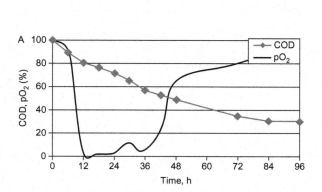
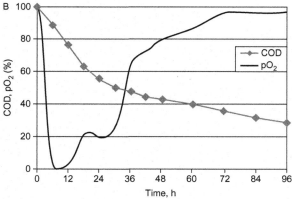

FIGURE 10.7   Biodegradation of process waters from the production of potato starch, batch profile (1) at 55°C (mixed culture): (A) hot *post* protein recovery wastewater (HPW); (B) cold potato wash water (PSW).

Once again, consumption of reducing substances was very high—on the order of 80–90%—and COD reductions were achieved in the range 60–70%. Increasing the dilution rate from 0.02/h to 0.05/h after 13 days did not appear to influence the biodegradation performance significantly. In all cases, the best performance was obtained at pH = 7. This justified carrying out most work at pH = 7.0.

Lasik et al. (2010) reported a comparative analysis of the aerobic thermophilic treatment effect on a high-strength (COD = 35 g-O_2/L) effluent from potato processing, using batch, repeated-batch (with cell recycle and medium replacement), and continuous mode operations. The analysis consisted of: (1) examining the extent of removal of the major parameters such as COD, TOC, TN, and TP, and (2) determining the impact of oxygen deficit on the formation and consumption of organic acids in the course of the three treatment modes. When use was made of the repeated-batch operation and mixed thermophilic population of *Bacillus* sp., the values of the COD and TOC removal rates were more than twice as high as those obtained with the continuous

process, and more than five times as high as those obtained with the batch process (Table 10.8). The results of the study (Table 10.8) showed that the rise in the dilution rates accounted for a noticeable increase not only in the removal rates of COD and TOC, but also in those of TN and TP.

In conclusion, we isolated mixed cultures with the capacity for high levels of biodegradation of both potato-processing wastes tested. The feasibility of carrying out this process in either batch or continuous operation under thermophilic conditions was established, using the wastewaters both separately and mixed together. Given the batch reaction times involved (more than 96 h) and continuous process retention times (20–50 h, corresponding to dilution rates of 0.05/h to 0.02/h, respectively), operation at thermophilic temperatures should produce a pasteurized product.

4.1.5 Bioconversion of Wheat Stillage

Aerobic biodegradation of wheat stillage was investigated at elevated temperature (45°C) (Krzywonos et al., 2010). The aim of this study was to assess how the separation of solids affects the course and efficiency of the batch process, using a mixed culture of bacteria of the genus *Bacillus*. Investigations with and without solids separation were carried out for 144 hours in a 5 L bioreactor, with aeration at 1.6 vvm, agitation at 550 rpm, and at constant pH 6.5. The results showed that separation of solids is not essential, because it had only a minor effect on the reduction of the substrate COD (SCOD) determined after solids separation. SCOD reduction was 88.25% for non-filtered and 92.85% for filtered stillage. Moreover, during biodegradation of the non-filtered stillage the bacterial consortium was able to remove more than 50% of the suspended solids present.

A comparison of the use of mixed cultures of thermo- and mesophilic bacteria of the genus *Bacillus* to treat starch-based distillery effluents at 45°C showed that COD removal varied from 82.6% for maize stillage to 93.7% for wheat-potato stillage. These values were

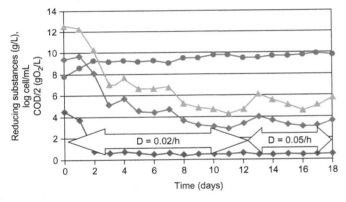

FIGURE 10.8 Continuous biodegradation of a mixture of HPW and PSW (mixed culture, HPW:PSW 1:1, 300 rpm, 60°C, pH 7.0). ○, reducing substances; ■, cell; △, COD with biomass; ◇, COD without biomass.

TABLE 10.8 Calculated Removal Rates (per Liter of Treated Wastewater per Hour)

	Batch Biodegradation				Repeated-Batch Operation			Continuous Biodegradation		
	45°C	55°C	60°C	62°C	12 h	6 h	4 h	D = 0.02/h	D = 0.05/h	D = 0.08/h
COD removal rate (mg-O_2/(L h)	463.5[*]	415.2[*]	366.7[*]	345.9[*]	885.6	1532.8	2075.6	412.2	747.5	983.1
TOC removal rate (mg-C/(L h)	120.9[*]	101.5[*]	97.5[*]	98.2[*]	253.5	452.4	608.4	108.4	220.1	254.6
TN removal rate (mg-N/(L h)	7.9[*]	13.6[*]	12.1[*]	13.3[*]	21.1	31.6	29.1	10.9	22.2	23.7
TP removal rate (mg-P/(L h)	2.6[*]	3.6[*]	2.8[*]	2.4[*]	3.6	5.4	4.6	1.8	3.1	3.4

Source: Lasik et al. (2010).
[*]Calculated for the first 50 h of the process (time required for total replacement of 1 L of wastewater being treated at D = 0.02/h).

obtained with HRTs of 112 h and 165 h for maize stillage and wheat-potato stillage, respectively (Cibis, 2002; Cibis et al., 2006; Krzywonos et al., 2008).

5. A NEW BIOREACTOR DESIGNED FOR THERMOPHILIC DIGESTION

Of the major factors that affect the efficiency of the thermophilic aerobic biodegradation process, temperature and oxygen availability deserve particular attention. The literature includes many references that substantiate the possibility of treating efficiently various high-strength wastewaters at increased temperatures using consortia of thermophilic and mesophilic microorganisms. Promising results have been reported for the biodegradation of potato slops from rural distilleries (Cibis et al., 2002, 2006; Krzywonos et al., 2002, 2008, 2009), the bioremediation of dairy wastes (Kosseva et al., 2001, 2003, 2007), the biodegradation of olive oil and lipid-rich wool scouring wastewater (Becker et al., 1999), and the utilization of the effluents from potato processing (Lasik and Nowak, 2007). However, regardless of the biotreatment method used, the microflora participating in aerobic thermophilic biodegradation utilize oxygen at a high rate, which is responsible for the frequent occurrence of insufficient oxygen supply. The major underlying causes of this insufficiency are: (1) the high oxygen demand of the rapidly growing thermophilic aerobic bacteria, (2) the limited solubility of oxygen at elevated temperature and high pollution load, and (3) poor aeration while attempting to keep process costs down (LaPara and Alleman, 1999; Vogelaar et al., 2000; Cibis et al., 2002, 2006; Ugwuanyi et al., 2005; Lasik and Nowak, 2007; Krzywonos et al., 2008). The supply of oxygen therefore remains one of the major challenges for thermophilic aerobic biotreatment processes.

Our work using thermophilic populations within conventional strirred-tank bioreactors led us to consider other designs of bioreactor to meet the challenge of thermophilic aerobic bioprocessing. One design with promising features was that of a novel gas-liquid reactor developed at the University of Birmingham: the Cocurrent Downflow Contactor (CDC) Reactor. The design of this reactor is simple, capable of high gas holdups and highly efficient gas-liquid mass transfer (Boyes et al., 1994), and involves low residence times (Box 10.1). It was modified for bioprocessing use, and a pilot-scale CDC was incorporated within a transportable rig.

5.1 General Layout and Operation System

The general layout of the Birmingham TAD system designed for the bioconversion of whey waste is shown in Figure 10.9. The main components are the Cocurrent Downflow Contactor (CDC) bioreactor, receiver, and pump along with pipework, fittings, and control and measurement systems. The CDC reactor is a Schott glass column, 0.1 m in diameter and 2.0 m in height with a volume of approximately 16 L (maximum pressure 4.0 bar and maximum temp 80°C). The receiver is a QVF glass vessel (0.3 m diameter and 0.5 m height) with a conical section at the bottom, which reduces to 0.1 m diameter. An additional reducer is installed, which allows the 0.1 m diameter glass cone outlet to be connected to the 0.08 m flanged inlet of the pump. The total volume of the receiver with the cone section is approximately 42 L. In addition to the volumes of the CDC reactor and receiver with cone section, the internal volume of the pump and the piping in the system necessitates a minimum total system working volume of 130 L. Heating is effected by a heating element, and data logging and measurement of temperature, pressure, dissolved oxygen, pH, gas input, and liquid flow rates are undertaken with suitable probes attached to the control unit and connected computer.

5.2 Bioreactor Concept and Description

The CDC evolved from a novel concept of contacting a liquid continuum and a dispersed phase. An intense shearing of the dispersed phase is induced with a minimum expenditure of energy compared with that required for motive power. Where the dispersed phase is a gas or another liquid, an enormous interfacial area is generated in a small containment volume. The interface is subjected to rapid surface renewal through repeated rupture and coalescence, resulting in intense mixing and highly efficient mass transfer. High interfacial areas are produced by exploiting a controlled hydrodynamic flow regime and do not require mechanical aids such as stirrers or baffles. Equipment engineering developments are often a trade-off between performance and operational and capital costs. In the case of the CDC, not only has the performance been improved but operational and capital costs have also been substantially reduced (Boyes et al., 1994).

Some other advantages over conventional gas-liquid contactors (such as stirred tank reactors) are:

- Simple design and operation
- Efficient gas utilization ($\sim$100%)
- High mass transfer coefficient ($k_L = 0.8 - 2.6 \times 10^{-4}$ m/s), and very close approach to equilibrium (97%) in short contact times (<10 s)
- High gas-liquid interfacial area (2000$-$6000 m^2/m^3) and gas holdup (50$-$65%)

BOX 10.1

BASIC FEATURES OF A COCURRENT DOWNFLOW CONTACTOR BIOREACTOR

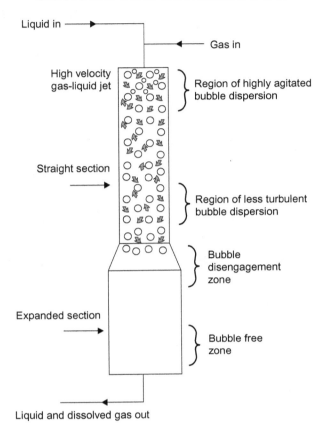

The bioreactor is a column with a gas-liquid contacting entry zone at its top. The entry zone has a gas inlet line connected by a T-piece to a high-velocity liquid inlet stream. The end of the liquid stream is constricted, allowing the formation of a high-velocity, fully submerged gas-liquid jet in the column. The constriction is effected simply by a hydrodynamically designed orifice. The kinetic energy of the stream entering the column produces intense turbulence and mixing. The balance between the rise velocity of the gas bubbles and the liquid downflow velocity produces a constant, rapidly renewed bubble matrix and high-efficiency mass transfer between the gas and liquid phases in the body of the column (WRK Design and Services Ltd, UK).

- Low power consumption and small operating volume
- No internal moving parts, and easy maintenance.

5.3 Bioreactor Performance

Experimental results have shown that the CDC bioreactor system can function as an efficient biodegradation process. Our studies identified *Bacillus* sp. as the numerically dominant organisms within the thermophilic population. Treatment with such populations of cabbage wastes from a major international food processor within stirred reactors and the CDC bioreactor showed both population increases and the maintenance of temperatures significantly in excess of 60°C.

When treating a non-pasteurized "sweet" whey feedstock, the CDC bioreactor rig achieved a 75−80% reduction of COD in approximately 160 h of operation (approximately 120 h after inoculation with *Bacillus* sp. culture), and supported real growth of biomass. An observed temperature drop of mostly 3 to 4°C between the receiver vessel and column could have been due to a combination of cooling on contact with

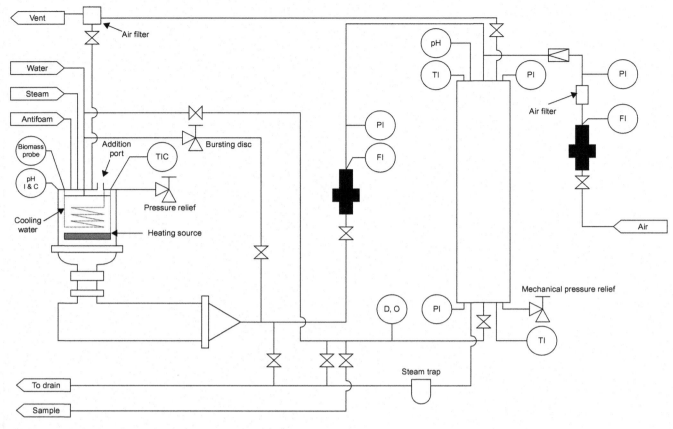

FIGURE 10.9   The general layout of the Birmingham TAD system.

the airstream and heat loss from the system. However, a minimum temperature of 55 to 56°C had been maintained for more than 100 h. Compared with the European Commission's recommendation on biological treatment of biowaste, that aerobic biological sanitation should maintain the material at a minimum of 55°C for 24 h without interruption, this means that the products from this rig, treated as above, should be considered as sanitized (i.e., pathogenfree) (Kent, 2004).

6. FEED PRODUCTION FROM FOOD INDUSTRY WASTES

It has been proposed that TAD, as a vehicle for protein enrichment, may be applied to waste intended for upgrading and recycling as components of animal feed, given also that several agricultural and food industry wastes are currently being studied for upgrading and recycling, particularly in solid-state fermentations such as ensilaging (Ugwuanyi et al., 2006). The emphasis has been on exploiting the capacity of thermophilic populations to rapidly degrade carbohydrates and lipids (with loss of carbon as carbon dioxide) while accumulating N as microbial protein and

the selective conservation of waste protein under appropriate digestion conditions (elevated temperature) where nitrification is unlikely to occur.

White potato peel waste has been digested in a batch 12 L working volume CSTR (Ugwuanyi et al., 2008), operating at 55°C, over a range of aeration rates and pH. The predominant thermophilic populations active in the digestion process were identified as members of the genus *Bacillus*. The majority of the thermophiles were identified as *B. coagulans* and *B. licheniformis* with growth temperature optima at 55°C and growth temperature maximum not exceeding 60°C. Together these accounted for over 60% of thermophilic populations isolated. Other participating populations were identified as *B. stearothermophilus* with growth temperature optimum at 65°C and maximum extending up to and above 70°C. The population types and relations of the thermophiles did not appear to be significantly influenced by pH or aeration rate, even though total population declined slightly at pH 9.5 relative to the lower pH values studied. In general, amino acid content (protein quality) of the wastes decreased as the pH of digestion increased, and this became more pronounced above pH 8.0. This was in addition to the significant drop in the content of crude protein as the digestion pH increased above neutral, particularly at pH 9.0 and above. Comparing digestions at 0.5 and 1.0 vvm aeration rates, the quantity of amino acids was generally higher at the higher aeration rate. Therefore, the variation in amino acid content of wastes with digestion pH and aeration rates suggests that the potential exists for improvement in protein quality of digest with process optimization. Comparison of the quality of the digested waste protein with the Food and Agricultural Organization (FAO) standard in respect of some essential amino acids indicated that the protein was just comparable to that recommended for use in supplementation of animal feed. The digestion conditions (e.g., temperature) employed were such as to inactivate and discourage the growth of any pathogenic and enteric populations that may have been present in or entered the digesting waste. The ability of TAD to achieve sterility of treated biomass is a principal reason for the growth and acceptability of the process. However, issues related to the digestibility of the microbial biomass might be a concern and will need to be handled as for similar SCP products. This study demonstrated the basis for, and capability of, TAD to achieve protein enrichment and upgrading of carbohydrate-rich low cost waste for possible application/utilization in animal feeding.

A further point to note is that the AT process has twofold higher μ_m and higher K_s values than mesophilic processes (Abeynayaka and Visvanathan, 2011). Under higher loading rates, a higher degradation rate of the thermophilic process can be achieved. The thermophilic aerobic process also has a tenfold higher k_d. Under substrate-limited conditions higher decay causes low net biomass production. Higher decay rates in the thermophilic aerobic process release nitrogen back to the wastewater as organic nitrogen. The process has considerable potential for low cost reprocessing for animal feed use of a wide variety of food/agricultural wastes, much of which attract considerable cost penalties for treatment and disposal by conventional methods and would otherwise have no economic value (Ugwuanyi et al., 2006). Another example is cheese whey. Treated via our thermophilic technology with mixed *Bacillus* sp., this effluent can be utilized in pig farming (Lammas Resources Ltd, UK). The feeding process is more efficient compared with the use of raw cheese whey, because the concentration of indigestible lactose is dramatically reduced; also, the pasteurized material is enriched with microbial protein and amino acids. A further reason to use TAD-derived products as feed supplements is that the microbial biomass produced at the end of the process contains plenty of spores. They could serve as a probiotic component in the animal feed after licensing and evaluation of the safety of individual *Bacillus* strains as well as species present in the digester, on a case-by-case basis.

Moreover, bacterial spore formers, mostly of the genus *Bacillus*, are among the large number of probiotic products in use today (for example, BioGrow®; http://www.provita.co.uk) (Hong et al., 2004). Understanding the nature of this probiotic effect is complicated, not only because of the complexities of understanding the microbial interactions that occur within the gastrointestinal tract (GIT), but also because *Bacillus* species are considered allochthonous microorganisms (ones that have a bimodal life cycle of growth and sporulation in the environment as well as within the GIT). The review of Hong et al. (2004) summarizes the commercial applications of *Bacillus* probiotics. Specific mechanisms for how *Bacillus* species can inhibit gastrointestinal infections are covered, as well as the safety and licensing issues that affect the use of *Bacillus* species for commercial development.

7. CONCLUSIONS

After the isolation, adaptation, and screening of a substantial collection of thermophilic microorganism cultures and mixed populations, our research consortium has at its disposal robust populations capable of high levels of aerobic biodegradation over a range of temperatures (in excess of 60°C). RAPD-PCR, Biolog phenotypic profiles, Rep-PCR, and ERIC-PCR have identified most of the organisms in these populations as *Bacilli*, largely clustered into groups of similar genetic composition to *B. licheniformis*, *B. subtilis*, and *B. pumilis*. We have investigated biodegradation profiles and sequences on each target waste, using mixed populations and single species, and have built up an extensive knowledge base useful for assisting in process design and model development.

The potential of some of our target wastes as sources of added-value products has been confirmed, with the successful production of xanthan gums from cheese whey (Papagianni et al., 2001), and polyhydroxyalkanoates from a number of carbon sources including cheese whey (Pantazaki et al, 2003).

A novel reactor configuration has been designed and constructed, based upon Cocurrent Downflow Contactor (CDC) technology, the CDC bioreactor system. This is capable of enhanced oxygen mass transfer and gas utilization and very high gas phase hold-up. As a result of testing the rig, we have confirmed its capability of effectively biodegrading cheese whey, with more than 75% reduction in COD having been achieved.

Combining the advantages of low biomass yields and rapid kinetics associated with high temperature operation and stable process control of aerobic systems, AT processes are valuable options for the treatment of FIW, but clearly they have been under-used. Having the potential of both producing pathogen-free products and the generation of energy out of the process, TAD could be extensively used for the recovery of valuable products, such as animal feed, fertilizer, biopolymers, biosurfactants, organic acids, and others.

References

Abeynayaka, A., Visvanathan, C., 2011. Mesophilic and thermophilic aerobic batch biodegradation, utilization of carbon and nitrogen sources in high-strength wastewater. Bioresour. Technol. 102, 2358−2366.

Beaudet, R., Gagnon, C., Bisaillon, J.G., Ishaque, M., 1990. Microbial aspects of aerobic thermophilic treatment of swine waste. Appl. Environ. Microbiol. 56, 971−976.

Becker, P., Kostner, D., Popov, M.N., Markossian, S., Antranikian, G., Markl, H., 1999. The biodegradation of olive oil and the treatment of lipid-rich wool scouring wastewater under aerobic thermophilic conditions. Water Res. 33, 653−660.

Boudouropoulos, I.D., Arvanitoyannis, I.S., 2000. Potential and perspectives for application of environmental management system (EMS) and ISO 14000 to food industries. Food Rev. Int. 16, 177−237.

Boyes, A.P., Raymahasay, S., Tilston, M.W., Lu, X-X., Sarmento, S., Chugtai, A., et al., 1994. The cocurrent downflow contactor—a novel reactor for gas-liquid-solid catalysed reactions. Chem. Eng. Tech. 17, 307−312.

Bruce, A.M., Pike, E.B., Fisher, W.J., 1990. A review of treatment process options to meet the EC sludge directive. J. IWEM 4, 1−13.

Cheunbarn, T., Pagilla, K.R., 2000. Aerobic thermophilic and anaerobic mesophilic treatment of sludge. J. Environ. Eng.-ASCE 126, 790−795.

Chiang, C.F., Lu, C.J., Sung, L.K., Wu, Y.S., 2001. Full-scale evaluation of heat balance for autothermal thermophilic aerobic treatment of food processing wastewater. Water Sci. Technol. 43, 251−258.

Cibis, E., Kent, C.A., Krzywonos, M., Garncarek, Z., Garncarek, B., Miskiewicz, T., 2002. Biodegradation of potato slops from a rural distillery by thermophilic aerobic bacteria. Bioresour. Technol. 85, 57−61.

Cibis, E., Krzywonos, M., Miskiewicz, T., 2006. Aerobic biodegradation of potato slops under moderate thermophilic conditions: effect of pollution load. Bioresour. Technol. 97, 679−685.

Couillard, D., Zhu, S., 1993. Thermophilic aerobic process for the treatment of slaughterhouse effluents with protein recovery. Environ. Pollut. 79, 121−126.

EC SCAR report, 2011. Sustainable food consumption and production in a resource-constrained world.

El Fantroussi, S., Agathos, S.N., 2005. Is bioaugmentation a feasible strategy for pollutant removal and site remediation? Curr. Opin. Microbiol. 8, 268−275.

EPA, 1992. Control of Pathogens and Vector Attraction in Sewage Sludge. Office of Research and Development, US Environmental Protection Agency, Washington, DC, EPA/625/R-92/013 December.

Ercolini, D., Hill, P.J., Dodd, C.E.R., 2003. Bacterial community structure and location in stilton cheese. Appl. Environ. Microbiol. 69, 3540−3548.

Fields, P.A., 2001. Review: protein function at thermal extremes: balancing stability and flexibility. Comp. Biochem. Physiol. 129, 417−431.

Haruta, S., Nakayama, T., Nakamura, K., Hemmi, H., Ishii, M., Igarashi, Y., et al., 2005. Microbial diversity in biodegradation and reutilization processes of garbage. J. Biosci. Bioeng. 99, 1−11.

Harwood, E.R. (Ed.), 1989. Bacillus. Plenum Press, New York.

Henze, M., Harremoes, P., la Cour Jansen, J., Arvin, E., 2002. Wastewater Treatment: Biological and Chemical Processes. Springer, Berlin.

Hong, H.A., Duc, L.H., Cutting, S.M., 2004. The use of bacterial spore formers as probiotics. FEMS Microbiol. Rev.

Juteau, P., 2006. Review of the use of aerobic thermophilic bioprocesses for the treatment of swine waste. Livest. Sci. 102, 187−196.

Kent, C.A., 2004. EC FP5 Quality of Life and Management of Living Resources. Final Report on "Enhanced, Intelligent Processing of Food and Related Wastes using thermophilic populations". European Commission, Brussels.

Kosseva, M.R., Kent, C.A., Lloyd, D.R., 2001. Thermophilic bioremediation of whey: effect of physico-chemical parameters on the efficiency of the process. Biotechnol. Lett. 23, 1675−1679.

Kosseva, M.R., Kent, C.A., Lloyd, D.R., 2003. Thermophilic bioremediation strategies for a dairy waste. Biochem. Eng. J. 15, 125−130.

Kosseva, M.R., Fatmawati, A., Palatova, M., Kent, C.A., 2007. Modelling thermophilic cheese whey bioremediation in a one-stage process. Biochem. Eng. J. 35, 281−288.

Kristjansson, J.K., 1992. Thermophilic Bacteria. CRC Press, Boca Raton.

Krzywonos, M., Cibis, E., Miskiewicz, T., Kent, C.A., 2008. Effect of temperature on the efficiency of the thermo- and mesophilic aerobic batch biodegradation of high-strength distillery wastewater (potato stillage). Bioresour. Technol. 99, 7816−7824.

Krzywonos, M., Cibis, E., Lasik, M., Nowak, J, Miskiewicz, T., 2009. Thermo- and mesophilic aerobic batch biodegradation of high-strength distillery wastewater (potato stillage)—utilisation of main carbon sources. Bioresour. Technol. 100, 2507−2514.

Krzywonos, M., Cibis, E., Ryznar-Luty, A., Miskiewicz, T., Borowiak, D., 2010. Aerobic biodegradation of wheat stillage (distillery wastewater) at an elevated temperature—effect of solids separation. Biochem. Eng. J. 49, 1−6.

LaPara, T.M., Alleman, J.E., 1999. Thermophilic aerobic biological wastewater treatment. Water Res. 33, 895−908.

LaPara, T.M., Konopka, A., Nakatsu, C.H., Alleman, J.E., 2000. Thermophilic aerobic wastewater treatment in continuous-flow bioreactors. J. Environ. Eng. 126, 739−744.

Lasik, M., Nowak, J., 2007. Effect of pollution load and oxygen availability on thermophilic aerobic continuous biodegradation of potato processing wastewater. Eng. Life Sci. 7, 187−191.

Lasik, M., Nowak, J., Krzywonos, M., Cibis, E., 2010. Impact of batch, repeated-batch (with cell recycle and medium replacement) and continuous processes on the course and efficiency of aerobic thermophilic biodegradation of potato processing wastewater. Bioresour. Technol. 101, 3444−3451.

Lim, B.R., Huang, X., Hu, H.-Y., Goto, N., Fujie, K., 2001. Effects of temperature on biodegradation characteristics of organic pollutants and microbial community in a solid phase aerobic bioreactor treating high strength organic wastewater. Water Sci. Tech. 43, 131−137.

Liu, S.G., Song, F.Y., Zhu, N.W., Yuan, H.P., Cheng, J.H., 2010. Chemical and microbial changes during autothermal thermophilic aerobic digestion of sewage sludge. Bioresour. Technol. 101, 9438−9444.

Liu, S., Zhu, N., Li, L.Y., 2012. The one-stage autothermal thermophilic aerobic digestion for sewage sludge treatment: stabilization process and mechanism. Bioresour. Technol. 104, 266−273.

Loll, U., 1976. Purification of concentrated organic wastewaters from the foodstuffs industry by means of aerobic-thermophilic degradation process. Prog. Water Technol. 48, 373−379.

Loperena, L., Mario Daniel Ferrari, M.D., Díaz, A.L., Ingold, G., Leticia Verónica Pérez, L.V., Carvallo, F., et al., 2009. Isolation and selection of native microorganisms for the aerobic treatment of simulated dairy wastewaters. Bioresour. Technol. 100, 1762–1766.

Metcalf & Eddy Inc., Tchobanoglous, G., Burton, F.L., Stensel, H.D., 2003. Wastewater Engineering: Treatment and Reuse. McGraw-Hill, Boston.

Nakano, K., Matsamura, M., 2001. Improvement of treatment efficiency of thermophilic oxic process for highly concentrated lipid wastes by nutrient supplementation. J. Biosci. Bioeng. 92, 532–538.

Niaounakis, M., Halvadakis, C.P., 2006. Olive Processing Waste Management. Elsevier Ltd., Amsterdam, Waste Management Series 5, pp. 238–244.

Pantazaki, A.A., Tamvaka, M.G., Langlois, V., Guerin, P., Kyriakidis, D.A., 2003. Polyhydroxyalkanoates (PHA) biosynthesis in *Thermus thermophilus*: purification and biochemical properties of the PHA synthase. Mol. Cell. Biochem. 254, 173–183.

Papagianni, M., Psomas, S.K., Batsilas, L., Paras, S., Kyriakidis, D.A., Liakopoulou-Kyriakides, M., 2001. Xanthan production by *Xanthomonas campestris* in batch culture. Process Biochem. 37, 73–80.

Pedro, M.S., Hayashi, N.R., Mukai, T., Mukai, T., Ishii, M., et al., 1999. Physiological and chemotaxonomical studies on microflora within a composter operated at high temperatures. J. Biosci. Bioeng. 88, 92–97.

Rockstrom, J., et al., 2009. Planetary boundaries: exploring the safe operating space for humanity. Ecol. Soc. 14, 32.

Ryckeboer, J., Mergaert, J., Coosemans, J., Deprins, K., SWings, J., et al., 2003. Microbiological aspects of biowaste during composting in a monitored compost bin. J. Appl. Microbiol. 94, 127–137.

Silva, M.T.S.L., Santo, F.E., Pereira, P.T., Roseiro, J.C.P., 2006. Phenotypic characterization of food waste degrading *Bacillus* strains isolated from aerobic bioreactors. J. Basic Microbiol. 46, 34–46.

Skjelhaugen, O.J., 1999. Thermophilic aerobic reactor for processing organic liquid wastes. Water Res. 33, 1593–1602.

Sonnleitner, B., Fiechter, A., 1983. Bacterial diversity in thermophilic aerobic sewage sludge. II. Types of organisms and their capacities. Appl. Microbiol. Biotechnol. 18, 174–180.

Sürücü, G.A., Chian, E.S.K., Engelbrecht, R.S., 1976. Aerobic thermophilic treatment of high strength wastewater. J. Wat. Pollut. Control Fed. 4, 669–679.

Suutari, M., Laakso, S., 1994. Microbial fatty acid and thermal adaptation. Crit. Rev. Microbiol. 20, 285–328.

Thompson, I.P., van der Gast, C.J., Cirio, L., Singer, A.C., 2005. Bioaugmentation for bioremediation: the challenge of strain selection. Environ. Microbiol. 7, 909–915.

Tilman, D., Fargione, J., Wolff, B., D'Antonio, C., Dobson, A., Howarth, R., et al., 2001. Forecasting agriculturally driven global environmental change. Science 292, 281–284.

Tolner, B., Poolman, B., Konings, W.W., 1997. Adaptation of microorganisms and their transport systems to high temperatures. Comp. Biochem. Physiol. 118, 423–428.

Tripathi, C.R., Allen, D.G., 1999. Comparison of mesophilic and thermophilic aerobic biological treatment in sequencing batch reactors treating bleached kraft pulp mill effluent. Water Res. 33, 836–846.

Ugwuanyi, J.O., Harvey, L.M., McNeil, B., 2005. Effect of aeration rate and waste load on evolution of volatile fatty acids and waste stabilization during thermophilic aerobic digestion of a model high strength agricultural waste. Bioresour. Technol. 96, 721–730.

Ugwuanyi, J.O., Harvey, L.M., McNeil, B., 2006. Application of thermophilic aerobic digestion in protein enrichment of high strength agricultural waste slurry for animal feed supplementation. J. Chem. Technol. Biotechnol. 81, 1641–1651.

Ugwuanyi, J.O., Harvey, L.M., McNeil, B., 2008. Diversity of thermophilic populations during thermophilic aerobic digestion of potato peel slurry. J. Appl. Microbiol. 104, 79–90.

Visvanathan, C., Nhien, T.T.H., 1995. Study on aerated biofilter process under high temperature conditions. Environ. Technol. 16, 301–314.

Vogelaar, J.C.T., Klapwijk, A., Van Lier, J.B., Rulkens, W.H., 2000. Temperature effects on the oxygen transfer rate between 20 and 55°C. Water Res. 34, 1037–1041.

Wiegel, J., Ljungdahl, L.G., Demain, A.L., 1985. The importance of thermophilic bacteria in biotechnology. CRC Crit. Rev. Biotechnol. 3, 39–108.

Zvauya, R., Parawira, W., Mawadza, C., 1994. Aspect of aerobic thermophilic treatment of Zimbabwean traditional opaque-beer brewery wastewater. Bioresour. Technol. 48, 273–274.

11

Modeling, Monitoring, and Process Control for Intelligent Bioprocessing of Food Industry Wastes and Wastewater

Maria R. Kosseva and C.A. Kent

1. INTRODUCTION

Mathematical modeling is the use of mathematical language to describe the behavior of a system. It has been widely utilized in science and engineering in order to improve understanding of the behavior of systems, explore new theoretical concepts, predict system performance, and aid in the solution of practical design problems. In the latter context, mathematical models offer the potential to reduce, or even replace, the need for physical experimentation when exploring new material and/or process options. Given the challenges and costs involved in conducting appropriate laboratory and pilot scale investigations, an increased ability to assess new process options through such modeling is to be welcomed (Mason, 2006).

Mathematical modeling is a cyclic process, which starts with setting up a conceptual (verbal) model; the next step is to develop a provisional mathematical model (MM), and then solve the equations within the chosen model boundaries. The procedures of parameter estimation, parameter sensitivity analysis and optimization, and experimental design for model testing and training follow. Finally, if the model cannot predict the main process responses and fit the experimental data created for verification of the model, the modeling cycle will start from the beginning.

2. MATHEMATICAL MODELS OF BIOREACTORS AND BIODEGRADATION PROCESSES

Mathematical model-based simulations of actual bioreactor runs predict how process variables, such as substrate and product concentrations, change and how nutrient feeding should be "tuned" with respect to time, pattern, concentration, and composition to elicit a desired response. Insights gained from modeling can guide us in the adjustment of a process, reducing the number of characterization cycles required. Furthermore, comparing actual experimental results with model predictions helps improve the models themselves. Figure 11.1 summarizes and depicts a non-exhaustive classification of model forms useful in bioreactor modeling and simulation involving nonlinear modeling classification and its techniques (Julien and Whitford, 2007).

Most of the modeling work in our laboratories (Kent et al., 2004) has used *explicit models*, in which there are explicit statements of the relationships between inputs and outputs. These relationships are frequently in the form of differential equations, representing material (and/or energy) balances for major components of the model, and often based upon stated key reaction schemes and assumed or determined kinetic forms. These may be also termed *deterministic* or *mechanistic* models.

Food Industry Wastes.
DOI: http://dx.doi.org/10.1016/B978-0-12-391921-2.00011-1

191

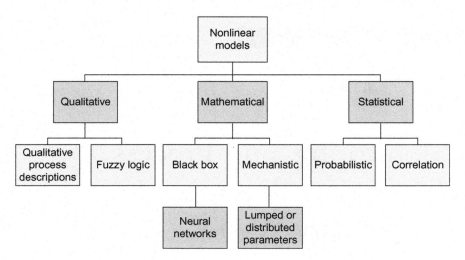

FIGURE 11.1 **Classification of nonlinear model forms used in bioreactor modeling.** *Source: Julien and Whitford (2007).*

Within our European Commission-funded thermophilic bioremediation project, two central scientific and technological objectives were to develop, test, and assess process models and knowledge-based control schemes to achieve robust, "optimal" populations and performance of specific aerobic thermophilic bioremediation processes, including cheese whey and potato stillage/distiller's slops. Elsewhere, mathematical models were also developed for dry anaerobic thermophilic digestion of chicken waste and sludge (Kosseva et al., 2010).

2.1 Modeling of Aerobic Biodegradation of Cheese Whey

In order to support our bioremediation technology, and enable its effective exploitation, at the University of Birmingham we developed a mathematical model to describe the one-stage aerobic biodegradation process described in Chapter 10, and also estimated optimal values of the model parameters. This model was intended for two areas of use: (a) to facilitate the rapid development of process conditions most suited to given remediation tasks by means of predictive simulation of bioreaction patterns, and (b) as the centre of a bioreactor control system, for example to regulate the feeding regime of a continuous or fed-batch bioreactor according to a given cost function related to the process aim. Because of the complexity of the reacting system—multiple substrates and multiple organism species—we considered that the most realistic approach was likely to be to develop a model that simplified the description of substrates or organisms or both and yet enabled reasonable approximations of process behavior.

Several of our models had their origins in the IAWQ Activated Sludge Models 1, 2, and 3 (Henze

et al., 2000) and are mass-balance-based models. Their structure was simplified to accommodate the food industry wastes targeted within our European project (presented in Chapter 10).

Biodegradation patterns of the one-stage whey bioremediation process showed that lactose was the first substrate to be degraded by the LAB and yeast, with lactate and acetate production, respectively, followed by consumption of lactate and acetate by *Bacillus* sp. at 55°C and 60°C (as described in Chapter 10). We considered two modeling approaches: a "full model" Approach I, in which all three substrates are consumed, and a simplified Approach II.

2.1.1 Approach I

We developed a mathematical model for aerobic biodegradation of whey based on the following approximations.

- A mixed culture of LAB, "lactic" yeast, and *Bacillus sp.* was used in practice, but biomass was lumped into one "equivalent culture" with a single, equivalent specific growth rate, μ.
- The carbon sources in whey were expressed as three "equivalent substrates": 1, lactose; 2, lactate; 3, acetate.
- Separate specific growth rates, one on each "substrate": μ_1 for LAB/yeasts (on lactose), μ_2 for thermophiles (on lactate), μ_3 for thermophiles (on acetate).
- Multiple Monod-type kinetics, for every carbon source.

A glossary of terms is given in Box 11.1.
Biomass growth:

$$\frac{dX}{dt} = \mu \cdot X - K_d \cdot X \qquad X(0) = X^0 \qquad (1)$$

BOX 11.1

GLOSSARY OF TERMS IN MATHEMATICAL MODEL FOR AEROBIC BIODEGRADATION OF WHEY

μ	specific growth rate (h^{-1})	X	lumped biomass concentration (g/L^{-1})
C_1	concentration of lactose (g/L^{-1}), experimental data	$Y_{X/C1}$, $Y_{X/C2}$, $Y_{X/C3}$	yield coefficients ((g biomass)/ (g substrate))
C_2	concentration of lactate (g/L^{-1})	t	duration of the experiment (h)
C_3	concentration of acetate (g/L^{-1})		
K_{1C}, K_{2C}, K_{3C}	saturation constants (g/L^{-1})		
K_d	decay coefficient (h^{-1})		

Carbon substrate consumption:

$$\frac{dC_1}{dt} = -\frac{1}{Y_{X/C1}}\mu_1 \cdot X \qquad C_1(0) = C_1^0 \qquad (2)$$

$$\frac{dC_2}{dt} = -\frac{1}{Y_{X/C2}}\mu_2 \cdot X + \frac{1}{Y_{X/C1}}\mu_1 \cdot X \qquad C_2(0) = C_2^0 \qquad (3)$$

$$\frac{dC_3}{dt} = -\frac{1}{Y_{X/C3}}\mu_3 \cdot X + \frac{1}{Y_{X/C1}}\mu_1 \cdot X \qquad C_3(0) = C_3^0 \qquad (4)$$

Biomass growth kinetics:

$$\mu = \mu_1 + \mu_2 + \mu_3 \qquad (5)$$

$$\mu_1 = \mu_{1max}\frac{C_1}{K_{1C} + C_1} \qquad (6)$$

$$\mu_2 = \mu_{2max}\frac{C_2}{K_{2C} + C_2} \qquad (7)$$

$$\mu_3 = \mu_{3max}\frac{C_3}{K_{3C} + C_3} \qquad (8)$$

Figure 11.2 (A−D) shows "best fit" solutions of the mathematical model, related to biomass (A), lactose (B), lactic acid (C), and acetic acid (D) concentrations.

Values of yield coefficients calculated for each experimental run (at different temperatures and DO levels) and from the model are shown in Table 11.1. Specific growth rates were estimated using biomass concentrations obtained from experiments and from the model; these are also shown in the table. The average specific growth rate estimated from the experimental data was approximately 0.09/h, whereas that found from the model data was approximately 0.08/h. The highest specific growth rates, in the range 0.12/h to 0.16/h, were observed at DO=40% and temperatures of 60°C and 65°C.

A similar trend was observed for the yield coefficients on lactic acid: the highest values were observed at 55°C and 65°C and DO = 40%. The average yield coefficients are given in Table 11.1. These were 0.35 g/g on lactose substrate, 0.43 g/g on lactic acid substrate,

and 0.86 g/g on acetic acid substrate. This applied to all three groups of microorganisms. At 55°C, the highest yields on lactose and growth rates were observed at DO = 20%. This is consistent with the properties of LAB, which are microaerophilic bacteria. Their yield and specific growth rate declined with an increase in DO. At 60°C, growth of thermotolerant yeast on lactose predominated. This could be an explanation of the highest yield at that temperature being at DO = 80%. An increase in yields on acetic acid with a rise in DO at 55°C can be explained by increased production of yeast when they shift their metabolism to acetic acid production under aerobic conditions, also with the aerobic growth of thermophilic biomass on acetic acid produced by yeast.

It has been claimed that thermophilic aerobic treatment may be limited by a poor oxygen transfer rate owing to lower oxygen solubility at higher temperatures. However, Boogerd et al. (1990) found that the oxygen transfer rate showed only minor changes in the temperature range 15−70°C in a mechanically mixed bioreactor. Vogelaar et al. (2000) showed that the reduced oxygen saturation concentration in thermophilic sludge was offset by the increased overall oxygen transfer coefficient, and that the oxygen transfer rate was only slightly affected by liquid temperature in the range 20−55°C. We performed a statistical assessment on our data of the impact of DO on Y and μ, using the correlation data analysis tool in Microsoft® Excel. Correlation coefficients varied from −0.39 to −0.20 for $Y_{X/C1}$ and $Y_{X/C2}$ (Y_1 and Y_2), which demonstrated a small negative correlation. The correlation coefficient obtained for $Y_{X/C3}$ (Y_3) was 0.50, showing a medium positive correlation between DO and yield coefficient on acetate, as acetate is produced by yeast under aerobic conditions. The correlation coefficient determined for μ was −0.52—a negative correlation between DO and specific growth rate of biomass. These statistics were in general agreement with the trends observed and explained above. Therefore, in our work

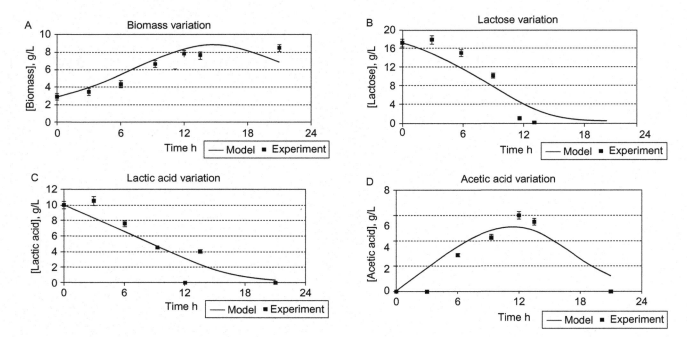

FIGURE 11.2   Simulation results (Approach I) and experimental data for 55°C, DO=80%. Assessments of levels of error in analytical measurements (shown in this figure and in Figures 11.3 and 11.4) were made on the basis of triplicate samples, for biomass dry weight determinations, and manufacturer's information, for HPLC determinations of concentrations of lactose, lactic acid, and acetic acid (Kosseva et al., 2007).

TABLE 11.1   Values of the Yield Coefficients and the Specific Growth Rates Calculated for Each Experimental Run and from the Model (Subscripts m)

Exptl. run	T (°C)	DO (%)	Y_1 (g biomass/ g lactose)	Y_{1m} (g biomass/ g lactose)	Y_2 (g biomass/ g lactate)	Y_{2m} (g biomass/ g lactate)	Y_3 (g biomass/ g acetate)	Y_{3m} (g biomass/ g acetate)	μ (h^{-1})	μ_m (h^{-1})
1	55	20	0.443	0.422	0.352	0.323	0.751	0.60	0.11	0.086
2	55	40	0.355	0.274	0.496	0.30	0.857	0.812	0.094	0.06
3	55	80	0.327	0.319	0.494	0.570	0.865	1.0	0.087	0.09
4	60	40	0.275	0.281	0.473	0.462	0.941	1.18	0.12	0.08
5	60	80	0.356	0.331	0.383	0.496	0.90	1.466	0.056	0.061
6	65	20	0	0	0.422	0.439	0	0	0.087	0.078
7	65	40	0	0	0.530	0.508	0	0	0.16	0.145
8	65	60	0	0	0.372	0.454	0	0	0.017	0.028
9	65	80	0	0	0.32	0.14	0	0	0.06	0.05
Average			0.351	0.325	0.427	0.410	0.863	1.010	0.088	0.075

Source: Kosseva et al. (2007).

we can generally conclude that no clear influence of DO levels on the biomass parameters studied (Y and μ) was shown.

2.1.2 Approach II

The biodegradation profile at 65°C showed no evidence of lactose degradation and acetate production, so the following simplifications were applied to the model of Approach I. Growth rates of LAB and yeast were neglected: $\mu_1 = 0$ (for LAB) and $\mu_2 = 0$ (for yeasts) and the model was modified as shown below.

$$\frac{dX}{dt} = \mu \cdot X - K_d \cdot X \quad X(0) = X^0 \qquad (9)$$

$$\frac{dC_1}{dt} = 0 \quad C_1(0) = C_1^0 \qquad (10)$$

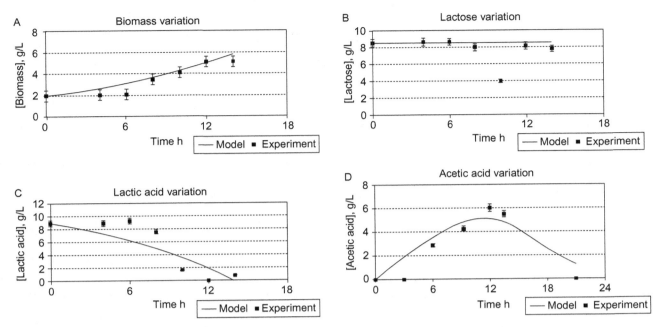

FIGURE 11.3 Simulation results (Approach II) and experimental data for 65°C, DO = 20%. *Source: Kosseva et al. (2007).*

$$\frac{dC_2}{dt} = -\frac{1}{Y_{X/C2}} \mu_2 \cdot X \qquad C_2(0) = C_2^0 \qquad (11)$$

$$\frac{dC_3}{dt} = 0 \qquad C_3(0) = C_3^0 \qquad (12)$$

$$\mu = \mu_2 = \mu_{2max} \frac{C_2}{K_{2C} + C_2} \qquad (13)$$

Graphical "best fit" solutions for this approach are given in Figure 11.3 (A–D) for biomass (A), lactose (B), lactic acid (C), and acetic acid (D) concentrations. If we apply the original model (Approach I), the results of the simulation are given in column 5 of Table 11.2. It may be seen that the values of μ_{1max} and μ_{3max} approach zero, which was our assumption for the simplified model applied at 65°C (Approach II, shown in column 6). This suggests that our model can be simplified at 65°C and should be valid over a range of temperatures.

Experimental verification of the model (Approach I) was carried out, and one of the three independent experimental results is shown in Figure 11.4 for 60°C and 80% DO.

Reasonably good fits to process data were obtained using these models over a range of temperatures, including those within the thermophilic region. Values of "best fit" model parameters were generated to predict biomass-specific growth rates. The average specific growth rate calculated was 0.097/h at 55°C, while the experimental one was 0.079/h. At 65°C the calculated average specific growth rate was 0.075/h, while the experimental one

was 0.089/h. The results obtained suggest that temperature may have exerted a larger influence on the biodegradation process than DO, as the composition of the microbial community changed with temperature over the range 55–65°C. The average biomass yields generated were 0.350 g/g (on lactose substrate), 0.430 g/g (on lactic acid substrate), and 0.86 g/g (on acetic acid substrate), whereas yields calculated using the model were 0.325 g/g on lactose substrate, 0.410 g/g on lactic acid substrate, and 1.01 g/g on acetic acid substrate. Our investigations suggest that modeling of complex bioreaction systems via grouping key substrates and microbial species into a limited number of "equivalent clusters" is worthy of consideration as a possible means of facilitating rapid process development and practical process operation.

2.2 Modeling of the Biodegradation of Potato Stillage/ Distiller's Slops

The models for this wastewater were developed by the Wroclaw group in our EC-funded project, and are modifications of ASM2, based upon data from potato slops biodegradation experiments. Working with continuous culture experiments, it was assumed that organic substances in potato slops are removed following two different types of kinetics: reducing substances and glycerol are consumed with the first type of kinetic, whereas the main organic acids are assimilated with the second type of kinetic.

III. IMPROVED BIOCATALYSTS AND INNOVATIVE BIOREACTORS FOR ENHANCED BIOPROCESSING OF LIQUID FOOD WASTES

TABLE 11.2   Values of the Model Parameters Calculated for Different Experimental Runs, Using Approach I and Approach II

Parameter	55°C, DO = 80%	55°C, DO = 40%	60°C, DO = 80%	65°C, DO = 20%	65°C, DO = 20%
K_{1C} (g/L^{-1})	6.274	7.99	17.158	6.11	–
K_{2C} (g/L^{-1})	9.617	8.07	3.586	0.001	0.0013
K_{3C} (g/L^{-1})	5.452	3.34	5.9813	3.98	–
μ_{1max} (h^{-1})	0.158	0.218	0.2091	0.014	–
μ_{2max} (h^{-1})	0.144	0.095	0.0972	0.111	0.096
μ_{3max} (h^{-1})	0.0515	0.065	0.0579	0.0457	–
K_d (h^{-1})	0.0855	0.094	0.0949	0.046	0.0179
$Y_{X/C1}$ (g/g^{-1})	0.319	0.286	0.7085	0.537	–
$Y_{X/C2}$ (g/g^{-1})	0.570	0.108	0.3874	0.565	0.531
$Y_{X/C3}$ (g/g^{-1})	0.865	0.120	0.1713	0.551	–

Source: Kosseva et al. (2007).

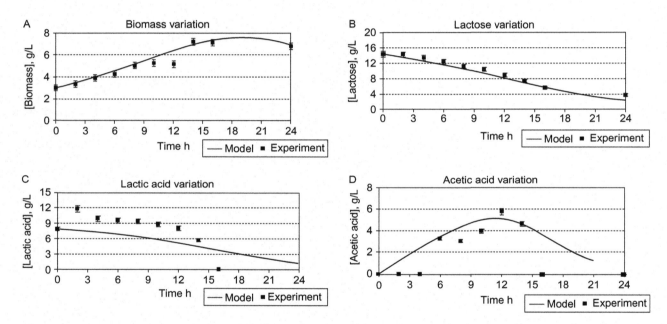

FIGURE 11.4   Simulation results (Approach I) and verification: experimental data for 60°C, DO = 80%. *Source: Kosseva et al. (2007).*

However, in batch experiments, it was observed that, except for acetic acid, all the carbon sources were assimilated roughly simultaneously. Acetic acid was produced during the first stage of the biodegradation process but was totally consumed later. This resulted in the following treatment of substrates and biomass.

- Biomass was considered as one, combined "representative species", with lumped concentration and growth kinetic properties.
- Substrates were divided into two groups:
  - *In the continuous model*: (i) organic acids, and (ii) reducing substances and glycerol.

- *In the batch model*: (i) acetic acid, treated as both a substrate and a product, and (ii) all other carbon sources, treated as a lumped substrate.
- Biomass growth following Moser kinetics was assumed, with no oxygen limitation.

On this basis, two models were developed:

- A version for continuous biodegradation: using (i) organic acids (C_1), and (ii) reducing substances and glycerol (C_2), as substrates.
- A version using (i) acetic acid, treated as both a substrate and a product (P), and (ii) all other carbon sources, treated as a lumped substrate (S).

2.2.1 Version for Continuous Biodegradation

$$\frac{dX}{dt} = \mu \cdot X - D \cdot X \tag{14}$$

$$\frac{dC_1}{dt} = -\frac{1}{Y_{X/C_1}} \cdot \mu_1 \cdot X - m_{C_1} \cdot X + D \cdot (C_{10} - C_1) \tag{15}$$

$$\frac{dC_2}{dt} = -\frac{1}{Y_{X/C_2}} \cdot \mu_2 \cdot X - m_{C_1} \cdot X + D \cdot (C_{20} - C_2) \tag{16}$$

$$\frac{dN}{dt} = -\frac{1}{Y_{X/N}} \cdot \mu \cdot X + D \cdot (N_0 - N) \tag{17}$$

$$\frac{dP}{dt} = -\frac{1}{Y_{X/P}} \cdot \mu \cdot X + D \cdot (P_0 - P) \tag{18}$$

$$\mu_1 = \frac{\mu_{1max} \cdot (C_1 - C_{1R})}{K_{1C} + C_1 - C_{1R}} \cdot \frac{N - N_R}{K_N + N - N_R} \cdot \frac{P - P_R}{K_P + P - P_R} \tag{19}$$

$$\mu_2 = \frac{\mu_{2max} \cdot (C_2 - C_{2R})}{K_{2C} + C_2 - C_{2R}} \cdot \frac{N - N_R}{K_N + N - N_R} \cdot \frac{P - P_R}{K_P + P - P_R} \tag{20}$$

$$\mu = \mu_1 + \mu_2 \tag{21}$$

where: X is lumped biomass concentration; μ is specific growth rate; D is dilution rate; C_1 and C_2 are concentrations of lumped carbon substrates (C_1, organic acids; C_2, the sum of reducing substances and glycerol); C_{10} and C_{20} are concentrations of lumped carbon substrates in the inflow; C_{1R} and C_{2R} are concentrations of remainder of lumped carbon substrates, treated as parameters that should be estimated; N is concentration of lumped nitrogen substrate, for example total nitrogen; N_0 is concentration of lumped nitrogen substrate in the inflow; N_R is concentration of remainder of lumped nitrogen substrate, treated as a parameter that should be estimated; P is concentration of lumped phosphorus substrate, for example total phosphorus; P_0 is concentration of lumped phosphorus substrate in the inflow; P_R is concentration of remainder of lumped phosphorus substrate, treated as a parameter that should be estimated; $Y_{X/C1}$, $Y_{X/C2}$, $Y_{X/N}$, $Y_{X/P}$ are yield coefficients; K_{C1}, K_{C2}, K_N, K_P are saturation constants; m_{C1}, m_{C2} are maintenance coefficients.

2.2.2 Version for Batch Biodegradation

$$\frac{dX}{dt} = \mu_1 \cdot X + \mu_2 \cdot X - L \cdot X \tag{22}$$

$$\frac{dS}{dt} = -\frac{1}{Y_{X/S}} \cdot \mu_1 \cdot X - B \cdot X \cdot S - M \cdot X \tag{23}$$

$$\frac{dP}{dt} = A \cdot X \cdot S - \frac{1}{Y_{X/P}} \mu_2 \cdot X \tag{24}$$

$$\mu_1 = \frac{\mu_{1max} \cdot S^n}{K_S + S^n} \tag{25}$$

$$\mu_2 = \frac{\mu_{2max} \cdot P^k}{K_P + P^k} \tag{26}$$

where: X is number of cells; S is concentration of lumped carbon substrate (expressed as total COD of the liquid phase less COD of acetic acid); P is concentration of acetic acid; μ_{1max}, μ_{2max} are maximum specific growth rates; L is autolysis coefficient; $Y_{X/S}$, $Y_{X/P}$ are yield coefficients; K_S, K_P are saturation constants; M is maintenance coefficient (includes also substrate forming caused by autolysis); A is coefficient of acetic acid production; B is coefficient of substrate consumption for acetic acid production.

2.2.3 Comparison of Model Output and Experimental Data

Figure 11.5 compares the batch model output (simulated on MATLAB®/SIMULINK®) (curves) with experimental plots (data symbols) of the state variables from a biodegradation of Stillage II (see Table 10.4) at pH 8.0, 900 rpm.

The agreement between calculated and experimental data is good, with correlation coefficients of model fit for cell numbers, acetic acid, and substrate given as 0.9022, 0.9177, and 0.9830, respectively. This suggests that the model should be suitable for use with this biodegradation process, assuming it is not oxygen-limited.

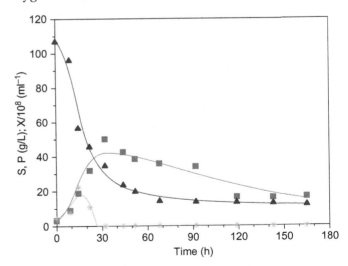

FIGURE 11.5 Batch model compared with experimental data for Stillage II. Curves show batch model output, symbols show experimental data: continuous line and triangles, COD of acetic acid; broken line and squares, number of cells; dotted line, acetic acid.

2.3 Modeling of Anaerobic Digestion (AD)

Many of the models of this type of process use kinetics-based approaches to fit experimental behavior, incorporating inhibition terms when needed and leading sometimes to a large number of parameters with questionable mechanistic interpretation. Some of these models have inherited their structures from the ASM family of models (Henze et al., 2000), originally developed for aerobic systems that are more clearly kinetically controlled, provide high energy yields, and proceed far from thermodynamic equilibrium. Anaerobic processes, in contrast, provide low energy yields and their conversions can proceed very closely to thermodynamic equilibrium (Kleerebezem and Stams, 2000). This suggests that anaerobic processes may be controlled thermodynamically rather than kinetically. In anaerobic digestion, the step of carbohydrate fermentation (acidogenesis) by fermenting microbial populations leads to a variety of products that are subsequently methanized by other microbial populations. It is also known that the types of fermentation products obtained are influenced by the operational conditions (Horiuchi et al., 2002).

Batstone et al. (2006) presented a review on extensions, applications, and critical analysis of the ADM1 model, assessing future requirements for standardized anaerobic process modeling. An important limitation was recognized by Kleerebezem and van Loosdrecht (2006), in that stoichiometry is based only on catabolic reactions. Anabolic reactions source COD from the substrate, but in cases of carbon limitation, they source excess carbon from carbon dioxide. This is unrealistic for most organisms (except hydrogen utilizers), especially those that do not produce bicarbonate (e.g., butyrate oxidizers). Because biomass yields in general are relatively small, and because most reactions produce carbon dioxide during catabolism, the overall impact is small. However, when the ADM1 is used as a basis for assessing metabolic products, the updated basis proposed by Kleerebezem and van Loosdrecht (2006) should be used (Batstone et al., 2006).

Parameters for the ADM1 have also been applied to a range of systems, including an ASBR treating winery wastewater (Batstone et al., 2004) and a thermophilic manure digester (Batstone et al., 2003). Generally, modifications of kinetic parameters have been of the order of 20–50%, which is relatively small compared with the large variations found in some parameters used for activated sludge systems.

2.3.1 Case Study 1: Anaerobic Treatment of Chicken Wastes

Applying anaerobic thermophilic digestion to chicken waste (CW), one can combine the advantages

of pasteurizing the effluent waste at elevated temperatures with the use of ammonia released as a sanitation agent. Running thermophilic biogas processes at high ammonia levels also produces a residue with a high fertilizer value. Such processes provide a method of bacterial sanitation without a preceding pasteurization of the incoming organic waste. Thermophilic dry AD of dehydrated waste-activated sludge (DWAS) with 80% water content showed that the release and stripping of ammonia prior to AD can prevent the inhibition of methane production by a two-stage process (Nakashimada et al., 2008). This two-stage AD process can also decrease the cost of further treatment of digested sludge and can recover a high concentration of ammonia nitrogen effectively. Stable dry AD has also been carried out by recycling ammonia-stripped biogas, using CW containing a large amount of nitrogen (Abouelenien et al., 2009). The characteristics of raw CW and digested seed sludge used as inoculum (Ozu sludge) are shown in Table 11.3 (Abouelenien et al., 2009).

At Hiroshima University, we also developed a dynamic model for the above anaerobic thermophilic treatment of CW and seed sludge, based on the ASM1 model. The experimental system, procedures, and analytical methods used are described by Abouelenien et al. (2010). We applied a similar mass-balance-based modeling approach and model assumptions to those used for the aerobic thermophilic treatment of cheese whey (Kosseva et al., 2007).

2.3.1.1 MODEL ASSUMPTIONS

- The flow pattern is approximated to a completely mixed pattern.
- The mixed culture of seed sludge microorganisms is treated as one "equivalent culture" (biomass).

TABLE 11.3 Initial Characteristics of Raw Chicken Waste and Digested Seed (Ozu) Sludge

Parameter	Unit	Chicken Waste	Ozu Sludge
Water content	% w/w	75	80
TS	% w/w	25	20
VS	% TS	58	53
TOC	g-C/kg-TS	380	268
TKN	g-N/kg-TS	87	32
TAN	g-N/kg-TS	10.5	3.2

Source: Abouelenien et al. (2009).
TAN, total ammonia-nitrogen; TKN, total Kjeldahl nitrogen; TOC, total organic carbon; TS, total solids; VS, volatile solids.

- Separate specific growth rates for each class of organism, one for each substrate (μ_1, μ_2, μ_3, μ_4).
- Total (net) specific growth rate, μ, is made up from separate growth rates.
- Monod-type growth kinetics for each "equivalent" C-source.
- Based on chemical analyses, carbon sources in CW are expressed as nine "equivalent substrates": C_1 protein, C_2 lipids, C_3 cellulose, C_4 uric acid, C_5 long chain fatty acids (LCFA), C_6 sugars, C_7 acetate, C_8 propionate, C_9 methane.

A glossary of terms used in the model is given in Box 11.2. Biomass growth kinetics is expressed by the sum of specific organisms' growth rates:

$$\mu = \mu_1 + \mu_2 + \mu_3 + \mu_4 \qquad (27)$$

Four main steps during the AD process were identified: hydrolysis; fermentation or acidogenesis; acetogenesis; and methanogenesis (as depicted in Figure 7.1). Knowing the CW initial characteristics and based on HPLC and GC analysis of raw CW as well as intermediate products of AD in the closed system, we can write down the main reactions in the STR bioreactor. The following scheme represents a conceptual theoretical model of our process, which accounts only for carbon flux in the system, assuming that full hydrolysis of proteins, lipids, and cellulose will precede fermentation of uric acid and monosaccharides, oxidation of LCFA, and methane production.

2.3.1.2 MAIN REACTIONS ASSUMED IN THE MODEL

1. Hydrolysis:

Proteins → Amino Acids:
$$C_5H_7NO_2 + H_2O \rightarrow C_5H_9O_3N$$

Lipids → LCFA:
$$C_{57}H_{104}O_6 + 3H_2O \rightarrow CH_2OHCHOHCH_2OH + 3CH_3(CH_2)_{16}COOH$$

Cellulose → Carbohydrates:
$$(C_6H_{10}O_5)_n + H_2O \rightarrow n/4C_6H_{12}O_6 + n/4C_5H_{10}C_5 + n/2C_{12}H_{22}O_{11}$$

2. Fermentation of amino acids and monosaccharides:

$$C_5H_9O_3N + 3H_2O \rightarrow CH_3CH_2COOH + 2CO_2 + 3H_2 + NH_3$$

Uric Acid:
$$C_5H_4N_4O_3 + 5.36H_2O \rightarrow 0.82CH_3COOH + 4NH_3 + 3.36CO_2 + 0.1H_2$$

BOX 11.2

GLOSSARY OF TERMS FOR DYNAMIC MODEL OF ANAEROBIC THERMOPHILIC TREATMENT OF CHICKEN WASTE AND SEED SLUDGE

C_1	Protein concentration (g_{COD}/kg^{-1})	X	Lumped biomass concentration (g_{COD}/kg^{-1})
C_2	Lipid concentration (g_{COD}/kg^{-1})		
C_3	Cellulose concentration (g_{COD}/kg^{-1})	Y_1, Y_2, Y_3, Y_4	Yield coefficients (g_{VS}/g^{-1}_{COD})
		μ	Total specific growth rate (d^{-1})
C_4	Amino acid (uric acid) concentration (g_{COD}/kg^{-1})	$\mu_{max1}, \mu_{max2}, \mu_{max3}, \mu_{max4}$ (d^{-1})	Maximum specific growth rates of the seed sludge components consisting of the following microorganisms:
C_5	LCFA concentration (g_{COD}/kg^{-1})		
C_6	Monosaccharide concentration (g_{COD}/kg^{-1})		1—bacteria responsible for fermentation of amino acids, uric acid degrader—a newly-isolated bacterium;
C_7	Acetic acid concentration (g_{COD}/kg^{-1})		
C_8	Propionate concentration (g_{COD}/kg^{-1})		2—acetogenic bacteria responsible for oxidation of LCFA;
C_9	Methane concentration (g_{COD}/kg^{-1})		3—bacteria responsible for fermentation of sugars;
K_{h1}, K_{h2}, K_{h3}	Hydrolysis constants (d^{-1})		4—acetoclastic methanogenic bacteria
K_{1S}, K_{2S}, K_{3S}	Saturation constants (g_{COD}/kg^{-1})		
Kd	Decay coefficient (d^{-1})		

Sugars:

$$C_6H_{12}O_6 + 4H_2O \rightarrow 2CH_3COO- + 2HCO_3^- + 4H^+ + 4H_2$$

3. Oxidation of LCFA:

$$LCFA \rightarrow Acetic\ \ Acid + Hydrogen$$

4. Synthropic Acetogenic reactions:

Propionate Acetogenesis:

$$CH_3CH_2COO^- + 3H_2O \rightarrow CH_3COO^- + HCO_3^- + H^+ + 3H_2 \quad \Delta G^\circ = +76.1\ kJ\ mole^{-1}$$

Propionic Acid → Acetic Acid + Hydrogen (ignored in the model)

$$2CO_2 + 4H_2 \rightarrow CH_3COO^- + H^+ + 2H_2O$$

5. Methane production from acetic acid:

Acetic Acid → Methane

$$CH_3COO^- + 2H_2O \rightarrow CH_4 + HCO_3^-$$

When combining several steps together, some dynamic information can be lost. As H_2 utilization is relatively fast compared with oxidation of LCFA, the loss will not be significant. Acetate production from H_2 is also quite a fast reaction; furthermore there was no H_2 detected in our gas samples, so it was ignored in this version of the model. Propionate acetogenesis was also ignored in our model; because of the positive Gibbs free energy change of this reaction ($\Delta G^\circ = +76.1 kJ/mol$), also the concentration of propionic acid did not significantly vary in the HPLC analyses of fermentation samples.

Owing to thermodynamic constraints, the acetate-utilizing methanogenic reaction proceeds much better at elevated temperatures. The thermodynamic equilibrium model, suggested by Oh and Martin (2007), shows that the methanogenesis step requires thermal energy and electrons, so that anaerobic digestion may achieve high substrate degradation and high conversion to methane. We found that an optimum temperature of 55°C suited both the fermentation of CW and stripping of NH_3 produced, so that methane yields reached 155 mL/g_{VS}, and the level of total ammonia-nitrogen was maintained below 2 g_N/kg wet sludge in the bioreactor in the pH range 8.5–9.

The following kinetic expressions were used for hydrolysis reaction of proteins, lipids, and cellulose.

1. Hydrolysis of (a) proteins, (b) lipids, (c) cellulose:

$$r_{h1} = \frac{dC_1}{dt} = -K_{h1}C_1; \quad C_1(0) = C_1^0; \quad (28a)$$

$$r_{h2} = \frac{dC_2}{dt} = -K_{h2}C_2; \quad C_2(0) = C_2^0; \quad (28b)$$

$$r_{h3} = \frac{dC_3}{dt} = -K_{h3}C_3; \quad C_3(0) = C_3^0 \quad (28c)$$

2. Biomass growth and decay:

$$\frac{dX}{dt} = r_x - r_d = \mu X - K_d X \quad X(0) = X^0 \quad (29)$$

3. Fermentation of uric acid:

$$\frac{dC_4}{dt} = -\frac{r_{x1}}{Y_1} = -\frac{\mu_1 X}{Y_1}; \quad C_4(0) = C_4^0; \quad \mu_1^1 = \frac{\mu_{max1} X C_4}{Ks_1 + C_4} \cdot \frac{P}{Kp + P} \quad (30)$$

4. Oxidation of LCFA:

$$\frac{dC_5}{dt} = -\frac{r_{x2}}{Y_2} = -\frac{\mu_2 X}{Y_2} = -\frac{\mu_{max2} X C_5}{(Ks_2 + C_5)Y_2} \quad C_5(0) = C_5^0 \quad (31)$$

5. Fermentation of glucose:

$$\frac{dC_6}{dt} = -\frac{r_{x3}}{Y_3} = -\frac{\mu_3 X}{Y_3}; \quad C_6(0) = C_6^0;$$
$$\mu_3 = \frac{\mu_{max3} X C_6 Ks_3}{Ks_3 + C_7 + (K_3 ia^* C_6/C_7)} \quad (32)$$

6. Acetoclastic methane production:

$$\frac{dC_9}{dt} = \frac{r_{x4}}{Y_4} = \frac{\mu_4 X}{Y_4}; \quad C_9(0) = C_9^0;$$
$$\mu_4 = \frac{\mu_{max4} X C^2_7}{C^2_7 + Ks_4} \cdot \frac{1}{1 + (P/K_i)} \quad (33)$$

The following reactions of hydrolysis and fermentation lead to acetic acid production from:

a. Uric acid production and consumption:

$$\frac{dC_4}{dt} = Kh_1 C_1 - \frac{\mu_1 X}{Y_1} \quad (34)$$

b. LCFA production and consumption:

$$\frac{dC_5}{dt} = Kh_2 C_2 - \frac{\mu_2 X}{Y_2} \quad (35)$$

c. Glucose production and consumption:

$$\frac{dC_6}{dt} = Kh_3 C_3 - \frac{\mu_3 X}{Y_3} \quad (36)$$

So production and consumption of acetic acid can be summarized by the following equation:

$$\frac{dC_7}{dt} = \frac{dC_4}{dt} + \frac{dC_5}{dt} + \frac{dC_6}{dt} - \frac{dC_9}{dt} \quad (37)$$

Table 11.4 shows kinetics of biomass growth and hydrolysis.

As an example, the rough proportion of input substrates from CW was taken as: proteins (13 wt%), lipids (2 wt%), cellulose (60 wt%), uric acid (6 wt%), glucose (4 wt%), and acetic acid (15 wt%), based on our own analytical data and the following sources: Henuk and Dingle (2003); Gelegenis et al. (2007).

TABLE 11.4 Kinetics of Biomass Growth and Hydrolysis for Dynamic Model of Anaerobic Thermophilic Treatment of Chicken Waste and Seed Sludge

Reaction	Kinetics of Growth	Reference
Hydrolysis	$\frac{dC}{dt} = -K_h C$	Angelidaki and Sanders (2004)
Fermentation of uric acid	$\mu_1 = \frac{\mu_{max1} X C_4}{Ks_1 + C_4} \cdot \frac{P}{Kp + P}$ Commensalistic model of growth + production	Moser (1988)
Oxidation of LCFA	$\mu_2 = \frac{\mu_{max2} X C_5}{Ks_2 + C_5}$	Moser (1988)
Fermentation of glucose	$\mu_3 = \frac{\mu_{max3} X C_6 Ks_3}{Ks_3 + C_7 + (K_3 ia * C_6 / C_7)}$ Competitive inhibition by VFA	Moser (1988)
Acetoclastic methane production	$\mu_4 = \frac{\mu_{max4} X C^2_7}{C^2_7 + Ks_4} \cdot \frac{1}{1 + (P/K_i)}$ Allosteric inhibition; noncompetitive inhibition by product, NH3	Hill (1983)

Various kinetic expressions (Gavala et al., 2003) have been tested for the specific biomass growth rate of individual bacteria and are reported in Table 11.4. Initial values of variables were calculated based on COD analyses, and thermophilic kinetic constants were taken from our data and references, including Angelidaki and Sanders (2004). For example, μ_{max4} of acetoclastic methanogenic bacteria was taken as the maximum specific growth rate of *Methanosarcinaceae*, which is reported to dominate at high VFA and NH$_3$ concentrations at elevated temperatures in manure digesters (Karakashev et al., 2005).

According to our dynamic model, simulated on MATLAB®/SIMULINK® (Eqs 27–37), higher values of acetic acid concentrations and CH$_4$ production are achieved during batch process compared with the experimental degradation of total COD contained in CW (Figure 11.6). This may be explained by the additional reaction of the hydrolysis of cellulose (Eq 28c), which is included in the theoretical model of C-fluxes but does not exist in the real system of CW lacking preliminary pretreatment. In the process diagram (Figure 11.8) of total organic carbon (TOC) digestion, it is expressed as insoluble or unreacted TOC. The results obtained by simulation of the continuous biodegradation process (Figure 11.7) also show that this model requires further elaboration and possibly more complexity. However, our investigations suggest that modeling of complex bioreaction systems via the combination of key substrates and microbial species into a limited number of "equivalent clusters" is worthy of consideration as a possible means of facilitating rapid process development and practical process operation. Further work on the models reported here will be needed to improve some of their predictions, including allowance for time-varying specific growth rates. Although, in cases where approximate predictions are all that is required, such models could have much to commend them.

In practical terms, we used a theoretical SIMULINK®-based model (Eqs 27–37) combined with a direct mass balance over the digestion processes of TOC and total nitrogen (TN = TKN + TAN) content, setting the stream flows and elemental compositions/reaction stoichiometry into a Microsoft Excel® environment. The calculated results show that a cumulative production of approximately 104.5 L of methane gas can be achieved, when treating 4 kg of raw CW (72% wet) added in portions (from 0.12 kg to 0.18 kg) semi-continuously to 3.5 kg of seed sludge (80% water). Under the above dry conditions 1 kg CW was converted to CH$_4$ and NH$_3$ over approximately 40–45 days or after 8 semi-batch runs.

When calculating TN mass-balance, the assumption is made that 70% of N-compounds in the crude proteins of CW goes to nitrogen in uric acid (UA), and the following UA fermentation reaction is effected by the new bacteria isolated:

$$C_5H_4N_4O_3 + 5.36H_2O \rightarrow 0.82CH_3COOH$$
$$+ 4NH_3 + 3.36CO_2$$

Simultaneously, about 2 kg of fertilizer (ammonium sulfate) are produced from 4 kg raw CW, trapping the ammonia gas released after fermentation. The amount of sulfuric acid (4N H$_2$SO$_4$) required for reaction with NH$_3$ is about 1.5 kg.

$$2NH_3 + H_2SO_4 = (NH_4)_2SO_4$$

The results, calculated from mass balancing and stoichiometry of the reactions on the basis of TOC and TN evaluation, are very similar to those achieved experimentally.

This study also demonstrated that thermophilic dry anaerobic digestion including ammonia stripping is an effective treatment and sanitation technique for chicken waste as well as DWAS.

As a further development of the ASM family, the mass-balance-based models presented are suitable for predictive simulation of bioreaction patterns. Combined with the stoichiometry and thermodynamics of the processes, they could serve as a design tool and a basis for a bioreactor control system.

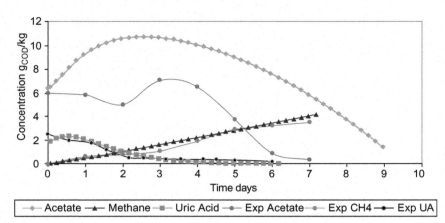

FIGURE 11.6 Comparison of the theoretical model output for batch methane production (△), acetic acid variation (◇), and uric acid variation (□) with experimental plots (·).

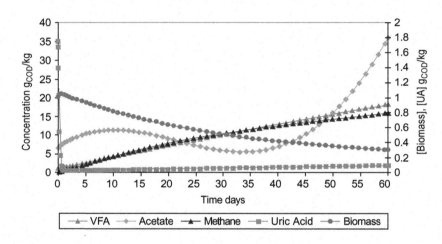

FIGURE 11.7 Modeling result for continuous CW degradation process (dilution rate $D = F/V$ $= 0.033/h$). F = Feed flow rate through the reactor, L/h; V = Reactor volume, L. [△], VFA; [◇], acetic acid; [△], methane; [□], uric acid; [•], biomass.

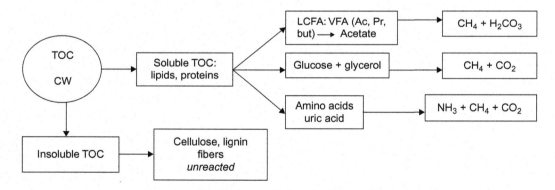

FIGURE 11.8 Process diagram of total organic carbon (TOC) digestion of chicken waste.

2.4 Modeling of an Autothermal Thermophilic Aerobic Digester (ATAD)

Gomez et al. (2007) presented a new biochemical model for aerobic digestion that, in addition to mass balances, introduces energy balances to dynamically predict the temporary evolution of the temperature in an ATAD. It includes 20 variables within liquid and gas phases; however, it does not include temperature-dependent kinetic parameters. The model proposed by Kovacs et al. (2007) consists of a mass balance, which includes temperature-dependent kinetic parameters, but it lacks an energy balance. These two models have advantages and disadvantages and they are complementary, but neither is fully comprehensive. That said, the model presented by Kovacs et al. (2007) contains only 9 variables and is considerably less complex than the model proposed by Gomez et al. (2007).

2.4.1 Mass Balance

Rojas et al. (2010) combined the mass balance and kinetic parameters presented by Kovacs et al. (2007) with a minor modification of the energy balance presented by Gomez et al. (2007). A mass balance is needed to determine variations in the VS content of the sludge (as a function of time), which directly influences the stabilization degree, as well as biological heat generation, thus indirectly influencing the pasteurization process. The mass balance suggested by Kovacs et al. (2007) is an extension of ASM1 applied at thermophilic temperatures. This model was built to reproduce results of laboratory experiments in which temperature was maintained constant throughout the reaction. One of the main modifications of this model is the separation of the biomass into two categories: mesophilic and thermophilic biomass. Another modification is the inclusion of an activation process for the thermophilic biomass. The structure of the model, including hydrolysis, growth, and decay processes, is illustrated in Figure 11.9 (Rojas et al., 2010).

The generic mass balance equation of the extended model is:

$$\frac{d(VX_i)}{dt} = \sum_j v_{ij}\rho_j V + In(X_i) - Out(X_i) \quad (38)$$

where V is the working reactor volume (L), X_i the concentration of component i (g/L), v_{ij} the stoichiometric coefficient of process j affecting component i, ρ_j represents the process rate (g/L/day), and $In(X_i)$ and Out (X_i) are the inlet and outlet transport terms (g/day), respectively.

The specific reaction rates can be obtained from the Petersen matrix shown in Table 11.5. Kovacs et al. (2007) have also determined the values of the kinetic parameters at different temperatures.

2.4.2 Energy Balance

An energy balance is needed in order to account for temperature variations during the reaction. These directly affect the pasteurization process and the kinetics of the reaction, and thereby have an indirect effect upon the stabilization process. The energy balance, taken from the work of Gomez et al. (2007), makes use of gas and liquid–solid phases, which interact with each other via energy and mass transfer. In contrast, Rojas et al. (2010) modified this energy balance to account only for sensible heat flux through the sludge stream, sensible and latent heat flux through the gas stream, heat loss to the surroundings, biological heat generation, and heat input through mixing, as depicted in Figure 11.10.

The elimination of the gas phase by Rojas et al. (2010) significantly simplifies the model, which was justified in the same reference. Additionally, it is assumed that mass transfer between gas and liquid phases can be neglected.

On the basis of the previous assumptions, the energy balance of the liquid phase is expressed by Eq (39):

$$\frac{dH_l}{dt} = In(H_l) - Out(H_l) + P_{bio} + P_{mix} - P_{wall} + Q_s^{in} - Q_s^{out} + Q_{lat}^{in} - Q_{lat}^{out} \quad (39)$$

where H_l is the total enthalpy of the liquid phase (MJ). $In(H_l)$ and $Out(H_l)$ are the sensible heat gain and loss through the sludge stream, respectively. P_{bio} is the biological heat production (kW), P_{mix} the heat coming from the mixing (kW), and P_{wall} the heat loss through the reactor walls (kW). Q_s^{in} and Q_s^{out} are the sensible heat gain and loss through the gas stream, respectively. Finally, Q_{lat}^{in} and Q_{lat}^{out} represent the latent heat gain and loss through the gas stream (kW).

Specific formulae for $In(H_l)$, $Out(H_l)$, P_{bio}, P_{mix}, and P_{wall} can be found in Gomez et al. (2007). The sensible heat gain and loss through the gas stream can be determined via Eqs (40) and (41), respectively:

$$Q_s^{in} = m_{in}^g C_p^g T_{in}^g \quad (40)$$

$$Q_s^{out} = m_{out}^g C_p^g T^g \quad (41)$$

where m_{in}^g and m_{out}^g are the mass flow rate of the influent and effluent gas streams (kg/day), respectively. C_p^g is the specific heat capacity of the gas (kJ/kg°C). T_{in}^g and T^g are the influent and effluent gas temperatures (°C).

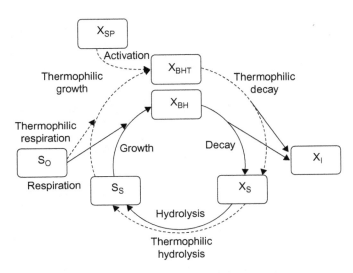

FIGURE 11.9   Structure of extended ASM1 conceptual model at thermophilic temperatures. *Source: Adapted from Kovacs et al. (2007).*

TABLE 11.5 Petersen Matrix of the Extended ASM1 Model at Thermophilic Temperatures

j Processes	1 S_I	2 S_S	3 X_I	4 X_S	5 X_{SP}	6 $X_{B,H}$	7 $X_{B,H,T}$	8 X_P	9 S_0	Process rates, ρ_j
1 Aerobic growth of mesophiles		$-1/Y_H$				1			$-(1-Y_H)/Y_H$	$\mu_H \frac{S_S}{K_S+S_S}\cdot\frac{S_0}{K_0+S_0} X_{B,H}$
2 Aerobic growth of thermophiles		$-1/Y_{H,T}$					1		$-(1-Y_{H,T})/Y_{H,T}$	$\mu_{H,T}\frac{S_S}{K_{S,T}+S_S}\cdot\frac{S_0}{K_{0,T}+S_0} X_{B,H,T}$
3 Mesophilic decay				$1-f_P$		-1		f_P		$b_H X_{B,H}$
4 Thermophilic decay				$1-f_{P,T}$			-1	$f_{P,T}$		$b_{H,T} X_{B,H,T}$
5 Mesophilic hydrolysis of slowly biodegradable substrate		1		-1						$k_H \frac{X_S/X_{B,H}}{K_X+(X_S/X_{B,H})}\cdot\frac{S_0}{K_0+S_0} X_{B,H}$
6 Thermophilic hydrolysis of slowly biodegradable substrate		1		-1						$k_{H,T}\frac{X_S/X_{B,H,T}}{K_{X,T}+(X_S/X_{B,H,T})}\cdot\frac{S_0}{K_{0,T}+S_0} X_{B,H,T}$
7 Thermophilic activation					-1		1			$\mu_{SP}\frac{S_0}{K_{0,T}+S_0} X_{B,H,T}$
	Dissolved inert organic matter (mg-COD/L)	Readily biodegradable substrate (mg-COD/L)	Particulate inert organic matter (mg-COD/L)	Slowly biodegradable substrate (mg-COD/L)	Inactive biomass (mg-COD/L)	Active mesophilic biomass (mg-COD/L)	Active thermophilic biomass (mg-COD/L)	Inert organic matter from decay (mg-COD/L)	Dissolved oxygen (mg-O$_2$/L)	

Y_H ($Y_{H,T}$): mesophilic (thermophilic) yield constant f_P ($f_{P,T}$): inert fraction of mesophilic (thermophilic) biomass

μ_H ($\mu_{H,T}$): maximal specific growth rate of mesophiles (thermophiles) b_H ($b_{H,T}$): maximal specific decay rate of mesophiles (thermophiles) K_0 ($K_{S,T}$, $K_{0,T}$): saturation constant of substrate and oxygen k_H, K_X ($k_{H,T}$, $K_{x,T}$): maximal specific hydrolysis constant and hydrolysis saturation constant μ_{SP}: specific constant rate of thermophilic activation

Source: Adapted from Kovacs et al. (2007).

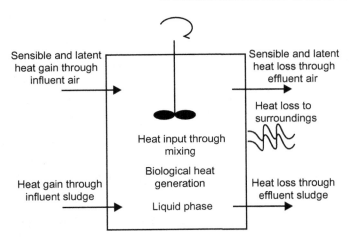

FIGURE 11.10 Conceptual energy balance of ATAD reactor accounting for sensible and latent heat flux through gas streams but without gas phase. *Adapted from Rojas et al. (2010).*

The latent heat gain and loss can be calculated according to Eqs (42) and (43), respectively:

$$Q_{lat}^{in} = m_{in}^g w_{in}^g L_{in}^g \qquad (42)$$

$$Q_{lat}^{out} = m_{out}^g w_{out}^g L_{out}^g \qquad (43)$$

where w_{in}^g and w_{out}^g are the specific humidity of ambient air and effluent air (kg water/kg dry air), respectively. L_{in}^g and L_{out}^g are the latent heat of evaporation of the influent and effluent gas (kJ/kg), respectively. This formulation allows for the condensation or evaporation of water accompanying the gas phase, depending on gas inlet and reactor conditions. The specific humidity is calculated via Eqs (44) and (45):

$$w_i = \frac{18.015}{2896} \frac{P_i}{P_{atm} - P_i} \qquad (44)$$

with

$$P_i = RH_i P_i^{sat} = RH_i \times 10^{8.07 - (1730.63/233.42 - T_i)} \qquad (45)$$

where P_{atm} is the atmospheric pressure (760 mmHg), P_i is the actual water vapor pressure (mmHg), and P_i^{sat} the saturation vapor pressure (mmHg) determined using the Antoine equation. The latent heat of evaporation (kJ/kg) is calculated as shown in Eq (46):

$$L_i = 2501 - 2.361 T_i \qquad (46)$$

Rojas and Zhelev (2011) used the above model to minimize the energy requirement of an ATAD plant.

2.5 Modeling of Wastewater Treatment Plants (WWTPs)

Very complex models have recently been built that combine aerobic and anaerobic digestion processes in WWTPs. De Gracia et al. (2009) present a mathematical model developed to reproduce the performance of a generic sludge digester working under either aerobic or anaerobic operational conditions. The digester has been modeled as two completely mixed tanks associated with gaseous and liquid volumes. The conversion model has been developed based on a plant-wide modeling methodology (PWM) and comprises biochemical transformations, physicochemical reactions, and thermodynamic considerations. The model predicts the reactor temperature and the temporary evolution of an extensive vector of model components, which are completely defined in terms of elemental mass fractions (C, H, O, N, and P) and charge density. The methodology, developed and verified by Grau et al. (2007a), allows the direct connection of models for different processes and makes it possible to establish a systematic mathematical procedure for the automatic estimation of influent characteristics in WWTPs. This procedure was developed in parallel to complement the PWM methodology, as it requires a large number of model components (Grau et al., 2007b). The construction of a PWM consists of three consecutive steps: (i) the construction of the Plant Transformations Model, (ii) the construction of the Unit Process Model, and (iii) the construction of the integrated PWM (De Gracia et al., 2009).

This model has been calibrated using experimental data from different facilities (i.e., laboratory biodegradability reactors, anaerobic digestion pilot plants, and a full-scale ATAD). Real data from four different facilities and a straightforward calibration have been used to successfully verify the model predictions in the cases of mesophilic and thermophilic anaerobic digestion as well as ATAD (De Gracia et al., 2009).

2.5.1 Steady-State Models of WWTPs

Steady-state models are useful for design of wastewater treatment plants (WWTPs) because they allow reactor sizes and interconnecting flows to be simply determined from explicit equations in terms of unit operation performance criteria (Ekama, 2009). Once the overall WWTP operations are estimated, dynamic models can be applied to the connected unit operations to refine their design and evaluate their performance under dynamic flow and load conditions. To model anaerobic digestion (AD) within plant-wide WWTP models, not only COD and nitrogen (N) but also carbon (C) fluxes entering the AD need to be defined. Current plant-wide models, like benchmark simulation model No2 (BSM2 described by Grau et al., 2007b), impose a C-flux at the AD influent. Ekama (2009) extended the COD and N mass balance steady-state models of activated sludge organics degradation, nitrification/denitrification, and AD and aerobic digestion of wastewater sludge. He linked them with bioprocess transformation stoichiometry to form C, H, O, N, COD, and charge mass-balance-based models, so that C (and H and O) can be tracked through the whole WWTP. By assigning a stoichiometric composition (x, y, z, and a in $C_xH_yO_zN_a$) to each of the five main influent wastewater organic fractions and ammonia, these, and the products generated from them via the biological processes, are tracked through the WWTP. The model has been applied to two theoretical case study WWTPs treating the same raw wastewater to the same final sludge residual biodegradable COD. It was demonstrated that much useful information can be generated with relatively simple steady-state models to aid WWTP layout design and track different products leaving the WWTP via the solid, liquid, and gas streams, such as aerobic versus anaerobic digestion of waste activated sludge, N loads in recycle streams, methane production for energy recovery, and greenhouse gas (CO_2, CH_4) generation. To reduce trial and error usage of WWTP simulation software, Ekama (2009) has recommended that they be extended to include preprocessors based on mass balance steady-state models to assist with WWTP layout design, unit operation selection, reactor sizing, option evaluation and comparison, and wastewater characterization before dynamic simulation. Steady-state models of WWTP unit operations are therefore a very useful complement to the dynamic simulation ones.

3. PROCESS ANALYTICAL TECHNOLOGY

Efficient monitoring and control of aerobic and anaerobic digestion processes have to be carried out in order to enhance their performance. These operations are required to stabilize and optimize biological production, maximize production rate and capacity, and enhance productivity without running the risk of process instability or inhibition. By applying the principles of process analytical technology (PAT), the above issues can be addressed effectively (Junker and Wang, 2006). The biorefinery sectors also can absorb advanced process control involving PAT instrumentation, likewise the technology-leading pharmaceutical sector (Holm-Nielsen, 2008). PAT is primarily directed towards the pharmaceutical industry, but the initiative also covers other industries. In 2004 the American Food and Drug Administration (FDA) completed a "Guidance for Industry", where PAT was defined as: "a system for designing, analyzing, and controlling manufacturing through timely measurements during processing of critical quality and performance parameters of raw and process intermediates". The goal is to monitor and control the process on-line as early as possible in the process and in real time at strategically

selected process locations, with steps that ensure the quality of the final product. The term Analytical Technology in PAT refers to analytical chemical, physical, microbiological, mathematical, data, and risk analyses conducted in an integrated manner. The term "Quality" of a product means the final quality of various industrial processes. It can be either the concentration, purity, strength, or similarity of the processed products. "Quality cannot be tested into products; it should be built-in or it should be by design" (FDA PAT guidance for industry, 2005), so product quality/quantity has to be optimized during the ongoing process (Holm-Nielsen, 2008).

Recent advances in PAT allow complex bioconversion processes to be monitored and deciphered, using, for example, spectroscopic and electrochemical measurement principles. In combination with chemometric, multivariate data analysis, these emerging process monitoring approaches have the potential of bringing AD process monitoring and control to a new level of reliability and effectiveness. Classical process parameters include simple-to-measure parameters such as pH, temperature, and redox potential. These parameters are all relatively easy to quantify on-line using commercially available technology. For at-line or off-line determinations, there is no need for extensive training of the personnel performing the analysis. However, these parameters provide only very limited insight into the complex biochemical network of coupled reactions that occur in an AD process. Comprehensive insight into the dynamics of the microbiological process can be obtained if the monitoring program is augmented to also include volatile fatty acids (VFA), ammonia/ammonium, and off-gas composition. Conventional determination of these parameters requires technically more complex analytical hardware and highly skilled personnel. Recent advances in development of (multivariate) process sensors have made it possible to substitute these maintenance-heavy methods with rugged and reliable process sensors, the sensor-analyte quantification being based on multivariate calibration (Martens and Naes, 1991). There is still a need for specialist intervention, as most (multivariate) process analyzers require extensive calibration and validation. However, once validated, the process analyzers provide the plant operator with detailed information about the state of the process. Madsen et al. (2011) suggest the monitoring principle outlined in Box 11.3 as the optimal solution for detailed deciphering of process dynamics in practically any chemical or biochemical reactor.

A common assumption in chemical engineering is to presume homogeneous mixtures, allowing for easier process modeling. In reality, however, biotechnological processes such as AD are far from being homogeneous systems. Since most of the species of interest are

BOX 11.3

THE MONITORING PRINCIPLE

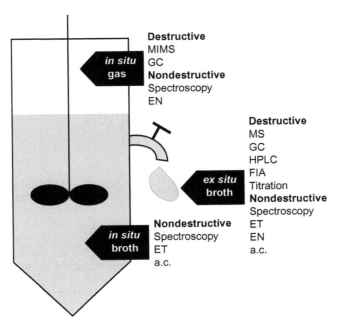

A general PAT monitoring scheme applicable for any chemical, environmental, or biochemical process. The *in situ* interfaces to both the liquid and the gas phase can be accomplished with commercial hardware. Adding a battery of (destructive) reference modalities for at-line analysis further increases the knowledge level pertaining to virtually any process (Madsen et al., 2011).

dissolved in the liquid phase, filtration is often applied to get rid of the heterogeneous, interfering material such as particulate matter. Moreover, formation of gas bubbles in the broth is the nightmare of spectroscopists. Usually these issues are counteracted in industry using "de-bubblers" and macerators, where appropriate. A clear, particle-free, and gas bubble-free liquid is much preferred in conventional spectroscopy. It is necessary to face these issues firmly in the future and to incorporate from the design phase the principles disseminated by the Theory of Sampling (TOS) (Esbensen and Paasch-Mortensen, 2010). TOS presents a complete methodology for evaluating the total sampling error associated with both static and process sampling. The process of taking a sample—characterizing a complex, large system based on a small part thereof—is not an easy one. The distinction between the accuracy of the sampling process (the *bias*) and the precision of the analysis is crucial for any appreciation of TOS.

Food wastes form a highly complex and heterogeneous suspension because of a significant proportion of solids: a high "dry matter" percentage of lignocellulosic

fiber particles and macromolecule aggregates such as starch, sugars, and proteins. Sampling from such systems consequently will introduce a significant sampling bias, if not properly counteracted. Esbensen's study dealt exhaustively with all relevant sampling issues using a "complete TOS toolbox" (Esbensen and Paasch-Mortensen, 2010). Holm-Nielsen et al. (2006) discussed the benefits and drawbacks of recycling lines and methods for assuring representative samples from complex geometries such as reactors. Holm-Nielsen (2008) also presented a new holistic view on monitoring the complex AD process in particular and of bioconversion processes in general.

By introducing the type of reliable on-line PAT sensor systems delineated in Box 11.4 in addition to the at-line analysis approaches, it should be possible to deliver fully comprehensive monitoring data for improved control of AD processes. Appropriate sensors and physical sampling means should be deployed at critical control points: (1) at raw materials and feeding inlet/outlet, (2) in recurrent loops positioned strategically in the process, and (3) at posttreatment

BOX 11.4

ON-LINE PAT SENSOR SYSTEMS AND TOS CONCEPT

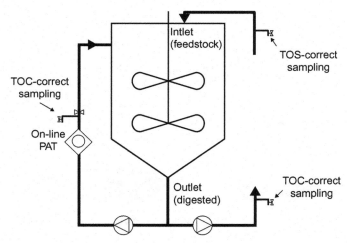

On-line PAT measurement and TOS-sampling concept in a full-scale AD plant context. (Holm-Nielsen et al., 2006).

locations. These developments "will contribute towards the goal of an integrated process and product monitoring management and regulation system, orders of magnitude more reliable than the current state of the art", according to Holm-Nielsen et al. (2006).

The application of numerous monitoring techniques for quantification of the key metabolites and their general principles of operation, including the way that sensors can provide insight into ongoing bioprocesses, has been described by Vanrolleghem and Lee (2003). Three different uses for sensors can be found: for monitoring (operator support), in automatic control systems, and as tools for plant auditing/optimization/modeling by consultants.

Generally, monitoring and control of wastewater treatment plants rely on four building blocks: (1) insight into the process as summarized by a process model, (2) sensors that provide on-line data, (3) adequate monitoring and control strategies, and (4) actuators that implement the controller output.

4. CONTROL STRATEGY DEVELOPMENT

Industrial bioprocesses are well known for their complex natures, which makes the application of modern automation techniques in this area a challenging task. Specifically, aqueous waste biotreatment processes based on the Activated Sludge process possess a

number of characteristic features, which must be considered during the control design process (Ohtsuki and Kawazoe, 1998):

- Unsteadiness: the inflow volume and concentration do not remain constant
- Nonlinearity: the reactions of the activated sludge process often reach pseudo-stability when substrates, nutrients, or oxygen are limited (often represented by a Monod-type equation)
- Complexity of the process
- Poor process understanding and dependence on empirical knowledge (phenomena such as sludge bulking and foaming are poorly understood)
- Lack of kinetic information on bacteria
- Lack of industrially viable sensors for on-line measurement of key process variables.

For these reasons there has developed a tradition in both the academic and industrial communities to address the challenging task of automation of industrial bioprocesses by the use of various advanced approaches, ranging from model-based to artificial intelligence (AI)-based techniques. In this context, knowledge-based control systems (KBCS) capable of integrating various paradigms (both AI-based and model-based) offer a useful tool for the implementation of advanced control in the WWTP application area (Hrnčiřík et al., 2002). Bioprocess control decisions are often made on the basis of both physical, chemical, and microbiological principles (*described by deterministic algorithms*) and qualitative process knowledge (*represented in various AI paradigms*). As a result, one of the key problems in knowledge-based control systems design is the development of an architecture able to manage efficiently and synergistically all the different elements of the available quantitative and qualitative process knowledge. A possible solution to the above problem is the application of multi-agent systems architecture (Lažanský et al., 2001), where decisions are based on processing of information from various sources diverse in nature and in different forms.

The application of knowledge-based approaches to bioprocess operation has been reviewed by Shioya et al. (1999). The review covered fuzzy inference and other knowledge-based methodologies, such as artificial neural networks (ANN), expert systems, and genetic algorithms, and provided an understanding of the outline of bioprocess control by a knowledge-based approach.

In contrast to conventional control strategies operating in closed loops with respect to the cell environment, the "physiological state" (or even "metabolic state") control strategy is in a closed loop with respect to the cell state. Consequently, the environment is not a goal but a tool for manipulating cell physiology. From all the subtasks of a general physiological state control strategy, the task of on-line recognition of the physiological state of the microorganism being cultivated is of key importance. The classification schemes involved in this approach are usually based to some extent on expert knowledge representation, and are frequently implemented using various artificial intelligence techniques (Konstantinov, 1996).

For process control itself, model-based control techniques are playing an increasingly important role. Advanced models of the processes in a wastewater treatment plant, based upon stoichiometry of key overall reaction schemes and component or elemental mass balancing over the bioreactor, such as the ASM1, 2, and 3, have rarely been used for practical control schemes owing to their complexity. They nevertheless have provided a very useful tool for the evaluation of different control strategies (Lindberg, 1997). Therefore, many reduced models and methods for model simplification have been suggested (e.g., using simplifying assumptions or process knowledge that reduces the model (Jeppson, 1996; Samuelsson, 2001) by "lumping together" all microbial or biochemical species present into one "representative, average species"). Besides those based on physical modeling, it is also possible to model the process using black-box methods (e.g., ANNs (Samuelsson, 2001)).

In contrast, Rosen et al. (2005) have described the implementation of ADM1 in MATLAB®/SIMULINK® as an integrated part of the Benchmark Simulation Model No2 (BSM2), which was further developed and used in control strategies for ATAD technology (Vrecko et al., 2006; Zambrano et al., 2009).

4.1 Fuzzy Logic Control

Since the introduction of fuzzy logic control by Zadeh (1965), this technique has evolved into an established practice for the control of biotechnological processes (Venkateswarlu and Naidu, 2000). Unlike crisp set theory, fuzzy set theory allows a transition from the classic bivalent concept of truth to the gradual and multivalent truth concept. Instead of crisp values (binary yes/no choices), the fuzzy theory deals with linguistic variables that are expressed by corresponding fuzzy sets. The distinct, numerical input values, received from the measuring devices, are assigned to linguistic terms and evaluated by distinct rule bases representing the expert knowledge of the process in the form of "if…then" rules (Krause et al., 2011). This approach could be especially appropriate to the complex nature and lack of process knowledge of waste biotreatment processes.

Honda and Kobayashi (2011) focused their work on the characteristics of fuzzy control systems and the usefulness for the automatic control of bioprocesses of fuzzy control, which can be categorized into two types. The first type utilizes direct inferencing, in which fuzzy inferencing directly determines the outputs from a knowledge base and online data. This method allows bioprocess control to be easily automated using the knowledge of expert operators, and it simply transfers the operators' know-how into the control system. The second type is indirect inferencing, in which fuzzy inferencing is first used to estimate the culture phase or physiological state, after which empirical control strategies are employed in each phase or state (Konstantinov and Yoshida, 1990; Shimizu et al., 1995).

A schematic representation of a control strategy in which process variables were determined directly from fuzzy inference is shown in Box 11.5 (Honda and Kobayashi, 2011). In this case, the control strategy consisted of feedforward estimation and feedback control. Assuming that the cellular yield from glucose did not vary during cultivation, the glucose feed rate, F^*, could be estimated by feedforward means, in accordance with the increase of cell concentration:

$$F* = \mu XV/Y_{x/s} So \qquad (47)$$

In this work, specific growth rate was determined from the cell concentration, which was determined from turbidity values measured online by a turbidometer. Glucose concentration in the broth was measured every 5 minutes by an on-line enzymatic analyzer. The value of $Y_{x/s}$ was calculated from serial cell concentration data and glucose consumption within 20 min. The feedforward determination of glucose feed rate F^* was just a rough approximation. To keep the glucose concentration precisely at a set value, the feed rate had to be corrected by feedback

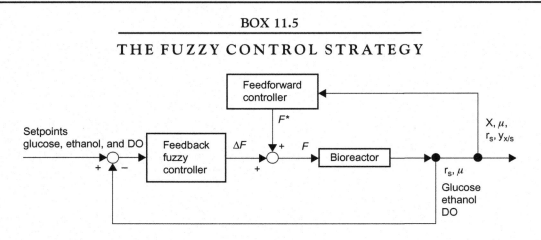

BOX 11.5

THE FUZZY CONTROL STRATEGY

Schematic representation of a fuzzy control strategy:

F^*, normal glucose feed rate; F, real glucose feed rate; ΔF, corrected glucose feed rate; X, cell concentration; μ, specific growth rate; r_s, specific glucose consumption rate; $Y_{x/s}$, cellular yield; DO, dissolved oxygen concentration. (Honda and Kobayashi, 2011).

control ($F = F^* + \Delta F$). For this correction, the value of ΔF was determined by the fuzzy control algorithm.

Many linguistic rules have been developed by skilled operators for the precise control of bioprocesses, and some industrial processes have been operated under fuzzy control: for example, mevalotin precursor production (Sankyo Co.), vitamin B$_{12}$ production (Nippon Roche Co.), and sake mashing (Sekiya Brewing Co., Gekkeikan Co., and Ozeki Co.).

4.2 Control Strategy Development for Food Wastes

The control strategies we developed within our thermophilic bioremediation project fall into three distinct categories:

- supervisory control strategies
- physiological state classification strategies (or indirect inferencing)
- direct control strategies.

Most of these were incorporated, as individual software agents, into the supervisory or knowledge-base level of BIOGENES II, the Knowledge Based Control System (KBCS) developed by our partners in the Institute of Chemical Technology Prague (ICTP) (Hrnčiřík et al., 2002).

4.2.1 Supervisory Control Strategies

Every strategy in this group outlined a set of process conditions to be maintained in order to achieve

optimal operation of the corresponding bioprocess. Specifically, this was done by defining the sets of controlled variables, their corresponding set point values, and manipulated variables for a given bioprocess. One example of supervisory control strategies is a bioconversion of whey into alternative products. The thermophilic treatment of cheese whey offers the potential for a number of different processing strategies as shown in Figure 11.11.

Other strategies covered include those developed in Birmingham, INETI Lisbon, ICTP, Wrocław, and Aristotle University of Thessaloniki, for glycerol and potato slops biodegradation, "population-focused" processing using flow cytometry to detect and follow physiological changes in subpopulations of a culture, process control to maintain an optimal physiological state of the culture, and process phase definition in xanthan gum production.

4.2.2 Physiological State Classification Strategies

This second group of strategies includes:

- rule-based strategies
- fuzzy, neural network, and fuzzy/neural strategies
- case-based reasoning strategies.

These are process-independent, designed for application with an arbitrary bioprocess for which a physiological process model is available. The classification strategies may be based on expert knowledge, machine learning, or both.

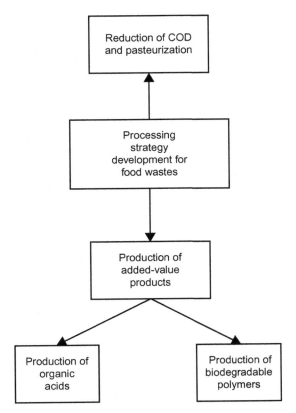

FIGURE 11.11 Strategy for bioconversion of cheese whey into alternative products.

4.2.3 Direct Control Strategies

These implement on the process level the process-specific control recipes formulated by the supervisory control strategies. These strategies are process-independent except for the model-based ones, for which an appropriate process model of the corresponding process application must be defined. The strategies include the following that have been incorporated into BIOGENES II:

- fuzzy logic controller
- model-based predictive control
  - based on a mechanistic process model
  - based on a fuzzy process model
- adaptive control using a statistical criterion
- direct digital control using a PSD controller.

4.2.4 Development of the KBCS

One of the key problems in KBCS design is the development of architecture able to manage efficiently and synergistically the different elements of available process knowledge. In this respect, multi-agent architecture offers a useful tool. For this reason, a multi-agent approach was used for the design and implementation of BIOGENES II. In this way it was possible to design the KBCS as an open system, capable of controlling several different degradation and productive bioprocesses demanding different control approaches and allowing for an easy integration of new process knowledge, process models, and control strategies.

BIOGENES II consists of two levels. All the real-time process control tasks are implemented on the basic system level in a compact programmable logic controller (PLC) with Momentum I/O units. This level also includes the human–machine interface (HMI), which is implemented through InTouch process visualization software and run on a PC. The higher, supervisory or knowledge-based level is represented by individual agents, created in various programming environments, realizing the specific tasks of process supervisory control. The two system levels are mutually interconnected via a real-time database, which can be accessed by both levels in order to write and read all the necessary data.

5. CONCLUSIONS

Process modeling is both a science and an art because a good dose of creativity is required to make assumptions for a computationally simple yet predictive model. Modeling inherently involves a compromise between accuracy (complexity) and the cost and effort involved in developing a model (Julien and Whitford, 2007). We have developed several mass-balance-based mathematical models covering our target wastes and found that the IAWQ ASM "family" was a useful starting point. Good fits to process data were obtained using modifications to the ASM's concepts of "lumping" mixed populations and mixed substrates into a small number of "clusters" of "equivalent" substrate or biomass. We have further developed design tools, such as process models, and operating strategies to facilitate the development and use of efficient processes for the thermophilic, aerobic treatment of a range of food-industry-related waste streams.

Applying a similar mass-balance-based modeling approach, a mathematical model of the dry anaerobic thermophilic treatment was further developed to predict successful production of methane and ammonia from chicken wastes and seed sludge.

To ensure optimal product quality of bioprocesses, it is necessary also to develop intelligent control systems with integrated monitoring of key parameters. However, the inherent complexity (time-varying behavior and nonlinearity) of biological processes presents many challenges to the control system. We have designed a number of control and operating strategies, including staged processes combining mesophilic and thermophilic conditions, waste blending as the manipulated variable in a simple dissolved oxygen control loop, identification of optimal process termination time, and "physiological state" and "population state" determinations from data record analyses.

We have also constructed a modular, adaptable multi-agent KBCS with an open structure that will allow it to run with a wide range of hardware and software operating systems. The "knowledge base" is structured to incorporate much of our process knowledge, our models and our control and operating strategies, to carry out routine control and management operations and to provide sophisticated process supervision through its "expert system" shell. Although designed with the bioprocessing of our target wastes in mind, its modular structure can be readily customized to a specific application, whether waste treatment or another biotransformation process.

Acknowledgement

Dr. Maria Kosseva is grateful to the Japan Society for the Promotion of Science for their financial support. Professors Naomichi, Nakashimada, and Dr. Abouelenien are acknowledged for their kind collaboration.

References

Abouelenien, F., Kitamura, Y., Naomichi, N., Nakahsimada, Y., 2009. Dry anaerobic ammonia—methane production from chicken manure. Appl. Microbiol. Biotechnol., 757—764.

Abouelenien, F., Fujiwara, W., Namba, Y., Kosseva, M., Nishio, N., Nakashimada, Y., 2010. Improved methane fermentation of chicken manure via ammonia removal by biogas recycle. Bioresour. Technol. 101, 6368—6373.

Angelidaki, I., Sanders, W., 2004. Assessment of the anaerobic biodegradability of macropollutants. Rev. Environ. Sci. Bio/Technol. 3, 117—129.

Batstone, D.J., Pind, P.F., Angelidaki, I., 2003. Kinetics of thermophilic, anaerobic oxidation of straight and branched chain butyrate and valerate. Biotechnol. Bioeng. 84, 195—204.

Batstone, D.J., Torrijos, M.J., Ruiz, C., Schmidt, J.E., 2004. Use of an anaerobic sequencing batch reactor for parameter estimation in modelling of anaerobic digestion. Water Sci. Technol. 50, 295—303.

Batstone, D.J., Keller, J., Steyer, J.P., 2006. Variable stoichiometry with thermodynamic control in ADM1. Water Sci. Technol. 54, 1—10.

Boogerd, F.C., Bos, P., Kuenen, J.G., Heijnen, J.J., Van der Lans, R.G.L.M., 1990. Oxygen and carbon dioxide mass transfer and the aerobic autotrophic cultivation of moderate and extreme thermophiles. Biotechnol. Bioeng. 35, 1111—1119.

De Gracia, M., Grau, P., Huete, E., Gomez, J., Garcıa-Heras, J.L., Ayesa, E., 2009. New generic mathematical model for WWTP sludge digesters operating under aerobic and anaerobic conditions: model building and experimental verification. Water Res. 43, 4626—4642.

Ekama, G.A., 2009. Using bioprocess stoichiometry to build a plant-wide mass balance based steady-state WWTP model. Water Res. 43, 2101—2120.

Esbensen, K.H., Paasch-Mortensen, P., 2010. Theory of sampling—the missing link in process analytical technologies (PAT). In: Bakeev, K.A. (Ed.), Process Analytical Technology, Second ed. Wiley, Chichester, UK, pp. 37—80.

FDA PAT Guidance for industry, December 2005. Available at: < http://www.fda.gov/cder/guidance/6419fnl.htm>.

Gavala, H.N., Angelidaki, I., Ahring, B.K., 2003. Kinetics and modeling of anaerobic digestion process. Adv. Biochem. Eng. Biotech. 81, 57—93.

Gelegenis, J., Georgakakis, D., Angelidaki, I., Mavris, V., 2007. Optimization of biogas production by co-digesting whey with diluted poultry manure. Renewable Energy 32, 2147—2160.

Gomez, J., De Gracia, M., Ayesa, E., Garcıa-Heras, J.L., 2007. Mathematical modelling of autothermal thermophilic aerobic digester. Water Res. 41, 959—968.

Grau, P., Vanrolleghem, P.A., Ayesa, E., 2007a. BSM2 plant-wide model construction and comparative analysis with other methodologies for integrated modeling. Water Sci. Technol. 56, 57—65.

Grau, P., Beltran, S., de Gracia, M., Ayesa, E., 2007b. New mathematical procedure for the automatic estimation of influent characteristics in WWTPs. Water Sci. Technol. 56, 95—106.

Henuk, Y.L., Dingle, J.G., 2003. Poultry manure: source of fertilizer, fuel and feed. World's Poultry Sci. J. 59, 350—360.

Henze, M., Gujer, W., Mino, T., van Loosdrecht, M.C.M., 2000. Activated sludge models ASM1, ASM2, ASM2d and ASM3. IWA Publishing, London, Scientific and Technical Report No. 9.

Hill, D.T., 1983. Design parameters and operating characteristics of animal waste anaerobic digestion systems swine and poultry. Agric. Wastes 5, 157—178.

Holm-Nielsen, J.B., 2008. Process analytical technologies for anaerobic digestion systems: robust biomass characterisation, process analytical chemometrics, and process optimisation. Ph.D. thesis, Aalborg University, Aalborg, Denmark.

Holm-Nielsen, J.B., Dahukl, C.K., Esbensen, K.H., 2006. Representative sampling for process analytical characterization of heterogeneous bioslurry systems—a reference study of sampling issues in PAT. J. Chemolab. 83, 114—126.

Honda, H., Kobayashi, T., 2011. Fuzzy control of bioprocess. In: Moo-Young, M. (Ed.), Comprehensive Biotechnology, Second ed. vol. 2. Elsevier, Netherlands, pp. 863—873.

Horiuchi, J.-I., Shimizu, T., Tada, K., Kanno, T., Kobayashi, M., 2002. Selective production of organic acids in anaerobic acid reactor by pH control. Bioresour. Technol. 82, 209—213.

Hrnčiřík, P., Náhlík, J., Vovsík, J., 2002. The BIOGENES system for knowledge-based bioprocess control. Expert Syst. Appl. 23, 145—153.

Jeppson, U., 1996. Modelling aspects of wastewater treatment processes. PhD thesis, Lund Institute of Technology, Sweden.

Julien, C., Whitford, W., 2007. Bioreactor monitoring, modeling, and simulation, In: Bioreactors, Chapter 1. BioProc. Int. Suppl. pp. 10—17.

Junker, B.H., Wang, H.Y., 2006. Bioprocess monitoring and computer control: key roots of the current PAT Initiative. Biotechnol. Bioeng. 95, 227—261.

Karakashev, D., Batstone, D.J., Angelidaki, I., 2005. Influence of environmental conditions on methanogenic compositions in anaerobic biogas reactors. Appl. Environ. Microbiol. 71, 331—338.

Kent, C.A., Roseiro, J.C., Rychtera, M., Náhlík, J., Miśkiewicz, T., Nowak, J., Liakopoulou-Kyriakides, M., 2004. Commission of the European Communities Fifth FRAMEWORK Programme, Project QLK3-CT-1999-00004: Enhanced, Intelligent Processing of Food and Related Wastes Using Thermophilic Populations: Final Report.

Kleerebezem, R., Stams, A.J.M., 2000. Kinetics of syntrophic cultures: a theoretical treatise on butyrate fermentation. Biotechnol. Bioeng. 67, 529—543.

Kleerebezem, R., van Loosdrecht, M.C.M., 2006. Critical analysis of some concepts proposed in ADM1. Water Sci. Technol. 54, 51—57.

Konstantinov, K.B., 1996. Monitoring and control of the physiological state of cell cultures. Biotechnol. Bioeng. 52, 271—289.

Konstantinov, K.B., Yoshida, T., 1990. An expert approach for control of fermentation processes as variable structure plants. J. Ferment Bioeng. 70, 48–57.

Kosseva, M.R., Fatmawati, A., Palatova, M., Kent, C.A., 2007. Modelling thermophilic cheese whey bioremediation in a one-stage process. Biochem. Eng. J. 35, 281–288.

Kosseva, M.R., Nakashimada, Y., Nishio, N., 2010. Report to Japan Society for the Promotion of Science. Mathematical modelling of dry ammonia-methane fermentation of chicken manure with dehydrated waste-activated sludge waste.

Kovacs, R., Mihaltz, P., Csikor, Z., 2007. Kinetics of autothermal thermophilic aerobic digestion—application and extension of activated sludge model no.1 at thermophilic temperatures. Water Sci. Technol. 56, 137–145.

Krause, D., Birle, S., Hussein, M.A., Becker, T., 2011. Bioprocess monitoring and control via adaptive sensor calibration. Eng. Life Sci. 11, 402–416.

Lažanský, J., Štěpánková, O., Mařík, V., Pěchouček, M., 2001. Application of the multi-agent approach in production planning and modelling. Eng. Appl. Artif. Intell. 14, 369–376.

Lindberg, C.F., 1997. Control and estimation strategies applied to the Activated Sludge process. PhD thesis, Uppsala University, Sweden.

Madsen, M., Holm-Nielsen, J.B., Esbensen, K.H., 2011. Monitoring of anaerobic digestion processes: a review perspective. Renewable Sustainable Energy Rev. 15, 3141–3155.

Martens, H., Naes, T., 1991. Multivariate Calibration, First ed. John Wiley and Sons, Great Britain.

Mason, I.G., 2006. Mathematical modelling of the composting process: a review. Waste Manage. 26, 3–21.

Metcalf and Eddy, Inc., 2003. Wastewater Engineering: Treatment and Reuse, fourth ed. McGraw-Hill, New York.

Moser, A., 1988. Bioprocess Technology: Kinetics and Reactors, Springer–Verlag.

Nakashimada, Y., Ohshima, Y., Minami, H., Yabu, H., Yuzaburo Namba, Y., Nishio, N., 2008. Ammonia–methane two-stage anaerobic digestion of dehydrated waste-activated sludge. Appl. Microbiol. Biotechnol. 79, 1061–1069.

Oh, S.T., Martin, A.D., 2007. Thermodynamic equilibrium model in anaerobic digestion process. Biochem. Eng. J. 34, 256–266.

Ohtsuki, T., Kawazoe, T., 1998. Intelligent control system based on blackboard concept for wastewater treatment processes. Water Sci. Technol. 37, 77–85.

Rojas, J., Zhelev, T., 2011. Energy efficiency optimisation of wastewater treatment: Study of ATAD. Comput. Chem. Eng. (online).

Rojas, J., Zhelev, T., Bojarski, A.D., 2010. Modelling and sensitivity analysis of ATA. Comput. Chem. Eng. 34, 802–811.

Rosen, C., Vrecko, D., Gernaey, K.V., Jeppsson, U., 2005. Implementing ADM1 for benchmark simulations in Matlab/Simulink.

Samuelsson, P., 2001. Modelling and control of Activated Sludge processes with nitrogen removal. PhD thesis, Uppsala University, Sweden.

Shimizu, H., Miura, K., Shioya, S., Suga, K., 1995. On-line state recognition in a yeast fed-batch culture using error vectors. Biotechnol. Bioeng. 41, 165–173.

Shioya, S., Shimizu, K., Yoshida, T., 1999. Knowledge-based design and operation of bioprocess systems. J. Biosci. Bioeng. 87, 261–266.

Vanrolleghem, P.A., Lee, D.S., 2003. On-line monitoring equipment for wastewater treatment processes: state of the art. Water Sci. Technol. 47, 1–34.

Venkateswarlu, C., Naidu, K.V.S., 2000. Dynamic fuzzy model based predictive controller for a biochemical reactor. Bioprocess. Biosyst. Eng. 23, 113–120.

Vogelaar, J.C.T., Klapwijk, A., Van Lier, J.B., Rulkens, W.H., 2000. Temperature effects on the oxygen transfer rate between 20 and 55°C. Water Res. 34, 1037–1041.

Vrecko, D., Gernaey, K., Rosen, C., Jeppsson, U., 2006. Benchmark simulation model No.2, in Matlab-Simulink: towards plantwide WWTP control strategy evaluation. Water Sci. Technol. 54, 65–72.

Zadeh, L.A., 1965. Fuzzy sets. Inf. Control 8, 338–353.

Zambrano, J.A., Gil-Martinez, M., Garcia-Sanz, M., Irizar, I., 2009. Benchmarking of control strategies for ATAD technology: a first approach to the automatic control of sludge treatment systems. Water Sci. Technol. 60, 409–417.

PART IV

ASSESSMENT OF WATER AND CARBON FOOTPRINTS AND REHABILITATION OF FOOD INDUSTRY WASTEWATER

CHAPTER

12

Accounting for the Impact of Food Waste on Water Resources and Climate Change

Ashok K. Chapagain and Keith James

1. BACKGROUND

Feeding the 9 billion people expected to inhabit our planet by 2050 will be an unprecedented challenge, which is further complicated by uncertainties and damaging changes in climate and other environmental factors. Much of the debate revolves around increasing food production. Despite the fact that global food production is currently more than enough, on a per capita basis, to adequately feed the global population, nearly a billion people remained food insecure in 2010. One cause, among many, of this deficiency is a poor distribution system exacerbated by societal attitudes towards food consumption globally, leading to over-consumption and wastages in some places and hunger and malnutrition in others. The recent trend to globalization has, on the one hand, helped to facilitate global distribution to some extent but, on the other, it has made the supply chain more complex, making it difficult to understand the true environmental cost of the food system and to formulate better food policy at both ends of the food chain.

Much agricultural production produces excessive amounts of greenhouse gases, uses unsustainable amounts of resources, and pollutes the environment. It not only contributes to the phenomenon of global warming by the emission of greenhouse gases (GHG), but also has detrimental environmental impacts such as deforestation, rapid loss in biodiversity, water scarcity (both in terms of quantity and quality), limiting economic growth, and hampering much of the aquatic ecosystem and services that accrue from water-based systems. We are consuming the earth's resources much faster than they can be replenished and are destroying the very systems on which our food supply depends; some two thirds of our ecosystems, including our forests, oceans, rivers, and lakes, are in decline (WWF, 2010).

The food we buy accounts for 23% of our ecological footprint—a measure of our environmental impact on the world. However, not all of this food is consumed. The comprehensive WRAP report *Household Food and Drink Waste in the UK* (Quested and Johnson, 2009) highlighted the importance of understanding the connection between wastage rates and resources used in the production process of these goods. The report classified household food waste in the UK based on avoidability (*avoidable, possibly avoidable, unavoidable*), disposal routes, and reasons for disposal. "Avoidable waste" is classified as the food and drink thrown away that was, at some point prior to disposal, edible (e.g., milk, lettuce, fruit juice, meat (excluding bones, skin, etc.)). "Possibly avoidable waste" is classified as the food and drink that some people eat and others do not (e.g., bread crusts) or that can be eaten when a food is prepared in one way but not in another (e.g., potato skins). "Unavoidable food waste" is classified as the waste arising from food and drink preparation that is not, and has not been, edible under normal circumstances (e.g., meat bones, egg shells, pineapple skin, tea bags).

The WRAP report estimated that 8.3 million tonnes of food and drink were wasted by households in the UK in 2008 (Quested and Johnson, 2009). The average UK household wastes around 22% of total food and drink purchases, and the proportion of waste deemed avoidable or possibly avoidable prior to disposal

Food Industry Wastes.
DOI: http://dx.doi.org/10.1016/B978-0-12-391921-2.00012-3

217

amounts to 81% of the total food and drink wasted. The report concludes that reducing the considerable amount of household food and drink waste generated in the UK saves households money while reducing our environmental impact. With the publication of WWF-UK's report on the water footprint of the food and fiber consumption in the UK and its impact on global water resources (Chapagain and Orr, 2008), quantification of water use by food consumption in the UK became possible.

Unnecessary wasting of food represents a direct waste of precious water resources, though to date there has been little quantification of water associated with food waste (Lundqvist et al., 2008). Recent studies have been undertaken to quantify the volume of water waste related to a single fruit industry (mango) in Australia (Ridoutt et al., 2009) and beef, potatoes, and tomatoes in the UK (Langley et al., 2010; Lewis, 2010). In the USA it has been reported that wasted calories account for about a quarter of the country's freshwater consumption (The Economist, 2009).

In addition to water, the UK food economy is also responsible for a significant quantity of greenhouse gas emissions in the UK and abroad. Several studies have attempted to quantify the footprint of specific food and drink products, but to date no research has been published on the national impacts of UK household food waste. Given the fact that a large proportion of food is wasted, it is also relevant to assess the equivalent carbon footprint of household food waste.

The quantification of water and carbon footprints of food and drink waste is potentially of interest to a range of stakeholders such as consumers, food retailers, suppliers, producers, NGOs, environmental agencies, water managers, and national and regional governments. The information can be used in a variety of contexts such as:

• identifying foods with high and low environmental impacts;
• identifying where to focus efforts to reduce the environmental impact of food production and to improve management of natural resources;
• understanding the way in which changes to the food supply chain can contribute to wider environmental policy objectives; and
• supporting activity in preventing food waste.

It is important to recognize that food production is, in most parts of the world, an important economic activity that provides benefits to many people. This is particularly the case in developing countries where agriculture is often the primary source of income for poor rural communities. Equally, mismanagement of natural resources such as water can have adverse impacts on the poorest people.

This chapter is useful in understanding the wider impacts of food wasted in the UK and catalyzing a discussion on the implications of these and formulating necessary policy interventions at appropriate levels. This is done, first, by quantifying the water footprint of food waste and its potential impacts, especially in water-scarce regions; then by linking water footprint data with the greenhouse gas impacts of wasted food, accounting for the whole life cycle. The chapter also identifies where in the world water is used to produce the part of the food being wasted in the UK.

The specific objectives of this study are to:

• quantify the water footprint of the food wasted by UK households in total and by country of origin;
• establish the linkages of wastage to locations where water resources are used and, in doing so, shed light on potential impacts on freshwater ecosystems;
• establish the carbon footprint of food waste by UK households in total and by item; and
• classify the footprints by waste categories such as avoidable food waste.

The underlying hypothesis behind this chapter is that, if the food and drink were properly "managed" (i.e., better decisions made during purchasing, storage, etc.), then wasted food and drink would not have had to be produced and the environmental impacts from all stages of the supply chain would be reduced. In other words, the resources employed to produce food and drink subsequently wasted would be available for other uses. In this study we have focused on the avoidable and possibly avoidable fraction of food waste only, as the impact associated with unavoidable food waste is allocated to food that has not been wasted (e.g., banana skins from bananas that have been eaten, tea bags used to make cups of tea), and from which unavoidable food waste is a natural consequence.

The chapter does not attempt to paint a comprehensive picture of every environmental, social, or economic impact—positive and negative—of food production, consumption, and waste.

2. DEFINING WATER FOOTPRINTS

The water footprint is an indicator of freshwater use that looks not only at the direct water use of a consumer or producer, but also at the indirect water use (Hoekstra and Chapagain, 2008). The water footprint of a product comprises three color-coded components (Hoekstra et al., 2011), which are green water (water evaporated from soil moisture supplemented by rainfall), blue water (water withdrawn from ground or surface water sources), and gray water (the polluted volume of blue water returned after production) (Box 12.1).

BOX 12.1

COMPONENTS OF WATER FOOTPRINT

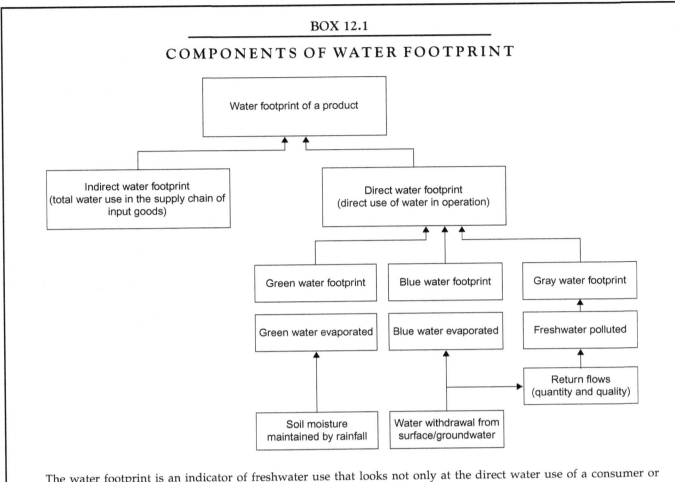

The water footprint is an indicator of freshwater use that looks not only at the direct water use of a consumer or producer, but also at the indirect water use (Hoekstra & Chapagain 2008). The water footprint of a product comprises three color-coded components (Hoekstra *et al.* 2011), which are green water (water evaporated from soil moisture supplemented by rainfall), blue water (water withdrawn from ground or surface water sources), and grey water (the polluted volume of blue water returned after production).

Green water is the soil water derived from rainfall. It is often seen as "free" and its use unproblematic, but this is not necessarily always the case. In the majority of situations, interception of rainfall by the plant canopy of cash crops is less than that of natural vegetation; this leads to the scenario where replacing natural vegetation by cash crops can increase groundwater and river flows (Scanlon et al., 2007). More recent work on green water has shown that it is an important dimension in global food and water security (Chapagain et al., 2006; Chapagain and Orr, 2009; Aldaya et al., 2009; Chapagain and Hoekstra, 2010; Mekonnen and Hoekstra, 2010). The inclusion of green water is again being questioned since some researchers contend that "green water (and other resources) is only accessible through access to and occupation of land" and that "the consumption of green water in agri-food product life

cycles is better considered in the context of land use" (SABMiller and WWF, 2009; Ridoutt and Pfister, 2009). The inclusion of green water within the volumetric water footprint is a useful auditing tool, but the need to consider water use within the wider context of land use and environmental impact has exposed its limitations.

The blue water footprint is an indicator of consumptive use of surface water (lakes and rivers) or groundwater (Hoekstra et al., 2011). The term consumptive refers to the volume that meets any of the three conditions of either being evaporated, or incorporated into the product, or water diverted into another catchment (not available in the same place and time). Generally, this is the part of the water abstracted by farmers for irrigation or by water companies to supply the general population that meets either of the above three conditions of being consumptive use.

The gray water footprint concept has grown out of the recognition that the size of water pollution can be expressed in terms of the volume of water that is required to dilute pollutants such that they become harmless (Chapagain, 2006; Hoekstra and Chapagain, 2008). The idea of expressing water pollution in terms of a water volume needed to dilute the waste is not new; it is based on the quality and quantity of the polluted return flows (effluent) and is quantified on the basis of the size and existing water quality standards of the receiving water bodies. Food processing, whether it is simply washing prior to sale (e.g., carrots) or more complicated preparation (e.g., preparing a pizza with multiple toppings) uses large quantities of water. This is normally blue water which, once used, is generally discharged back to surface waters. Although most of this is "non-evaporative use", the returned water is usually of a lower quality than the abstracted blue water, and additional blue water may be required to dilute or assimilate emissions (pollution) to the freshwater ecosystem from the production process. In the absence of accurate information on the assimilation capacity of freshwater ecosystems in the majority of places, gray water footprint accounting can be quite difficult and controversial.

As argued by Hoekstra et al. (2011), a water footprint is not a measure of the severity of the local environmental impact of water consumption and pollution. The local environmental impact of a certain amount of water consumption and pollution depends on the vulnerability of the local water system and the attributes of water consumers and polluters making use of the same system. Water footprint accounts give explicit information on how water is appropriated for various human purposes in a specific time and location (Box 12.2). A water footprint can inform the discussion about sustainable and equitable water use and allocation and also form a good basis for a local assessment of environmental, social, and economic risks and impacts.

All components of a total water footprint are specified both in time and space (Box 12.2). While it is widely accepted that blue water resources are limited and their exploitation can have obvious effects, green water is often seen as water that could be exploited with limited adverse impacts on freshwater ecosystems. However, taking the use of soil moisture for granted has immensely undervalued the importance of green water in managing water resources wisely (Rockström, 2001; Falkenmark, 2003). Green water may also be scarce, and in the context of food waste it represents a potentially significant opportunity cost; if it were not used to grow food that was subsequently wasted, it could be used for an alternative crop that might have significant economic and/or nutritional value. In agriculture, green water can be substituted for blue water and vice versa, so both must be accounted for to obtain a full picture.

Agricultural production uses large amounts of water; for example, the recent WWF report suggests that imported food and fiber account for 62% of the UK's total water footprint (Chapagain and Orr, 2008). In countries where water stress is less extreme, such as the UK, the impact of water use is generally concentrated in certain areas (such as East Anglia) and is restricted to certain times of the year. Although water abstraction for agriculture is less than 1% of total blue water abstraction in the UK, in some catchments and at peak times it can exceed abstraction for domestic water supply. However, in countries where water stress is common, blue water abstraction can have much more severe impacts.

In this study the scope of the water footprint is limited to the agricultural production phase, which is the stage with the largest water footprint in the whole supply chain. It is assumed that the quality of return flows is just enough to meet local norms and standards, although this inevitably underestimates the gray water footprints. In addition, the calculation does not include gray water footprints arising from other stages in the whole life cycle of the food.

2.1 Defining Carbon Footprints

Carbon footprint accounting can be carried out at a variety of levels (e.g., national, per person, product, service, etc.). Despite the high level of interest in carbon footprinting, there is a surprising lack of agreed upon definitions as to what a carbon footprint is. The *Guide to PAS 2050* (BSI, 2008) suggests that: "The term 'product carbon footprint' refers to the greenhouse gas emissions of a product across its life cycle, from raw materials through production (or service provision), distribution, consumer use and disposal/recycling. It includes the greenhouse gases carbon dioxide (CO_2), methane (CH_4) and nitrous oxide (N_2O), together with families of gases including hydrofluorocarbons (HFCs) and perfluorocarbons (PFCs)". In this study we've used the Carbon Footprint as a measure of the total amount of GHG emissions that are directly and indirectly caused by an activity or are accumulated over the life stages of a product. This includes activities of individuals, populations, governments, companies, organizations, processes, industry sectors, etc. Products include goods and services. In any case, all direct (on-site, internal) and indirect emissions (off-site, external, embodied, upstream, and downstream) need to be taken into account.

Greenhouse gas emissions arise at every point in the life of a product. Emissions may be direct (from animals, fertilizer application, fuel use) or indirect (e.g., from electricity generation). Each GHG has a different

BOX 12.2

PRESENTING THE WATER FOOTPRINT OF A SUPPLY CHAIN

All components of a total water footprint are specified both in time and space. Of these three components, the inclusion of green water is the most commonly debated. While it is widely accepted that blue water resources are limited and their exploitation can have obvious effects, green water is often seen as water that could be exploited with limited adverse impacts on freshwater ecosystems. However, taking the use of soil moisture for granted has immensely undervalued the importance of green water in managing water resources wisely (Rockström, 2001; Falkenmark, 2003). Green water may also be scarce, and in the context of food waste it represents a potentially significant opportunity cost; if it was not used to grow food that was subsequently wasted, it could be used for an alternative crop that might have significant economic and/or nutritional value. In agriculture, green water can be substituted for blue water and vice versa, so both must be accounted for to obtain a full picture.

potential to increase atmospheric temperature. To enable them to be discussed in a common language, characterization factors are applied to the gases against a standard radiative effect (the act of emitting or causing the emission of radiation). The characterization model for climate change, as developed by the Intergovernmental Panel on Climate Change (IPCC), contains a series of internationally recognized characterization factors. Factors are expressed as global warming potential (GWP) for a time horizon of 100 years (GWP100), in kg carbon dioxide equivalent (CO_2 eq)/kg emission. For a calculation of lifetimes and a full list of greenhouse gases and their global warming potentials please refer to Solomon et al. (2007).

In this study, all emissions are viewed from a consumption perspective. This means that emissions associated with cultivating and transporting food destined for the UK but grown elsewhere are included, and in turn a proportion of emissions from UK agriculture are allocated to food exported from the UK, and these are not included herein. Several studies have highlighted that imports of all goods account for around a third of the UK/European greenhouse gas emissions from a consumption perspective (Wiedmann et al., 2008; Brinkley and Less, 2010; Davis and Caldeira, 2010).

Unlike a water footprint, there are no local or regional interpretations of the impact of carbon emissions; a kilogram of carbon dioxide emitted in one country contributes to climate change in the same way as a kilogram emitted elsewhere (Forster et al., 2007). In this document, the objective is to account for the total carbon footprint of food waste and visualize where most emissions occur.

We've assumed that the greenhouse gas emissions associated with growing the same crop in different countries remain constant. This is a significant limitation, as from studies by Mila i Canals et al. (2007) and others it is known that the emissions associated with growing a foodstuff vary by season and location. The use of single figures in this analysis does not allow for illustration of the varying emissions associated with wasting food at different times of the year or from different sources. There is a range of data gaps for different food products at present. In the short to medium term we anticipate that this will be filled through a wider adoption of product carbon footprinting using PAS 2050, the WRI/WBCSD GHG Protocol, and the upcoming ISO Standard on product carbon footprinting.

3. ACCOUNTING CARBON FOOTPRINT

Greenhouse gas emissions have been calculated in two ways. Firstly, estimates have been made at a national level based on emissions from whole sectors (e.g., agriculture, food and drink manufacturing) and other top-level sources. Greenhouse gas emissions data has been taken from a variety of sources to derive an overview of the impact of food and drink across a range of sectors from agriculture to households and waste management. The emissions associated with food have been allocated to the final quantity that is consumed by UK households, net of exports and food and drink that is sold via the hospitality/catering sector. The factors therefore account for waste along the supply chain associated with the production of food for household consumption, although they exclude emissions from managing this waste. Secondly, as part of this chapter, individual carbon factors for different products have been identified. The available figures center on unprocessed (fruit, vegetables, raw meat) rather than processed foods (e.g., ready meals, sauces, prepared sandwiches, carbonated drinks) and do not cover all food products wasted. These are significant gaps and mean that the sum of the figures outlined herein will not add up to the average of all foods.

Data that complies with either the International Standard on life cycle assessment, ISO14040, or the Standard on carbon footprinting, PAS 2050, has been used wherever possible with some exceptions defined where applicable. Defra (2009a)[1] and DECC (2010) publish data on emissions associated with the food, drink, and tobacco sector in the UK and associated with food imported. However, while applicable at a top-down level, this information cannot be used at a product level. In lieu of this, product-specific reports have been used where possible. One study (ERM, 2009) has been identified that contains information on processing pork. Here, the figures associated with pork have been applied to all meats. Processing data from Yoshikawa (Yoshikawa et al., 2008) have been used for stoned fruit, broccoli, cabbage, carrots, potatoes, spinach, and sprouts, Beccali et al. (2009) for citrus fruit, Ruini and Marino (2009) for pasta, Sakaorat et al. (2009) for rice, Zespri et al. (2009) for Kiwi fruit, Cadbury (2008) for chocolate, and Tesco (2008) for orange juice.

Watkiss et al. (2005) identify that approximately 9 million tonnes of CO_2 emissions were associated with the import of 28 million tonnes of food to the UK. Although this figure is CO_2 only, it has been used in this study as a rough approximation for the impact of importing one tonne of food, with an emission factor of 0.320 tonnes CO_2 per tonne of imported food used for emissions that occur abroad in transporting food to the UK. An average packaging emission factor has been applied to all food items, based on data collected for the Courtauld Commitment. This includes primary and secondary packaging. The emissions factors used are identified in James (2010).

Tassou et al. (2009) identified emissions associated with storage, distribution, and retail of 11 specimen food products in a study to test and inform the development of PAS 2050. The emissions derived accounted for the time products spend on shelf, the temperature at which they are kept, and the volume of the product. Here, these emissions factors have been applied to similar food items. For example, emissions factors for bread have been applied to crackers, pastries, and flour. Most fruit have been classified as fresh apples. Soft berries, cucumbers, leafy salad, and lettuce have been allocated the same retail distribution center (RDC), transport, and retail emissions as strawberries. Emissions associated with drink products have been estimated based on WRAP estimates in discussion with industry.

Pretty et al. (2005) calculated that UK household food shopping involves around 8 km of car travel per household per week (in addition to travel by bus, bicycle, and on foot). The food expenditure survey suggests that food consumption is about 12 kg per person per week. With an average UK household size of 2.32 persons, this equates to 28 kg per household per week.

---

[1]The UK Government's Department for Environment, Food and Rural Affairs.

Defra (2010a) provide emissions factors for a range of vehicles. Based on a lower medium size passenger car, and assuming that the sole purpose of the journey is food shopping, this equates to 0.228 kg CO_2 eq per km. This equates to 0.07 kg CO_2 eq per kilogram of food transported home, or 0.07 tonnes CO_2 eq per tonne of food.

As with processing, limited food-specific emission factors are available for home storage and preparation of food. Information on drink storage is based upon WRAP estimates in discussion with industry, squash has been taken from Tesco (2008), peas have been taken from Defra (2009b), pasta from Ruini and Marino (2009), and meat from ERM (2009). At waste stage, an average emission factor for all foods has been applied (Table 12.1). Information on agriculture and fertilizer manufacture is based upon UK agriculture and fertilizer use only.

TABLE 12.1 The Carbon Factors Used to Estimate the Average Impact of Food

Life Cycle Stage	Total (million tonnes CO_2 eq)	Per Tonne Food Waste (tonnes CO_2 eq)[*]	Data Source
Agriculture	52.7	1.22	Defra (2006)
Fertilizer manufacture	3.5	0.08	Composition (AIC, 2008) Impacts (Davis and Haglund, 1999)
Food and drink manufacturing, UK	13.3	1.09	DTI (2002)
Food and drink manufacturing, non-UK	47.2	1.09	Defra (2009a)
Packaging	6.6	0.17	WRAP[†]
Transport (including overseas)	18.4	0.48	Defra (2005)
Food exports	− 8.0	− 0.18	Defra (2009a)
Retail	5.0	0.13	Tassou et al. (2009)
Home related	16.1	0.42	Brook Lyndhurst (2008)
Waste emissions (avoidable food only)	2.4	0.41	WRAP[**]
Total	130[*]	3.8	

[*]Total and per tonne factor is for food purchased by households only.
[†]WRAP calculation based on Courtauld Methodology and packaging data.
[**]Calculation based upon 62% to landfill, 8% to energy from waste, 22% to sewer, and 8% to home composting.

Of the 5.3 million tonnes of avoidable food and drink waste, 17% is drink and 83% food. The bottom-up calculation takes these relative proportions into account in obtaining the overall carbon factor for food and drink (3.8). Defra estimates that greenhouse gas emissions from the UK food chain are equivalent to about 180 million tonnes CO_2 eq, excluding waste. This suggests an average figure of 4.2 tonnes CO_2 eq per tonne of food, excluding waste management impacts. The results from both assessments are similar in the magnitude of impact.

Data on food and drink brought into the home (from retail, takeaways, and sources of free food) come from Defra's Family Food Survey for the calendar year 2007 (Defra 2008a). The food types used in this chapter were aligned as closely as possible with the classification of food and drink in the Family Food dataset, and then estimates for amounts of purchases at the food group level were obtained. It estimates that UK households purchase approximately 38.5 million tonnes of food and drink per annum, with a further 4.8 million tonnes consumed outside of the home.

3.1 Land Use Change

Land Use Change is responsible for 18% of global greenhouse gas emissions, principally from deforestation (Herzog, 2009). The FAO (2007) estimates that 58% of deforestation is due to commercial agriculture Land Use Change. Conventionally, when considering emissions from Land Use Change, accounting methods ascribe these to the products grown on the land recently converted to agriculture. However, if we consider the whole system, it is the level of demand for a product that drives expansion in agricultural land. In 2009 FCRN (Food Climate Research Network) and WWF published *How Low Can We Go?*, a study that quantified the emissions associated with Land Use Change attributable to UK demand for foodstuffs (Audsley et al., 2010). The study estimated that 87 million tonnes CO_2 equivalent can be attributed to deforestation related to the UK food economy and provides figures for a variety of foods.

These figures have been used to report separately on Land Use Change impacts. It is important to note that that estimates of CO_2 eq from Land Use Change are still subject to large uncertainties. The IPCC 4th Assessment Report cites error ranges of up to ± 2933 million tonnes CO_2 at the global level in the 1990s (Forster et al., 2007).

4. DATA

The production statistics are taken from FAO (FAOSTAT data, 2008), international trade data are

TABLE 12.2 Shares of Household Food Waste Based on Ability to be Avoided in the UK (million tonnes per year)

Categories	Avoidable	Possibly Avoidable	Unavoidable	Total
Food	4.5	1.5	1.1	7.1
Drink	0.8	0.0	0.4	1.2
Total	5.3	1.5	1.5	8.3

TABLE 12.3 Total Water Footprint of Household Food Waste in the UK (million m^3 per year)

	Avoidable	Possibly Avoidable	Total
Internal water footprint	1473	339	1812 (29%)
External water footprint	3895	555	4450 (71%)
Total water footprint	5368	894	6262 (100%)
Total percentage	86%	14%	100%

retrieved from International Trade Centre (ITC, 2006, 2009), and the water resources withdrawal data are taken from FAO (2003a,b). The virtual water content data for agricultural products are taken from Chapagain and Hoekstra (2004). The various sources of other data on climate, crop coefficients, and crop periods that are used in the calculations of water footprint are listed in volume 2 of the WWF-UK report (Chapagain and Orr, 2008). Based on primary ingredient, the virtual water flow results are regrouped to match with the list of household waste products in the WRAP database. Data on carbon has been drawn from various sources as listed in the earlier section on carbon footprint.

Table 12.2 presents a short summary of the household food waste in the UK, split by ability to avoid. The unavoidable part in the table is presented as additional information for comparison among different categories of household food waste. A full list of household food waste data is available from WRAP (Quested and Johnson, 2009).

As the chapter is based on data retrieved from a variety of sources, it is inevitable that any errors in these sources can influence the result of this analysis. Every effort has been made to cross-check these data sources with various other independent sources, and the selection of datasets used in this study is made on the basis of the scope and the degree of precision achievable within the scope and limitations of this study.

5. RESULTS OF WATER FOOTPRINT ACCOUNTING

The total water footprint of food waste in UK households is 6262 million m^3 per year, of which 5368 million m^3 per year is attributed to avoidable food waste, and a further 894 million m^3 to possibly avoidable waste. These figures represent 8% and 1% of the UK's total food water footprint respectively. In per capita terms, the water footprint of total avoidable and possibly avoidable household food waste in the UK is 284 liters per person per day. By comparison, the daily

average household water use in the UK (i.e., water from the tap) is about 150 liters per person per day (Defra, 2008b). Out of the total household food waste, 243 liters per person per day (86%) is completely avoidable and the remainder is possibly avoidable (Table 12.3).

A large part (71%) of the avoidable food waste in the UK is from imported products. Please note that the table doesn't show the water footprint of unavoidable food waste, as this will be counted towards water footprint of actual food consumption. The rationale for this is that unavoidable waste (e.g., banana skins, bones) is an integral part of food consumption. The share of the UK's External Water Footprint (EWF) for agricultural products (excluding textiles) is about 60%, whereas the EWF of household food waste is 71%. It shows that the imported food products are more water intensive (m^3 per tonne) than the products from the UK itself.

Though the total food waste by quantity is about 22% by weight, it is only 14% in terms of equivalent water footprint (Box 12.3). This is because the wasted food has relatively low water content per tonne of products. The Internal Water Footprint of wasted food is relatively bigger for livestock products (i.e., meat and dairy) compared to that for crop products. However, for imported products, it is the crop product component that has higher EWF.

The products with the largest share of the water footprint of household food waste are presented in Figure 12.1. It is seen that beef and cocoa products are the top two products in the list of water footprints of household food waste. They also rank highly for external water footprint among agricultural products in the UK as reported in the WWF-UK report (Chapagain and Orr, 2008). It is to be noted that for complex products with more than one ingredient, where possible an estimate of the composition has been made. Where this has not been possible, the waste has been assumed to consist of the single ingredient with the most significant contribution to the total water

BOX 12.3

INTERNAL AND EXTERNAL WATER FOOTPRINT OF HOUSEHOLD FOOD WASTE OF CROP AND LIVESTOCK PRODUCTS

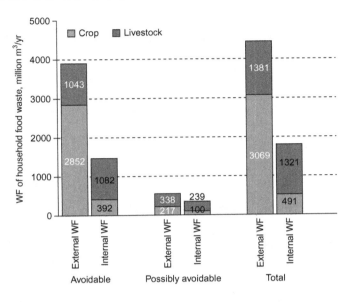

Though the total food waste by quantity is about 22% by weight, it is only 14% in terms of equivalent water footprint. This is because the wasted food has relatively low water content per tonne of products. The Internal Water Footprint (IWF) of wasted food is relatively bigger for livestock products (i.e., meat and dairy) compared to that for crop products. However, for imported products, it is the crop product component that has higher External Water Footprint (EWF).

footprint of the product. For example, cakes have been assumed to be made of wheat, and the water footprint for cocoa has been used for chocolate, although other ingredients (e.g., sugar) are also used within this product.

The countries with the highest external water footprints related to UK household food waste are presented in Table 12.4. A complete list of these locations and the size of the UK's water footprint is presented in Chapagain and James (2011). Although Ghana, Brazil, and the Ivory Coast feature at or near the top of the list of EWFs, it is important to note that the products originating in these locations are mainly rain fed and so exert limited pressure on blue water resources in these locations. A complete map of the water footprint of the UK's household food waste is presented in Figure 12.2.

Impact assessments of these water footprints were conducted using hydrological attributes of regions where water is used in food cultivation and processing (Box 12.4). Water stress in these locations is calculated as the ratio of actual blue water withdrawal to the net blue water available after taking into account environmental flow requirements. Following the scheme of impact categorization suggested in WWF-UK (Chapagain and Orr, 2008), the various countries are then grouped on the basis of the severity of the impacts, using the schematic presented here.

A shortcoming of this approach is that it doesn't take into account green water availability in these locations, nor the impact of gray water. Thus, calculations of stress on hydrology, based only on the blue water availability in these locations, are incomplete. From this exercise it is found that Egypt, Israel, Pakistan, India, Thailand, and Spain are examples of countries falling into Group D, where water stress is very high and the external water footprint of the UK's household food waste is also high. In contrast, countries falling into Group B have relatively lower water stress, and the water footprint of UK household food waste in these countries is relatively low. Although Ghana and Brazil, both in Group A, support a large part of the external water footprint of household food waste in the UK, water stress in these countries is low.

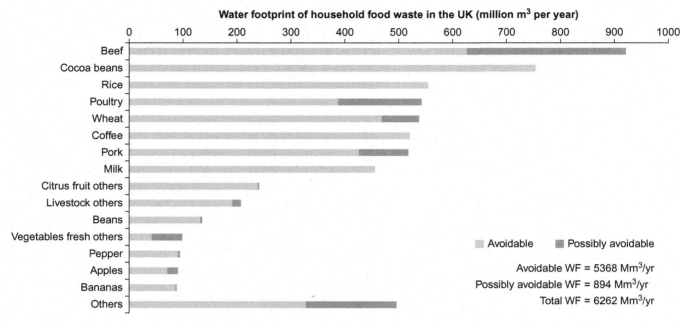

FIGURE 12.1   Total water footprint of household food waste in the UK for major food categories.

6. RESULTS OF CARBON FOOTPRINT ACCOUNTING

It is estimated that the total carbon footprint of food and drink consumed in the UK is 130 million tonnes CO_2 eq per year, which is approximately equivalent to a fifth of UK territorial emissions, or 2 tonnes of CO_2 eq per person per year. Excluding emissions from wasted items, the average impact of a tonne of food and drink purchased is 3.4 tonnes CO_2 eq, rising to 3.8 tonnes CO_2 eq per tonne of food alone.

This chapter uses two approaches to quantify the carbon footprint of food waste. The first of these is a top-down approach, which gives the total carbon footprint of the UK's household food and drink waste as 25.7 million tonnes CO_2 eq, of which 20 million tonnes CO_2 eq is associated with avoidable waste.

The top-down results (Box 12.5) suggest that, on average, approximately one quarter of the impact of food is associated with growing/rearing the crops and animals that enter the food chain, one quarter is associated with food processing, and one eighth is associated with home-related impacts (e.g., cooking). Waste management and degradation accounts for one tenth of emissions.

Secondly, the carbon footprint of household food and drink waste has been constructed from the bottom up, to allow allocation of emissions to the country of origin. Due to data gaps for specific foods, it should be noted that the carbon emissions attributed to specific foods do not add up to the top-down average. For example, no specific farm emission data was identified for approximately 10% of avoidable food waste by weight, and 20% of possibly avoidable food waste. No data to allocate specific emissions from regional distribution centers was identified for 6% of avoidable food waste by weight, and 20% of possibly avoidable food waste. Where food waste has not been identifiable, the carbon emissions associated with this have not been allocated to any specific country.

Out of the total carbon footprint of the UK's household waste of food and drink (Table 12.5), 78% is related to waste under the "avoidable" category and 22% under the "possibly avoidable" category. The average carbon footprint of avoidable household food waste is 330 kg CO_2 eq per person per year. This is equivalent to approximately one third of the emissions of CO_2 (rather than CO_2 eq) associated with household electricity use per person in the UK (DECC, 2010).

The products with the greatest share in the carbon footprint of household food waste are presented in Figure 12.3. Avoidable food waste and possibly avoidable food waste are responsible for 6,092,000 tonnes and 1,538,000 tonnes of CO_2 equivalent per year, respectively. Data limitations mean that few processed foods were able to be identified, and subsequently limited emissions have been attributed to these. The complete list of household food wastes and associated carbon footprints is available in Chapagain and James (2011).

TABLE 12.4 Countries with the Highest External Water Footprint Related to UK Household Food Waste

Locations	EWF (million m³/year) Avoidable	Possibly Avoidable	Total	Top Products and EWF (million m³/year)
Ghana	423	0	423	Cocoa beans 413, Coffee 5, Bananas 4, Pineapples 1
Brazil	271	64	336	Beef 148, Coffee 89, Poultry 47, Livestock others 35, Cocoa beans 5
India	263	22	284	Rice 165, Pepper 54, Beans dry 30, Oilseeds others 18, Coffee 5
Ireland	175	71	246	Beef 177, Poultry 27, Pork 26, Livestock others 12, Wheat 1
Netherlands	168	50	218	Pork 92, Poultry 74, Livestock others 24, Beef 15, Vegetables fresh others 6
Thailand	176	21	197	Rice 109, Poultry 62, Livestock others 14, Citrus fruit others 6, Vegetables fresh others 3
Ivory Coast	171	2	173	Cocoa beans 148, Coffee 16, Bananas 4, Stone fruit 2, Oilseeds others 2
France	129	38	166	Poultry 43, Pork 25, Wheat 25, Maize 13, Livestock others 13
Denmark	128	29	156	Pork 131, Poultry 10, Livestock others 9, Beef 3, Wheat 1
USA	128	7	135	Rice 55, Beans dry 49, Wheat 12, Apples 4, Stone fruit others fresh 3
Italy	102	15	118	Rice 37, Beef 19, Livestock others 12, Citrus fruit others 8, Wheat 7
Pakistan	114	0	115	Rice 111, Beans dry 3, Fruit fresh others 1
Nigeria	113	0	114	Cocoa beans 113
Spain	87	17	104	Rice 22, Pork 8, Poultry 8, Beef 8, Fruit fresh others 6
Canada	88	10	98	Wheat 71, Beans dry 26, Beef 1
Others	1359	207	1565	
Total	3895	555	4450	

In addition to the direct emissions associated with the life cycle of food, we also estimate that avoidable food waste generated in the UK is responsible for emissions associated with Land Use Change totaling 7.6 million tonnes CO_2 eq per annum (Box 12.6). The impact has not been split by external and internal Land Use Change. All figures used to calculate Land Use Change are global averages rather than nation-specific. As such they are not necessarily representative of emissions arising within the UK as a consequence of Land Use Change.

Inclusion of emissions associated with Land Use Change would increase the average carbon footprint of avoidable food and drink waste by approximately one fifth. As a proportion of Land Use Change emissions associated with UK consumption of agricultural products, it is 7%. The impact of Land Use Change is discussed further in the case studies, conclusions, and recommendations of this chapter. In recent years there has been an increase in the UK's fruit and vegetable consumption even as production of these crops in the UK is decreasing (FAOSTAT data, 2010). Major increases in agricultural product flows from the Mediterranean region, South Africa, South America, and elsewhere have made up the difference. To illustrate the connection between use of natural resources overseas and household food waste in the UK, case studies are set out below of three food products (wheat, tomato, and beef).

7. CASE STUDIES

Case A: Wheat

The carbon footprint of household waste of wheat products in the UK is 1,556,000 tonnes CO_2 eq per year. The internal part of the total carbon footprint is 1,448,000 tonnes CO_2 eq per year and the rest is due to activities at external locations. Almost 87% of the carbon footprint is composed of avoidable food waste rather than possibly-avoidable food waste. The greatest contribution to the footprint is made by emissions from agriculture and processing (e.g., into bread, cakes, and pastries). Packaging (transit and primary) accounts for approximately 10% of the carbon footprint of the product, and the emissions from waste management is the next most significant source.

The total water footprint of wheat consumption (e.g., bread, cakes) in the UK is 7483 million m³ per year (Chapagain and Orr, 2008). The water footprint

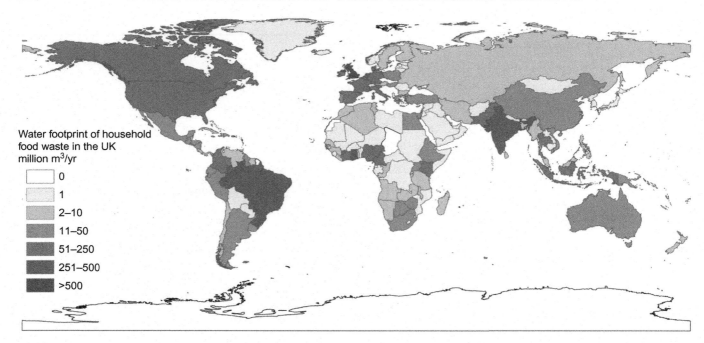

FIGURE 12.2   Map of the external water footprint of UK household food waste.

of household waste of wheat products in the UK is equal to 143 million m^3 per year (2% of total UK wheat water footprint), all of which is classed as avoidable. The major regions where the UK's water footprint from household waste of wheat products falls are shown in Figure 12.4.

As wheat is mostly rain fed in the UK, the impact on blue water resources in the UK is negligible. However, this does vary by region. The Environment Agency (2008) notes that, although farmers use less than 1% of the total amount of water abstracted in England and Wales for spray irrigation, this can reach 20% in East Anglia, and that on occasion more water is used on a hot dry day for spray irrigation than for public water supply. Nearly all of this water is lost by evaporation and can therefore represent a significant contributor to the internal water footprint of wheat production.

The EWF of waste of wheat products in the UK is presented in Figure 12.4. The darker the area, the larger is the EWF of UK household wheat waste in these areas. Note that wheat is rainfed in some of these regions and irrigated in others.

Case B: Tomato

The carbon footprint of household waste of tomato products in the UK is 853,000 tonnes per year. In contrast to wheat, it is mostly due to activities outside of the UK. The external component is 81% of the total emissions in this case. Almost 99% of the carbon footprint derives from avoidable waste of tomato products. The carbon footprint of tomatoes can vary significantly depending on whether they are grown in field or in greenhouse and upon how greenhouses are heated (Wiltshire et al., 2009). However, any variations in the emissions associated with the manner in which tomatoes are grown do not alter the dominant role of production in the carbon footprint.

Total water use for tomato production in the UK is only 0.8 million m^3 per year. However, as an importer of tomatoes, the UK's water footprint in relation to tomato consumption is 13.9 million m^3 per year. Out of the total water footprint of tomato consumption in the UK (including tomatoes, cook-in sauces, and ketchup), the waste at household contributes to 9.6 million m^3 per year, of which 9.4 million m^3 per year is classed as avoidable. This is about 68% of the total water footprint of tomato consumption in the UK. Most of the water footprint of household tomato wastage falls overseas (Figure 12.5). Currently the UK imports mainly from Spain, followed by Italy, the Netherlands, Morocco, Turkey, and Portugal.

A significant share of the total waste (57%) is related to the household waste of tomatoes imported from Spain. The countries mentioned above together

BOX 12.4

HOT-SPOTTING OF LOCATIONS WITH HIGHEST IMPACT ON WATER RESOURCES

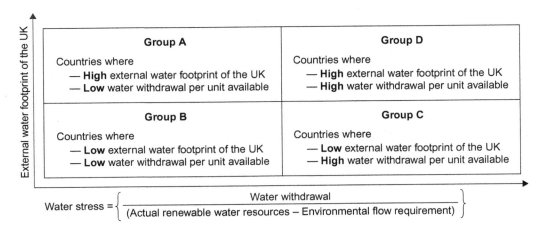

Impact assessments of these water footprints were conducted using hydrological attributes of regions where water is used in food cultivation and processing. Water stress in these locations is calculated as the ratio of actual blue water withdrawal to the net blue water available after taking into account environmental flow requirements. Following the scheme of impact categorization suggested in WWF-UK (Chapagain & Orr 2008), the various countries are then grouped based on the severity of the impacts using the schematic presented here.

contribute a total share of 97% of the total external water footprint of tomato waste in the UK household. As tomatoes are mostly irrigated and grown in greenhouses in the western hemisphere, the impacts are notable from carbon as well as blue water availability perspectives.

A recent study (Chapagain and Orr, 2009) shows that the EU consumes 955,000 tonnes of Spanish fresh tomatoes annually, which evaporates 71 million m^3 of water per year and would require a further 7 million m^3 of water per year to dilute leached nitrates in Spain. The main tomato-producing regions in Spain are in the Ebro valley (Navarra, Rioja, and Zaragoza) and Guadiana valley (Extremadura), in the south-east catchments of the Júcar, Segura, and Sur (Valencia, Alicante, Murcia, and Almería), and Canary Islands. The majority of fresh tomato imports to the UK originate from the southern Spanish mainland and the Canary Islands. These sites are among the most significant in Spain in terms of water stress. Other than water consumption, the main environmental issues associated with tomato cultivation are water pollution, soil pollution, and erosion, with habitat loss from expanding cultivation in some areas. The over-exploitation of aquifers from horticulture exports has affected water quantity and quality, including water salinization and

declining water tables, with additional loss of biodiversity, ecological value, and landscape amenity across the Mediterranean area (Martínez-Fernández and Selma, 2004). Current water use, in Almeria for example, is around 4–5 times more than annual rainfall and is mainly obtained from deep wells with high salinity of water, limiting the possibilities for water reuse. Almeria also has the largest poly-tunnel concentration in the world, with around 40,000 ha of greenhouse crops grown predominantly under flat-roof greenhouses. Ensuring sustainable, equitable, and productive use of water resources in these regions presents a challenge to all companies and organizations with a stake in these fresh produce supply chains.

Case C: Beef

The carbon footprint of household waste of beef in the UK is 1,176,000 tonnes CO$_2$ eq per year. Two thirds of the total direct emissions are made up of avoidable household waste of beef. About 67% of the total direct emissions are due to activities outside of the UK border. However, this excludes the impact of Land Use Change. Based on the average global impacts of Land Use Change to facilitate cattle farming, this would add

BOX 12.5

CONTRIBUTION TO THE AVERAGE CARBON FOOTPRINT OF HOUSEHOLD FOOD AND DRINK WASTE USING THE TOP-DOWN APPROACH (TONNES CO_2 EQ PER TONNE)

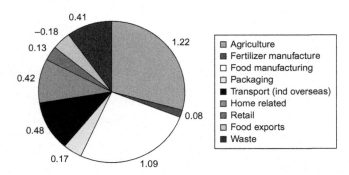

The top-down results suggest that, on average, approximately one quarter of the impact of food is associated with growing/rearing the crops and animals that enter the food chain, one quarter is associated with food processing, and one eighth is associated with home-related impacts (e.g., cooking). Waste management and degradation accounts for one tenth of emissions.

3.4 million tonnes CO_2 eq to the emissions attributable to avoidable and possibly-avoidable household waste of beef, roughly three times higher than direct emissions. The impact of Land Use Change is discussed further in the case study on Brazil.

As beef is a complex product for which to trace the final water footprint, the headline water footprint in these exporting countries does not reveal the full impact on water resources as a result of water used to grow cattle feed. As well as variation by country (Figure 12.6), the farming practices employed (e.g., extensive "grazing", "industrial" farming) will also influence the water footprint, and a detailed location-specific analysis is needed before interpreting the results. The total external water footprint per country should be further distinguished between beef from internal feed and imported feed in these locations.

8. DISCUSSION AND CONCLUSION

This study demonstrates the water and carbon footprints of food and drink waste generated at household level within the UK. For the first time, the carbon and water footprints of food waste have been identified by country of origin.

Much of the produce that carries a significant water footprint is also significant in terms of carbon footprint. Milk, beef, pork products, poultry, coffee, rice, and apples are identified as large contributors to both

indicators. Some notable differences between the carbon and water footprint data can be seen in the cases of bananas, citrus fruit, peppers, and cocoa, which have relatively high avoidable water footprints and relatively low avoidable carbon footprints; and potatoes, tea, and cucumbers and gherkins, which have relatively high avoidable carbon footprints but low avoidable water footprints.

Chocolate is another example of contrasting impacts. The quantity of chocolate thrown away is relatively small (24,000 tonnes of chocolate bars plus 7,000 tonnes in hot chocolate). Its contribution to the carbon footprint of food and drink waste is also small (117,000 tonnes CO_2 equivalent), yet it significantly contributes to the water footprint of food waste, accounting for over 750 million m^3 of water. Dairy and meat products constitute a relatively low proportion of food and drink waste by tonnage but are significant in terms of both the water and carbon footprint. This is because of the amount of feed required to support livestock, as well as enteric emissions from the animals themselves during their lives. This highlights the need to consider more than one issue when identifying priority products to be addressed by waste prevention initiatives.

The assessment of the carbon footprint excludes the impact of Land Use Change associated with demand for certain foodstuffs. For some products this makes little difference. For example, Land Use Change associated with demand for mushrooms (which are subsequently wasted) contributes 9 tonnes CO_2 eq per year. However, the demand for land to grow wheat for bread is responsible for an additional 270,000 tonnes CO_2 eq per year. In the case of beef and lamb, emissions associated with Land Use Change are actually higher than direct emissions associated with avoidable waste. This highlights the need to understand the boundaries of any assessment of environmental impact and the issues that one is in a position to influence.

In using the information contained in a water or carbon footprint, a number of issues need to be considered. The amount of water used or carbon emitted does not take account of how efficiently crops are grown and food produced within the local environment. In addition, the footprint does not consider the level of competition for water resources and local issues such as access to water. There is therefore a need also to consider scarcity of water in a given catchment area, and also how effectively it is being managed. In considering these issues from a policy perspective, it is important to take account of a range of economic, social, and environmental factors, many of which have not been discussed in this study.

Egypt, Israel, Pakistan, India, Thailand, and Spain are examples of countries where water stress is very high and the external water footprint of the UK's food waste is also relatively high. In such countries an appropriate policy response from UK stakeholders should include consideration of how to play a constructive role in improving information flows, stakeholder engagement, data availability, and institutional capacity in order to support water management regimes that adequately balance the needs of different

TABLE 12.5 Total Carbon Footprint of Household Food Waste in the UK (thousand tonnes per year)

	Avoidable Waste	Possibly Avoidable Waste	Total
Internal emissions	14,002	3223	16,230
External emissions	6138	696	6833
Unattributed emissions		1658	1658
Total carbon footprint	20,140	5577	25,717

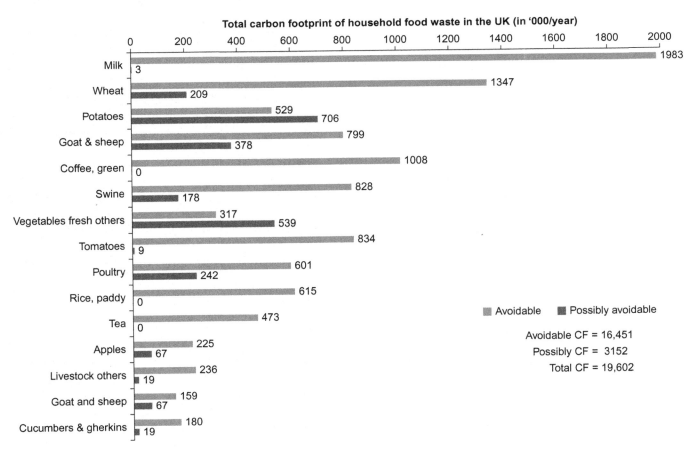

FIGURE 12.3 Total carbon footprint of household food waste in the UK for major food categories.

BOX 12.6

CARBON FOOTPRINT OF AVOIDABLE AND POSSIBLY AVOIDABLE HOUSEHOLD FOOD AND DRINK WASTE IN THE UK FOR DIRECT EMISSION AND EMISSIONS FROM LAND USE CHANGES ('000 TONNES PER YEAR)

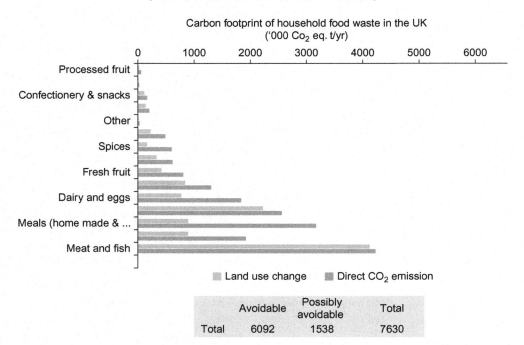

Carbon footprint of household food waste in the UK ('000 CO_2 eq. t/yr)

	Avoidable	Possibly avoidable	Total
Total	6092	1538	7630

In addition to the direct emissions associated with the life cycle of food, we also estimate that avoidable food waste generated in the UK is responsible for emissions associated with Land Use Change totaling 7.6 million tonnes CO_2 eq per annum. The impact has not been split by external and internal Land Use Change. All figures used to calculate Land Use Change are global averages rather than nation-specific. As such they are not necessarily representative of emissions arising within the UK as a consequence of Land Use Change.

users and ensure the continued viability of the freshwater ecosystems.

Conversely, though Ghana and Brazil support a large part of the external water footprint of household food waste in the UK, water stress in these countries is relatively low (with local exceptions). It may be that other environmental issues are more pressing and that positive socioeconomic impacts from sourcing produce from these countries might outweigh the negative externalities related to hydrology.

The work presented here could be further developed and improved in a number of ways. Information on the water footprint of food and drink is presented herein as a total figure. This could be disaggregated to highlight the contribution of green, blue, and gray water to a particular water footprint. The water footprint could also be disaggregated within a nation, to highlight particular areas of water stress and surplus to inform decisions regarding food production. At

present, the impact of Land Use Change on the carbon footprint is discussed separately from the emissions directly associated with the food supply chain. Further analysis could be used to highlight particular "hot spots" that would require further attention, as well as highlighting the interlinked nature of a range of issues, such as agriculture and forestry.

To address the water and carbon footprints of food and drink, and food and drink waste, WWF-UK (Macdiarmid et al., 2011) and others (e.g., Stehfest et al., 2009) have advocated a focus on reducing or altering patterns of consumption, which could be complementary to objectives around a healthy diet. However, a quick win in the UK would be to reduce the sheer volume of food waste at household level. It would bring positive changes in terms of impacts from the use of fertilizers and feedstuffs in the food industry overall. Another way of dealing with food waste is to reduce its creation along the full supply chain. Packaging protects

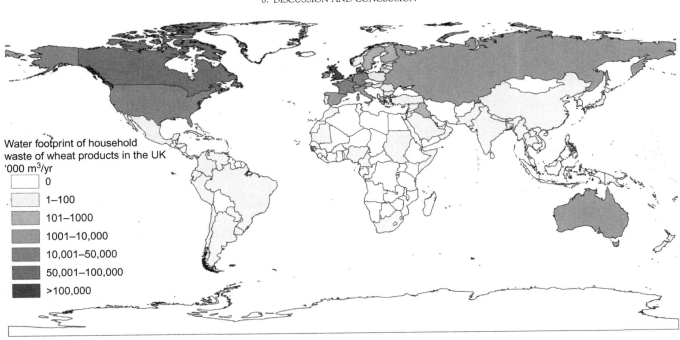

FIGURE 12.4   External water footprint of UK household waste of wheat products.

FIGURE 12.5   External water footprint of UK household waste of tomato products.

food from damage during its transportation from farms and factories via warehouses to retailing, as well as preserving its freshness upon arrival. Like other waste, food waste can be sent to landfill, but some food waste can also be fed to animals (typically swine), or it can be biodegraded by composting or anaerobic digestion, and reused to enrich soil.

Demand for food is increasing, and with a projected global population of 9 billion by 2050, there is a need to ensure that the food we do produce is effectively grown, stored, distributed, prepared, and consumed, with minimal waste. Beyond this the research shows the value of considering a range of issues when tackling waste, and highlights the effect of considering water

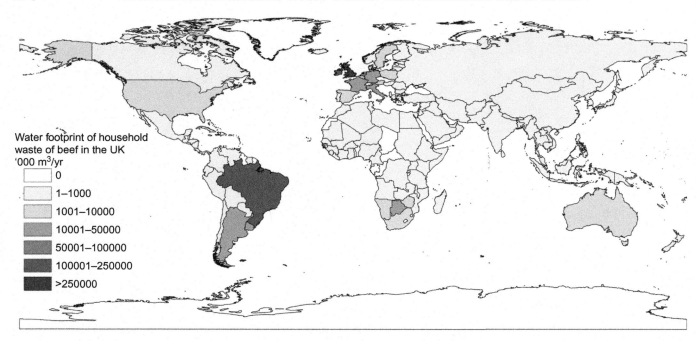

FIGURE 12.6   External water footprint of UK household waste of bovine products.

and greenhouse gas emissions in informing priorities. While it will never provide the answer by itself, reducing food waste could contribute toward reducing serious environmental problems associated with increasing scarcity of water resources and with climate change. Consumers, manufacturers, retailers, and those in the hospitality sector can all contribute to minimizing food waste, an action that is particularly relevant in the context of the global food security debate.

The way food is stored, packaged, labeled, and marketed, and our expectations of what good food looks like and when we can purchase it, can all influence the amount we waste. It is rather less wasteful if we align consumption to availability and adapt to the seasonal variability in supplies. A way forward to reduce waste is possible if waste generation is decoupled from economic growth as argued by Sjöström and Östblom (2010).

Reducing food waste requires action by all sectors of society. Those involved in farming can use the information contained in this and the referenced reports to reflect on the crops that they grow and their suitability given water availability and efficiency of production. Businesses can act in a number of ways, such as reviewing their supply chains and the processes and technologies they employ. Businesses can also help consumers by providing information on products, altering pack sizes to reduce waste, implementing standards, or giving advice on how to reduce waste within the home. Households can act to reduce food waste through a range of simple measures, as advocated

through the Love Food Hate Waste campaign, supported by work undertaken by other sectors of society. Love Food Hate Waste continues to be a valuable initiative in highlighting the magnitude of household food waste and ways in which this can be tackled.

This study identifies the water footprint associated with meeting the UK's food and drink needs, and the carbon footprint across the supply chain that delivers this, from farm to plate, including emissions associated with Land Use Change. Avoidable food waste represents approximately 6% of the UK water footprint, equivalent to almost twice the direct household water use of the UK. Just the internal water footprint from household food waste in the UK is equivalent to the direct household water use of 33 million people. Avoidable food waste also represents approximately 3% of the UK's domestic greenhouse gas emissions, with further emissions abroad. Excluding Land Use Change, emissions associated with avoidable food waste are equivalent to the annual greenhouse gas emissions from 7 million cars.

The case studies highlight the need to consider context when interpreting water footprint data. It is not necessarily the total volume of the water footprint that is critical; rather, it is the local context within which that water is used to produce food that determines the impacts on the environment and on other water users. The impacts in areas suffering water stress (e.g., excessive extraction of water) are normally far more significant than in areas with relatively plentiful water supplies.

This research could be extended in future to consider the sources of water used in different countries (green, blue) and the impacts of effluent (gray water) across additional stages in the supply chain. The carbon footprint data could also be developed to provide nation-specific factors. Even without this additional detail the key conclusion, that reducing food waste can make a significant contribution to addressing the water and carbon footprints of the UK, is clear. The research provides a framework around which to discuss how food waste could be reduced to provide the greatest environmental benefit.

Acknowledgement

This chapter is adapted from a report jointly published by WWF-UK and WRAP in 2011. We would like to thank both organizations for their kind permission to use a large part of their report in this book.

References

AIC, 2008. Fertiliser Statistics 2008 Report. AIC, Peterborough.

Aldaya, M.M., Allan, J.A., Hoekstra, A.Y., 2009. Strategic importance of green water in international crop trade. Ecol. Econ. 69 (4), 887–894.

Audsley, E., Brander, M., Chatterton, J., Murphy-Bokern, D., Webster, C., Williams, A., 2010. How Low Can We Go? An Assessment of Greenhouse Gas Emissions from the UK Food System and the Scope to Reduce Them by 2050. How Low Can We Go? WWF-UK, and the Food Climate Research Network.

Beccali, M., Cellura, M., Iudicello, M., Mistretta, M., 2009. Resource consumption and environmental impacts of the agrofood sector: life cycle assessment of Italian citrus-based products. Environ. Manag. 43 (4), 707–724.

Brinkley, A., Less, S., 2010. Carbon Omissions Consumption-Based Accounting for International Carbon Emissions. Policy Exchange, London.

BSI, 2008. Guide to PAS 2050: How to Assess the Carbon Footprint of Goods and Services. British Standards Institution, London.

Cadbury, 2008. Dairy Milk Carbon Footprint. Available at: <http://www.cadbury.com/ourresponsibilities/purplegoesgreen/Pages/CarbonFootprint.aspx> (accessed 15.02.08.).

Chapagain, A.K., 2006. Globalisation of Water; Opportunities and Threats of Virtual Water Trade. Balkema, Leiden.

Chapagain, A.K., Hoekstra, A.Y., 2004. Water Footprints of Nations. UNESCO-IHE, Delft, the Netherlands, Value of Water Research Report Series No. 16.

Chapagain, A.K., Hoekstra, A.Y., 2010. The Green, Blue and Grey Water Footprint of Rice From Both a Production and Consumption Perspective. UNESCO-IHE, and University of Twente, the Netherlands, Value of Water Research Report Series No. 40.

Chapagain, A.K., James, K., 2011. The Water and Carbon Footprint of Household Food and Drink Waste in the UK. WRAP and WWF-UK, UK.

Chapagain, A.K., Orr, S., 2008. UK Water Footprint: The Impact of the UK's Food and Fibre Consumption on Global Water Resources. WWF-UK, Surrey.

Chapagain, A.K., Orr, S., 2009. An improved water footprint methodology linking global consumption to local water resources: a case of Spanish tomatoes. J. Environ. Manag. 90 (2), 1219–1228.

Chapagain, A.K., Hoekstra, A.Y., Savenije, H.H.G., Gautam, R., 2006. The water footprint of cotton consumption: an assessment of the impact of worldwide consumption of cotton products on the water resources in the cotton producing countries. Ecol. Econ. 60 (1), 186–203.

Davis, J., Haglund, C., 1999. Life Cycle Inventory (LCI) of Fertiliser Production. Fertiliser Products Used in Sweden and Western Europe. Chalmers University of Technology, Gothenburg.

Davis, S., Caldeira, K., 2010. Consumption-based accounting of CO_2 emissions. Proc. Natl. Acad. Sci. 107 (12), 5687–5692.

DECC, 2010. Local and Regional CO2 Emissions Estimates for 2005–2008. DECC, Available at: <http://www.decc.gov.uk/en/content/cms/statistics/climate_change/gg_emissions/uk_emissions/2008_local/2008_local.aspx>.

Defra, 2005. Food Transport: The Validity of Food Miles as an Indicator of Sustainable Development. Defra, Available at: <http://statistics.defra.gov.uk/esg/reports/foodmiles/default.asp>.

Defra, 2006. Animal Health and Welfare Strategy Indicators: Headline Indicator 4B. Defra. Available at: <http://www.defra.gov.uk/foodfarm/policy/animalhealth/eig/indicators/h4b.htm> (accessed 09.09.09.).

Defra, 2008a. Family Food in 2007: A Report on the 2007 Expenditure and Food Survey. Defra, London.

Defra, 2008b. Future Water: The Government's water strategy for England. Defra Report. UK.

Defra, 2009a. Environmental Indicators In Your Pocket 2009. Available at: <http://www.defra.gov.uk/evidence/statistics/environment/eiyp/pdf/eiyp2009.pdf>.

Defra, 2009b. Understanding the GHG Impacts of Food Preparation and Consumption in the Home. Defra. Available at: <http://randd.defra.gov.uk/Document.aspx?Document = FO0409_8192_FRP.pdf>.

Defra, 2010a. 2010 Guidelines to Defra/DECC's GHG Conversion Factors for Company Reporting. Defra. Available at: <http://www.defra.gov.uk/environment/business/reporting/conversion-factors.htm>.

DTI, 2002. Energy Consumption In The UK. DTI, London, UK.

Environment Agency, 2008. Water Resources in England and Wales—Current State and Future Pressures. Environment Agency, UK.

ERM, 2009. Scoping LCA of Pork Production. BPEX, Kenilworth, UK.

Falkenmark, M., 2003. Freshwater as shared between society and ecosystems: from divided approaches to integrated challenges. Phil. Trans. R. Soc. B: Biol. Sci. 358 (1440), 2037–2049.

FAO, 2003a. AQUASTAT 2002. Food and Agriculture Organization of the United Nations, Rome.

FAO, 2003b. AQUASTAT 2003. Food and Agriculture Organization of the United Nations, Rome.

FAO, 2007. State of the World's Forests 2007. Food and Agriculture Organization of the United Nations, Rome.

FAOSTAT data, 2008. FAO Statistical Databases. FAO. Available at: <http://faostat.fao.org/default.jsp> (accessed 10.01.08.).

FAOSTAT data, 2010. Food Balance Sheets. FAO. Available at: <http://faostat.fao.org/site/291/default.aspx> (accessed 10.01.10.).

Forster, P., Ramaswamy, V., Artaxo, P., Berntsen, T., Betts, R., Fahey, D.W., et al., 2007. Changes in atmospheric constituents and in radiative forcing. In: Solomon, S., Qin, D., Manning, M., Chen, Z., Marquis, M., Averyt, K.B., et al., (Eds) Climate Change 2007: The Physical Science Basis. Contribution of Working Group I to the Fourth Assessment Report of the Intergovernmental Panel on Climate Change. Cambridge University Press, Cambridge, UK and New York, NY, USA.

Herzog, T., 2009. World Greenhouse Gas Emissions in 2005. World Resources Institute, Washington DC.

Hoekstra, A.Y., Chapagain, A.K., 2008. Globalization of Water: Sharing the Planet's Freshwater Resources. Blackwell Publishing Ltd, Oxford, UK.

Hoekstra, A.Y., Chapagain, A.K., Aldaya, M.M., Mekonnen, M.M., 2011. The Water Footprint Assessment Manual: Setting the Global Standard. Earthscan.

ITC, 2006. PC-TAS Version 2000–2004 in HS or SITC, CD-ROM. International Trade Centre, Geneva.

ITC, 2009. PC-TAS Version 2004–2008 in HS or SITC, CD-ROM. International Trade Centre, Geneva.

James, 2010. Methodology for assessing the climate change impacts of packaging optimisation under the Courtauld Commitment Phase 2. WRAP, Banbury.

Langley, J., Yoxall, A., Heppell, G., Rodriguez-Falcon, E., Bradbury, S., Lewis, R., et al., 2010. Food for thought?—A UK pilot study testing a methodology for compositional domestic food waste analysis. Waste Manag. Res. 28 (3), 220–227.

Lewis, E., 2010. Uncovering Water Losses in UK Household Food Waste: A Critical Assessment of the Water Offsetting Potential Arising from the Waste of Beef, Potatoes and Tomatoes. Kings College of London.

Lundqvist, J., de Fraiture, C., Molden, D., 2008. Saving Water: From Field to Fork—Curbing Losses and Wastage in the Food Chain. SIWI, SIWI Policy Brief.

Lyndhurst, B., 2008. London's Food Sector GHG Emissions: A Report for the Greater London Authority. London.

Macdiarmid, J, Kyle, J., Horgan, G., Loe, L., Fyfe, C., Johnston, A., et al., 2011. Livewell—A Balance of Healthy and Sustainable Food Choices. WWF-UK, Surrey.

Martínez-Fernández, J., Selma, M.A., 2004. Assessing the Sustainability of Mediterranean Intensive Agricultural Systems Through the Combined Use of Dynamic System Models, Environmental Modelling and Geographical Information Systems. Edward Elgar Publishing, Cheltenham.

Mekonnen, M.M., Hoekstra, A.Y., 2010. A global and high-resolution assessment of the green, blue and grey water footprint of wheat. Hydrol. Earth Syst. Sci. 14 (7), 1259–1276.

Milà i Canals L., Hospido, A., Clift, R., Truninger, M., Hounsome, B., Edwards-Jones, G., 2007. Environmental effects and consumer considerations of consuming lettuce in the UK winter. LCA in Foods 5th International Conference. Sweden.

Pretty, J.N., Ball, A.S., Lang, T., Morison, J.I.L., 2005. Farm costs and food miles: an assessment of the full cost of the UK weekly food basket. Food Pol. 30 (1), 1–19.

Quested, T., Johnson, H., 2009. Household Food and Drink Waste in the UK. WRAP, Banbury.

Ridoutt, B.G., Pfister, S., 2009. A revised approach to water footprinting to make transparent the impacts of consumption and production on global freshwater scarcity. Global Environ. Change 20 (1), 113–120.

Ridoutt, B.G., Juliano, P., Sanguansri, P., Sellahewa, J., 2009. Consumptive water use associated with food waste: case study of fresh mango in Australia. Hydrol. Earth Syst. Sci. Discuss. 6, 5085–5114.

Rockström, J., 2001. Green water security for the food makers of tomorrow: windows of opportunity in drought-prone savannahs. Water Sci. Technol. 43 (4), 71–78.

Ruini, L., Marino, M., 2009. LCA of Semolina Dry Pasta. EC, Brussels.

SABMiller & WWF, 2009. Water Footprinting: Identifying & Addressing Water Risks in the Value Chain. SABMiller, and WWF-UK, Surrey.

Sakaorat, K., Woranee, P., Phanida, S., Harnpon, P., 2009. Life cycle assessment of milled rice production: case study in Thailand. Eur. J. Sci. Res. 30 (2), 195–203.

Scanlon, B.R., Jolly, I., Sophocleous, M., Zhang, L., 2007. Global impacts of conversions from natural to agricultural ecosystems on water resources: quantity versus quality. Water Resour. Res. 43 (3), W03437.

Sjöström, M., Östblom, G., 2010. Decoupling waste generation from economic growth : an analysis of the Swedish case. Ecol. Econ. 69, 1545–1552.

Solomon, S., Qin, D., Manning, M., Chen, Z., Marquis, M., Avery, K.B., et al., 2007. Contribution of Working Group I to the Fourth Assessment Report of the Intergovernmental Panel on Climate Change. Cambridge University Press, Cambridge, UK.

Stehfest, E., Bouwman, L., van Vuuren, D., den Elzen, M., Eickhout, B., Kabat, P., 2009. Climate benefits of changing diet. Climatic Change 95 (1), 83–102.

Tassou, S.A., De-Lille, G., Ge, Y.T., 2009. FO0405 Greenhouse Gas Impacts of Food Retailing. Defra, London.

Tesco, 2008. Our Carbon Label Findings. Available at: <http://www.tesco.com/assets/greenerliving/content/documents/pdfs/carbon_label_findings.pdf> (accessed 16.10.10.).

The Economist, 2009. A Hill of Beans: America's Food-Waste Problem is Getting Worse. Available at: <http://www.economist.com/node/14960159> (accessed 26.11.2009.).

Watkiss, P., Smith, A., Tweddle, G., McKinnon, A., Browne, M., Hunt, A., et al., 2005. The Validity of Food Miles as an Indicator of Sustainable Development. Defra, London.

Wiedmann, T., Wood, R., Lenzen, M., Minx, J., Guan, D., Barrett, J., 2008. Development of an Embedded Carbon Emissions Indicator—Producing a Time Series of Input–Output Tables and Embedded Carbon Dioxide Emissions for the UK by Using a MRIO Data Optimisation System. Defra, London.

Wiltshire, J., Wynn, S., Clarke, J., Chambers, B., Cottrill, B., Drakes, D., et al., 2009. Scenario building to test and inform the development of a BSI method for assessing greenhouse gas emissions from food. Defra, London.

WWF, 2010. Living Planet Report 2010. WWF-International, Gland, Switzerland.

Yoshikawa, N., Amano, K., S. K., 2008. Evaluation of environmental loads related to fruit and vegetable consumption using the hybrid LCA method: Japanese case study. Proceedings of Life Cycle Assessment VIII. Seattle, Washington.

Zespri, New Zealand Ministry of Agriculture & Forestry, Landcare Research, Massey University, Plant & Food Research & AgriLINK, 2009. Carbon Footprint of Kiwi Fruit. Zespri, Mount Maunganui, New Zealand.

Electrical Energy from Wineries—A New Approach Using Microbial Fuel Cells

Sheela Berchmans, A. Palaniappan, and R. Karthikeyan

1. INTRODUCTION

Wineries are found worldwide, and more than 60 million tonnes of grapes are produced annually (Devesa-Rey et al., 2011). Historically, the species *Vitis vinifera* has been cultivated all over the world for wine making. Apart from being a popular drink served along with food, red wine is now considered a heart-healthy drink (Shrikhande, 2000). The disposal of winery and distillery effluents is one of the main environmental problems related to wine and alcohol-producing industries around the globe. Winery wastewater contains remains of grape pulp, skins, seeds, leaves, dead yeast cells, cell fragments, and various organic compounds used in filtration, precipitation, and cleaning processes. Thorough studies of the composition of soluble compounds of winery wastewater have shown that ethanol and to a lesser extent sugars (fructose and glucose) constitute more than 90% of the organic load of winery effluent. It is considered worthwhile to recover the organic load of winery wastewater rather than dissipating it into sludge and CO_2. Further, ethanol is a compound that is easy to extract (Bustamante et al., 2005).

2. WINERY WASTEWATER TO ELECTRICITY—CONCEPTUAL APPROACH

Traditional winery effluent treatment concentrates mainly on bringing down the biological oxygen demand (BOD) of wastewater by various techniques, before using the water either for irrigation or for soil conditioning. Recently, it has been demonstrated that recovery of H_2, methane, useful antioxidants (dietary supplements), and pullulan are possible from winery wastewater (Yu et al., 2002; Riaño et al., 2011). Also, winery waste sludge was shown to be an effective adsorbent for the adsorption of heavy metals from aqueous solutions (Arvanitoyannis et al., 2006). Apart from these scattered reports, the concept of "waste to wealth" is still at infancy in the wine industry. On the wealth side too, so far no one, other than our group, has reported methods to produce electricity from winery waste. With ever increasing demand for clean energy, scientists around the world are looking at ways to tap energy from every possible resource.

In this context, the last decade has witnessed a revival of interest in the development of microbial fuel cells (MFCs) for innovative applications as a source of energy. Microorganisms are able to perform the dual duty of electricity generation and biodegradation simultaneously. By combining wastewater treatment appropriately with current generation, recovery of electricity is feasible. Electricity recovery from biomass or production of hydrogen from biomass will lead to energy recovery without net carbon emissions into the ecosystem. This is an unexplored area in the field of treatment of winery wastewater. Hence this chapter describes viable ways of recovering electricity, along with biodegradation of winery wastewater, using MFCs.

3. MICROBIAL FUEL CELLS

The "central nervous system" of an MFC is electrochemistry. From the history of electrochemistry it is clear that the oldest and most significant application of electrochemistry the world has witnessed is the storage and conversion of energy. In a galvanic cell/battery, the stored chemical energy is converted into electrical energy through spontaneous redox reactions. The compounds

Food Industry Wastes.
DOI: http://dx.doi.org/10.1016/B978-0-12-391921-2.00013-5

undergoing spontaneous redox reactions are contained within the device itself. If these compounds (fuels) are supplied through an external source as and when they are depleted, then the energy conversion devices are called fuel cells. The MFC is one variant of many available fuel cells.

The classical way of producing electrical energy using fossil fuels has resulted in global warming and consequently increasing sea levels because of melting ice. To protect Mother Nature and as an alternative to fossil fuels, the energy available in naturally occurring biomass is converted into biofuels or into electricity after gasification (which involves Carnot limitation). On the other hand, the degradation of biomass by microorganisms does not involve Carnot limitation, resulting in better energy conversion. Scientists searching for alternative renewable energy sources have investigated the generation of electrical energy from microorganisms.

MFCs offer a solution whereby energy in chemical bonds is converted to electrical energy through catalytic reactions of microorganisms under anaerobic conditions. MFCs have also been studied for applications as biosensors, such as sensors for biological oxygen demand measurements. Presently, real-world applications of MFCs are restricted because of their low power density (several W/m^2).

3.1 What is Special about Electrochemical Energy Conversion?

During electrochemical energy conversion, in any fuel cell, spontaneous oxidation/de-electronation takes place at an anode (electron sink) and reduction/electronation takes place at a cathode (electron source). The current produced can be discharged through an external load. The electrochemical reactions at the anode and cathode (oxidation and reduction) bring about the conversion of chemical energy (Gibb's free energy) into electrical energy. Hence, one might expect complete transfer of the energy difference between anode and cathode compartments to external load. However, in practical applications this is not true because of the following.

The change in free energy in a reaction is given by

$$-\Delta G = W_{rev} - P\Delta V \quad (1)$$

where W_{rev} include all types of work (i.e., electrical, surface, gravitational, etc.) and QUOTE corresponds to the work of expansion. The change in free energy exclusive of work of expansion can be written as

$$-\Delta G = W_{rev} \quad (2)$$

The general expression for the free energy change in an electrochemical reaction can be written in the form

$$-\Delta G = nFE^o \quad (3)$$

where n is the number of electrons transferred per mole of the reactant, F is the Faraday constant in coulombs per equivalent, and E is the thermodynamic equilibrium potential in volts. Hence the electrochemical work carried out during conversion of electrical energy is

$$W_{rev} = nFE^o \quad (4)$$

All the available free energy can be converted into electrical work. In this sense it is said that in an electrochemical energy conversion, the ideal maximum efficiency is 100%. All the available energy will intrinsically be converted into electricity. However, not quite all the energy difference between the reactants and products of an electrochemical reaction can be made available even during an electrochemical reaction. This is because some of the energy will be lost due to ordering and disordering (entropy loss or gain) of molecules. Therefore intrinsic maximum efficiency of electrochemical conversion is

$$\varepsilon_{max} = \frac{\Delta G}{\Delta H} = \frac{-nFE^o}{\Delta H} \quad (5)$$

Comparison of ΔG, ΔH, and E^o values of typical fuel cell reactions suggest that maximum intrinsic efficiency of electrochemical energy conversion is in the range of 90% (Bockris and Reddy, 2002).

During electrochemical energy conversion, the maximum cell potential value E^o is obtained only under thermodynamic equilibrium when no current is drawn from the cell. As soon as the current is discharged through external load, E^o changes to E:

$$\varepsilon_o = \left\{\frac{nFE^o}{\Delta H}\right\}\left(\frac{E}{E^o}\right) \quad (6)$$

$$\left(\frac{E}{E^o}\right) \quad (7)$$

The above equation is true only if the reactants are completely converted into products without any losses. The equation should be modified by a Faradaic efficiency factor ε_f to take into account any incomplete conversion of reactants to products:

$$\varepsilon_o = \varepsilon_{max}\varepsilon_f\left(\frac{E}{E^o}\right) \quad (8)$$

The factor that mainly determines the efficiency is the actual cell potential E. The overall efficiency of electrochemical energy conversion depends on how the overall cell potential varies with the current density that the cell is producing. The cell has to consume a fraction of total cellular energy available to get its two electrode reactions to take place at the desired rate. The remaining fraction of the total energy is available for useful work. The actual realizable cell voltage during discharge is mainly influenced by three factors: ohmic, activation, and concentration overpotentials. The cell potential E and resulting efficiency of electrochemical energy conversion are determined by the activation overpotential, by the electrolyte conductance, and by mass transfer. At low current density, the influence of activation overpotential is dominant. At medium and high current densities, the dominant factors are ohmic overpotential and concentration overpotential, respectively.

3.2 Working Principle of MFCs

Conventional MFCs consist of biological anodes and abiotic cathodes. Abiotic cathodes are generally catalysts or electron transfer mediators that are used to attain high electron transfer rates. Biocathodes use microorganisms to assist cathodic reactions. In the case of aerobic biocathodes, with oxygen as the terminal electron acceptor, electron mediators, such as iron and manganese, are first reduced at the cathode (abiotically) and then reoxidized by bacteria. Anaerobic biocathodes directly reduce terminal electron acceptors, such as nitrate and sulfate, by accepting electrons from a cathode electrode during microbial metabolism (He and Angenent, 2006). Hence, from the above discussion we understand that the vital parts of the MFC are the anode, cathode, and electrolyte. In an MFC, the microorganisms are fed with fuels like glucose, alcohols, etc., which act as electron donors. Microorganisms catalyze the oxidation of reduced substrates/electron donors and the process of oxidation goes through the respiratory electron transport chain; releasing some of the electrons produced from cell respiration to the anode. The electrons released at the anode flow through an external circuit to the cathode and generate current. For each electron that is produced, a proton must be conducted to the cathode through the electrolyte (the aqueous solution) to maintain the current. The electrons and protons react with oxygen at the cathode to form water. Besides oxygen, ferricyanide or $KMnO_4$ can also be used, which will result in greater overall potentials. A cation exchange membrane is used to separate the catholyte and the anolyte and also prevents chemicals and materials other than protons from reaching the cathode compartment. Boxes 13.2 and 13.3 illustrate biological oxidation of manganese and iron ions, respectively.

When the electron donor undergoes oxidation along the respiratory chain, a large free energy change is produced when the oxidation is coupled with the reduction of the terminal electron acceptor oxygen. Microorganisms store energy in the form of adenosine triphosphate (ATP). In some bacteria, reduced substrates are oxidized

and electrons are transferred to respiratory enzymes by NADH, the reduced form of nicotinamide adenine dinucleotide (NAD). These electrons flow down a respiratory chain and finally the electrons are released to a soluble terminal electron acceptor, such as nitrate, sulfate, or oxygen. The maximum potential of the process is ~1.2 V, on the basis of the potential difference between the electron carrier (NADH) and oxygen under standard conditions. When the natural terminal acceptor is replaced by an electrode having a similar energy level, the available free energy in the electron transport chain can be trapped by the electrode. The electrons released in the electron transport chain are channeled towards the electrode instead of reaching oxygen when the conditions remain anaerobic. During the electron flow in the respiratory chain a series of enzymes move protons across an internal membrane, creating a proton gradient. The protons flow back into the cell through the enzyme ATPase, creating 1 ATP molecule from 1 adenosine diphosphate for every 3−4 protons. Box 13.1 lists redox potentials of electron donors/acceptors in the electron transport chain.

The possible types of electron transfer between a microbe and an electrode are (a) direct electron transfer, (b) direct electron transfer through pili, (c) mediated electron transfer, and (d) through oxidation of secondary metabolites.

a. *Direct electron transfer.* The microorganisms can have direct electrical communication with the electrode through the electroactive center of certain proteins. In this case, the electron transfer rate can be very low due to the insulation of the active site of the enzyme in the protein environment, which is buried inside the bacterial membrane. On the other hand, there are a few microorganisms that can communicate directly with the electrode due to the presence of redox proteins in the outer membrane, referred to as exoelectrogens. These microorganisms have their electron transferring centers with the proper orientation at the periphery of the redox enzyme that faces towards the external medium or towards the electrode. Microorganisms such as *Geobacter sulfurreducens, Hansenula anomala, Acetobacter aceti, Gluconobacter roseus*, and *Shewanella oneidensis* exhibit direct electron transfer (Busalmen et al., 2008; Fricke et al., 2008; Karthikeyan et al., 2009; Prasad et al., 2007).

b. *Direct electron transfer through pili.* Another direct electron transfer path involves biological nanowires 2−3 nm long called pili, made of fibrous protein structures. These thin protruding "wires" most likely assist direct electron transfer between the microbe and the electrode.

c. *Mediated electron transfer.* This type of electron transfer usually proceeds at much faster rates.

BOX 13.1

REDOX POTENTIALS OF ELECTRON DONOR–ACCEPTOR SYSTEM IN THE ELECTRON TRANSPORT CHAIN AND REDOX POTENTIALS OF SOME TERMINAL ELECTRON ACCEPTORS AT PH 7.2 VERSUS NORMAL HYDROGEN ELECTRODE (NHE)

Redox systems	E^0 (V)	Observation
$NAD^+/NADH + H^+$	−0.32	
$FAD^+/FADH_2$	−0.22	Metabolic redox
Cytochrome b (+3)/Cytochrome b (+2)	0.07	reaction
Ubiquinone (ox)/Ubiquinone (red)	0.10	
Cytochrome c (+3)/Cytochrome c (+2)	0.22	Probable e^- exit
$[Fe(CN)_6]^{3-}/[Fe(CN)_6]^{4-}$	0.36	
MnO_2/Mn^{2+}	0.60	
$MnO_2/MnOOH$	0.64	Terminal e^-
Fe(+3)/Fe(+2)	0.77	acceptors for
$\frac{1}{2}O_2 + 2H^+/H_2O$	0.82	cathodic reaction
$KMnO_4/MnO_{2\,(s)}$	1.23	

The energy derived from the redox reactions occurring in the respiratory chain during metabolism provides energy for the survival of bacteria. The electrons released in the respiratory chain are accepted by terminal acceptors. In the case of an MFC, the terminal electron acceptors are present in the cathode compartment, which collects the electrons received at the anode during metabolism. The anolyte should be free from any electron-accepting reagents.

Mediators, often referred to as electron transfer shuttles, are redox active compounds, which are stable in the oxidized and reduced states. Mediators are able to quickly diffuse in and out of the enzymatic channels, thus effectively shuttling electrons from the enzyme active site to the electrode surface. Electron transfer rates are very high in this case. However, some of the mediators can have a toxic effect on the microorganisms, and the use of mediators is not compatible with the use of MFCs for water purification.

d. *Oxidation of secondary metabolites.* Another promising mechanism involves the direct oxidation at the anode of exported catabolites by the microbes, such as dihydrogen or formate (Schaetzle et al., 2008).

3.3 Electrochemical and Bacterial Losses During Energy Conversion

As mentioned earlier, the actual electrochemical energy conversion depends upon the net voltage realized by the cell under working conditions during current discharge. The cell voltage is influenced by three factors—ohmic, concentration, and activation overpotentials—as mentioned previously. In the case of MFCs, efficiency is further influenced by microbial metabolism and electron transfer mechanism.

LOSSES DUE TO OHMIC OVERPOTENTIAL The ohmic losses in an MFC arise due to the resistance to the flow of electrons through the electrodes and interconnections, and the resistance to the flow of ions through the catholyte, anolyte, and the ion exchange membrane. Ohmic losses can be minimized by reducing the electrode spacing, increasing the conductivity of the electrolyte to a level tolerable to microorganisms, using a membrane with low resistivity, and using better interconnections.

LOSSES DUE TO ACTIVATION OVERPOTENTIAL Sluggish electron transfer kinetics give rise to activation overpotentials. This may occur during electron channeling from microorganism to the anode or at the microorganism to mediator/electrode interface. Slow electron acceptance rate at the cathode also causes activation overpotentials. Activation losses can be minimized by increasing the surface area, using better catalytic electrodes, and using enriched biofilms.

BOX 13.2

BIOLOGICAL MN (II) OXIDATION

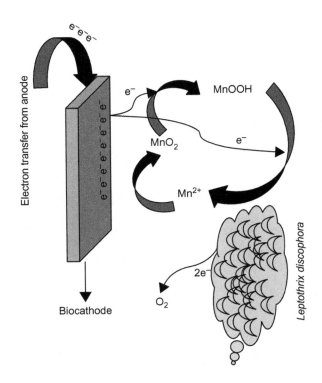

The ability of the bacterial species *Leptothrix discophora* to oxidize Mn^{2+} to MnO_2 can be utilized to regenerate the electron acceptor MnO_2 that can be used as a cathode material in the MFC. In the first step, which is an abiotic reaction, MnO_2 is reduced to an intermediate product, $MnOOH$, by accepting one electron from the cathode. This is followed by a further reduction of $MnOOH$ to Mn^{2+} through the acceptance of another electron. The second step is accomplished by manganese-oxidizing bacteria such as *Leptothrix discophora*, which oxidizes Mn^{2+} to MnO_2 by releasing two electrons to oxygen.

LOSSES DUE TO CONCENTRATION OVERPOTENTIAL Concentration losses occur mainly at high current densities due to limited mass transfer of chemical species to the electrode surface by diffusion. At the anode, concentration losses are caused by a limited supply of electron donors toward the electrode.

INTRACELLULAR LOSSES DURING BACTERIAL METABOLISM To generate metabolic energy, bacteria transport electrons from an electron donor at a low potential (through the electron transport chain) to the final electron acceptor (such as oxygen or nitrate) at a higher potential. In an MFC, the anode is the final electron acceptor and its potential determines the energy gain for the bacteria. Two kinetic processes are involved in intracellular potential losses from electron donor to outer membrane. First, bacteria oxidize the electron donor, producing intracellular reducing power in the form of NADH. NADH is then oxidized by transferring its electrons into the membrane proteins that constitute the electron transport chain. Ultimately the electron reaches the outer membrane before terminating at the anode, during which a certain amount of extracellular potential loss also occurs.

3.4 Electrochemical Techniques Generally Used in MFC Studies

Having learned the basic components of MFCs, the electron transport mechanism, and maximum possible energy conversion, we will look at a few

BOX 13.3

BIOLOGICAL MN (II) OXIDATION

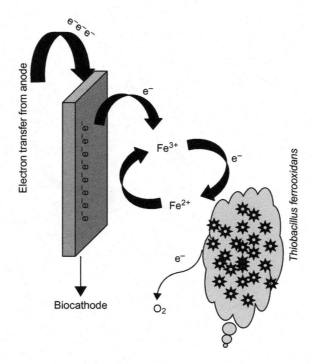

Iron compounds can undergo biological oxidation to a greater extent than manganese compounds, and this reaction can be used to regenerate ferric ions and hence can be used as a biocathode in a microbial fuel cell. In the first step, ferric ions accept electrons from the cathode through abiotic reaction and are converted into Fe^{2+} ions. The microbial activity of *Thiobacillus ferrooxidans* can oxidize Fe(II) back to Fe(III). Hence iron compounds can be biologically recycled by accepting electrons from electrodes via electrochemical reactions and releasing electrons to the terminal electron receptor.

electrochemical techniques that are generally used in MFC studies.

3.4.1 MFC Polarization Studies

The first technique involves what are called polarization studies, where the polarization curves are plots of electrode potential (or MFC voltage) as a function of current or current density. Such plots can be obtained with a potentiostat or a variable external resistance load or electronic load for highly reproducible and precise measurements. There are four methods for the measurement of MFC polarization curves: (a) constant resistance discharge, measured by connecting different resistors to the MFC and determining the resulting currents and voltages; (b) potentiodynamic polarization, in which current is measured at a slow voltage scan rate (e.g., 1 mV/s); (c) galvanostatic

discharge, in which current is maintained constant and the resulting voltages are measured; and (d) potentiostatic discharge, in which the voltage is maintained constant and the resulting currents are measured. Constant potential and current data are useful when MFCs are designed as power supplies for practical systems; constant resistance values do not yield useful information for MFC studies, especially when MFCs of different configurations and dimensions are being compared (Zhao et al., 2009).

3.4.2 Electrode Polarization Techniques

To get information about the suitability of anodes and cathodes for MFCs, polarization curves need to be recorded for individual electrodes (anode or cathode) by introducing a reference electrode in the respective compartment. With the help of three electrode configurations, many measurement techniques, such as potentiodynamic, galvanodynamic, and potentistatic, can be performed to assess single electrode performance. However study of half-cell reactions using anodes/cathodes of smaller area (<1 cm^2) is ideal for the evaluation of single electrode performance.

Tafel plots are more often used to characterize reactions of interest in chemical fuel cells. The Tafel equation describes the relationship between activation overpotential η_{act} and exchange current density i_0. The Tafel equation is the high-field approximation of the Butler–Volmer equation.

$$\eta_{act} = b \log_{10}\left(\frac{i}{i_0}\right) \qquad (9)$$

where i is the current density and b is Tafel slope (mV/dec).

Tafel plots are normally chosen to explore uncomplicated electron transfer processes that have very well-defined stoichiometry. They have been utilized to a lesser extent to investigate direct electron transfer properties of microorganisms, because of the more complex nature of the microbial electron transfer processes and due to their ill-defined stoichiometry. A few researchers have shown Tafel plots recorded using MFCs for the anode and with a cathode of equal area as the counter electrode. Such experiments were found not to suffer from counter electrode limitations (Rabaey et al., 2010). However, this factor needs supporting data from other researchers as well.

3.4.3 Current Interruption Technique

This technique is used extensively to measure the internal ohmic resistance of chemical fuel cells and is currently employed for MFCs. The basic principle of the technique is to interrupt the current flow and to observe the resulting voltage transients. On current interruption, the ohmic overpotential effects are initially seen, as it is an instantaneous process. Other voltage losses involve higher relaxation times and they appear later in the voltage transients. The main disadvantage is that the measurements need to be performed within a short duration ($<10\,\mu s$) for precise and accurate determinations. Further, it is difficult to separate the effects of concentration and activation overpotential.

3.4.4 Electrochemical Impedance Spectroscopy (EIS)

The electrochemical impedance spectroscopy technique is used to characterize the complete MFC and to determine the electrochemical properties of the individual electrodes. In addition the internal resistance of the MFC can also be measured using EIS.

In the case of two-compartment MFCs separated by a cation exchange membrane, a separate reference electrode has to be placed in each compartment. This design permits the measurement of potentials of the individual electrodes unaffected by ohmic resistance of the membrane and the conductivity of the electrolyte present in the two compartments.

There are two common graphical representations used in EIS: Nyquist plots and Bode plots. A small amplitude (e.g., 10 mV) alternating current (AC-sine wave) perturbation signal is applied to the MFC (a) to ensure nonlinear harmonic effects are not interfering with the data collection and (b) to prevent damage to the biofilm attached to the electrode surface. To avoid nonlinear responses when measuring EIS spectra at different currents or potentials, it is important that the direct current (DC) polarization curves are examined first so that the correct polarization potential can be worked out. Compared with the current interruption method, EIS is a sensitive technique offering detailed information about kinetic parameters, reaction mechanism, electrode/electrolyte conductivities, and biofilm behavior.

3.4.5 Cyclic Voltammetry

Cyclic voltammetry (CV) offers a fast and recognized method to distinguish whether bacteria use mobile redox shuttles to transfer their electrons, or pass the electrons "directly" through membrane-associated compounds (Logan et al., 2006). However, the method of recording cyclic voltammograms for investigating MFC reactions is to be considered with care. Certain researchers suggest that to record CV curves, a reference electrode must be placed in the anode chamber of the MFC close to the anode (working electrode); the counter electrode (e.g., platinum wire) is preferably placed in the cathode chamber but can also be placed in the anode chamber (Zhao et al., 2009). The option of using a Pt counter electrode in the anode compartment itself is better, provided that a counter electrode with a larger area than that of the anode is used. Using Pt wire as the counter electrode in the cathode compartment is not a good choice, as the electrolyte used is different in the cathode compartment. Further, if the Pt counter electrode is not larger than the anode, there is a chance of competitive reactions occurring in the cathode compartment at the counter electrode. It is better to carry out the CV technique using half-cell anodic reactions to identify electron transfer mechanisms and for the calculation of kinetic parameters. Low scan rates between 25 and 10 mV/s are preferred to carry out the CV techniques pertaining to MFC reactions. Many control experiments need to be performed to identify the true redox response of the biofilm.

3.5 Basic Performance Parameters of MFCs

Organic loading rate (OLR) (kg/m^3/day). Fuel supply to the MFC is calculated in terms of OLR:

$$OLR = \frac{W}{V \times T} \qquad (10)$$

where W is amount of COD (kg), V is volume (m^3), and T is time (day).

Power calculation (W). Voltage across the external resistance or load can be measured by using a multimeter. The derived current can be calculated from Ohm's law. Power will be the product of voltage and current:

$$P = VI \qquad (11)$$

where P is power (W), I is current (A), and V is voltage.

Power density (W/m^2 or W/m^3).

$$P.D = P/A \text{ or } P/Vol \qquad (12)$$

where $P.D$ is power density, A is area of the anode (m^2), and Vol is volume of the anolyte (m^3).

Theoretical coulombs (C_t). In a defined fuel system,

$$C_t = FbSV/MW \qquad (13)$$

where F is Faraday's constant, 96,485 C/mol of electrons, b is mol of electrons available for removal per mol of substrate, M is substrate concentration (mol/L), S is substrate concentration (g/L), and MW is molecular weight of the substrate.

In an undefined fuel system,

$$C_t = Ff\Delta S_{COD}V \qquad (14)$$

where f is factor of 1 mol of electrons/8 g-COD, and ΔS_{COD} is substrate concentration (g-COD/L).

Produced Coulombs (Cp):

$$C_p = \int_0^t It \qquad (15)$$

Coulombic efficiency (CE %):

$$CE\% = \frac{C_p}{C_t} \times 100 \qquad (16)$$

3.6 Construction of MFCs

3.6.1 Evolution of MFC Configurations and Designs

The evolution of MFC configuration has passed through several stages. Initially a two-compartment cell separated by a cation exchange membrane, resembling a proton exchange membrane (PEM) fuel cell, was constructed. Similar to PEM fuel cells, oxygen reduction was investigated as the cathode reaction by many researchers. To avoid mass transport limitation of oxygen supply to the cathode, air cathodes were introduced, which paved the way towards the development of single-chamber MFCs. The air cathode was exposed to air on one side and was bonded to the proton exchange membrane on the other side. Such a configuration decreased the internal resistance and increased the power output. As a next stage, membraneless MFCs were developed that exhibited better performance at low cost (Figure 13.1). In the absence of the membrane, significant oxygen crossover to the anode chamber takes place and the cathode gets contaminated by microbes. This reduces the coulombic efficiency as the available fuel now undergoes oxidation by consuming oxygen. The bacterial colonization of the cathode may also decrease long-term operation. In a few cases cathode fouling was avoided by using a nanoporous polymer membrane, based on nylon, cellulose, or polycarbonate, that inhibits bacterial growth.

A widely used and inexpensive design is a two-chamber MFC with the conventional "H" shape, made up of two bottles connected by a cation exchange membrane (CEM) such as Nafion or salt bridge. The salt bridge MFC, however, produces little power due to the high internal resistance. H-shaped systems are acceptable for basic parameter research, such as examining power production using new materials, or understanding different types of microbial communities that arise during the degradation of specific compounds, but they typically produce low power densities. Larger power densities have been achieved using oxygen as the electron acceptor when aqueous cathodes are replaced with air cathodes. Several changes of these basic designs have evolved in an effort to increase power density or supply of continuous flow of fuel

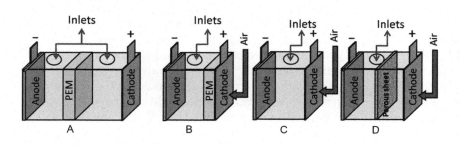

FIGURE 13.1 Evolution of different stages of microbial fuel cells: (A) conventional two-compartment cell; (B) single-chamber cell with air cathode; (C) membraneless MFC; (D) anode and cathode compartments separated by a nanoporous membrane.

through the anode chamber in contrast to the batch model. Systems have been designed with concentric circular compartments with either the inner or outer compartment acting as anode/cathode chamber. Another variation is to design the MFC like an upflow fixed-bed biofilm reactor, with the fluid flowing continuously through porous anodes toward a membrane separating the anode from the cathode chamber. MFCs have been designed to resemble hydrogen fuel cells, wherein a cation exchange membrane is sandwiched between the anode and cathode. To increase the overall system voltage, MFCs can be stacked with the cells shaped as a series of flat plates or linked together in series (Logan et al., 2006) (Figure 13.2).

4. MICROBIAL FUEL CELLS AND WINERIES—A CASE STUDY

In wineries, an activated sludge process is commonly used to treat the effluents. The activated sludge process can be replaced by MFCs since they can concurrently degrade the waste and produce electricity, as discussed earlier. It is essential to identify the required optimal MFC technology to have an advantage over the activated sludge process. The factors to be considered for the development of optimal MFC technology are to construct MFCs that can maintain low internal resistance in spite of an increase in the volume of the reactor, to establish efficient methods to separate the anode and cathode compartments, and to optimize reactor designs. Further, in the activated sludge process, diverse microbial consortia are used for the treatment. However, the use of microbial fauna that proliferate in wineries would also be a good choice for the investigation of treatment of winery effluents (Karthikeyan and Berchmans, 2012).

In wine, because of the harsh physicochemical conditions and low nutrient content, only a few bacterial species can grow. Of those few bacteria, acetic acid bacteria (AAB) and lactic acid bacteria (LAB) are most often observed in wine. The growth and metabolism of AAB and LAB are often associated with a change in the sensory attributes of the wine. *Gluconobacter oxydans* was the

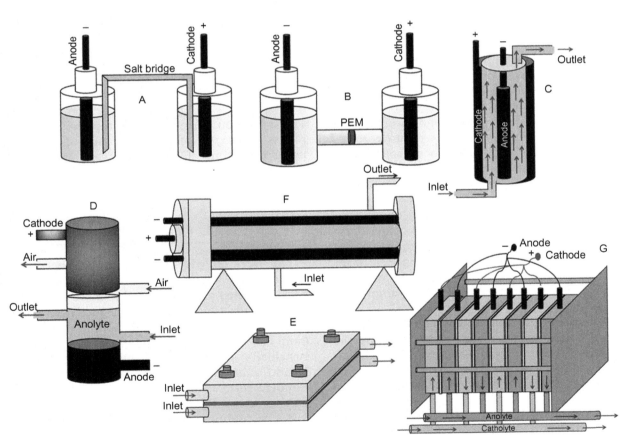

FIGURE 13.2 Evolution of various designs of MFC: (A) two-compartment MFC separated by a salt bridge; (B) two-compartment MFC separated by a cation exchange membrane; (C) and (D) upflow cylindrical MFCs; (E) MFC similar to a sandwich type PEM fuel cell; (F) two-compartment MFC with concentric circular anode and cathode compartments; (G) stacked MFC assembly.

main representative of acetic acid bacteria on sound, unspoiled red or white grapes. However, *Acetobacter aceti* and, to a lesser extent, *A. pasteurianus* became more widespread as the grapes became spoiled. These two species accounted for 75–85% of the acetic acid bacteria on the infected grapes. Acetic acid bacteria were present at all stages of wine making, from the mature grape through vinification to conservation. Low levels of *A. aceti* stay in the wine and they exhibited rapid proliferation on short exposure of the wine to air and caused significant increases in the concentration of acetic acid. Higher temperature of wine storage and higher wine pH favored the development and metabolism of *A. aceti* and *A. pasteurianus* species. *Acetobacter aceti* and *Gluconobacter roseus* belong to gram-negative prokaryotic species. Gram-negative prokaryotic cells have respiratory redox proteins located in the cell membrane that are accessible from the periplasm. The outer membrane contains porins, which makes them permeable to a wide variety of low molecular weight charged mediators. The periplasmic membrane-bound pyrroloquinoline quinone (PQQ)-containing enzymes of these genera provide fast and highly efficient oxidation of a wide variety of substrates.

Alcohol dehydrogenase (ADH) of acetic acid bacteria, consisting of the genera *Acetobacter* and *Gluconobacter*, catalyzes the first step of acetic acid production (i.e., the oxidation of ethanol to acetaldehyde). However, the presence of the quinoheme protein cytochrome C complex bound to the periplasmic side of the cytoplasmic membrane suggests that these microorganisms are likely to undergo direct electron transfer at the electrode surface. *Acetobacter aceti* is capable of oxidizing acetic acid to carbon dioxide. Hence the combination of *A. aceti* and *G. roseus* can be utilized for complete biodegradation of a fuel/carbon source. The presence of the membrane-bound built-in mediators and the ability of *Acetobacter* sp. to break acetic acid to CO_2 are the key points for considering the above two microorganisms for current generation. The efficiency of current generation using *A. aceti* and *G. roseus* was evaluated separately and as a mixture in a two-compartment cell separated by a Nafion membrane type MFC.

The influence of different anode materials (graphite felt, nickel foam, nickel foam modified with compounds like polyaniline, chitosan, and titanium carbide) and different substrates (ethanol, acetate, and glucose) on current generation have been analyzed. The suitability of the two microorganisms *A. aceti* and *G. roseus* for use in wineries is reflected in the results obtained from our research work (Karthikeyan et al., 2009; Karthikeyan and Berchmans, 2012). The results clearly show that by using the mixed culture of *A. aceti* and *G. roseus*, a maximum coulombic conversion

efficiency of 45% is achieved with spoiled wine as the fuel in MFCs. The current output of the fuel cells varies between $45 \mu A/cm^2$ and $250 \mu A/cm^2$. The maximum power output derived from the mixed biocatalysts was $18.8 W/m^3$. Also, the MFCs fabricated with the mixed catalysts show a sustainable operating voltage of 0.535 V with a current output of 0.89 mA. Hence the power output will be $470 \pm 5 \mu W$. This power will be 500 times higher than the power required to run a wall clock, which requires only $8–10 \mu W$ at minimum voltage of $1.3 \pm 0.2 V$. When two MFCs are connected in series, the power output obtained is sufficient to run devices like wall clocks normally run with 1.5 V dry cells. A power management system (PMS) that enables a sediment microbial fuel cell (SMFC) to operate a remote sensor consuming 2.5 W of power (Donovan et al., 2001) has been designed. By using a custom PMS to microbial energy in capacitors and use the stored energy in short bursts using two DC/DC converters and a digital logic circuit to convert low-level power from a SMFC (maximum $0.38 V \times 64 mA = 24 mW$ and an average continuous power of 3.4 mW) to 2.5 W power for 5 s. Similar electronic circuitry can be developed to deliver power output of the order of watts from our MFC. It is difficult to make a comparison of power outputs derived from other MFCs as there are number of variables, such as volume of anolyte and electrode materials (nature of materials used for the preparation of anodes and cathodes and their geometric and surface areas), involved in the design, which vary from MFC to MFC. The maximum current density reported in the recent past is $3.0 mA/cm^2$ for the oxidation of 5 g/L glucose using inoculums of heat-treated soil with an electrocatalyst like tungsten carbide (Rosenbaum et al., 2006).

5. CONCLUSIONS

Our work on MFCs is the proof of concept of using winery waste for electrical current generation, and we have identified areas that need to be researched before commercialization. To promote a successful scaling-up of MFCs, focus should be laid on the development of more efficient MFC configurations with low cost and sustainable materials. More importantly, the microbial kinetics must be improved to make MFC technology competitive to commercial high-rate anaerobic digestion systems. Recovery of valuable products from MFCs, in addition to the generation of electricity, will add economic value to the technology (Fornero et al., 2010). Engineered microorganisms can also offer solutions for the improvement of MFC technology. (Fishilevich et al., 2009).

In general, while wining and dining improves the social life of mankind, the luxury of living should be passed on to the coming generations without causing damage to our planet. Considering the greener aspects of electricity recovery without net carbon emissions, MFC technology definitely needs attention. Suitable innovations (in terms of low cost technology or recovering valuable by-products or synthetic biological techniques) are required to replace the existing activated sludge process with MFCs. More research efforts are required to provide innovative solutions to overcome the existing bottlenecks and to make the use of MFCs in winery treatment a realistic prospect.

References

Arvanitoyannis, I.S., Ladas, D., Mavromatis, A., 2006. Potential uses and applications of treated wine waste: a review. Int. J. Food Sci. Tech. 41, 475−487.

Bockris, J.O.M., Reddy, A.K.N., 1973. Modern Electrochemistry, second ed. vol. 2. Plenum Press, New York.

Busalmen, J.P., Nunez, A.E., Berna, A., Feliu, J.M., 2008. C-type cytochromes wire electricity-producing bacteria to electrodes. Angew. Chem. Int. Ed. 47, 4874−4877.

Bustamante, M.A., Paredes, C., Moral, R., Caselles, J.M, Espinosa, A. P., Murcia, M.D.P., 2005. Uses of winery and distillery effluents in agriculture: characterisation of nutrient and hazardous components. Water Sci. Technol. 5, 145−151.

Devesa-Rey, D., Vecino, X., Alende, J.L.V, Barral, M.T., Cruz, J.M., Moldes, A.B., 2011. Valorization of winery waste vs. the costs of not recycling. Waste Manage. 31, 2327−2335.

Donovan, C., Dewan, A., Peng, H., Heo, D., Beyenal, H., 2001. Power management system for a 2.5 W remote sensor powered by a sediment microbial fuel cell. J. Power Sources 196, 1171−1177.

Fishilevich, S., Amir, L., Fridman, Y., Aharoni, A., Alfonta, L., 2009. Surface display of redox enzymes in microbial fuel cells. J. Am. Chem. Soc. 131, 12052−12053.

Fornero, J.J., Rosenbaum, M., Angenent, L.T., 2010. Electric power generation from municipal, food, and animal wastewaters using microbial fuel cells. Electroanalysis 22, 832−843.

Fricke, K., Harnisch, F., Schröder, U., 2008. On the use of cyclic voltammetry for the study of anodic electron transfer in microbial fuel cells. Energy Environ. Sci. 1, 144−147.

He, Z., Angenent, L.T., 2006. Application of bacterial biocathodes in microbial fuel cells. Electroanalysis 18, 2009−2015.

Karthikeyan, R., Berchmans, S., 2012. Simultaneous degradation of bad wine and electricity generation with the aid of the coexisting biocatalysts Acetobacter aceti and Gluconobacter roseus. Biores. Technol. 104, 388−393.

Karthikeyan, R., Sathish kumar, K., Murugesan, M., Berchmans, S., Yegnaraman, V., 2009. Bioelectrocatalysis of Acetobacter aceti and Gluconobacter roseus for current generation. Environ. Sci. Technol. 43, 8684−8689.

Logan, B., Hamelers, B., Rozendal, R., Scröder, U., Keller, J., Ferguia, S., et al., 2006. Microbial fuel cells: methodology and technology. Environ. Sci. Technol. 40, 5181−5192.

Prasad, D., Arun, S., Murugesan, M., Padmanaban, S., Satyanarayanan, R.S., Berchmans, S., et al., 2007. Direct electron transfer with yeast cells and construction of a mediatorless microbial fuel cell. Biosens. Bioelectron. 22, 2604−2610.

Rabaey, K., Angenent, L., Schröder, U., Keller, J., 2010. Importance of Tafel plots in the investigation of bioelectrochemical systems. In: Lowy, D.A (Ed.), Bioelectrochemical Systems. IWA publishing, London, pp. 153−183.

Riaño, B., Molinuevo, B., González, M.C.G., 2011. Potential for methane production from anaerobic co-digestion of swine manure with winery wastewater. Biores. Technol. 102, 4131−4136.

Rosenbaum, M., Zhao, F., Schröder, U., Scholz, F., 2006. Interfacing electrocatalysis and biocatalysis with tungsten carbide: a high-performance, noble-metal-free microbial fuel cell. Angew. Chem. Int. Ed. 455, 6658−6661.

Schaetzle, O., Barrière, F., Baronian, K., 2008. Bacteria and yeasts as catalysts in microbial fuel cells: electron transfer from microorganisms to electrodes for green electricity. Energy Environ. Sci. 1, 607−620.

Shrikhande, A.J., 2000. Wine by-products with health benefits. Food Res. Int. 33, 469−474.

Yu, H., Zhu, Z., Hu, W., Zhang, H., 2002. Hydrogen production from rice winery wastewater in an upflow anaerobic reactor by using mixed anaerobic cultures. Int. J. Hydrogen Energ. 27, 1359−1365.

Zhao, F., Slade, R.C.T., Varcoe, J.R., 2009. Techniques for the study and development of microbial fuel cells: an electrochemical perspective. Chem. Soc. Rev. 38, 1926−1939.

CHAPTER

14

Electricity Generation from Food Industry Wastewater Using Microbial Fuel Cell Technology

Wen-Wei Li, Guo-Ping Sheng, and Han-Qing Yu

1. INTRODUCTION

Microbial fuel cell (MFC) technology offers a promising sustainable solution to meet the increasing needs of energy and wastewater treatment (Logan, 2009; Rozendal et al., 2008). In an MFC, the chemical energy stored in organics of wastewater is directly transformed to electrical energy via a series of electrochemical reactions catalyzed by microorganisms. Extracting energy from wastewater is considered one of the most promising future schemes of sustainable energy production and wastewater treatment (Angenent et al., 2004). In this respect, several biotechnologies have been extensively investigated, including anaerobic fermentation for methane and biohydrogen production (Hallenbeck and Ghosh, 2009). While biohydrogen technology is still to be demonstrated at an industrial scale, biogas technology has seen widespread application worldwide (Martins das Neves et al., 2009). MFC is a relatively new technology. At present, the energy efficiency of MFC, in most cases even at lab scales, still cannot compete with the anaerobic fermentation processes (Foley et al., 2010). One of the most important reason is that the electrochemically active microorganisms (EAMs) at the anode cannot effectively utilize the complex organic substrates in wastewater and cannot outcompete most fermenters.

Nevertheless, there are several distinct advantages of MFC over these conventional bioprocesses. First, MFC enables more efficient and complete removal of pollutants and thus ensures a better effluent quality. The effluent from anaerobic fermentation processes usually contains volatile fatty acids (VFAs) and chemical oxygen demand (COD) of up to several hundred mg/L (Yu et al., 2002); thus a subsequent aerobic treatment is needed to meet the discharge limits. However, this leads to significant energy consumption for aeration and subsequent sludge treatment and increases the operating cost (Cusick et al., 2010). In addition, in the conventional anaerobic–aerobic treatment process, some hazardous by-products like sulfide may be released and damage the facilities (Hamelin et al., 2011). In comparison, a high COD removal efficiency can be achieved by MFC even at low influent concentration (Kim et al., 2010), sulfide can be effectively utilized as an electron donor and contribute to electricity generation (Rabaey et al., 2006), and most of all, deriving net energy from this process is possible (Fornero et al., 2010a; Pant et al., 2011). Furthermore, the presence of the electrode in an MFC anodic chamber provides a platform to trigger multiple biological and electrochemical reactions to facilitate simultaneous removal of multiple pollutants (Venkata Mohan et al., 2010). Second, from the energy recovery and utilization point of view, electricity is apparently a more desirable option than methane or hydrogen, because the electricity can be utilized *in situ* and there are no problems such as gas-product separation, purification, and incineration (if for power generation). Because of the many advantages, MFC is considered a highly pursuable technology to capture energy from various organic wastewaters. Nevertheless, MFC is not likely to completely replace anaerobic fermentation processes because of their different implementation scope and process characteristics. Instead, one promising strategy

Food Industry Wastes.
DOI: http://dx.doi.org/10.1016/B978-0-12-391921-2.00014-7

249

would be to complement MFC with anaerobic fermentation for both efficient wastewater treatment and energy recovery (Oh et al., 2010), as we shall discuss in detail in this chapter.

Food industry wastewaters (FIWs) are considered an ideal substrate for electricity generation because of their rich organic content, high biodegradability, and abundant availability (Digman and Kim, 2008). To date, a wide variety of FIWs have been investigated for electricity generation using MFC, including waste streams from brewery, winery, dairy, and canteen, and effluent from starch, whey, vegetable, meat, fish, and other food-processing industries. During the last decade, there have been considerable improvements in the power density (PD) and coulombic efficiency (CE) of MFC as well as the overall cost of such processes, attributed to significant progress in MFC design, engineering, and operation and a better understanding and manipulation of the microbe—electrode interactions. Implementing MFC to harvest energy from energy-rich wastewater is still a highly pursuable future direction of MFC application. Because of the food-derived nature of FIWs, COD removal in MFC is usually not a problem for such wastewater (Velasquez-Orta et al., 2011). In fact, the biggest difference between MFC and conventional anaerobic treatment processes lies not in the degradation of organics, but in the metabolic pathway and the electron acceptors that consume the released electrons. Thus, in this chapter we focus mainly on the electricity generation of FIW-fueled MFCs.

Because of the high energy density and abundance of FIWs, FIW-fueled MFCs may have the greatest potential to be scaled up for practical power generation in the near future. On the other hand, however, the high variability of FIW in composition and volume and their complex nature make the MFC bioconversion process difficult. In this chapter, we summarize the recent research efforts in improving the power density and energy efficiency of MFCs from various FIWs, evaluate the major factors influencing such processes, and discuss the challenges for a practical implementation of the technology. In particular, a case study of electricity generation from molasses wastewater using a stacked MFC is presented to offer a detailed analysis of the advantages and limitations of this technology. Finally, the future possibilities of this technology toward large-scale implementation are evaluated, and several future research directions to address the present challenges are proposed. This chapter gives an overview of the current situation of MFCs that use FIW as a substrate and assesses the feasibility of this technology for commercialized power generation application.

2. CURRENT STATUS OF ELECTRICITY GENERATION FROM FOOD INDUSTRY WASTEWATERS

FIWs generally contain a high content of degradable organics that can be readily utilized for electricity generation. There are many types of FIWs with composition, strength, and solution chemistry varying substantially depending on their sources; thus their potentials for electricity generation also vary significantly (Pant et al., 2010). Here, the advantages and limitations of several typical wastewaters as substrates for MFCs are introduced. A list of some typical processes is given in Table 14.1.

Waste streams from breweries, with COD concentrations typically in the range 3000—5000 mg/L, have been most frequently used as substrates to power MFCs (Feng et al., 2008; Wang et al., 2008; Wen et al., 2009). Compared with domestic wastewater, brewery wastewater contains a higher amount of easily degradable carbohydrate and thus has high energy intensity (Pant et al., 2010). Feng et al. (2008) investigated electricity generation from brewery wastewater in an air-cathode MFC at fed-batch mode. A high PD (205 mW/m^2 normalized to the anode area, or 5.1 W/m^3 normalized to the effective volume of the anodic chamber, hereinafter the same) was obtained in this system when full-strength raw brewery wastewater was used. Katuri and Scott (2010) designed a hybrid upflow MFC to treat brewery wastewater, and a stable PD of 51.5 mV/m^2 was continuously generated.

Wastewater from wineries based on the use of cane molasses is also a suitable feedstock for MFC. The usually very high COD and the presence of a high content of inhibitory substances, such as sulfur compounds and salt, somehow lower energy efficiency and even suppress the microbial activity. Thus, winery wastewaters usually need to be appropriately diluted or treated by fermentation prior to being utilized by EAMs (Zhong et al., 2011). Cusick et al. (2010) reported a 40.9% higher power output in an air-cathode MFC fed with organic-rich winery wastewater than with domestic wastewater.

Starch-processing wastewaters have also been recently explored as substrates for MFCs (Jamuna and Ramakrishna, 1989; Kaewkannetra et al., 2011). An earlier study by Kim et al. (2004) demonstrated that starch-processing wastewater, at an initial COD concentration of 1700 mg/L, can be utilized by microbial consortia at an MFC anode to continuously generate electricity. Kaewkannetra et al. (2011) operated a 30 L two-chamber MFC using full-strength wastewater

TABLE 14.1 Electricity Generation Performance of MFCs Fed with Various FIWs

Wastewater Type	COD Load* (mg/L)	(kg/(m^3day))	Maximum PD[†] (mW/m^2)	(W/m^3)	Maximum CE (%)	Reference
Brewery wastewater	1,168	–	–	5	3.6	(Fornero et al., 2010a)
Brewery wastewater	2,239	–	483	12	20	(Wang et al., 2008)
Brewery wastewater	627	–	264	9.5	–	(Wen et al., 2009)
Brewery wastewater	2,240	–	528	13.2	–	(Feng et al., 2008)
Brewery wastewater	430	–	330	18	–	(Katuri and Scott, 2010)
Brewery wastewater	1,501	–	669	24.1	–	(Wen et al., 2010)
Brewery wastewater	2,800	–	–	5.0	3.6	(Fornero et al., 2010a)
Winery wastewater	656–2,555	–	124	0.27		(Huang et al., 2011)
Winery wastewater	2,200	–	–	6.7	18	(Cusick et al., 2010)
Cassava mill wastewater	16,000	–	1,771	18.2	20	(Kaewkannetra et al., 2011)
Winery wastewater	–	3.20	115.5	1.01	1	(Zhong et al., 2011)
Starch-processing wastewater	4,852	–	239.4	–	8	(Lu et al., 2009)
Cereal wastewater	595	–	81	–	–	(Oh and Logan, 2005)
Rice mill wastewater	2,200–2,250	–	–	2.3	–	(Behera et al., 2010)
Dairy wastewater	4,440	–	–	1.10	–	(Venkata Mohan et al., 2010)
Dairy wastewater	1,200	–	–	5.7	–	(Ayyaru and Dharmalingam, 2011)
Dairy wastewater	1,562	–	150	–	–	(Nimje et al., 2011)
Cheese whey wastewater	730	–	18.4	–	1.9	(Antonopoulou et al., 2010)
Yogurt waste	8,169	–	92	–	–	(Cercado-Quezada et al., 2010)
Chocolate wastewater	1,459	–	1,500	–	–	(Patil et al., 2009)
Meat-packing wastewater	5,540	–	139	–	5.2	(Heilmann and Logan, 2006)
Seafood wastewater	5,200	–		16.2	15	(You et al., 2010)
Potato-processing wastewater	2,100	–	217	–	21.1	(Kiely et al., 2011)
Vegetables extract	–	0.70	57.38	–	–	(Mohan et al., 2010)
Canteen wastewater	–	1.74	5.09	–	–	(Goud et al., 2011)
Palm oil mill wastewater	–	8–10.5	44.6	–	–	(Cheng et al., 2010)
Pre-fermented winery wastewater	–	–	1,410	–	–	(Zhang et al., 2009a)
Pre-fermented cereal wastewater	5,516–5,635	–	371	–	–	(Oh and Logan, 2005)
Pre-fermented canteen wastewater	–	1.74	98.8	–	–	(Goud and Mohan, 2011)
Pre-fermented canteen wastewater	5,300	–	240.3	–	–	(Choi et al., 2011)
Corn stover hydrolysate	1,000	–	367	–	30	(Zuo et al., 2006)
Wheat straw hydrolysate	2,000	–	123	–	37.1	(Zhang et al., 2009c)

CE, Coulombic efficiency; COD, chemical oxygen demand; PD, power density.
*COD load refers to soluble COD in wastewater that is directly fed into the MFC anodic chamber and when a maximum PD is achieved.
[†]PD is normalized to the projected surface area of the anode electrode or to the anodic chamber volume.

from a cassava mill (COD = 16,000 mg/L) as the substrate for electricity generation. After enrichment and acclimation, the MFC yielded a very high PD of 1771 mW/m^2.

Dairy wastewaters offer another potential source of renewable energy. However, such wastewaters also frequently contain a high concentration of proteins and lipids (Fang and Yu, 2000). These compounds are less biodegradable and can adversely affect the microbial activity in anaerobic degradation processes when present at high concentrations (Demirel et al., 2005). Free ammonia released during protein degradation and long-chain fatty acids from liquid hydrolysis have all been demonstrated to inhibit anaerobic microorganisms (Ramsay and Pullammanappallil, 2001). Thus, there is greater difficulty in the treatment and energy extraction from dairy wastewaters than with other carbohydrate-rich FIWs. Nevertheless, several recent studies show that such wastewater seems to less affect the microbial activity when being utilized for power generation in MFCs (Nimje et al., 2011; Venkata Mohan et al., 2010).

In addition to the above wastewaters, some waste streams from more complex sources have also been explored as MFC feedstock, including meat-processing (Heilmann and Logan, 2006; Katuri et al., 2012) and fish-processing wastewaters (You et al., 2010), vegetable wastewater (Durruty et al., 2011; Kiely et al., 2011), and even canteen-based food wastewater (Goud et al., 2011). Mohan et al. (2010) reported the use of extract from composite vegetable wastes, after mastication and filtration, for bioelectricity generation in a non-catalyzed air-cathode MFC, which yielded a maximum PD (57.38 mW/m^2) at OLR of 0.7 kg/(m^{3}d). Patil et al. (2009) tested the feasibility of deriving electricity from chocolate industry wastewater in a dual-chamber MFC. More excitingly, Goud et al. (2011) obtained a maximum PD of 107.9 mW/m^2 from composite canteen food wastewater.

Today, a growing variety of substrates are being brought into MFC systems by researchers and engineers, with substantially enhanced performance. In general, FIWs are suitable substrates for MFCs, but power generation performances can vary significantly with the characteristics of the treated wastewater. One primary reason for the lower PD of MFCs with complex substrates compared with those using glucose and carboxylic acids is the substantial loss of energy during microbial metabolism and electron transfer. It is well known that polymers such as polysaccharides and proteins cannot be directly utilized by most EAMs. Thus, they need to be converted to simple forms such as glucose and VFAs in the first place (Fornero et al., 2010a). To achieve this, the anodic microbes have to sustain a fermentative population in order to degrade the complex organics, which generally leads to lower PD and CE than with VFA- or glucose-fed MFCs (Liu et al., 2004; Rabaey et al., 2005b). However, regarding the complex nature of FIWs and the usually large fractions of less-degradable organics, the presence of diverse microorganisms, especially acidogenic bacteria, is of critical importance. In practical treatment processes, the complex contents such as polysaccharides, proteins,

and lipids are first converted to sugars, amino acids, and VFAs by acidogenic bacteria, which can then be easily degraded by EAMs for electricity generation (Choi et al., 2011). Thus, the presence of acidogenic bacteria is necessary to break down the complex substrates, but the number of competitive fermentative bacteria should be minimized as much as possible to favor a high PD. This has led to the concept of pre-fermentation pretreatment. Up to now, there are already several successful cases of using pre-acidified FIWs as substrates for MFC, which generally show better power production than raw wastewater (Oh and Logan, 2005; Durruty et al., 2011; Goud and Mohan, 2011; Kiely et al., 2011). Despite the many advantages, however, such a two-step configuration also increases the process complexity and cost, and there arise other new problems such as how to smoothly connect the two parts without affecting individual processes (Li and Yu, 2011). Thus, a balanced consideration of these aspects and a further process optimization are needed in future studies.

3. FACTORS AFFECTING ANODIC PERFORMANCE

The production of electricity in an MFC involves complex processes of biological and electrochemical reactions and charge transfer. Thus, the substrate availability, microbial community, reactor design, and operating parameters all exert significant influences on an MFC. Given that various organics in FIWs are mainly degraded and utilized at the anodic chamber of MFCs, the anodic performance is critical for such FIW-fueled MFCs.

3.1 Wastewater Properties

Wastewater property is one of the most critical factors that directly affects the microbiology and conversion performance of an MFC. In fact, many industrial wastewaters with high toxic or recalcitrant contents are usually out of consideration as MFC feedstock. FIW is in general a favorite substrate for MFC, but the PD and CE vary significantly depending on the wastewater type and accordingly its properties, such as composition, conductivity, and pH.

Previous studies show that MFC performance is closely associated with the wastewater composition. First, simple and easily degradable organics like soluble sugars and VFAs are preferred as substrates in MFCs over higher-molecular compounds such as cellulose, starch, and proteins. This is in line with the fact that FIW-fed MFCs mostly show lower PD and CE than those with pre-fermented (Zuo et al., 2006) or glucose substrate (Logan et al., 2007). Second, the influences of inhibitory substances in wastewater, such as free ammonia (Nam et al., 2010b; Kim et al., 2011) and long-chain fatty acids (Lalman and Bagley, 2001) formed during organics degradation, could also severely inhibit the function of EAMs, especially at higher strengths. Lastly, the various oxidative substances such as nitrate and sulfate in wastewater, and even the oxygen leaked from the cathode side, would all compete with the anode electrode for electrons.

Despite the high energy content of FIWs, the low ion conductivity usually presents a substantial drawback of such wastewaters for MFC operation. A negative correlation between wastewater conductivity and internal resistance of MFCs has been documented (Cheng and Logan, 2007; Fornero et al., 2010b). A higher conductivity is beneficial for transport of ions through the anode biofilm and in the bulk phase. As shown in Table 14.2, raw FIWs typically have relatively low conductivity (mostly below 6 mS/cm) compared to many other waste streams. For example, the solution conductivity is 50 mS/cm for sea water (He et al., 2007) and 34 mS/cm for landfill leachates (Greenman et al., 2009). FIWs with higher conductivities generally favor more electricity generation than the low-conductive streams with similar COD strength (Velasquez-Orta et al., 2011). A dramatic improvement in PD (up to 245%) was achieved by Huang and Logan (2008) by directly

raising the wastewater conductivity from 0.8 mS/cm to 10.2 mS/cm. In addition, the conductivity is also closely associated with the organic strength. A dilution of brewery wastewater from a COD concentration of 2,240 to 200 mg/L was found to decrease solution conductivity from 3.23 to 0.12 mS/cm (Feng et al., 2008). As a consequence, the internal resistance dramatically increased from 595 to 4340 Ω, while PD dropped dramatically from 205 to 29 mW/m^2. These studies clearly indicate a positive relationship between ionic conductivity and power generation. However, an overly high salinity and ion conductivity may pose detrimental effects on the anodic microorganisms (Digman and Kim, 2008). Thus, there exists a suitable range of wastewater conductivity for MFCs (Oh and Logan, 2006; Zuo et al., 2006).

Low pH buffering capacity is another challenge for FIWs as substrates for MFCs. This is especially true for brewery or pre-fermented FIWs, which typically have high VFA contents and a poor pH buffering capacity (Goud et al., 2011). A low value of anolyte pH not only suppresses microbial activity but also decreases electron discharge from the biocatalyst, thus a higher wastewater pH is generally found to favor MFC operations by neutralizing the protons produced at the anode (Behera et al., 2010; Zuo et al., 2006). For example, a significant increase of PD (158%) was achieved in a recent study by adding 200 mM phosphate buffer into the brewery wastewater at anodic chamber (Feng et al., 2008). However, this buffer addition strategy could not be a practical option, due to economic limitations and environmental considerations such as salt-induced secondary pollution.

TABLE 14.2 Typical Values of Conductivity and pH for Some Raw FIWs

Wastewater	Conductivity (mS/cm)	pH	Reference
Dairy wastewater	2.92	8.7	(Velasquez-Orta et al., 2011)
Bakery wastewater	1.238	5.9	(Velasquez-Orta et al., 2011)
Brewery wastewater	0.702	3.5	(Velasquez-Orta et al., 2011)
Brewery wastewater	3.23	6.5	(Feng et al., 2008)
Molasses wastewater	42.7	4.2	(Zhong et al., 2011)
Cassava wastewater	2.30	5.5	(Kaewkannetra et al., 2011)
Meat-packing wastewater	5.8	–	(Heilmann and Logan, 2006)
Starch-processing wastewater	5.6	6.5	(Lu et al., 2009)
Cassava wastewater	2.3	5.5	(Kaewkannetra et al., 2011)
Potato wastewater	5.2	6.1	(Kiely et al., 2011)
Dairy wastewater	2.7	8.3	(Kiely et al., 2011)

3.2 Anodic Microbiology

Electricity generation in MFCs arises from the process that microorganisms metabolize organic substrates to generate electrons and subsequently transfer the released electrons to an electrode, either via direct contact or by using soluble electron shuttles (Logan, 2009). Therefore, the anodic microorganisms play a critical role in the bioconversion and energy transfer process. Many EAMs, such as various *Shewanella*, *Geobacter*, and *Proteobacter* (Lovley, 2008), along with a growing number of other species like *Enterococcus* (Fornero et al., 2010a) and *Planctomyces* (Patil et al., 2009), have been extensively found in MFC systems and show high capabilities to metabolize simple carbohydrates and transfer the realeased electrons to the electrode. However, most of these EAMs cannot directly degrade complex organics. Thus, although excellent performance can be obtained by a pure culture system using simple substrates, the treatment of complex wastewater usually requires a mixed culture. Especially for real FIW treatment, the substrate degradation and conversion efficiency is heavily dependent on a synergetic cooperation between multiple species (Velasquez-Orta et al., 2011). At the same time, however, the presence of other electron-consuming bacteria such as methanogens and nitrate- and sulfate-reducing species means competition for the substrate and decreases the number of electrons diverted toward the electrode (Figure 14.1). Thus, the development of a well-balanced microbial ecosystem is of critical importance for this system.

Attributed to a higher complexity of the FIW composition, anodic microbial communities of MFCs fed with FIW generally show higher diversity than those fed with pure substrate, and the community composition has been found to significantly affect the electricity recovery (Fornero et al., 2010a; Kiely et al., 2011). The cooperation of microbial consortia for enhanced hydrolysis and utilization of the complex organic matter to favor current generation has been extensively demonstrated (Kiely et al., 2011). In addition, many anodic microorganisms also exhibit syntrophic interactions in improving the overall electron transfer via secreting electron mediators (Rabaey et al., 2005a). Therefore, establishing a diverse and electrochemically active microbial community while minimizing competitive species is of vital importance for the development of high-performance MFCs. To achieve this, efforts should be taken to optimize the inoculum, reactor design, and operating parameters, because all these factors can significantly affect the microbial dynamics. First of all, the choices of inoculum source and appropriate enrichment methods are key parameters to enrich desirable anodic consortia for treatment of specific wastewater. To shorten the start-up process and improve adaptability and long-term operating stability, a common practice is to inoculate with indigenous species. For example, power generation from yogurt wastewater was recently investigated in an MFC with various sources of inoculum (Cercado-Quezada et al., 2010). As expected, the MFC inoculated with compost-leachate showed higher maximum PD ($92 \, mW/m^2$) than the MFC inoculated with anaerobic sludge ($54 \, mW/m^2$). Likewise, the MFC inoculated with marine sediment was demonstrated to tolerate high salinity and osmosis pressure of sea water, and to enable a high power density of the MFC when seafood wastewater was used as the fuel (You et al., 2010). Secondly, the growth conditions of microorganisms are important. It has been well recognized that anodic microbial ecosystems are also significantly affected by engineering factors such as wastewater properties and operating conditions (Rodrigo et al., 2009). Thus, appropriate pretreatment and conditioning of the wastewater and an optimized, dynamic control of the operating parameters could both be useful approaches to maximizing energy capture from FIWs.

In addition to community composition, the biofilm thickness can also significantly influence the substrate metabolism and electron transfer. Complex substrate and small hydraulic shearing generally favor a thick biofilm and consequently a higher biodegradation rate of substrates due to the presence of more biomass. However, a thick biofilm also increases the resistance of substrate diffusion toward inner microorganisms (Picioreanu et al., 2007; Aelterman et al., 2008). Moreover, the faster-growing fermenters are likely to outcompete the EAMs and dominate in a thick biofilm, which usually leads to gas pockets within the biofilm and thus increases the resistance (Fornero et al., 2010a). Thus, a more viable choice would be to maintain a moderately thin biofilm while increasing the overall area of electrode surface. In this respect, design and engineering strategies such as application of three-dimensional electrodes, simplification of the wastewater through pre-fermentation to favor slow-growing EAMs, and offering moderate hydraulic shearing would all be helpful.

3.3 Reactor Design Parameters

Reactor design is an important aspect that is directly related to substrate—microbe—electrode interactions and governs the efficiency and scalability of an MFC. Reactor configuration, electrode material, and separator selection are three major parameters of MFC design that significantly affect the energy losses and electron-quenching reactions at the anode (Clauwaert et al., 2008; Pham et al., 2009; Rozendal et al., 2008).

To date, a wide variety of MFC configurations have been reported, and their categorization also varies

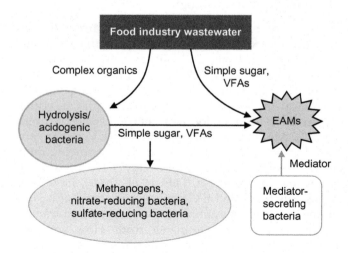

FIGURE 14.1 Interactions of anodic microbial consortia in FIW-fed MFCs.

a lot. For FIW treatment, the hydrodynamics, which control the carbon and nutrient diffusion to the anodic microorganisms and the shearing force to the biofilm (Shah et al., 2010), could be a vital factor affecting substrate—microbe contact and biofilm formation (Pham et al., 2008). To date, various enhanced-flow reactors have been tested for FIW treatment, showing different performance and economics. One simple approach is to install baffles into the anodic chamber to enable continuously forced flow (Feng et al., 2010b; Rabaey et al., 2005c). Another attractive way to enhance fluid mixing is to incorporate a high-rate anaerobic bioreactor into the MFC anodic chamber design (Deng et al., 2010). An upflow membraneless MFC was applied by Cheng et al. (2010) for treatment of high-strength palm oil mill effluent, with enhanced mass transfer and maximum PD of 44.6 mW/m^2 achieved. High efficiencies of treatment and electricity recovery were also achieved for brewery wastewater treatment in an improved upflow MFC that consisted of a bottom anaerobic digestion zone, a middle single-chamber MFC zone, and an upper clarifier zone (Katuri and Scott, 2010). To further improve the electrode surface area and mixing, an anaerobic fluidized-bed—MFC integrated design was also explored by adding biofilm carrier materials such as granular carbon or graphite into the anodic chamber and applying effluent circulation (Kong et al., 2011).

To justify practical MFC implementation, one of the biggest challenges is to find low-cost, efficient, and sustainable electrode materials. For the anode electrode, carbon-based material in various forms have been extensively investigated, such as paper, cloth, granule, graphite fiber, and even brush (He et al., 2006; You et al., 2007; Liu et al., 2010). However, these carbon-based materials mostly have relatively poor conductivity and mechanical durability compared with metal materials. To address this, one present trend in MFC electrode design is to combine metal and carbon materials to constitute a composite electrode (Logan et al., 2007). Another important trend is electrode surface modification through physicochemical pretreatment or coating conductive materials (Cheng and Logan, 2007; Feng et al., 2010a). Electrochemical pre-oxidation has been suggested as a useful tool to create carboxyl-containing functional groups and favor EAM enrichment (Torres et al., 2009; Cercado-Quezada et al., 2011; Tang et al., 2011). In addition, modification with nano-materials is also gaining increasing popularity in recent years (Zou et al., 2008; Sun et al., 2010; Liang et al., 2011). All this progress in electrode materials and treatment techniques has significantly improved the technological and economical feasibility of MFC for FIW treatment.

One of the most difficult parts of MFC design is the separator between anode and cathode. To use or not to use a separator usually presents a dilemma in reactor design because of the multiple advantages and limitations of a separator for MFCs (Li et al., 2011). On the one hand, a separator isolates the substrate, microbial consortia, and environmental conditions of anode and cathode so that the potential difference between electrodes can be maximized. On the other, it also inevitably aggravates the pH splitting between anolyte and catholyte and increases the internal resistance of the MFC. Moreover, separators also usually account for a considerable part of MFC cost, and there is the problem of membrane fouling during long-term operation. Thus, regarding separator design, there are now two different development directions, each with certain degrees of success.

Most researchers choose to improve the performance of a separator by developing novel materials with superior properties. The conventional proton exchange membranes (PEM) have many drawbacks such as limited proton diffusion, high internal resistance, poor mechanical strength, and high cost. In recent years, a variety of novel separator materials have been investigated as a substitute to PEM, including anion exchange membranes, bipolar membranes, ultrafiltration/microfiltration membranes, microporous fabricates, composite separators, and even salt bridges (Patil et al., 2009; Tang et al., 2010; Li et al., 2011). Compared with fine-pore polymer membranes, various coarse-porous materials, due to their cost advantages and better proton transfer performance, are attracting more interest as separators, such as canvas cloths (Zhuang et al., 2009), glass fibers (Zhang et al., 2009b), and earthen pots (Behera et al., 2010).

Meanwhile, efforts are underway on another frontier. Some researchers have tried to completely abandon the separator in MFC design while seeking other ways for proton transfer, such as adopting full-loop mode (Clauwaert et al., 2009; Li et al., 2009; Lefebvre et al., 2011). You et al. (2010) designed a U-shaped MFC with anoxic/oxic architecture for seafood wastewater treatment, and no separator was adopted in this system (Figure 14.2). In this design, the arc-shaped tube connecting the anoxic zone (which serves as the anodic chamber) and oxic zone (which serves as the cathodic chamber) can enable a smooth flow of the anolyte to the cathode while effectively preventing the intrusion of oxygen to the anode. This advective flow, together with the high salinity of the wastewater and packed carbon granule electrode, contributes to a low internal resistance of 100 Ω despite load changes. With such a separator-free design, a PD of 16.2 W/m^3 and CE of 15% were obtained, accompanied by up to 80% COD removal. However, a possible negative

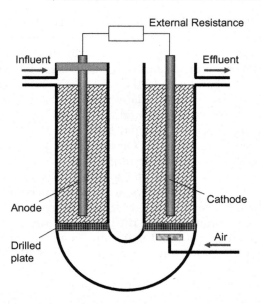

FIGURE 14.2 Schematic diagram of a U-shaped separator-free MFC for treatment of seafood wastewater. *Adapted from You et al. (2010)*

consequences of such a separator-free design is an increased substrate in the cathode chamber, which may result in growth of heterotrophic bacteria.

3.4 Operating Parameters

In MFC, the structure, composition, and activity of anodic biofilm are affected by the wastewater properties and operation conditions, such as organic loading rate (OLR), temperature, and dissolved oxygen (DO) concentration. Furthermore, as a bioelectrochemical system, the anodic microbial ecology and electron transfer are also closely associated with electrochemical parameters such as electrode potential.

Wastewater strength is a critical operating parameter that governs the efficiency and pathways of substrate metabolism in an MFC. Thus, the OLR of wastewater, either controlled in terms of COD concentration or hydraulic retention time, can significantly affect the electricity generation. One major advantage of FIW as a fuel for MFC is the high energy content per unit volume. Thus, FIW-fed MFCs generally show higher PD than those fed with domestic wastewater (Nam et al., 2010a). A moderately higher OLR also facilitates enhanced substrate diffusion into the anodic biofilm and decreases the concentration overpotential (You et al., 2010). However, an overly high OLR would adversely suppress the microbial activity and decrease the power generation (Goud et al., 2011; Goud and Mohan, 2011), and this turning point varies significantly with wastewater type and reactor configurations (Cercado-Quezada et al., 2011; Mohan et al., 2010).

Like conventional biological systems, there exists an optimal range of operating temperatures for MFCs. A mesophilic temperature of 30°C was found to significantly increase the PD and treatment efficiency of an MFC than lower temperature (23°C) (Ahn and Logan, 2010). This was in line with the result of another study (Kong et al., 2011), where an increase of temperature from 18°C to 30°C led to more power generation. However, a decrease in PD was observed when the temperature exceeded 30°C, possibly caused by a denaturation of the enzyme at such high temperature. Thus, to ensure a high-performance MFC, an appropriate mesophilic temperature should be maintained.

Anodic electrode potential can significantly influence the energy gain by microorganisms and thus alter the microbial community composition (Fornero et al., 2010a). As such, electrode potential has been frequently utilized as a valid electrochemical tool to accelerate start-up and enhance power generation of MFCs. From a thermodynamic perspective, a higher anode potential enables more metabolic energy gain for the bacteria (Logan et al., 2006), and this higher energy gain offers EAMs a competitive advantage over the non-EAM species (Borole et al., 2011; Sleutels et al., 2011). However, inconsistent results were reported by other researchers (Aelterman et al., 2008; Torres et al., 2009). Torres et al. (2009) found that at an anode potential of −150 mV (versus SHE, hereinafter the same), a denser and more conductive anodic biofilm and higher PD were obtained than at higher anode potential. The promotion effect of a moderate anode potential (at 0 mV) was recently evidenced by several researchers (Aelterman et al., 2008; Wei et al., 2010). All these results together indicate that appropriately increasing the anode potential could be beneficial for promoting EAM enrichment and accelerating MFC start-up, but this is not a case of the higher the better.

4. ELECTRICITY GENERATION FROM A SCALABLE MFC—A CASE STUDY

In a certain sense, the success of an MFC for practical application will depend on the degree to which it can be scaled up to treat real wastewater at high efficiency and be an acceptable investment. Thus, the efficiency, scalability, stability, and cost are all critical factors to be considered in judging its practical feasibility. In light of these, in this case study we chose a well-designed stacked MFC (Figure 14.3), which used molasses wastewater as substrate (Zhong et al., 2011). An introduction of this case aims to facilitate a better understanding of the possibilities and specific challenges for an MFC in treating practical FIWs.

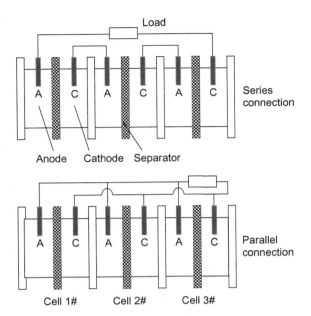

FIGURE 14.3 Stacked MFCs. The power output of a single MFC unit is limited despite optimization. To scale up an MFC, one viable strategy is to assemble multiple MFCs together, which has led to the development of stacked MFCs (Aelterman et al., 2006). Theoretically, MFCs can be stacked in series or parallel connection to achieve any desired voltage or current. However, voltage reversal may occur when the voltages in the individual cells are not matched, leading to reduced overall performance (Oh and Logan, 2007). Fuel starvation and insufficient microbial activity are considered two major causes of voltage reversal in stacked MFCs.

4.1 MFC Setup and Characteristics

A stacked air-cathode MFC with a total volume of about 2.7 L was used. It consisted of four cell units (U1–U4) separated by three upright overflow plates (Figure 14.4). Each unit chamber was further separated into two zones (i.e., a smaller downflow zone and a larger upflow zone) by a vertically hung baffle. A carbon felt anode was set at the middle of the upflow zone. Moreover, graphite granules were added into each cell to offer more anodic surface for biofilm growth and electron transfer. A wet-proofed and platinum-loaded carbon cloth was set at one side of each cell, and each cathode was bonded to a copper mesh that served as a current collector. No separator was employed in this system. The reactor was inoculated with sludge taken from the anaerobic digestion tank of a sewage treatment plant. Raw molasses wastewater from an alcohol factory in Guangxi Province, China, was used as the feedstock after adjusting the pH to 7.5 and appropriate dilution. The raw wastewater was characterized by high COD concentration (127,500 mg/L), high sulfate (7,616 mg/L), and deep brown color. During operation, the diluted wastewater was continuously pumped into the stacked MFC from one end of the reactor, flowed through the four units in

turn, and finally exited from the other end of the reactor. Three different connecting strategies of the electrode pairs were tested: isolated, series stacked, and parallel stacked.

This stacked design confers several advantages over conventional MFCs: (i) the stacked modules and continuous-flow operation suggest good scalability and applicability of the system to practical wastewater treatment; (ii) the vertical baffles favor high retention of sludge biomass and enable complete substrate degradation, and the addition of graphite granules further increases the substrate–microbe–electrode interfaces; (iii) the overflow mode enables the enrichment of different microbial consortia in individual units, thus the overall system can have high adaptability to various wastewater and operating fluctuations; (iv) the bonded copper screen on the cathode creates a point-surface contact between different electrodes and thus significantly decreases the ohmic losses caused by electrode connection; and (v) the continuous flow and high-strength wastewater is helpful in alleviating the cell reversal caused by fuel starvation that is commonly seen in stacked systems.

4.2 Power Generation Performance

The power generation varied considerably with the applied OLR. Under isolated electrode-pair condition, the average PD peaked (115.5 mW/m^2, or 1.01 W/m^3) at a moderate OLR (3.20 kg/(m^{3}day)). This pattern of PD versus OLR profile agreed well with most MFC studies, where moderate OLR was found to favor power generation whereas overly high strength suppressed the EAMs. This demonstrates again the importance of keeping an appropriate OLR range for an MFC. Interestingly, great differences of PD were also observed between different MFC units. Overly high U1 showed the best performance under all OLRs, followed by U3, U2, and U4 in turn. The highest PD in U1 was mainly attributed to the presence of a large amount of readily degradable organics in the influent. While the considerable degradation of complex substrates was achieved in the first two units, the high concentration of hydrolyzed organics contributed to a high PD in U3. Thus, because of the different substrate degradation and power utilization behaviors in different units, such a stacked system is supposed to better adapt to influent fluctuations in practical operation. Moreover, the system exhibited a voltage loss of only 8.1% when connected in series, indicating good scalability.

Overall, this baffled stacking MFC showed comparable power generation performance with other single-unit systems for winery wastewater treatment (Huang

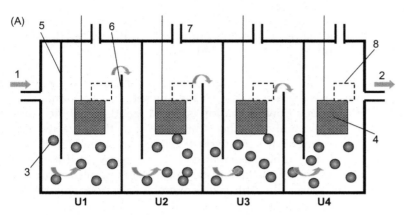

FIGURE 14.4 (A) Schematic diagram and (B) photograph of baffled stacking MFC. (1) water inlet; (2) water outlet; (3) graphite granules; (4) carbon felt anode; (5) baffle plate; (6) overflow plate; (7) gas outlet; (8) cathode (on one side of the reactor).

et al., 2011). However, the maximum PD of 1.01 W/m^3 is far from achieving any practical application. Moreover, it should be noted that the CE is pretty low (below 1%) in this system. This is mainly due to an overgrowth of non-EAMs and the many competitive reactions, such as sulfate reduction. In addition, because there were no separators in this configuration, the intrusion of oxygen might also account for a substantial portion of the electron loss and thus further lower the CE. Thus, these limitations and challenges should be taken into account in MFC design and operation. Moreover, pH neutralization of wastewater and application of Pt catalyst should be avoided to enable a more cost-effective system.

5. CONCLUSION

The past few years have witnessed rapid advances in MFC technology. Our knowledge of the microbial ecology, microbe—electrode interactions, and other factors influencing MFCs has dramatically improved. In addition, the amazing progress in respect to reactor design, material manufacturing, and process control has significantly lifted the achievable PD and lowered the cost of such systems. All these developments enable us to better exploit this technology and implement it into real-world scenarios. However, this

technology is far from mature. At the present stage, many technological and economic challenges are yet to be addressed before a commercialized application of this technology is possible. Several major hurdles include the low PD and energy efficiency, high cost of construction, difficulty in scaling-up, and fouling problems in long-term operation. Regarding power generation from FIWs, the high energy density and abundant availability of FIWs make them a highly pursuable substrate for MFC. However, because of their inherent complexity, relatively low conductivity, and usually the presence of fermenters within the wastewater, how to divert as many electrons as possible from the organics to the EAMs and electrode will be an especially critical issue to address in the future.

In summary, MFC-based technology for FIW treatment is a highly promising field that may see practical application in the near future, but there is still a long road ahead. Currently, there are significant challenges to its scaling-up and commercialization. The key to final success of this technology will depend on how we can address or minimize all these challenges to enable technologically viable and economically competitive processes. Especially, more breakthroughs will be required in microbial manipulation, system design, process control, and integration. All these warrant more comprehensive and in-depth investigations both in the laboratory and in the field.

Acknowledgements

The authors wish to thank the NSFC-JST Joint Project (21021140001) and the National Hi-Technology 863 Project (2008BADC4B18) for the support of this study.

References

Aelterman, P., Rabaey, K., Hai, P., Boon, N., Verstraete, W., 2006. Continuous electricity generation at high voltages and currents using stacked microbial fuel cells. Environ. Sci. Technol. 40, 3388–3394.

Aelterman, P., Freguia, S., Keller, J., Verstraete, W., Rabaey, K., 2008. The anode potential regulates bacterial activity in microbial fuel cells. Appl. Microbiol. Biotechnol. 78, 409–418.

Ahn, Y., Logan, B.E., 2010. Effectiveness of domestic wastewater treatment using microbial fuel cells at ambient and mesophilic temperatures. Bioresour. Technol. 101, 469–475.

Angenent, L.T., Karim, K., Al-Dahhan, M.H., Wrenn, B.A., Domiguez-Espinosa, R., 2004. Production of bioenergy and bio-chemicals from industrial and agricultural wastewater. Trends Biotechnol. 22, 477–485.

Antonopoulou, G., Stamatelatou, K., Bebelis, S., Lyberatos, G., 2010. Electricity generation from synthetic substrates and cheese whey using a two chamber microbial fuel cell. Biochem. Eng. J. 50, 10–15.

Ayyaru, S., Dharmalingam, S., 2011. Development of MFC using sul-phonated polyether ether ketone (SPEEK) membrane for electricity generation from waste water. Bioresour. Technol. 102, 11167–11171.

Behera, M., Jana, P.S., More, T.T., Ghangrekar, M.M., 2010. Rice mill wastewater treatment in microbial fuel cells fabricated using pro-ton exchange membrane and earthen pot at different pH. Bioelectrochemistry 79, 228–233.

Borole, A.P., Reguera, G., Ringeisen, B., Wang, Z.W., Feng, Y.j., Kim, B.H., 2011. Electroactive biofilms: current status and future research needs. Energy Environ. Sci. 4, 4813–4834.

Cercado-Quezada, B., Delia, M.L., Bergel, A., 2010. Testing various food-industry wastes for electricity production in microbial fuel cell. Bioresour. Technol. 101, 2748–2754.

Cercado-Quezada, B., Delia, M.L., Bergel, A., 2011. Electrochemical micro-structuring of graphite felt electrodes for accelerated for-mation of electroactive biofilms on microbial anodes. Electrochem. Commun. 13, 440–443.

Cheng, J., Zhu, X., Ni, J., Borthwick, A., 2010. Palm oil mill effluent treatment using a two-stage microbial fuel cells system integrated with immobilized biological aerated filters. Bioresour. Technol. 101, 2729–2734.

Cheng, S., Logan, B.E., 2007. Ammonia treatment of carbon cloth anodes to enhance power generation of microbial fuel cells. Electrochem. Commun. 9, 492–496.

Choi, J.D.R., Chang, H.N., Han, J.I., 2011. Performance of microbial fuel cell with volatile fatty acids from food wastes. Biotechnol. Lett. 33, 705–714.

Clauwaert, P., Aelterman, P., Pham, T., De Schamphelaire, L., Carballa, M., Rabaey, K., et al., 2008. Minimizing losses in bio-electrochemical systems: the road to applications. Appl. Microbiol. Biotechnol. 79, 901–913.

Clauwaert, P., Mulenga, S., Aelterman, P., Verstraete, W., 2009. Litre-scale microbial fuel cells operated in a complete loop. Appl. Microbiol. Biotechnol. 83, 241–247.

Cusick, R.D., Kiely, P.D., Logan, B.E., 2010. A monetary comparison of energy recovered from microbial fuel cells and microbial elec-trolysis cells fed winery or domestic wastewaters. Int. J. Hyd. Energy 35, 8855–8861.

Demirel, B., Yenigun, O., Onay, T.T., 2005. Anaerobic treatment of dairy wastewaters: a review. Process. Biochem. 40, 2583–2595.

Deng, Q., Li, X.Y., Zuo, J.E., Ling, A., Logan, B.E., 2010. Power gener-ation using an activated carbon fiber felt cathode in an upflow microbial fuel cell. J. Power Sources 195, 1130–1135.

Digman, B., Kim, D.S., 2008. Review: alternative energy from food processing wastes. Environ. Prog. 27, 524–537.

Durruty, I., Bonanni, P.S., González, J.F., Busalmen, J.P., 2011. Evaluation of potato-processing wastewater treatment in a micro-bial fuel cell. Bioresour. Technol.

Fang, H., Yu, H., 2000. Effect of HRT on mesophilic acidogenesis of dairy wastewater. J. Environ. Eng. 126, 1145–1148.

Feng, Y., Wang, X., Logan, B., Lee, H., 2008. Brewery wastewater treatment using air-cathode microbial fuel cells. Appl. Microbiol. Biotechnol. 78, 873–880.

Feng, Y., Yang, Q., Wang, X., Logan, B.E., 2010a. Treatment of carbon fiber brush anodes for improving power generation in air-cathode microbial fuel cells. J. Power Sources 195, 1841–1844.

Feng, Y.J., Lee, H., Wang, X., Liu, Y.L., He, W.H., 2010b. Continuous electricity generation by a graphite granule baffled air-cathode microbial fuel cell. Bioresour. Technol. 101, 632–638.

Foley, J.M., Rozendal, R.A., Hertle, C.K., Lant, P.A., Rabaey, K., 2010. Life cycle assessment of high-rate anaerobic treatment, microbial fuel cells, and microbial electrolysis cells. Environ. Sci. Technol. 44, 3629–3637.

Fornero, J.J., Rosenbaum, M., Angenent, L.T., 2010a. Electric power generation from municipal, food, and animal wastewaters using microbial fuel cells. Electroanalysis 22, 832–843.

Fornero, J.J., Rosenbaum, M., Cotta, M.A., Angenent, L.T., 2010b. Carbon dioxide addition to microbial fuel cell cathodes maintains sustainable catholyte pH and improves anolyte pH, alkalinity, and conductivity. Environ. Sci. Technol. 44, 2728–2734.

Goud, R.K., Babu, P.S., Mohan, S.V., 2011. Canteen based composite food waste as potential anodic fuel for bioelectricity generation in single chambered microbial fuel cell (MFC): bio-electrochemical evaluation under increasing substrate loading condition. Int. J. Hyd. Energy 36, 6210–6218.

Goud, R.K., Mohan, S.V., 2011. Pre-fermentation of waste as a strat-egy to enhance the performance of single chambered microbial fuel cell (MFC). Int. J. Hyd. Energy 36, 13753–13762.

Greenman, J., Gálvez, A., Giusti, L., Ieropoulos, I., 2009. Electricity from landfill leachate using microbial fuel cells: comparison with a biological aerated filter. Enzyme. Microb. Technol. 44, 112–119.

Hallenbeck, P.C., Ghosh, D., 2009. Advances in fermentative biohydro-gen production: the way forward? Trends Biotechnol. 27, 287–297.

Hamelin, L., Wesnaes, M., Wenzel, H., Petersen, B.M., 2011. Environmental consequences of future biogas technologies based on separated slurry. Environ. Sci. Technol. 45, 5869–5877.

He, Z., Shao, H., Angenent, L.T., 2007. Increased power production from a sediment microbial fuel cell with a rotating cathode. Biosens. Bioelectron. 22, 3252–3255.

He, Z., Wagner, N., Minteer, S.D., Angenent, L.T., 2006. An upflow microbial fuel cell with an interior cathode: assessment of the internal resistance by impedance spectroscopy. Environ. Sci. Technol. 40, 5212–5217.

Heilmann, J., Logan, B.E., 2006. Production of electricity from proteins using a microbial fuel cell. Water. Environ. Res. 78, 531–537.

Huang, J., Yang, P., Guo, Y., Zhang, K., 2011. Electricity generation during wastewater treatment: an approach using an AFB-MFC for alcohol distillery wastewater. Desalination 276, 373–378.

Huang, L., Logan, B., 2008. Electricity generation and treatment of paper recycling wastewater using a microbial fuel cell. Appl. Microbiol. Biotechnol. 80, 349–355.

Jamuna, R., Ramakrishna, S.V., 1989. SCP production and removal of organic load from cassava starch industry waste by yeasts. J. Ferment. Bioeng. 67, 126–131.

Kaewkannetra, P., Chiwes, W., Chiu, T.Y., 2011. Treatment of cassava mill wastewater and production of electricity through microbial fuel cell technology. Fuel 90, 2746–2750.

Katuri, K.P., Scott, K., 2010. Electricity generation from the treatment of wastewater with a hybrid up-flow microbial fuel cell. Biotechnol. Bioeng. 107, 52–58.

Katuri, K.P., Enright, A.M., O'Flaherty, V., Leech, D., 2012. Microbial analysis of anodic biofilm in a microbial fuel cell using slaughterhouse wastewater. Bioelectrochemistry. doi:10.1016/j.bioelechem.2011.12.002.

Kiely, P.D., Cusick, R., Call, D.F., Selembo, P.A., Regan, J.M., Logan, B.E., 2011. Anode microbial communities produced by changing from microbial fuel cell to microbial electrolysis cell operation using two different wastewaters. Bioresour. Technol. 102, 388–394.

Kim, B.H., Park, H.S., Kim, H.J., Kim, G.T., Chang, I.S., Lee, J., et al., 2004. Enrichment of microbial community generating electricity using a fuel-cell-type electrochemical cell. Appl. Microbiol. Biotechnol. 63, 672–681.

Kim, H.W., Nam, J.Y., Shin, H.S., 2011. Ammonia inhibition and microbial adaptation in continuous single-chamber microbial fuel cells. J. Power Sources 196, 6210–6213.

Kim, J.R., Premier, G.C., Hawkes, F.R., Rodríguez, J., Dinsdale, R.M., Guwy, A.J., 2010. Modular tubular microbial fuel cells for energy recovery during sucrose wastewater treatment at low organic loading rate. Bioresour. Technol. 101, 1190–1198.

Kong, W.F., Guo, Q.J., Wang, X.Y., Yue, X.H., 2011. Electricity generation from wastewater using an anaerobic fluidized bed microbial fuel cell. Ind. Eng. Chem. Res. 50, 12225–12232.

Lalman, J.A., Bagley, D.M., 2001. Anaerobic degradation and methanogenic inhibitory effects of oleic and stearic acids. Water Res. 35, 2975–2983.

Lefebvre, O., Shen, Y., Tan, Z., Uzabiaga, A., Chang, I.S., Ng, H.Y., 2011. Full-loop operation and cathodic acidification of a microbial fuel cell operated on domestic wastewater. Bioresour. Technol. 102, 5841–5848.

Li, W.W., Yu, H.Q., 2011. From wastewater to bioenergy and biochemicals via two-stage bioconversion processes: a future paradigm. Biotechnol. Adv. 29, 972–982.

Li, W.W., Sheng, G.P., Liu, X.W., Yu, H.Q., 2011. Recent advances in the separators for microbial fuel cells. Bioresour. Technol. 102, 244–252.

Li, Z., Zhang, X., Zeng, Y., Lei, L., 2009. Electricity production by an overflow-type wetted-wall microbial fuel cell. Bioresour. Technol. 100, 2551–2555.

Liang, P., Wang, H.Y., Xia, X., Huang, X., Mo, Y.H., Cao, X.X., et al., 2011. Carbon nanotube powders as electrode modifier to enhance the activity of anodic biofilm in microbial fuel cells. Biosens. Bioelectron. 26, 3000–3004.

Liu, H., Cheng, S., Logan, B.E., 2004. Production of electricity from acetate or butyrate using a single-chamber microbial fuel cell. Environ. Sci. Technol. 39, 658–662.

Liu, R.H., Sheng, G.P., Sun, M., Zang, G.L., Li, W.W., Tong, Z.H., et al., 2010. Enhanced reductive degradation of methyl orange in a microbial fuel cell through cathode modification with redox mediators. Appl. Microbiol. Biotechnol. 89, 201–208.

Logan, B.E., 2009. Exoelectrogenic bacteria that power microbial fuel cells. Nat. Rev. Microbiol. 7, 375–381.

Logan, B.E., Hamelers, B., Rozendal, R.A., Schrroder, U., Keller, J., Freguia, S., et al., 2006. Microbial fuel cells: methodology and technology. Environ. Sci. Technol. 40, 5181–5192.

Logan, B.E., Cheng, S.A., Watson, V., Estadt, G., 2007. Graphite fiber brush anodes for increased power production in air-cathode microbial fuel cells. Environ. Sci. Technol. 41, 3341–3346.

Lovley, D.R., 2008. The microbe electric: conversion of organic matter to electricity. Curr. Opin. Biotechnol. 19, 564–571.

Lu, N., Zhou, S.G., Zhuang, L., Zhang, J.T., Ni, J.R., 2009. Electricity generation from starch processing wastewater using microbial fuel cell technology. Biochem. Eng. J. 43, 246–251.

Martins das Neves, L.C., Converti, A., Vessoni Penna, T.C., 2009. Biogas production: new trends for alternative energy sources in rural and urban zones. Chem. Eng. Technol. 32, 1147–1153.

Mohan, S.V., Mohanakrishna, G., Sarma, P.N., 2010. Composite vegetable waste as renewable resource for bioelectricity generation through non-catalyzed open-air cathode microbial fuel cell. Bioresour. Technol. 101, 970–976.

Nam, J.Y., Kim, H.W., Lim, K.H., Shin, H.S., 2010a. Effects of organic loading rates on the continuous electricity generation from fermented wastewater using a single-chamber microbial fuel cell. Bioresour. Technol. 101, S33–S37.

Nam, J.Y., Kim, H.W., Shin, H.S., 2010b. Ammonia inhibition of electricity generation in single-chambered microbial fuel cells. J. Power Sources 195, 6428–6433.

Nimje, V.R., Chen, C.Y., Chen, H.R., Chen, C.C., Huang, Y.M., Tseng, M.J., et al., 2011. Comparative bioelectricity production from various wastewaters in microbial fuel cells using mixed cultures and a pure strain of Shewanella oneidensis. Bioresour. Technol.

Oh, S., Logan, B.E., 2005. Hydrogen and electricity production from a food processing wastewater using fermentation and microbial fuel cell technologies. Water Res. 39, 4673–4682.

Oh, S.E., Logan, B.E., 2006. Proton exchange membrane and electrode surface areas as factors that affect power generation in microbial fuel cells. Appl. Microbiol. Biotechnol. 70, 162–169.

Oh, S.E., Logan, B.E., 2007. Voltage reversal during microbial fuel cell stack operation. J. Power Sources 167, 11–17.

Oh, S.T., Kim, J.R., Premier, G.C., Lee, T.H., Kim, C., Sloan, W.T., 2010. Sustainable wastewater treatment: how might microbial fuel cells contribute. Biotechnol. Adv. 28, 871–881.

Pant, D., Van Bogaert, G., Diels, L., Vanbroekhoven, K., 2010. A review of the substrates used in microbial fuel cells (MFCs) for sustainable energy production. Bioresour. Technol. 101, 1533–1543.

Pant, D., Singh, A., Van Bogaert, G., Gallego, Y.A., Diels, L., Vanbroekhoven, K., 2011. An introduction to the life cycle assessment (LCA) of bioelectrochemical systems (BES) for sustainable energy and product generation: relevance and key aspects. Renew. Sustain. Energy Rev. 15, 1305–1313.

Patil, S.A., Surakasi, V.P., Koul, S., Ijmulwar, S., Vivek, A., Shouche, Y.S., et al., 2009. Electricity generation using chocolate industry wastewater and its treatment in activated sludge based microbial fuel cell and analysis of developed microbial community in the anode chamber. Bioresour. Technol. 100, 5132–5139.

Pham, H.T., Boon, N., Aelterman, P., Clauwaert, P., De Schamphelaire, L., Van Oostveldt, P., et al., 2008. High shear enrichment improves the performance of the anodophilic microbial consortium in a microbial fuel cell. Microb. Biotechnol. 1, 487–496.

Pham, T.H., Aelterman, P., Verstraete, W., 2009. Bioanode performance in bioelectrochemical systems: recent improvements and prospects. Trends Biotechnol. 27, 168–178.

Picioreanu, C., Head, I.M., Katuri, K.P., van Loosdrecht, M.C.M., Scott, K., 2007. A computational model for biofilm-based microbial fuel cells. Water Res. 41, 2921–2940.

Rabaey, K., Boon, N., Höfte, M., Verstraete, W., 2005a. Microbial phenazine production enhances electron transfer in biofuel cells. Environ. Sci. Technol. 39, 3401–3408.

Rabaey, K., Clauwaert, P., Aelterman, P., Verstraete, W., 2005b. Tubular microbial fuel cells for efficient electricity generation. Environ. Sci. Technol. 39, 8077–8082.

Rabaey, K., Ossieur, W., Verhaege, M., Verstraete, W., 2005c. Continuous microbial fuel cells convert carbohydrates to electricity. Water Sci. Technol. 52, 515–523.

Rabaey, K., Van de Sompel, K., Maignien, L., Boon, N., Aelterman, P., Clauwaert, P., et al., 2006. Microbial fuel cells for sulfide removal. Environ. Sci. Technol. 40, 5218–5224.

Ramsay, I.R., Pullammanappallil, P.C., 2001. Protein degradation during anaerobic wastewater treatment: derivation of stoichiometry. Biodegradation 12, 247–256.

Rodrigo, M.A., Cañizares, P., García, H., Linares, J.J., Lobato, J., 2009. Study of the acclimation stage and of the effect of the biodegradability on the performance of a microbial fuel cell. Bioresour. Technol. 100, 4704–4710.

Rozendal, R.A., Hamelers, H.V.M., Rabaey, K., Keller, J., Buisman, C.J.N., 2008. Towards practical implementation of bioelectrochemical wastewater treatment. Trends Biotechnol. 26, 450–459.

Shah, V., Borole, A.P., Hamilton, C.Y., 2010. Energy production from food industry wastewaters using bioelectrochemical cells. Emerging Environmental Technologies. Springer, Netherlands, pp. 97–113.

Sleutels, T.H.J.A., Darus, L., Hamelers, H.V.M., Buisman, C.J.N., 2011. Effect of operational parameters on Coulombic efficiency in bioelectrochemical systems. Bioresour. Technol. 102, 11172–11176.

Sun, J.J., Zhao, H.Z., Yang, Q.Z., Song, J., Xue, A., 2010. A novel layer-by-layer self-assembled carbon nanotube-based anode: preparation, characterization, and application in microbial fuel cell. Electrochim. Acta 55, 3041–3047.

Tang, X.H., Guo, K., Li, H.R., Du, Z.W., Tian, J.L., 2010. Microfiltration membrane performance in two-chamber microbial fuel cells. Biochem. Eng. J. 52, 194–198.

Tang, X.H., Guo, K., Li, H., Du, Z., Tian, J., 2011. Electrochemical treatment of graphite to enhance electron transfer from bacteria to electrodes. Bioresour. Technol. 102, 3558–3560.

Torres, C.I., Krajmalnik-Brown, R., Parameswaran, P., Marcus, A.K., Wanger, G., Gorby, Y.A., et al., 2009. Selecting anode-respiring bacteria based on anode potential: phylogenetic, electrochemical, and microscopic characterization. Environ. Sci. Technol. 43, 9519–9524.

Velasquez-Orta, S.B., Head, I.M., Curtis, T.P., Scott, K., 2011. Factors affecting current production in microbial fuel cells using different industrial wastewaters. Bioresour. Technol. 102, 5105–5112.

Venkata Mohan, S., Mohanakrishna, G., Velvizhi, G., Babu, V.L., Sarma, P.N., 2010. Bio-catalyzed electrochemical treatment of real field dairy wastewater with simultaneous power generation. Biochem. Eng. J. 51, 32–39.

Wang, X., Feng, Y.J., Lee, H., 2008. Electricity production from beer brewery wastewater using single chamber microbial fuel cell. Water Sci. Technol. 57, 1117–1121.

Wei, J., Liang, P., Cao, X., Huang, X., 2010. A new insight into potential regulation on growth and power generation of *Geobacter sulfurreducens* in microbial fuel cells based on energy viewpoint. Environ. Sci. Technol. 44, 3187–3191.

Wen, Q., Wu, Y., Cao, D., Zhao, L., Sun, Q., 2009. Electricity generation and modeling of microbial fuel cell from continuous beer brewery wastewater. Bioresour. Technol. 100, 4171–4175.

Wen, Q., Wu, Y., Zhao, L., Sun, Q., 2010. Production of electricity from the treatment of continuous brewery wastewater using a microbial fuel cell. Fuel 89, 1381–1385.

You, S., Zhao, Q., Zhang, J., Jiang, J., Wan, C., Du, M., et al., 2007. A graphite-granule membrane-less tubular air-cathode microbial fuel cell for power generation under continuously operational conditions. J. Power Sources 173, 172–177.

You, S.J., Zhang, J.N., Yuan, Y.X., Ren, N.Q., Wang, X.H., 2010. Development of microbial fuel cell with anoxic/oxic design for treatment of saline seafood wastewater and biological electricity generation. J. Chem. Technol. Biotechnol. 85, 1077–1083.

Yu, H.Q., Zhu, Z.H., Hu, W.R., Zhang, H.S., 2002. Hydrogen production from rice winery wastewater in an upflow anaerobic reactor by using mixed anaerobic cultures. Int. J. Hyd. Energy. 27, 1359–1365.

Zhang, B., Zhao, H., Zhou, S., Shi, C., Wang, C., Ni, J., 2009a. A novel UASB-MFC-BAF integrated system for high strength molasses wastewater treatment and bioelectricity generation. Bioresour. Technol. 100, 5687–5693.

Zhang, X., Cheng, S., Wang, X., Huang, X., Logan, B.E., 2009b. Separator characteristics for increasing performance of microbial fuel cells. Environ. Sci. Technol. 43, 8456–8461.

Zhang, Y., Min, B., Huang, L., Angelidaki, I., 2009c. Electricity generation and microbial community analysis of wheat straw biomass powered microbial fuel cells. Appl. Environ. Microbiol.

Zhong, C., Zhang, B., Kong, L., Xue, A., Ni, J., 2011. Electricity generation from molasses wastewater by an anaerobic baffled stacking microbial fuel cell. J. Chem. Technol. Biotechnol. 86, 406–413.

Zhuang, L., Zhou, S., Wang, Y., Liu, C., Geng, S., 2009. Membrane-less cloth cathode assembly (CCA) for scalable microbial fuel cells. Biosens. Bioelectron. 24, 3652–3656.

Zuo, Y., Maness, P.C., Logan, B.E., 2006. Electricity production from steam-exploded corn stover biomass. Energy Fuel 20, 1716–1721.

Zou, Y., Xiang, C., Yang, L., Sun, L.X., Xu, F., Cao, Z., 2008. A mediatorless microbial fuel cell using polypyrrole coated carbon nanotubes composite as anode material. Int. J. Hyd. Energy 33, 4856–4862.

ASSESSMENT OF ENVIRONMENTAL IMPACT OF FOOD PRODUCTION AND CONSUMPTION

CHAPTER

15

Life Cycle Assessment Focusing on Food Industry Wastes

Mónica Herrero, Adriana Laca, and Mario Díaz

1. INTRODUCTION

Life cycle assessment (LCA) is the environmental management tool most frequently used. The European Union has identified LCA as the best tool to evaluate the potential environmental impact of products. LCA is defined by UNE-EN ISO 14040 as a technique for assessing environmental aspects and potential impacts associated with a process, product, or service. This goal is achieved through the identification and quantification of raw materials, energy, and waste discharges into the environment. To undertake an LCA study, different stages should be established such as the goal and scope definition, the functional unit, the inventory or data collection, inventory analysis or data treatment, and the evaluation or impact assessment. LCA has wide-ranging applications, including decision making, product and process design, research and development, purchasing, information for defining company strategies, identification of areas of improvement, or selection of environmental indicators.

LCA-related terms have been coined with a wider perspective, seeking to establish straightforward criteria and guidelines. In this way, financial Life Cycle Cost integrates economic aspects in the LCA approach. This term may be used with two orientations: either including the total cost linked to the purchase, operation, and disposal of a product; or the cost of a product or service over its entire life cycle, considering also external costs. Life Cycle Thinking (LCT) integrates existing consumption and production strategies towards more coherent policy making affecting industry and consumers. LCT considers the whole life cycle, seeking to avoid the shifting of problems from one life cycle to another, from one geographic area to another, or from one environmental medium to another. Bearing in mind that the European

Landfill Directive (1999/31/EC) implies stringent regulations affecting wastes, Gentil et al. (2011) recently used LCT to evaluate the environmental consequences of waste prevention on waste management systems. Focusing on household waste, it was concluded that prevention of food waste had the highest environmental benefit in the context of municipal waste prevention.

Gustavsson et al. (2011) highlighted the existence of major gaps in knowledge of global food loss and waste, which in affluent economies is accounted for mainly by post-retail food waste (Parfitt et al., 2010), claiming that further research in this area is urgent. In this chapter, various LCA studies will be reviewed, focusing on valorization of food wastes by bioprocessing and also by nonbiological processing.

Minimization is placed at the top of the hierarchy of waste management strategies; hence, policies encourage industries, retailers, and householders accordingly. Waste prevention has the highest priority in waste policy in the European Union, as stated in Directive 2008/98/EC. Reducing food waste would provide significant financial and environmental benefits and improve resource efficiency. The importance of reducing waste at source wherever possible, rather than by end-of-pipe treatment or recycling, should be further encouraged (Henningsson et al., 2004). Significant progress can be achieved on resource efficiency in the food industry simply through improving dialogue between producers, retailers, and consumers (Henningsson et al., 2004). Recently, it was pointed out (Monier et al., 2010) that lack of knowledge and awareness regarding the heavy consequences of food discharged as waste can partly explain the problem. To promote LCT, environmental education and specific employment projects are viable options to reduce food waste at source. LCT/LCA can also help

Food Industry Wastes.
DOI: http://dx.doi.org/10.1016/B978-0-12-391921-2.00015-9

to reckon the necessity to improve technologies or to face innovative scenarios.

LCA can also identify critical points where the environmental management system can be improved. Williams and Wilkström (2011), in an LCA study, highlighted the importance of analyzing the risk of increasing food losses when changing packaging design to reduce packaging material, which is the main intention of the Packaging and Packaging Waste Directive of the European Union (Directive 94/62/EC).

Generation of liquid effluents with high organic content and large quantities of sludge and solid wastes is a problem common to all food industries. Regarding treatment of food waste and other wastes generated by food industries, different scenarios have been evaluated by LCA. In this chapter, various treatments for the biodegradable fraction or for nonbiodegradable wastes from the food industry analyzed from an LCA perspective will be reviewed (such as biogasification, composting, combustion, recycling, landfilling, or valorization as commodities). Many studies have covered these topics, frequently comparing different technologies. Nevertheless, it should be borne in mind that no single technology can solve waste flows entirely, and at the same time, environmental technologies generate residues or emissions that require further treatment. Moreover, careful examination is needed, since conclusions obtained from an LCA perspective may vary when different scopes, boundaries, or distances are taken into account.

With the help of an LCT perspective, the need for sustainable measures in the food industry and in the food distribution chain, implying the active cooperation of consumers, should launch a new endeavour for integrating food waste prevention at source and efficient waste management, under more ethical guidelines, in the current energy-saving context.

2. METHODOLOGY IN LIFE CYCLE ASSESSMENT

LCA allows us to evaluate (and quantify) the associated environmental burdens for a given product, process, or activity, taking into account its full course, from raw material acquisition to production, use, and disposal. This means that, with this perspective (i.e., from "cradle to grave"), environmental concerns are no longer limited to a particular industrial plant; it is also necessary to establish what portion of pollution comes from pre- and posttreatment of the product, including its transportation. Company responsibility thus extends upstream and downstream of the product chain, including the impact of the product behavior on its end use.

It is interesting to understand the significance and applications of an LCA study in order to assess the benefits of undertaking this task in a company's project. It is not only an instrument to protect the environment and natural resources. LCA also helps to support decision making in business. A distinction can be made between effect-oriented LCA—aimed at describing the consequences of changes (e.g., in a productive process (consequential))—and descriptive LCA (attributional), aimed at mapping the environmental impacts that the product investigated is made accountable for (Baumann and Tillman, 2004). As an environmental management tool for industries, the outcomes achieved can be used with internal and/or external scopes. Therefore, consideration should be given to the recipients of the information.

Internally, LCA is useful to identify points where substantial impacts are taking place; to promote goals, showing how resources can be invested intelligently; or to measure the progress of the company in promoting actions to reduce environmental burdens. LCA results may help the company to manage risk, by identifying opportunities and uncertainties in environmental performance, complementing other tools and methods. In addition, this methodology allows modeling of the environmental behavior of processes, establishing criteria to select those that minimize the environmental impacts identified. Currently, actions to reduce the environmental impacts of products are mainly focused on the production stage, being extended in some cases to their final treatment. Although a proper and initial eco-design will reduce the environmental impact of a product, inclusion of environmental aspects in the product design stage has not yet received adequate attention (Braungart and McDonough, 2002).

As an external tool, LCA results may help the company to transmit and communicate the environmental benefits achieved, to be used in marketing and benchmark performance for eco-labeling, environmental product declaration, etc. The present tendency is that companies are beginning to consider not only economical aspects but also social responsibilities and eco-efficient results.

The methodology for conducting an LCA commonly considers four phases: (a) Goal and scope definition, (b) Inventory analysis (LCI), (c) Impact assessment (LCIA), and (d) Interpretation. The International Standards ISO 14040-14044 provides principles, framework, and methodological requirements for conducting LCA studies. The aim of this section is to give a brief overview, since comprehensive methodology of LCA can be found elsewhere (Allen and Shonnard, 2002; Baumann and Tillman, 2004; Pennington et al., 2004).

(a) **Goal and scope definition.** In this phase, the purpose of the study, system boundaries, functional unit (FU), and assumptions are defined. This stage includes a proper definition of the system under study and its limits, along with the need for data collection, the assumptions made, and the objective and level of depth undertaken in the study. System boundaries determine which unit processes, connected by material and energy flows, are going to be included in the LCA study (Box 15.1). Definition of system boundaries is partly based on a subjective choice, made during the scope phase when boundaries are initially set. All of this should be compatible with the purpose of the study, defining geographic and temporal boundaries. Defining the functional unit enables the association of the inputs and outputs of the system to a particular reference unit.

(b) **Inventory analysis.** The LCI phase is a technical process that quantifies all inputs and outputs from the processes within the system boundaries, compiling material and energy flows. Energy and raw material consumptions are considered as inputs; emissions to air, water, and soil, solid wastes, and products as outputs. This phase is the most time-consuming and work intensive because of the need for data collection. This is a key point in any LCA study, since the inventory should be representative, containing good data quality. To this end, firstly, the process flow chart should be constructed in detail. Real data are more advisable, but,

when reliable data are not available, commercial databases can be useful for dealing with general data rather than specific products. In the latter case, site-specific data are required. There are many LCA databases available that can normally be bought together with LCA software (http://www.lefe-cycle.org). Another important point in LCI is allocation (i.e., assigning data correctly to each of the subsystems constituting the object of evaluation). A cause—effect relationship is established between raw materials consumption and generation of wastes and emissions in the product/process/service under analysis. In case of just one product or process being analyzed, assignment is direct. Problems may arise when one system generates more than one product (multifunction processes) or influences more than one life cycle (e.g., in open-loop recycling). In these cases, providing different product life cycles with different functions, decisions should be taken to allocate correctly environmental burdens to each product or service under study. The International Standards ISO 14041 establishes recommendations for allocation in LCI analysis, reviewed by Ekvall and Finnveden (2001), who focused also on the use of subdivision and system expansion in LCI practice.

(c) **Impact assessment.** The goal of the LCIA phase is to understand and evaluate environmental impacts based on LCI data (i.e., the inventory and impact assessment results are discussed together). In this phase, different elements are normally used.

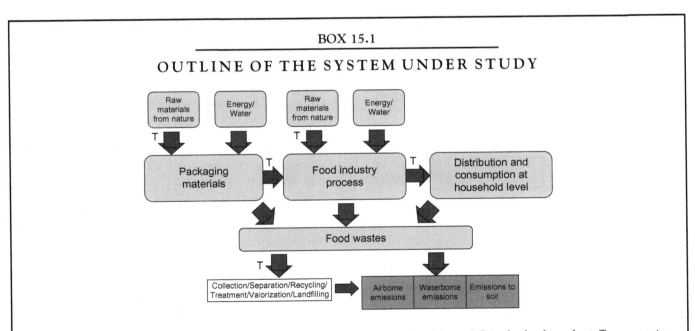

Schematic representation of subsystems, inputs, and outputs considered in an LCA of a food product. T represents transportation. At household level, energy/water consumption may be needed as well.

- *Classification.* The inventory inputs and outputs are assigned to the impact categories selected, which depend on the methodology chosen and are based on the expected types of impacts.
- *Characterization.* This provides a way to directly compare the LCI results within each category. Weighting factors are applied to unify all relevant substances within each impact category.
- *Normalization.* In order to establish a common reference to enable comparison of different environmental impacts, a reference quantity can be used to make the data "dimensionless" (optional).
- *Valuation.* Finally, the relative importance of the potential environmental impacts identified in the previous steps can be established by assigning them weighting (an optional step). Valuation is a controversial step, since it implies subjective statements to decide which category is more damaging and in what intensity in relation to the others, which may vary depending on local policies or other geographic considerations. This is a difficult step that usually receives little attention in current practice.

(d) **Interpretation.** The last phase of the process is the interpretation of the results obtained. By using this systematic technique, information from the previous stages is identified, quantified, checked, and evaluated. The final purpose of an LCA is to draw conclusions and recommendations useful for decision making. Significant environmental issues are identified for conclusions and recommendations, which should be consistent with the goal and scope of the study. These may include both quantitative and qualitative measures of improvement, such as changes in raw material use, product, industrial process, activity design, and consumer habits of consumption and waste management. In comparative studies, LCA enables identification of which of the compared alternatives has a better environmental performance.

3. UTILITY OF LCT/LCA TO PROMOTE LOWER-IMPACT HABITS IN CONSUMERS

Heller and Keoleian (2003) reported that effective opportunities to enhance the sustainability of the food system exist in changing consumption behavior, which will have benefits across agricultural production, distribution, and food disposition stages. Following this line, since to shift consumer habits requires information, scientific and accurate data should be available for decision-making strategies. Thus, LCA studies may well focus on the environmental consequences of changing consumer habits. As an example, using LCA, opportunities for reducing greenhouse gas (GHG) emissions as

a consequence of changing meat consumption patterns and the adoption of healthy and balanced diets have been reported (Roy et al., 2012). Similarly, by applying LCA, it was concluded that the overall environmental impact associated with a standard shopping basket is 10 times higher on average in a hypermarket than in a municipal market, and that if customers selected the least packaged products available in hypermarkets, each shopping basket could reduce significantly its associated environmental impact (Sanyé et al., 2011).

With an LCA approach, possibilities for increasing customer satisfaction while reducing the environmental impact from food packaging systems were explored using results based on consumers' demands on packaging (Williams et al., 2008). Data indicated that there were potentials to increase customer satisfaction while decreasing the environmental impact of packaging. Efficient packaging should be useful to reduce the impact of the food distribution chain. Although for decades the main issue in the packaging sector was reduction of material usage and improvement in recycling measures, and not particularly on reduction of food losses, the function of packaging should also be considered. A recent report on environmental impact of packaging and food losses with an LCA perspective (Williams and Wilkström, 2011) showed that, in some cases, it may be necessary to increase the environmental impact of packaging in order to reduce food waste. This may be especially true for items for which the environmental impact of the food is high relative to the packaging (such as cheese) and for food items with high losses (such as bread). In addition, it was reported that the transportation stage of the package is an important contributor to the environmental impact of the packaging systems (Madival et al., 2009). Comparisons of the impacts considering the same specific food product with different packaging systems have been carried out with an LCA perspective (De Monte at al., 2005; Hospido et al., 2006), taking into account also consumer behavior in recycling different containers (Calderón et al., 2010). In this work, along with the environmental charges involved in making each packaging material, differences in the weight of the packages were also considered in calculating the food product transportation impact. In the food industry, the use of energy for refrigeration results in one of the sector's most significant impacts (Henningsson et al., 2004). Reliance on high-energy processes in the food industry, such as canning and freezing/chill storage, may also need revision. LCA was useful to evaluate the impacts of some traditional or novel food preservation technologies (Pardo and Zufia, 2011). It was also applied to eco-design for the development of more efficient and sustainable food products (Zufia and Arana, 2008).

Despite the fact that, in recent years, presentation of environmental product information to consumers has been based on carbon footprint, Jungbluth et al. (2011) reported that this methodological approach may be insufficient for full environmental information, and thus the use of LCA has been recommended for this purpose, provided in a simplified form. With an LCT orientation, the authors highlighted that the environmental impacts of a product should be related to an overall environmental goal, similar to normalization. In the same study it was suggested that it was not sufficient to define environmental targets as a reduction of environmental impacts per gross domestic product. The time frame for achieving the critical burden was reckoned as a critical issue. In that work, the so-called "eco-time" (considering 20 years as a reasonable time frame within which the critical burden should be achieved) was proposed as a unit to be used in business directed to consumer communication. At the same time it was pointed out that communication of the respective LCA results to consumers in a simplified form is another important issue deserving attention.

Recently, Löfgren et al. (2011) reported that, to be effective, analysis methods intended to support improvement actions should also consider the decision makers' power to influence. The authors concluded that such an approach may help to focus on the environmental consequences of energy and material losses in manufacturing, rather than merely accounting for the contributions of individual stages of the life cycle to the overall environmental impact.

4. VALORIZATION OF WASTES BY BIOPROCESSING, FROM AN LCA PERSPECTIVE

Agricultural activities and the food industry generate a huge amount of organic wastes and by-products, which can be used as substrates in bio-transformations carried out by microorganisms or their enzymes. Resulting products can be employed as energy sources (e.g., biogas and biofuels), compost, or commercialized as high-value commodities used in other industries (food, dietetics, cosmetics, pharmacy, etc.).

4.1 Bioethanol

In order to implement policies on the mitigation of GHG emissions and reduce the dependence on fossil energy supplies, the increasing importance of biomass as a renewable energy resource is well recognized. Nowadays, the transport sector is almost fully dependent on fossil fuels. Hence, the potential use of biomass for the production of biofuel to be used in vehicles has received special attention. The use of biofuels seems to present environmental and socioeconomic advantages, although these sustainability credentials must be proved in a rigorous manner based on their LCA. Energy from biomass has an almost closed CO_2 cycle, but there are some GHG emissions in its life cycle (i.e., N_2O and CH_4 from fertilizer application, organic matter decomposition, and the employment of fossil fuels (Cherubini et al., 2009)).

The most common transportation biofuels are bioethanol and biodiesel, which can be produced from several sources. On a large scale, most bioethanol is nowadays produced from sugar or starch (mainly dedicated crops of sugar cane and corn). However, conflicts arise between food/feed and industrial uses of these products. Agricultural and forest wastes (e.g., straw, maize cob-stover, cotton stalks, sawdust, and forest thinning) as well as by-products from the food industry (e.g., sugar cane bagasse, barley hull, wheat barn, and rice husks) could be an interesting, less controversial alternative. The use of these lignocellulosic feedstocks as raw material usually involves an additional hydrolysis stage (prior to fermentation) with a potential for production of around 0.3 L of bioethanol per kg of dry mass (Singh et al., 2010). Nevertheless, this technical inconvenience can be outweighed by the advantages (i.e., they are in surplus, easily available, relatively cheap, and enable avoidance of the high impacts of dedicated crops production).

Various authors have employed LCA to analyze bioethanol production from different raw materials. Even for similar systems, different results were obtained depending on the type and management of raw materials, conversion, and end-use technologies. However, most studies have concluded that bioethanol can offer improved environmental performance in impact categories such as global warming, as well as a decrease in nonrenewable energy consumption. On the other hand, ethanol application frequently has adverse effects on other impact categories, such as photochemical oxidant formation, acidification, eutrophication, or human and ecological toxicity (von Blottnitz and Curran, 2007; Cherubini et al.; 2009, Singh et al., 2010; Laca et al., 2011; González García et al., 2012).

When lignocellulosic wastes are employed for energy production, a general conclusion is that fossil energy savings and GHG mitigation are increased. For biofuels, these GHG savings (with respect to fossil fuels) have been estimated to be higher than 80% if residues are used as feedstock. GHG emissions (g-CO_2-eq/km) are 25—50 for bioethanol from lignocellulose, 50—195 for bioethanol from dedicated crops, and 210–220 for gasoline. The ratio "nonrenewable energy input/energy output" is 0.15–0.45 for bioethanol from lignocellulose, whereas it is 1.2 for gasoline or diesel (Cherubini et al., 2009). Besides, benefits with respect to acidification, eutrophication, and human toxicity have been reported for bioethanol produced from waste bagasse (von Blottnitz and Curran, 2007).

Ethanol production from lignocelulosic biomass has the potential to produce a number of by-products, such as lignine and pentose sugars, with many uses after being concentrated or converted. The employment of lignin wastes for energy production to run biomass conversion plants increases fossil energy savings and GHG mitigation. Additionally, from residual sugars, many other products (e.g., organic acids, alcohols, 1,2-propandiol) can be manufactured along with bioethanol, which would improve the economic and environmental performance of the plant. Therefore, the use of lignocellulosic material could allow coproduction of valuable biofuels, chemical compounds, electricity, and heat. Some authors consider that this second generation ethanol from lignocellulosic feedstocks has the potential to become the major source of renewable energy in the word (Singh et al., 2010; González-García et al., 2012).

Some sugar-based food wastes or by-products can be also employed to produce bioethanol (e.g., fruit wastes, whey, or molasses). This is a good option in that it reduces the need for dedicated crops and at the same time valorizes an industrial waste. A recent work on molasses-based ethanol production in Nepal reported that avoided emissions are 76.6% when conventional gasoline is replaced by this fuel (Khatiwada and Silveira, 2011). However, very few studies focusing on the environmental aspects of bioethanol production from these materials are available, and further work from an LCA perspective should be developed to establish valid conclusions.

4.2 Biodiesel

Biodiesel is a vegetable oil- or animal fat-based diesel fuel consisting of fatty acid alkyl esters that can be produced from dedicated crops of edible sources (e.g., soybean, sunflower, rapeseed, palm, olive, corn, peanut). However, the constraints of the "Food versus Energy" crisis have motivated the development of technologies for the synthesis of biodiesel from nonedible sources (e.g., karanja, jatropha, neem, nagchampa, cotton), micro-algae, and waste sources (e.g., used cooking oil, waste engine oil, animal fats, sewage sludges). The employment of used cooking oil (UCO) as raw material reduces production costs and

helps to solve the environmental problems associated with the disposal of this waste. The limitation is that these UCOs are scarce. Waste animal fats are more abundant, but they require a more complex process. Municipal sewage sludge is also a potential source of lipid for biodiesel production, but the process is still quite costly (Siddiquee and Rohani, 2011).

LCA studies give different results depending mainly on the raw material and the process itself. Most LCA studies agree that biodiesel from virgin oils achieves 40–65% of the GHG emissions of conventional diesel (Cherubini et al., 2009). Annex V of the EU Renewable Energy Directive gives typical GHG emission savings for some biodiesel systems: 45% for rapeseed biodiesel, 62% for palm oil biodiesel, and 88% for UCO biodiesel. An environmental analysis of biodiesel production in Ireland gave a value of 54% GHG emission savings for animal fats biodiesel and a value of 69% for UCO biodiesel (Thamsiriroj and Murphy, 2011). In another study, GHG emission savings of 80, 72, 72, 75, 69, and 25% were calculated for waste vegetable oil, beef tallow, poultry fat, sewage sludge, soybean, and rapeseed biodiesels, respectively. It was reported that biodiesel fuels from wastes entail significantly lower impacts than conventional low-sulfur diesel for global warming, ozone layer depletion, and energy demand. Nevertheless, slightly increased values were found for acidification and eutrophication (Dufour and Iribarren, 2012). Processes using UCOs generally have a lower overall environmental impact than processes that use virgin oils, animal fats, or sludge.

In a recent work, three process design alternatives for biodiesel production from UCO have been considered: the alkali-catalyzed process including an FFA pretreatment, the acid-catalyzed process, and the supercritical methanol process using propane as co-solvent. Results showed that the last option is the most environmentally favorable alternative. The most relevant impact categories were in the cases of marine aquatic ecotoxicity and depletion of abiotic resources. When comparing the alkali- and the acid-catalyzed process, the former was the process with the lower overall environmental impact (Morais et al., 2010; Varanda et al., 2011).

From an environmental point of view, biodiesel fuels from wastes are a good alternative to both conventional diesel and first generation biodiesel. In particular, UCO arose as the most favorable lipid feedstock for biodiesel production. After 2017, European bioplants need to have a 60% GHG emission saving to be considered sustainable. UCO biodiesel meets the criteria, but it is recommended that animal fat biodiesel is processed with UCO to reach sustainability criteria (Thamsiriroj and Murphy, 2011; Dufour and Iribarren, 2012).

4.3 Biogas

Biogas obtained by anaerobic digestion of renewable resources can be used for heat or electricity production, and even as a transportation fuel. Several environmental benefits can be achieved, including a reduction in the amounts of biodegradable waste going to landfills. Among the raw materials available are sewage sludges, municipal organic wastes, agricultural harvesting residues, manure, ley crops, and also organic wastes from the food industry (e.g., whey, fruit and vegetable wastes, fish and meat wastes, molasses, coffee wastes, patisserie wastes, etc.) (Laca et al., 2011).

When fossil fuels are replaced by biogas systems, environmental improvements can normally be achieved. The energy input into biogas systems overall corresponds to 20–40% of the energy content in the biogas produced. It was reported that, depending on the biogas system, raw materials can be transported for different distances before the energy balance turns negative (e.g., 200 km for manure and up to 700 km for slaughterhouse waste) (Berglund and Börjesson, 2006). In addition to the direct environmental benefits, there are often indirect benefits of changed land use and handling of organic waste products. A study carried out on several biogas systems, including a system using food industry wastes as raw material, reported that GHG emissions decrease by approximately 75–90% when biogas-based fuel replaces fossil fuel to produce heat. CO_2 contributes to about 60–75% of these GHG emissions and CH_4 to 25–40%. Besides, it was estimated that the acidification and eutrophication potentials are reduced by up to 95% and the emission of particles by 30–70%. On the other hand, the photochemical oxidant creation potential increases typically by 20–70%. Similar environmental effects were observed when biogas was used for cogeneration of heat and power. When biogas replaces conventional transportation fuels, the contribution to the global warming potential is normally reduced by 50–80% and the contribution to the photochemical oxidant creation potential by approximately 20–70%. Similar benefits when biogas was used for heat or combined heat and power production were observed for acidification and eutrophication potentials (Börjesson and Berglund, 2007).

Another aspect that has been environmentally analyzed is the possibility of treating some packaging materials by anaerobic digestion. In the last decades, starch–polyvinyl alcohol (PVOH) blends have developed rapidly and are widely applied as packaging or agricultural mulch film. Under anaerobic digestion conditions, 58–62% biodegradation of starch–PVOH-based biopolymers was achieved. An LCA analysis of a starch-PVOH biolpolymer packaging system showed that atmospheric emissions released during the anaerobic digestion process and fuel combustion are the main contributors to acidification, eutrophication, global warming, and photochemical oxidation potentials, whereas abiotic depletion, ozone depletion, and toxic impacts are mainly caused by energy consumption and infrastructure requirements (Guo et al., 2011).

4.4 Compost

Recently, with an LCA approach, Giugliano et al. (2011) pointed out that the indicators for composting food waste were not environmentally convenient, for two reasons: firstly, the significant impact of collection-separation, and secondly, the small benefits associated with the substitution of peat and mineral fertilizers. The situation slightly improved when food and green wastes were considered for anaerobic digestion instead of composting, but the advantages were still modest in comparison with those of the recycling of packaging materials.

LCA was used in another study that analyzed different strategies for treatment of solid wastes in Sweden, including different fractions of municipal solid waste (MSW) (Finnveden et al., 2005). For the food waste fraction, composting, digestion, incineration, and landfilling were compared. In general, anaerobic digestion was preferable over composting and landfilling regarding energy use, emissions of greenhouse gases, and other factors in the study. It was observed that composting could be an interesting alternative if transport distances were kept low while they were longer for the other treatment alternatives. Large scale composting was therefore of limited interest, but as a general result, the advantages of composting were limited. It should be borne in mind that in this study it was assumed that residues from composting and digestion could be used as fertilizers whereas, as stated by the authors, this is not certain because of the risk of pollutants in the residues.

Lundie and Peters (2005), using LCA, showed that, for the impact categories considered, home composting was the best option for food waste management (among others analyzed, such as centralized composting and landfilling food waste with municipal waste as co-disposal). Emission of methane from the degradation of organic materials is a main reason for policies aiming to reduce landfilling of organic materials. The authors highlighted that, nevertheless, if operated without the required controlled aerobic conditions, home composting could greatly increase GHG emissions as a consequence of anaerobic methanogenesis.

Composting has been proposed for the recovery of stalk and dewatered wastewater sludge to produce a sanitized organic supplement for application in the vineyard, closing the organic matter cycle (Ruggieri et al., 2009). These two organic wastes have been traditionally incinerated or disposed of in landfill. In the work, the environmental impact of composting, and different alternatives to manage the high amount of organic wastes generated in wine production, were analyzed using LCA. *In situ* composting presented the best performance in most of the impact categories analyzed, while the energy balance showed that the composting systems involved less energy than systems based on mineral fertilizer consumption.

An important pitfall in the food industry and food distribution sectors is the huge range of oil-based polymers used for packaging. They are largely non-biodegradable and particularly difficult to recycle or reuse due to complex composites mixes. In recent years, development of bio-based packaging materials from renewable natural resources has received increasing attention. Despite their current higher costs compared with conventional plastics, the use of biopolymers—such as polylactide (PLA), thermoplastic starch (TPS), or others already in use—as alternatives for packaging seems to be feasible in the food sector. Bio-based packaging materials are potentially suitable for inclusion in the composting process or in anaerobic digestion. It has been reported that the most attractive route for treatment of bio-based packaging waste is domestic and/or municipal composting (Davis and Song, 2006), following the same behavior as organic matter within aerobic composting systems. Despite this, it was highlighted that many LCAs of bio-based and biodegradable materials neglect the post consumer waste treatment phase because of lack of consistent data, even though this stage of the life cycle may strongly influence the conclusions obtained (Hermann et al., 2011).

It has been reported that the use of disposable cutlery in the fast-food sector generates mixed heterogeneous waste (containing food waste and non-compostable plastic cutlery). This waste is not recyclable and it is currently disposed of in landfills or incinerated with or without energy recovery. Razza et al. (2009), with an LCA study whose functional unit was "serving 1000 meals", concluded that, using biodegradable and compostable plastic cutlery, an alternative management scenario was possible by valorization through composting. By shifting from the current to the alternative scenario, remarkable improvements were reported (an overall 10-fold energy saving and 3-fold GHG saving).

4.5 Other proposals

LCA has been used to evaluate the environmental impact of valorization of food wastes to obtain various commodities or to assess the impact of alternative

scenarios. Food industry by-products may be considered as wastes, negatively affecting the environment, or as resources, if appropriate valorization technologies are implemented. With a view to avoiding the environmentally high impact of raw materials in animal feed, LCA was applied to evaluate the incorporation of feed-use amino acids in the manufacturing of pig and broiler feeds (Mosnier et al., 2011). Interesting effects were reported, resulting in a reduced use of protein-rich ingredients such as soybean meal, recognized to have elevated contributions to environmental impacts associated with deforestation. For animal feeds, calculations were firstly made using only cereals or several protein-rich ingredients (soybean meal, rapeseed meal, and peas). As general results, incorporation of amino acids in various feed formulations had positive effects such as decreased values for eutrophication, terrestrial ecotoxicity, and cumulative energy demand, or reduction in climate change and acidification. Greater effects were especially found when this substitution was compared with high-environmental-impact raw ingredients such as soybean meal. Feeds specifically formulated to minimize GHG emissions had the lowest values for climate change and cumulative energy demand. In addition, it was suggested in the study that the costs of amino acids were not the limiting factor in their incorporation. So the use of food wastes as a resource for substituting high-impact raw materials in feedstocks, after adequate treatment to allay health concerns, deserves attention.

The sugar cane production process was reported from an LCA point of view (defining the daily sugar production of the mill as the functional unit) with the aim of identifying and quantifying the aspects with the largest environmental impact in the process, along with the analysis of alternatives for using by-products and wastes for valorization (Contreras et al., 2009). The first alternative represented conventional sugar production, implying the use of synthetic fertilizers, pesticides, bagasse combustion, and the use of molasses and agricultural wastes as animal feed. The other three alternatives incorporated the use of by-products and wastes for valorization. One alternative considered the use of wastewater, filter cake, and ashes to substitute for synthetic fertilizers. Another accounted for filter cake to be used for biogas production. The last one integrated alcohol and biogas production into the sugar production process. The major difference among the alternatives was found in the resource impact category. Advantages in producing alcohol, biogas, animal feed, and fertilizers from the by-products were achieved in the comparative study for resource savings.

Similarly, another LCA study (Gassara et al., 2011) dealt with apple pomace waste management and the repercussions of value addition of these wastes, in terms of their sustainability regarding GHG production. In the study, the functional unit defined was "the total production of apple pomace in Québec at 2007 of 16,209 tons". Different strategies for management of pomace waste were considered, comprising incineration, landfill, composting, solid-state fermentation to produce high-value enzymes (by *Phanerocheate chrysosporium*), and use as animal feed. Data indicated that, for the functional unit, among all the strategies, solid-state fermentation to produce enzymes was the most effective method for reducing GHG emissions, while apple pomace landfill resulted in higher emissions. LCA showed that the fermentation method to produce ligninolytic enzymes was the most environmentally sustainable process, as an aerobic process. The authors indicated that the absence of some data led them to consideration of some assumptions in order to calculate GHG emissions. The authors therefore remarked on the need for experimental studies to calculate GHG emissions coefficients during agro-industrial waste management.

5. VALORIZATION OF WASTES BY NONBIOLOGICAL PROCESSING OR DISPOSAL, FROM AN LCA PERSPECTIVE

5.1 Recycling of Packaging Materials (Plastics, Metal, Glass, Paper)

Recycling of this type of waste material has been analyzed from an LCA perspective in various studies over the past years. Recently, it was reported that in LCA evaluations, packaging materials recycling was always energetically and environmentally convenient, especially for metals, glass, homogeneous plastic, and paper (Giugliano et al., 2011).

It was observed (Finnveden et al., 2005) that a policy promoting recycling of paper and plastic materials, preferably combined with policies promoting the use of plastics to replace virgin materials, led to decreased use of total energy and emissions of gases contributing to global warming. In addition, if the waste could replace oil or coal as energy sources, and neither biofuels nor natural gas were alternatives, a policy promoting incineration of paper materials could be successful in reducing emissions of GHG.

Björklund and Finnveden (2005) analyzed data from the literature, based on life cycle comparisons of global warming impact and total energy use of waste management strategies. Their aim was to find out the extent to which previous reports agreed or contradicted, and whether generally applicable conclusions could be drawn. The analysis revealed key factors showing significant influence on the ranking between recycling, incineration, and landfilling of materials in

household wastes (excluding the biodegradable fraction). Firstly, producing materials from recycled resources (paper, cardboard) turned out to be often, but not always, less energy intensive, and caused less global warming impact, than production from virgin resources. For paper products, however, the savings of recycling were much smaller, and dependent on a number of factors such as paper quality, energy source avoided by incineration, and energy source at the mill. For nonrenewable materials (glass, metals, plastics) the savings were of enormous magnitude. Anyway, difficulty was encountered in drawing conclusions for a specific time and place, on the geographic and temporal boundaries.

A Life Cycle Costing approach was used to discuss an economic assessment of alternative scenarios involving different combinations of energy and materials recovery from MSW, adopting a welfare economic (i.e., social) perspective (Massarutto et al., 2011). Scenarios simulated various MSW management options, with different degrees of source separation and management of residuals. An interesting fact was that the aim of the study was not to individuate "the best" option, but rather to elucidate what were the assumptions that most influenced the feasibility of each scenario and determined the ranking of options. The study emphasized the need to consider waste management technologies as complementary parts of an integrated strategy, rather than alternatives. In this way, as stated, it was clear that both materials recycling and energy (and heat) recovery through incineration of residual waste were needed in order to effectively minimize the waste flow addressed to landfill.

It is important to note that some studies call attention to the difficulty in accurately assessing some environmental impacts. For example, information regarding toxicological impact of additives in plastics or paper, or regarding possible micropollutants that may occur in mixed wastes, is still scarce, though important efforts in research are being made. It was also claimed that the impact assessment methodology may not take into account the time and space of some interventions (Finnveden et al., 2005), so simplifications may hide, for example, obvious differences in the impacts from emissions at low levels in urban areas compared with emissions from high stacks in rural areas. As stated, these data gaps indicate that conclusions on an overall level should be drawn with caution. Hence, further research should be promoted.

5.2 Recovery of Combustion Energy

Electricity and heat can be generated using as fuel combustible biomass obtained from dedicated energy crops or organic residues (e.g., from agriculture, forestry, industries, and households). An important saving in fossil fuels can be achieved by using biomass as an energy source. For coal, oil, and natural gas, the generation of 1 MJ of electric energy implies a consumption of nonrenewable energy of between 1.7 and 4.2 MJ, whereas these values are in the range 0.1–0.4 MJ for biomass energy. In the case of heat energy, these values are between 1.1 and 1.5 for fossil fuels and only 0.01–0.15 for biomass. This renewable energy is normally evaluated as CO_2 neutral. However, considering the complete life cycle, there are some greenhouse gas emissions (i.e., CO_2, N_2O, and CH_4) due to cultivation, harvesting, processing, and transportation of the fuel. It has been reported that net GHG emissions from electricity generation can be reduced by more than 90% by employing biomass as raw material (Cherubini et al., 2009; Laca et al., 2011; Sebastian et al., 2011).

Two alternatives can be considered for large-scale biomass electricity generation: co-firing in an existing coal power plant or biomass-only fired power plants. Co-firing enables us to take advantage of the generally higher efficiency of very large-scale power plants. Besides, investment costs might be greatly reduced. As negative aspects of co-firing, more intensive particle conditioning is necessary and longer transport distances are generally required. A comparison of both kinds of power plants found that co-firing biomass in a coal power plant gives slightly lower GHG emissions than their conversion in a biomass-only fired plant. However, this fact depends enormously on certain parameters, such as power station efficiencies, the degree of pretreatment required for co-firing, and the influence of biomass particles in co-firing on the coal utility boiler efficiency and transport distances. This work also considered two sources for biomass (i.e., wheat straw and *B. carinata*), and in all cases agricultural residual biomass offered better results than the energy crop (Sebastian et al., 2011).

Fossil energy savings and GHG mitigation are increased if lignocellulosic wastes are used for energy production (Cherubini et al., 2009). These wastes are mainly generated in arable farms that consume a large amount of energy. Over the last few years, small-scale systems for energy generation based on biomass have progressed rapidly, and the number of applications suitable for farm use has increased. If the residual biomass is used on the farm of origin the environmental and economic costs of transport are avoided. A study has been developed to analyze a system based on the use of residual straw for total energy self-sufficiency on an organic arable farm. The straw feeds the straw furnace (for heat supply to the farm) and an ethanol production plant, which delivers ethanol fuel for field operations. The lignin separated out during ethanol production is used to produce process electricity and steam. This study

showed that it is possible for an organic farm to become self-sufficient in energy by using only its own residues. Nevertheless, low savings in GHG emission (9%) were achieved because of the impact on the soil carbon content of removing straw from the fields (Kimming et al., 2011).

Another possibility is to employ industrial wastes to replace fossil fuels in the industry that generates the residue. For example, sugar cane residues (e.g., excess bagasse and cane trash) can be employed as a bioenergy resource that is consumed in the sugar industry. This alternative provides some GHG reduction benefits (Nguyen et al., 2010).

Biomass fuels can be used to generate heat and electricity at ethanol plants. For example, in corn ethanol plants, dried distiller's grains with solubles, syrup, and corn stover can be used to produce electricity and process steam in order to reduce GHG emissions. An LCA of corn ethanol was conducted considering these biomass fuels and three different biomass conversion technologies/systems: process heat only systems (PH), combined heat and power (CHP) systems, and biomass integrated gasification combined cycle (BIGCC) systems. Much greater reductions in GHG emissions resulted for ethanol produced with biomass CHP or BIGCC compared with ethanol produced with gasoline or natural gas (in the case of BIGCC systems a 100% reduction was achieved) (Kaliyan et al., 2011).

5.3 Additional Recovery Proposals

Only scarce work has been focused, from a life cycle perspective, on the recovery of commodities from food industry wastes compared with the high number of LCA studies dealing with the impacts of different environmental technologies. Recovery of commodities from wastes may reduce environmental burdens while providing novel commercial products or the substitution of other nonrenewable materials towards more sustainable solutions. As an example, comparison of the environmental impacts of using chemical or, alternatively, meat meal fertilizers obtained from animal by-products (ABP, which comprises slaughter wastes not intended for human consumption) has been reported (Spångberg et al., 2011). Fertilizers are one of the major environmental issues in agricultural production. The authors noted that, when obtained from ABP in the form of meat and bone meal made in a production plant (commercialized as "Biofer product"), organic farms may use it as a sustainable source of plant nutrients, since only natural, renewable, and regenerative resources may be applied to the land in these cases. With an LCA approach, the study aimed to compare the environmental impact of two different systems for ABP handling and disposal. One system recovered the nutrient content of ABP by using it as fertilizer on arable land, replacing chemical fertilizer, and using animal fat (a by-product from the meat meal production) to replace fuel oil of fossil origin. The second system recovered the energy content of the ABP (co-combusted with a base fuel). The functional unit defined was "one kg of harvested spring wheat and treatment of 0.59 kg of ABP Category 2". Data analysis indicated that the system for nutrient recovery and chemical fertilizer replacement had lower emissions of GHG and acidification than the energy recovery system but had higher total use of energy and eutrophying emissions. The authors concluded that, overall, the results of the study greatly depended on the fuels replaced.

White cork granulate is mainly intended for products for beverage industries, such as stoppers for wine, beer, champagne, cider, etc. For this use, a pretreatment is required in order to clean the raw material before it is processed according to the health requirements of the food industry. It has been claimed that this eco-material could substitute other nonrenewable and more harmful materials. Recently, Rives et al. (2012) reported an LCA analysis of the production process of cork granulates, taking into account the operations after forest management. Production of cork granulate agglomerate (composed of crushed and joined cork by-products) generates a significant amount of dust (35% of the initial raw material that entered the system). However, since granulate cork is made from forestry or cork industry wastes, it could be considered as an example of raw material optimization in an industrial activity. Data obtained highlighted the potential of cork dust as a material that can be used as fuel to substitute other nonrenewable sources of energy such as diesel oil or electricity. However, the authors pointed out that the use of this waste as an energy source is still at an incipient point of implementation.

An interesting opportunity in the olive oil production industry is the exploitation of certain by-products obtained during the processing of olives for oil. Along with other pieces of work, LCA has been applied to a case study to obtain information on which strategic decisions can be made for optimizing the local olive oil production chain (Salomone and Ioppolo, 2011). Among the suitable strategies, alternative uses as fuels or fertilizers could be implemented.

5.4 Disposal in Landfills

In some European countries with scarce composting facilities, landfill continues to offer the cheapest waste management option. However, this will change as the landfill legislation is tightened and landfill sites diminish (Laca et al., 2011). Notably, an increasing number of landfills are only accepting inert wastes, so organic/degradable wastes have to be stabilized through biostabilization before they are accepted for landfilling

Conventional landfilling generates huge amounts of contaminant leachate and GHG emissions. Efforts have been made in recent years to better control these sites, advancing from simple open dumping places to modern environmentally engineered facilities. The importance of leachate and gas control measures in reducing the overall environmental impact from a conventional landfill has been assessed by LCA (Damgaard et al., 2011), identifying also cost-effective reduction measures. In that study, modeling was performed by applying the LCA model EASEWASTE. Data showed that, for the dump landfill, the main impacts were for spoiled groundwater, due to lack of leachate collection. However, it was demonstrated that leachate collection caused a slight increase in ecotoxicity and human toxicity via water since, despite leachate treatment, slight amounts of contaminants were released through emissions of treated wastewater to surface waters. The largest environmental improvement turned out to be a capping and leachate treatment system. Capping was reckoned to be cheap to establish, giving huge benefits in terms of lowered impacts. The leachate collection system, although expensive, rendered large benefits as well. Other gas measures were found to give further improvements, subject to a minor increase in cost.

It was reported that, when direct landfilling of raw food wastes was banned in Korea, efforts resulted in a highly effective recycling rate in food wastes (e.g., over 94% of food wastes recycled in the nation in 2006) (Kim and Kim, 2010). The common recycling types of food wastes were animal feeding, composting, and others (including anaerobic digestion or co-digestion with sewage sludge). Animal feeding from untreated food waste to animals has been prohibited in the EU since 2002, as a result of Transmissible Spongiform Encephalopathy disease. Kim and Kim (2010) analyzed by LCA the Global Warming Potential generated from 1 tonne of food wastes for each disposal system. It was observed in the study that, although feed manufacturing and composting were the common treatment methods employed, and were environmentally friendlier than other methods, at the same time they could negatively affect the environment if their by-products were not appropriately used as intended.

6. CONCLUSIONS

Even though LCA is a powerful tool to assess the environmental impacts of products, activities, and services, some limitations of it have been identified in recent years. The main difficulty is related to the LCA methodological approach, especially data quality and collection (such as the choice between average and marginal data or allocation problems), definition of the system, time boundaries, and process modeling. The huge amount of detailed data required for completing a full LCA can discourage some practitioners from using LCA as a decision-making support tool. Boundaries extension, quality, and availability of data (such as the still scarce toxicological information affecting the environment) influence the results significantly.

Nowadays, LCA-updated tools are required in order to give accurate answers to increasing demands in various areas for this kind of environmental information. Considering that it is not usual to have all necessary data to conduct an LCA, a large number of LCI databases and software tools have been released, including various methods for LCI compilation (Suh and Huppes, 2005). The most suitable choice depends on the specific features of each case, especially considering goal and scope, available resources, and time. Currently, commercial LCI databases of high quality, consistent and updated, are available. Nevertheless, LCI data for a particular material often does not coincide in different databases, and frequently the outcome of an LCA is dependent on the database used. The possibility of creating a common data exchange format that would further enhance data exchange is being discussed.

A weak point of this methodology is that the form in which LCA data are provided is extremely technical. With the objective of summarizing the LCA results into one single score, methods to produce input-related indicators have been developed (e.g., the ecological footprint, which measures how much land and water area a human population requires, and the carbon footprint, which considers all emissions of carbon dioxide caused by the process analyzed). These indicators can be used to represent the environmental consequences of human activities, but it must be taken into account that none of them covers all aspects considered by environmental policies (Galli et al., 2012). For strategic decision-making purposes, some authors are working on the development of tools that enable combination of the main footprints with the additional dimension of cost (De Benedetto and Klemes, 2009).

Despite the efforts to summarize LCA data into indicators, there still exists an obvious need to develop methods to present LCA results in a way that is accessible to consumers. More and more manufacturers are publishing LCA-based data on their products, but it is important to harmonize the way in which this information is published. It is expected that, in the future, environmental labels, corporate environmental reporting, and environmental guidebooks will be at consumers' disposal.

Finally, it is necessary to mention the usability of LCA methodology in different fields that require environmental information. So, elements of LCA may be used in environmental impact assessment and risk assessment. Several combinations of LCA with other environmental and non-environmental tools have already been reported (e.g. combinations of LCA and multi-criteria analysis, environmental performance indicators, emergy accounting, and even economic models (Hermann et al., 2007; Duan et al., 2011; Dandres et al., 2012)).

The topic treated in this book is not an exception, and the tools and tendencies in LCA discussed herein are applicable for the food industry, including the management of its wastes. It should be borne in mind that LCA is a relatively new environmental management tool, so advances continue to be made. To advance in its practical applications, additional international standardization of LCA will also be necessary (Laca et al., 2011).

7. CASE STUDY: LCA OF WASTE MANAGEMENT IN CIDER MAKING

A previous study analyzing industrial cider production with an LCA approach identified the main environmental loads and the contribution of the different subsystems to each impact category (Abad, 2010). In this section, results obtained in relation to waste management have been treated and analyzed to be presented as an example to illustrate the importance of waste management in this industrial sector. The system considered was a hypothetical factory sited in the north of Spain that produces sparkling cider. The functional unit (FU) selected was a cardboard box containing six bottles of cider (0.75 L per bottle). The system considered includes the whole life cycle with a cradle-to-grave perspective (see Box 15.2).

Input inventory data have been obtained from bibliographic sources, transport distance was calculated using maps, and output inventory data (emissions) were estimated from input data considering bibliographic information. A summary of these output data is shown in Box 15.3. As can be observed, most of the solid waste generated came from the glass of the bottles. Several subsystems have been considered, one of them being "waste management" (WM). This subsystem included the processes of composting, recycling, and landfill disposal, as indicated in Box 15.2. To complete the inventory and in order to carry out the life cycle impact assessment (LCIA), several databases

---

BOX 15.2

SYSTEMS BOUNDARIES USED IN THE CASE STUDY

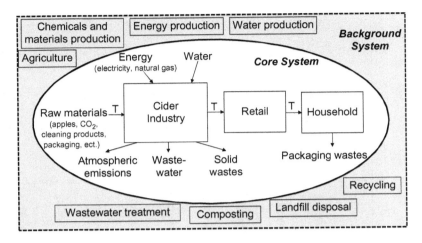

The system considered for LCA includes production of ingredients and materials (including agriculture phase), production of energy and water, transportation (T) of raw materials to the factory and distribution of final product to the consumer, product processing in the cider industry, and emissions from industry and household (to soil, airborne, waterborne; including wastewater treatment, composting, recycling, and landfill disposal).

BOX 15.3

SUMMARY OF OUTPUT INVENTORY DATA PER FUNCTIONAL UNIT (FU)—A CARDBOARD BOX CONTAINING 6 BOTTLES OF FINISHED PRODUCT READY TO BE CONSUMED

Form of waste	Management	Quantity per FU	Kind of wastes
Gases	Emission to the atmosphere	0.23 kg	Carbon dioxide
Liquids	Wastewater treatment	31.50 L	Wastewater
Solids	Composting	1.60 kg	Apple pressings
	Landfill disposal	1.15 kg	Organic wastes (yeast waste), glass, cork, cardboard
	Glass recycling	3.00 kg	Glass
	Cardboard recycling	0.30 kg	Cardboard

The gaseous emissions reported in this table only include carbon dioxide generated by fermentation (other emissions (e.g., from energy generation) were considered in other subsystems). It has been considered that wastewater generated in the industry is treated in a wastewater treatment plant. Solid wastes include wastes from the industrial process and also packaging wastes from households. Apple pressings are considered to be composted and the rest of solid wastes are recycled or disposed of in a landfill. Recycling percentages have been provided by Spanish nonprofit pro-recycling companies.

(Ecoinvent, Buwal 250, and LCA Food DK) were used online through the software tool SimaPro v7.2. The method applied was CML 2 Baseline 2000, which considers the following environmental impact categories: abiotic depletion (AD), acidification (A), eutrophication (E), global warming (GW), ozone layer

BOX 15.4

CONTRIBUTION OF WASTE MANAGEMENT TO TOTAL IMPACT

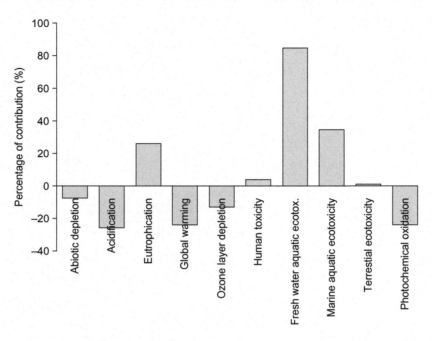

The contribution of waste management to the total impact for each category is shown. It must be noted that percentages shown in the graph were calculated for a 100% positive contribution. Negative contributions to the total impact indicate that waste management has a beneficial effect on that impact category.

depletion (OLD), human toxicity (HT), fresh water aquatic ecotoxicity (FWAE), marine aquatic ecotoxicity (MAE), terrestrial ecotoxicity (TE), and photochemical oxidation (PO). This method has been selected for the analysis because it includes ecotoxicity categories that are likely to be most affected by waste management.

Characterization results showed that the emission of carbon dioxide produced during apple must fermentation and wastewater treatment had a low contribution to total impact (lower than 5% in all the considered categories). However, the environmental loads associated with the subsystem WM were important, as can be observed in Box 15.4. WM had beneficial effects on AB, A, GW, OLD, and PO categories, due to the valorization processes considered (composting and recycling). On the contrary, environmentally damaging effects were observed for E, HT, FWAE, MAE, and TE categories. The subsystem WM turned out to be responsible for most of the total impact in the FWAE category (85%), showing also important contributions in the MAE and E categories (35% and 26%).

Taking into consideration results obtained after passing the normalization filter, the categories showing the highest environmental loads were MAE and FWAE, where WM plays an important role, particularly due to solid waste disposal in landfill. The other important contribution to MAE was the production of packaging material (mainly glass).

Considering normalization results, the method CMLS baseline indicated that the main contributions to the environmental impacts involved were originated by WM. LCIA results were strongly dependent on the treatment/valorization method considered for solid wastes. For example, if apple pressings were not composted, the total amount of wastes that went to landfill would have been more than doubled, and impacts associated with this activity would be considerably increased. It could be concluded that management of solid wastes is a key factor in the environmental impacts derived from cider-making production. Environmental improvement measures should be aimed at increasing the amount of wastes that are valorized and recycled, not only at industrial but also at household level. It has been reported (Pasqualino et al., 2011) that beverages packaging materials have a lesser impact on the environment if they are recycled rather than disposed of in landfills or incineration plants.

References

Abad, L.A., 2010. Life Cycle Assessment in a Sparkling Cider-Making Factory (In Spanish). Food Biotechnology Master (MBtA), University of Oviedo, Spain, Master Thesis.

Allen, D.T., Shonnard, D.R., 2002. Green Engineering. Environmental Conscious Design of Chemical Processes. Prentice-Hall, New Jersey, USA.

Baumann, H., Tillman, A.M., 2004. The Hitch Hiker's Guide to LCA. An orientation in life cycle assessment methodology and application. Studentlitteratur, Lund, Sweden.

Berglund, M., Börjesson, P., 2006. Assessment of energy performance in the life-cycle of biogas production. Biomass Bioenergy 30, 254–266.

Björklund, A., Finnveden, G., 2005. Recycling revisited—life cycle comparisons of global warming impact and total energy use of waste management strategies. Resour. Conserv. Recycl. 44, 309–317.

Börjesson, P., Berglund, M., 2007. Environmental systems analysis of biogas systems—Part II: the environmental impact of replacing various reference systems. Biomass Bioenergy 31, 326–344.

Braungart, M., McDonough, W., 2002. Cradle to Cradle. Remaking the Way We Do Things. North Point Press, New York.

Calderón, L.A., Iglesias, L., Laca, A., Herrero, M., Díaz, M., 2010. The utility of life cycle assessment in the ready meal food industry. Resour. Conserv. Recycl. 54, 1196–1207.

Cherubini, F., Bird, N.D., Cowie, A., Jungmeier, G., Schlamadinger, B., Woess-Gallasch, S., 2009. Energy- and greenhouse gas-based LCA of biofuel and bioenergy systems: key issues, ranges and recommendations. Resour. Conserv. Recycl. 53, 434–447.

Contreras, A.M., Rosa, E., Pérez, M., Van Langenhove, H., Dewulf, J., 2009. Comparative life cycle assessment of four alternatives for using by-products of cane sugar production. J. Cleaner Prod. 17, 772–779.

Damgaard, A., Manfredi, S., Merrild, H., Stensøe, S., Christensen, T. H., 2011. LCA and economic evaluation of landfill leachate and gas technologies. Waste Manage. 31, 1532–1541.

Dandres, T., Gaudreault, C., Tirado-Seco, P., Samson, R., 2012. Macroanalysis of the economic and environmental impacts of a 2005–2025 European Union bioenergy policy using the GTAP model and life cycle assessment. Renewable Sus. Energy Rev. 16, 1180–1192.

Davis, G., Song, J.H., 2006. Biodegradable packaging based on raw materials from crops and their impact on waste management. Ind. Crops Prod. 23, 147–161.

De Benedetto, L., Klemes, J, 2009. The environmental performance strategy map: an integrated LCA approach to support the strategic decision-making process. J. Cleaner Prod. 17, 900–906.

De Monte, M., Padoano, E., Pozzetto, D., 2005. Alternative coffee packaging: an analysis from a life cycle point of view. J. Food Eng. 66, 405–411.

Duan, N., Liu, X.D., Dai, J., Lin, C., Xia, X.H., Gao, R.Y., et al., 2011. Evaluating the environmental impacts of an urban wetland park based on energy accounting and life cycle assessment: a case study in Beijing. Ecol. Modell. 222, 351–359.

Dufour, J., Iribarren, D., 2012. Life cycle assessment of biodiesel production from free fatty acid-rich wastes. Renewable Energy 38, 155–162.

Ekvall, T., Finnveden, G., 2001. Allocation in ISO 14041—a critical review. J. Cleaner Prod. 9, 197–208.

Finnveden, G., Johansson, J., Lind, P., Moberg, G., 2005. Life cycle assessment of energy from solid waste—part 1: general methodology and results. J. Cleaner Prod. 13, 213–229.

Galli, A., Wiedmann, T., Ercin, E., Knoblauch, D., Ewing, B., Giljum, S., 2012. Integrating ecological, carbon and water footprint into a "Footprint Family" of indicators: definition and role in tracking human pressure on the planet. Ecol. Indic. 16, 100–112.

Gassara, F., Brar, S.K., Pelletier, F., Verma, M., Godbout, S., Tyagia, R.D., 2011. Pomace waste management scenarios in Québec—impact on greenhouse gas emissions. J. Hazard. Mater. 192, 1178–1185.

Gentil, E.C., Gallo, D., Christensen, T.H., 2011. Environmental evaluation of municipal waste prevention. Waste Manage. 31, 2371–2379.

Giugliano, M., Cernuschi, S., Grosso, M., Rigamonti, L., 2011. Material and energy recovery in integrated waste management systems. An evaluation based on life cycle assessment. Waste Manage. 31, 2092–2101.

González-García, S., Luo, L., Moreira, M.A., Feijoo, G., Huppes, G., 2012. Life cycle assessment of hemp hurds use in second generation ethanol production. Biomass Bioenergy 36, 268–279.

Guo, M., Trzcinski, A.P., Stuckey, D.C., Murphy, R.J., 2011. Anaerobic digestion of starch–polyvinyl alcohol biopolymer packaging: biodegradability and environmental impact assessment. Bioresour. Technol. 102, 11137–11146.

Gustavsson, J., Cederberg, C., Sonesson, U., van Otterdijk, R., Meybeck, A., 2011. Global food losses and food waste. Extent, causes and prevention. Food and agriculture organization of the United Nations (FAO).

Heller, M.C., Keoleian, G.A., 2003. Assessing the sustainability of the US food system: a life cycle perspective. Agr. Syst. 76, 1007–1041.

Henningsson, S., Hyde, K., Smith, A., Campbell, M., 2004. The value of resource efficiency in the food industry: a waste minimisation project in East Anglia, UK. J. Cleaner Prod. 12, 505–512.

Hermann, B.G., Kroeze, C., Jawjit, W., 2007. Assessing environmental performance by combining life cycle assessment, multi-criteria analysis and environmental performance indicators. J. Cleaner Prod. 15, 1787–1796.

Hermann, B.G., Debeer, L., De Wilde, B., Blok, K., Patel, M.K., 2011. To compost or not to compost: carbon and energy footprints of biodegradable materials' waste treatment. Polym. Degrad. Stab. 96, 1159–1171.

Hospido, A., Vazquez, M.E., Cuevas, A., Feijoo, G., Moreira, M.T., 2006. Environmental assessment of canned tuna manufacture with a life-cycle perspective. Resour. Conserv. Recycl. 47, 56–72.

Jungbluth, N., Büsser, S., Frischknecht, R., Flury, K., Stucki, M., 2011. Feasibility of environmental product information based on life cycle thinking and recommendations for Switzerland. J. Cleaner Prod. doi: 10.1016/j.jclepro.2011.07.016.

Kaliyan, N., Morey, R.V., Tiffany, D.G., 2011. Reducing life cycle greenhouse gas emissions of corn ethanol by integrating biomass to produce heat and power at ethanol plants. Biomass Bioenergy 35, 1109–1113.

Khatiwada, D., Silveira, S., 2011. Greenhouse gas balances of molasses based ethanol in Nepal. J. Cleaner Prod. 19, 1471–1485.

Kim, M-H., Kim, J-W., 2010. Comparison through a LCA evaluation analysis of foodwaste disposal options from the perspective of global warming and resource recovery. Sci. Total Environ. 408, 3998–4006.

Kimming, M., Sundberg, C., Nordberg, A., Baky, A., Bernesson, S., Norén, O., et al., 2011. Life cycle assessment of energy self-sufficiency systems based on agricultural residues for organic arable farms. Bioresour. Technol. 102, 1425–1432.

Laca, A., Herrero, M., Díaz, M., 2011. Life cycle assessment in biotechnology. In: second ed. Moo-Young, M., Butler, M., Webb, C., Moreira, A., Grodzinski, B., Cui, Z.F., Agathos, S. (Eds.), Comprehensive Biotechnology, 2. Elsevier, Amsterdam, pp. 839–851.

Löfgren, B., Tillman, A,M., Rinde, B., 2011. Manufacturing actor's LCA. J. Cleaner Prod. 19, 2025–2033.

Lundie, S., Peters, G.M., 2005. Life Cycle Assessment of food waste management options. J. Cleaner Prod. 13, 275–286.

Madival, S., Auras, R., Singh, S.P., Narayan, R., 2009. Assessment of the environmental profile of PLA, PET and PS clamshell containers using LCA methodology. J. Cleaner Prod. 17, 1183–1194.

Massarutto, A., de Carli, A., Graffi, M., 2011. Material and energy recovery in integrated waste management systems: a life-cycle costing approach. Waste Manage. 31, 2102–2111.

Monier, V., Mudgal, S., Escalon, V., O'Connor, C., Gibon, T., Anderson, G., et al., 2010. Preparatory Study on Food Waste Across EU 27. Contract 07.0307/2009/540024/ser/g4. Final report. European Commission DG ENV. Directorate C-Industry.

Morais, S., Mata, T.M., Martins, A.A., Pinto, G.A., Costa, C.A.V., 2010. Simulation and life cycle assessment of process design alternatives for biodiesel production from waste vegetable oils. J. Cleaner Prod. 18, 1251–1259.

Mosnier, E., Van Der Werf, H.M.G., Boissy, J., Dourmad, J.-Y., 2011. Evaluation of the environmental implications of the incorporation of feed-use amino acids in the manufacturing of pig and broiler feeds using life cycle assessment. Animal 5, 1972–1983.

Nguyen, T.L.T., Gheewala, S.H., Sagisaka, M., 2010. Greenhouse gas savings potential of sugar cane bio-energy systems. J. Cleaner Prod. 18, 412–418.

Pardo, G., Zufia, J., 2011. LCA of food-preservation technologies. J. Cleaner Prod. Available at: <http://dx.doi.org/10.1016/j.jclepro.2011.10.016>.

Parfitt, J., Barthe, M., Macnaughton, S., 2010. Food waste within food supply chains: quantification and potential for change to 2050. Phil. Trans. R. Soc. B 365, 3065–3081.

Pasqualino, J., Meneses, M., Castells, F., 2011. The carbon footprint and energy consumption of beverage packaging selection and disposal. J. Food Eng. 103, 357–365.

Pennington, D.W., Potting, J., Finnveden, G., Lindeijer, E., Jolliet, O., Rydberg, T., et al., 2004. Life cycle assessment part 2: current impact assessment practice. Environ. Int. 30, 721–739.

Razza, F., Fieschi, M., Innocenti, F.D., Bastioli, C., 2009. Compostable cutlery and waste management: an LCA approach. Waste Manage. 29, 1424–1433.

Rives, J., Fernandez-Rodriguez, I., Gabarrell, X., Rieradevall, J., 2012. Environmental analysis of cork granulate production in Catalonia—Northern Spain. Resour. Conserv. Recycl. 58, 132–142.

Roy, P., Orikasa, T., Thammawong, M., Nakamura, N., Xu, Q., Shiina, T., 2012. Life cycle of meats: an opportunity to abate the greenhouse gas emission from meat industry in Japan. J. Environ. Manage. 93, 218–224.

Ruggieri, L., Cadena, E., Martínez-Blanco, J., Gasol, C.M., Rieradevall, J., Gabarrell, X., et al., 2009. Recovery of organic wastes in the Spanish wine industry. Technical, economic and environmental analyses of the composting process. J. Cleaner Prod. 17, 830–838.

Salomone, R., Ioppolo, G., 2011. Environmental impacts of olive oil production: a life cycle assessment case study in the province of Messina (Sicily). J. Cleaner Prod. doi: 10.1016/j.jclepro.2011. 10.004.

Sanyé, E., Oliver-Solà, J., Gasol, C.M., Farreny, R., Rieradevall, J., Gabarrell, X., 2011. Life cycle assessment of energy flow and packaging use in food purchasing. J. Cleaner Prod. doi: 10.1016/j.jclepro.2011.11.067.

Sebastián., F., Royo, J., Gómez, M., 2011. Cofiring versus biomass-fired power plants: GHG (Greenhouse Gases) emissions savings comparison by means of LCA (Life Cycle Assessment) methodology. Energy 36, 2029–2037.

Siddiquee, M.N., Rohani, S., 2011. Lipid extraction and biodiesel production from municipal sewage sludges: a review. Renewable Sus. Energy Rev. 15, 1067–1072.

Singh, A., Pant, D., Korres, N.E., Nizami, A.S., Prasad, S., Murphy, J.D., 2010. Key issues in life cycle assessment of ethanol production from lignocellulosic biomass: challenges and perspectives. Bioresour. Technol. 101, 5003–5012.

Spångberg, J., Hansson, P.A., Tidåker, P., Jönsson, H., 2011. Environmental impact of meat meal fertilizer vs. chemical fertilizer. Resour. Conserv. Recycl. 55, 1078–1086.

Suh, S., Huppes, G., 2005. Methods for life cycle inventory of a product. J. Cleaner Prod. 13, 687–697.

Thamsiriroj, T., Murphy, J.D., 2011. The impact of the life cycle analysis methodology on whether biodiesel produced from residues can meet the EU sustainability criteria for biofuel facilities constructed after 2017. Renewable Energy 36, 50–63.

Varanda, M.G., Pinto, G., Martins, F., 2011. Life cycle analysis of biodiesel production. Fuel Process. Technol. 92, 1087–1094.

von Blottnitz, H., Curran, M.A., 2007. A review of assessments conducted on bio-ethanol as a transportation fuel from a net energy, greenhouse gas, and environmental life cycle perspective. J. Cleaner Prod. 15, 607–619.

Williams, H., Wilkström, F., 2011. Environmental impact of packaging and food losses in a life cycle perspective: a comparative analysis of five food items. J. Cleaner Prod. 19, 43–48.

Williams, H., Wilkström, F., Löfgren, M., 2008. A life cycle perspective on environmental effects of customer focused packaging development. J. Cleaner Prod. 16, 853–859.

Zufia, J., Arana, L., 2008. Life cycle assessment to eco-design food products: industrial cooked dish case study. J. Cleaner Prod. 16, 1915–1921.

Food System Sustainability and the Consumer

Monika Schröder

1. INTRODUCTION

In the world today, economic growth and affluence are not equally shared, as hunger and famines continue to coexist with "unprecedented opulence" (Sen, 1999: xi; 160). In a globalized economic system, rich countries are able to "export" environmental impacts, which means that poor countries are made to bear some of the costs of prosperity elsewhere (Allan, 2011; Foresight, 2011). Meanwhile, population growth continues apace with a global middle class that is projected to expand to some 4 billion people by 2030 from under half a million in 1960 (UN, 2010). Increased wealth from economic development leads to greater economic entitlement for citizens (Sen, 1999:39) and raised expectations regarding consumption. These developments place increasing strains on the natural environment as economic growth fuels resource demand—including demand for space to dispose of waste. The global ecological footprint (i.e., the biologically productive land and water required to provide the renewable resources people use) has doubled since 1966 (WWF, 2010). The carbon footprint, which refers to the greenhouse gas (GHG) emissions attributable to the consumption of goods and services, has increased 11-fold worldwide since 1961 (EU, 2010). Trends in water footprints too give considerable cause for concern (EU, 2010; Allan, 2011) and, in the world's poorest countries, biodiversity has decreased by 60% since 1970 (WWF, 2010). Most pressingly, economic activity is associated with climate change whose consequences, again, are not shared equitably across the world (Stern, 2007:31). In breaking carbon footprint down by country and category, Hertwich and Peters (2009) provide striking evidence of the precise locations of this inequity. Food is an example of an ecosystem service, alongside fibers for clothing, clean water, pollination, carbon capture from the atmosphere, fertile soils, and recreation (Rodriguez et al., 2006). There is increasing demand on the earth's land surfaces and water bodies to provide these services, especially food, and this is associated with significant GHG emissions (Garnett, 2008, 2011). In the UK, the largest impact is due to farming, but with sizeable variations between different types of food, for example, grain-fed versus hill cattle (Cabinet Office, 2008). Meat and dairy products, through the livestock sector, are a major source of GHG, and production in these is expected to double by 2050 (FAO, 2006). About 30% of globally produced food is either lost or wasted (Foresight, 2011) and this means "wasted" GHG emissions in the production and distribution of such food (Garnett, 2011). It is therefore not surprising that food waste has emerged as a prominent issue of public morality. The reason may be that food is essential to survival, yet is not available in sufficient quantity and quality for everyone; or perhaps this waste serves as a daily reminder of the excesses of contemporary consumer society with its "modern-day prodigals" (Sen, 1999:269). In any case, a more efficient, less wasteful food system would significantly reduce its pressure on the environment. For consumers there are far-reaching implications beyond mere resource efficiency, as they are being asked to address the role of GHG-intensive foods, especially meat in their diet (Friel et al., 2009; Oxfam, 2009; Foresight, 2011; Macdiarmid et al., 2011). Consumers therefore have a key role as agents of change in reducing the amount of food that is produced, as well as influencing the type of food that is produced and the efficiency with which it is produced.

The global agenda towards sustainable consumption has gathered a great deal of momentum in recent years although, from the beginning, with article 55 of the Charter of the United Nations (UN), there has been a strong focus on living standards as a crucial underpinning of global equity and justice (UN, 1945). Since 1945, three pillars of sustainable development

Food Industry Wastes.
DOI: http://dx.doi.org/10.1016/B978-0-12-391921-2.00016-0

have become firmly established, namely, economic and social development and environmental protection (UN, 1972, 1992, 2005). The Brundtland Report, "Our Common Future" (WCED, 1987), highlights intergenerational equity and has been widely circulated. Similarly important, the Johannesburg Plan of Implementation (UN, 2005) called for "changing unsustainable patterns of production and consumption" in order to reduce "resource degradation, pollution and waste". The Plan also promotes lifecycle analysis (LCA) as the method of choice when assessing resource degradation, pollution, and waste as integral stages of product and process design (Hertwich, 2005). The Kyoto Protocol (UN, 1998), the first international treaty to set GHG-emission targets, was not ratified by all of the major large GHG-emitting countries for a variety of reasons (Helm, 2009), but, with the Durban agreement (UN, 2011), progress has now been made, with clear implications for the consumer policy of signatories. The Stern Review (2007:ix; 603) identified economically efficient policy options for a low-carbon global economy, one of its recommendation being that deforestation to produce agricultural land should be curbed. The global sustainability agenda has informed the activities of regional, national, and local government. Sustainable consumption is a priority for the European Commission, which, under its Europe 2020 economic strategy, has identified resource efficiency as a flagship initiative (EU, 2011). Here there is explicit reference to waste minimization, food as a natural resource, and the need for more resource-efficient consumption patterns. This signals a wealth of opportunity for innovation, especially for businesses as the economic engines of change (Hart, 2007) but also for households, both as initiators and as potential adopters of change.

Consumer societies are characterized by consumerism as a materialistic, cultural orientation (Bauman, 2007:28; Worldwatch Institute, 2010:3), with the accumulation of goods and the sites for their purchase and consumption (Featherstone, 2007:13). Consumerism drives business as it demands that business starts out with the needs and the values of the consumer (Drucker, 2007:16) while government is concerned with creating a more sophisticated consumer/demand as an underpinning of sustainable economic growth (Cabinet Office, 2011). But with the proliferation of consumer choices and the responsibility placed on consumers to exercise them wisely, a discourse around the "tyranny of choice" has taken hold (Schwartz, 2004). Szmigin et al. (2009) find a new kind of "conscious" consumer, characterized as one whose purchases include ethical choices but who nevertheless requires a degree of flexibility, who cannot be entirely consistent in terms of sustainable behaviors. Generally, as consumers are becoming more mindful about the consequences of their choices, they are growing sensitive to

the global environmental challenges in particular. Surveys of pro-environmental behaviors in seventeen countries, conducted annually by National Geographic (2010), demonstrate increases in such behaviors, with Brazil, China, and Mexico as top scorers. While consumers in emerging economies are most likely to be concerned about global environmental and social issues, consumers in the UK, France, and Spain have significantly increased their consumption of local foods. More specifically, expenditure on green goods and services has increased by 18% in the UK in the two years to 2010 (Co-operative Bank, 2010), while Fairtrade certified foods have shown sustained growth for several years (Fairtrade Foundation, 2011). By joining in Earth Hour 2010 (WWF, 2010), 1 billion people are said to have shown their commitment to more sustainable consumption patterns. This chapter starts with a brief overview of the food supply chain in modern economies, with a particular focus on private households. Theories and current issues in consumer behavior and behavioral change are examined and provide a platform for the ultimate purpose of this chapter, namely an exploration of the role of consumers in the sustainability of the food system with particular reference to food waste.

2. FOOD SUPPLY CHAIN AND WASTE

The modern food industry comprises extended commodity chains with global reach at one end of the scale and short local food networks, designed to reconnect farmers with their markets, at the other. Certainly, consumer demand for greater transparency and simplicity has increased in this respect, most notably in the wake of bovine spongiform encephalopathy (BSE) (Smith et al., 1999). A model of the food supply chain is shown in Figure 16.1. This implies not only material flows but also communication flows, and the latter are naturally two-directional: where product information travels downstream and consumer requirements upstream. Also, while municipal services appear as the customers of households, they may be conceptualized with equal justification as service providers to households.

The food retailing and service industries are served directly by a range of sectors including agriculture, horticulture, capture fisheries and aquaculture, and manufacture (MAFF, 1999), and these in turn receive inputs from industries such as packaging and (fine) chemicals (Schröder, 2003:26). Value chains link these players and include households as an example of a "buyer value chain" (Porter, 1985:3, 35). Of the 18.4 million tonnes of total waste arising from the UK food and drink supply chain, approximately 27% originated from manufacturing, 0.5% from distribution, 7% from retailing, and 65% from households (WRAP, 2010a).

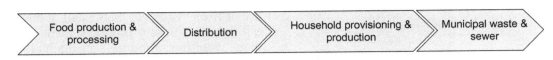

FIGURE 16.1   Food supply chain.

Value stream mapping serves to characterize supply chain processes and waste (Addy, 2011); and road mapping adds to this specified environmental targets, for example, the Milk Roadmap (Defra, 2008). The UK Food Industry Sustainability Strategy (FISS) Champions Group on Waste identified examples of avoidable food waste (IGD, 2007). These included under/overweight products and trimmings (i.e., crusts and tomato ends); technical errors with machinery; inconsistency of processes, such as cooking times and temperatures; and market-imposed waste such as 'take-back' systems and last-minute order cancellations. However, food waste occurs all the way along the food chain, from production/harvesting to processing/manufacturing and retailing. A "Designing out Waste" consortium was established under the auspices of the Green Alliance (2011) with several high-profile food industry players participating and focusing both on the economic and the environmental consequences of waste. Increasing energy and materials costs and the costs of transportation and waste disposal are all important drivers of waste reduction and increased efficiencies in the food industry (Wiltshire et al., 2008), and approaches such as lean manufacturing (Lehtinen and Torkko, 2005) and demand forecasting (Hennel, 2011) are increasingly utilized to achieve it. A study by Oakdene Hollins (2010) provides a snapshot of the amount and geographical distribution of food and packaging waste arising across 149 manufacturing sites and how this waste is being managed. It found that in 2009 these sites produced a total of 481,000 tonnes of waste, of which 9% was sent to landfill and 90.3% was recovered or recycled in some manner, a significant improvement on earlier surveys in 2006 and 2008. Waste segregation at source had been an important factor in this achievement. The European Directive on Waste (EU, 2008a) proposes a hierarchy of waste that has waste "prevention" as its primary objective, followed by "preparing for reuse", "recycling", "other recovery" (e.g., energy recovery), and finally "disposal". Its ultimate goal is to move Europe closer to a "recycling society" with "a high level of resource efficiency".

Through their choice of diet and other food-related routines, households influence the environmental impacts of the food supply chain overall. Households themselves generate ever more direct waste, but there are also trends towards recycling and a diversification of waste streams (OECD, 2002). The OECD acknowledges that while household consumption patterns worldwide have long been unsustainable, the drivers of these patterns are still not fully understood. Households are the basic unit for collecting consumption statistics (ONS, 2010), and in social and behavioral terms, they are groups of people who share resources, expenditures, and activities (Casimir and Tobi, 2011). "Household production" refers to the production of goods and services by the members of a household for their own consumption (Ironmonger, 2000). The food provisioning lifecycle for households is shown in Figure 16.2. Breaking this down into its component parts allows the critical points with regard to food waste to become more salient and hence more tangible in terms of approaches to the reduction or avoidance of such waste.

Acquisition comprises both food purchases—and home grown and foraged foods—and food in either short-term or long-term household storage. In fact, food preparation in modern households tends to draw more on foods already stored in the home than on freshly acquired ones. This of course has implications for the generation of food wastes in the home. The use phase of food provisioning covers food preparation, food service, and the actual eating event (meal, snack), all of them critical points in terms of waste generation and management. Food is purchased from a variety of retail sources, and shopping patterns focus on both food sources (local, global) and categories (fresh produce, prepared foods). If a household habitually grows its own food, it will have developed the capacity for dealing with seasonal gluts even if this just means recycling through the compost heap. Ending up with overstocks from food shopping is another matter as food is actively marketed at the point of sale, to which consumers are liable to succumb (Which, 2005) without, perhaps, any immediate plans for dealing with any excess. On the other hand, as long as they avoid stockpiling, households may consciously buy now for use at some point in the future, and this is one of the activities of household production. Excess food can be frozen, some directly, while some needs to be cooked, but here the requisite food skills need to be present in a household. The acquisition phase is also significant in the sense that it is here that consumers, knowingly or unknowingly, support embedded food waste or waste associated with its production. This calls for appropriate labeling. Labels also serve to render issues that may not be fully salient at the time of purchase more so. Food reaching the household will

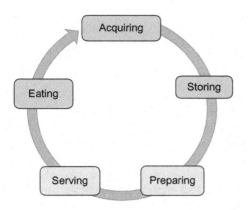

FIGURE 16.2 **Household food provisioning lifecycle.**

be allocated to stores, be they ambient, in the refrigerator, or in the freezer. Regular food inventories, implementation of stock rotations and shopping lists, and generally keeping only simple, essential, and versatile stores are common approaches towards reducing waste at this point. Waste created as part of cooking and serving food relates mostly to portion control and, in the case of the latter, service style. In the private realm such issues are obviously readily negotiable. However, lessons can perhaps be learned from research on plate waste in institutional settings, for example, the advantages and disadvantages of offer versus service provision, self-service, and so on (Buzby and Guthrie, 2002). WRAP (2007) found that over 80% of households in Great Britain did their main food shopping on a weekly basis within a single supermarket. Slightly more than half checked on household stock before shopping, and a similar percentage used a shopping list. Just under half succumbed to impulse purchases, especially in relation to promotions. Nearly every household undertook some home cooking at least two to three times a week. Food was most likely to be thrown away after the date mark had expired, even in the absence of any perceptible spoilage. Respondents also reported significant amounts of plate waste, and this was the result of poor portion management whether in home cooking or the purchasing of pre-prepared meals. Interestingly, nearly all respondents were initially adamant that there was little food waste in their household. This may be an aspect of the mindlessness associated with food, and other routine, consumption more generally.

In 2009, WRAP published a major report that estimated that 8.3 million tonnes per year of food and drink waste is generated by households in the UK, just over 6 kg per household per week, and two thirds of it is avoidable (WRAP, 2009). This waste is mostly due to spoilage or date mark expiry with slightly less being due to too much food having been "cooked, prepared or served". Nearly a quarter of arisings by weight

were of fresh vegetables and salad, followed by drink, fresh fruit, and bakery products. Some other European Union countries produce considerably more food waste per household than the UK, although others generate much less (EU, 2008b). In an exercise that linked food purchase data with household waste data, Defra (2010) found that 15% of UK food and drink purchases that could have been eaten were wasted. The most commonly wasted food was bread, followed by potatoes and vegetable wastes. Water used to produce food that householders in the UK waste represents 6% of the UK's water requirements; the same wasted food also represents 3% of the UK's domestic GHG emissions (WRAP, 2011a). As demonstrably the weakest link in the chain, it is not surprising then that a great deal of attention has been directed towards households and consumer practices and competences. In particular, the unexpectedly high levels of avoidable fruit and vegetables waste prompted a further investigation into home storage practices in this respect (WRAP, 2008). This found that fresh produce was usually disposed of after becoming perceptibly "off". The role of consumers in maintaining the integrity of refrigeration chains is further emphasized in a subsequent WRAP (2010b) report. The amounts and composition of food waste generated over a 4-week period in a representative sample of households in England were examined by Jones et al. (2008). They found that a fifth of the waste disposed of by households entered nonmunicipal waste disposal routes, including composting. Some 23% of what they termed "putrescible" kitchen waste was disposed of in this manner. WRAP's development of their consumer-facing "Love Food Hate Waste" campaign was informed by consumer research that included an exploration of the potential for change (WRAP, 2007).

3. CONSUMER BEHAVIOR AND BEHAVIORAL CHANGE

An understanding of consumer behavior draws on a range of theoretical perspectives from the three core behavioral disciplines of economics, psychology, and sociology (Figure 16.3). Actual consumer behavior is observable and expressed as "demand" in economics, but mainstream economic analysis of consumer behaviors, with its emphasis on a rational individual with stable tastes, has sometimes been challenged as too naive (Lewin, 1996). This may explain recent interest in behavioral economics, in particular, the consideration of "nudging" in the context of promoting and reinforcing socially desirable behaviors (Thaler and Sunstein, 2008). Cognitive and social psychology address themselves, respectively, to the mental states

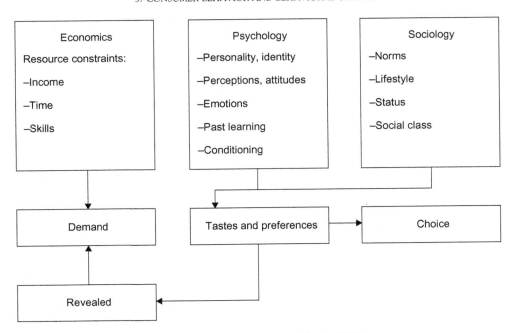

FIGURE 16.3 **Theoretical foundations of consumer studies.** *Adapted from Schröder (2003).*

and social factors that underlie any given consumer behavior or choice. Perceptions, beliefs, attitudes, intentions, motivations, judgments, and social influences cannot be observed directly, unlike demand, but must instead be elicited, either by interviewing consumers or by making inferences from observed actions (Lewin, 1996; Schröder, 2003:11). Attitudes and beliefs enter as constructs into various consumer choice models. Perhaps the best known among these are the theory of reasoned action (TRA) (Ajzen and Fishbein, 1980) and an extended version of this, the theory of planned behavior (TPB) (Ajzen, 1985). These models build on the assumption that volitional behavior is rational and predictable. However, it is now thought that consumers may actually "construct their preferences" at the very time of choosing, and are greatly influenced by available cues (Bettman et al., 2006). Gigerenzer (2000:242) refers to "human biases, fallacies and errors in (probabilistic) reasoning" that characterize individual consumer choices. He advocates heuristics ("rules-of-thumb") that can be developed to serve as shortcuts towards efficient decision making. Settled, rule-based behaviors are considered to be efficient for consumers because they do not burden the decision maker with "case-by-case cost–benefit calculations" (Prelec and Herrnstein, 1991:321). This raises the possibility that routine food waste creation in households may be efficient for those households as they must weigh time and effort being spent shopping against a certain level of inefficiency in use. Rules are also useful in building and reinforcing skills of resistance to some of the cues associated with over-

consumption and mischoosing. Limitations in people's cognitive abilities are among key factors that reinforce default and impulsive choices and patterns of choice (Which, 2005; Just et al., 2007). Distracted, thoughtless shopping behaviors are likely to result in excess food being brought into a household, food that may not be used in time, especially short-shelf life, ready-to-eat items. Examples of personal heuristics include buying just in time, buying only essential items for stock, not visiting certain supermarket aisles, sticking to a shopping list, and buying online. Social marketing approaches may also draw on heuristics being marketed to consumers by government and other agents. For example, the UK Cabinet Office (2010a) promotes the so-called MINDSPACE approach, where "S" stands for "salience". The role of salience is then explained in another policy document published that year (Cabinet Office, 2010b:16), which, in the context of diet choice, observes that "people show a strong tendency to anchor to an object or number that we are primed with, such as how much fruit to eat". The particular heuristic this brings to mind is the 5-A-DAY recommendation regarding fruit and vegetable consumption as published by the Department of Health (2003). In a similar vein, Oxfam (2009) promote a 4-a-week rule designed to help consumers reduce their meat consumption. This has been complemented by social psychological approaches such as the promotion of a meat-free day each week in the city of Ghent (Anon, 2009), drawing attention to GHG and climate change. "Reduce-Reuse-Recycle" derives from the European Directive on Waste (EU, 2008a) and is a heuristic that is being actively promoted by local councils throughout the UK.

Holbrook's (1999:12) typology of consumer value characterizes the value derived from any act of consumption as either extrinsic or intrinsic. Extrinsic value is experienced when a choice is simply a means to an end, and here Holbrook lists "efficiency", "excellence", "status", and "esteem". Intrinsic value is experienced when consumption takes place for its own sake, as with "play", "aesthetics", "ethics", and "spirituality". However, any given consumption event is likely to deliver combinations of these values and these may also bridge the extrinsic versus intrinsic categorization. In this context, the term "conspicuously ethical consumption" can be traced to Bauman (2007:108). Holbrook's typology is particularly useful in enabling visualization of a situation where intrinsic value, such as ethics, may once have been status value but over time has become integral to a person's self-concept (Schröder, 2003:20). However, each type of value in itself represents many bundles of further differentiated values. For example, a study by Jack et al. (1997) explored the "efficiency" of fruit as a snack. Here aspects of efficiency ("convenience") included storability, predictability of eating quality, and absence of waste and mess. Among the fruit, bananas and apples were perceived as more convenient than oranges and kiwi fruit. It is therefore important to understand exactly what prompts a particular consumer purchase. A bowl of fresh fruit may represent aesthetic value besides food value and therefore, if some of that fruit is finally not eaten, it may be inappropriate to categorize the situation as "waste". Consumers routinely engage in value trade-offs: for example, choice of animal welfare-friendly meat against budget constraints, access to nutritious food in relation to the latter, and so on (Schröder and McEachern, 2004). Such trade-offs will be sensitive to the task complexity, choice overload, time pressure, framing of alternatives in a given choice set, and other contextual factors. People strive to interpret contextual cues in a coherent manner in order to gain congruence between personal values and specifically to avoid cognitive dissonance when there has been an apparent transgression (Festinger, 1957). A range of strategies will be drawn on in an attempt to rationalize ostensibly irrational behaviors, prominent among them "neutralization", an approach that was first applied in the context of attitude—behavior gaps in delinquency (Sykes and Matza, 1957; Chatzidakis et al., 2007). Individuals engaged in neutralizing cognitive dissonance will deny responsibility, or injury, or the existence of a victim; or they will condemn the condemners or appeal to higher loyalties.

There are at least two conditions that must be satisfied for behavioral change to be successfully initiated and sustained: first, the motivation to act must be sufficiently strong; and, second, there has to be the capacity to act in line with motivations. The capacity to act is reliant both on external resources, such as physical access and access to key information, and on internal resourcefulness. A simple representation of behavioral change, as it builds on motivation and capacity, is given in Figure 16.4. The figure draws attention to the fact that motivation has both cognitive and affective elements. Knowledge needs first to be acquired, then gain saliency, and finally resonate with someone's personal value system. This idea may also be linked to Holbrook's (1999) intrinsic types of consumer value. Consumers, once motivated towards a behavioral change, will need the capacity to turn motivation into action. This implies, on the one hand, information being received from the supply chain and including waste disposal services and systems as to the nature of products and processes. On the other hand, people will use their generic consumer skills to develop efficient, personalized heuristics (search and decision rules) that they can deploy to embed target behaviors. Ideally, consumers will be supported by choice architectures in which the easy choice (default) coincides with the "rational"/"virtuous" choice. However, people may feel themselves to be "locked into" patterns of consuming (Jackson, 2006), and this can strengthen resistance to behavioral change. As an example, households may feel tied down by physical structures such as kitchen appliances and other household items that are replaced only infrequently. Switching to an entirely green lifestyle in a comprehensive manner would be difficult for most consumers and households as behaviors become socially embedded. When households embark upon major behavioral change, for example switching from a once-a-week supermarket trip to a pattern of shopping for local food and more frequently, reallocations of time and various trade-offs need to be implemented. These include planning for meals and items for stock, shopping routines, managing household food stocks, cooking, serving, the eating event itself, and facilities for food waste. As it is individual food provisioning patterns that lead to food waste, whether directly by households or generally through the types of food people buy, consumers might be held individually responsible and be expected to both initiate and implement behavioral change. However, a view that has been gaining in currency is that too much of the burden of solving social issues has been placed on individuals and the need for individual behavioral change (Nuffield Council on Bioethics, 2007). Among those in support, Kotler and Lee (2006:211) favor approaches that tackle infrastructures and other environmental factors, including government regulation. According to this view, pro-social and pro-environmental changes

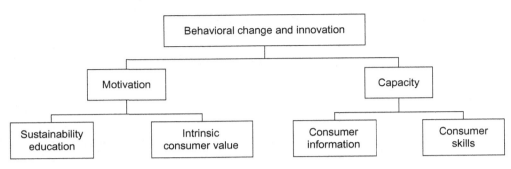

FIGURE 16.4 Motivation and capacity as preconditions for behavioral change and innovation.

should be made easy, the obvious or default choice. Thaler and Sunstein (2008) argue that an architecture of choice can be constructed to amplify the likelihood of specific choices over others. They state that if certain objects are made visible and salient (cues), people's behavior can be affected. For example, where there is provision for separate food waste collection, even when people live in flats, the amount of food waste a household generates is made more salient and this may serve to nudge behavioral change. Focus on user-friendly choice architectures, for example, where fast food does not become the default option for a time-pressed parent with limited food skills, should be considered as a way forward. This might be seen as a pertinent example of the "enabling" approach favored by the UK Government (Cabinet Office, 2008).

4. NEW PRODUCT DEVELOPMENT AND INNOVATION

Innovation, the successful exploitation of new ideas (DTI, 2003), is one of the core activities of business (Drucker, 2007:16). It is traditionally targeted either towards technologies and processes (present product versus new product) or demand (present customer versus new customer) (Ansoff, 1957). Demand focus is exemplified not only by radically innovative products but, more commonly, by adapting or (re)positioning existing products for new markets. Innovation is concerned with both materials (foods, packaging) and service elements, and, as part of the latter, marketing (consumer) communications, such as labeling, are especially important. Innovations targeting consumers are normally associated with, and frequently aim explicitly to bring about, consumer behavioral change. A current convention is to view innovation as a process of diffusion in social systems (Rogers, 2003:24) and sometimes as social contagion (Gladwell, 2002). Where Rogers' discourse focuses on the different categories of adopter that characterize this diffusion (Rogers, 2003:11, 212, 215), Gladwell draws particular attention to so-called "tipping points", critical stages when a new product or

idea begins to really resonate with a wider audience. But all innovations do not diffuse with equal speed. Many niche markets remain as such for a long time, and some never achieve mainstream popularity. For example, regarding organic food, one might argue that a long-established mass market (centuries-old traditional food production) had at one point shrunk to be left to Rogers' "laggards", here in terms of the industrialization of the food industry. Interestingly, when there was a change of paradigm, which criticized certain aspects of the industrialization of food, those same laggards found themselves transformed into modern-day innovators. As the term "motivation" (Figure 16.4) implies, motivation to act relies on motives and, to move action along, feelings. These can be negative, as in fear, anger, and disgust; or they can be positive, as in love, sympathy, and a consequent longing for fellowship. However, those wishing to motivate consumers to adopt more pro-social and pro-environmental behaviors need to beware of the possible effects of how they frame their message. While "nudging" is a much debated approach in this context (Thaler and Sunstein, 2008), it may only work effectively where it is pushing against doors that are already open in the sense that a nudge could cue a consumer, at the moment of choice, to adhere to established behavioral intentions. Those not persuaded of the behavior being promoted suiting them might be put off even further. The implications for the rate of diffusion of a given behavioral innovation may be that, while innovators and early adopters have their attitudes and behaviors reinforced, progress towards mass adoption may actually be delayed. Thøgersen and Ölander (2003) found a tendency for individual-level pro-environmental consumer behaviors to "spill over" (diffuse) gradually into new domains.

Lifecycle analysis (LCA) is a tool for calculating the environmental impacts, as emissions and resources use, during the production, distribution, use, and disposal of a product, as promoted in the Johannesburg Plan of Implementation (UN, 2005). In terms of product lifecycle, it has become popular to try to consider "cradle-to-cradle" approaches where the waste outputs of a given process become the inputs for other processes—as has always been one of the key features of organic farming. Picking up on Porter's (1985) observation that value can be added anywhere along the stages of a given value chain, Munasinghe et al. (2009) identify innovation levers across the lifecycle of products such as orange juice, crisps, bread, and milk. Having the impacts broken down along the supply chain highlights areas for improvement. LCA is equally applicable to household practices, and it is in the context of households that Hertwich (2005) explores LCA as a possible evaluation tool for sustainable consumption. Many consumers have little idea about the provenance of the food they buy, especially in terms of its environmental impact as well as other aspects of sustainability. Allan (2011:141) highlights this with particular reference to the water footprints of different foods and production systems. In most situations, consumers have nothing but the food label to answer their questions and potential concerns. Food product development today takes the approach of an integrated product design (IPD) (Gerwin and Barrowman, 2002) with the use of generic quality management tools such as quality function deployment (QFD). At the heart of QFD is the House of Quality (HoQ), a matrix that first captures market needs verbatim and then helps to translate them into specified product attributes (Chin et al., 2005). Cross-functional teams are a key feature of IPD, and QFD assists in efficiently allocating the requisite team skills and responsibilities (Schröder, 2003:34). QFD is equally useful for designing desirable features into a product as it is for designing out negative features, of which waste is an example. Hofmeister (1991) provides a step-by-step illustration of building the HoQ for a cake mix development. This shows in particular how consumer requirements ("whats") are captured verbatim and how the "voice of the consumer" is subsequently translated into technical specifications. Potential customers for a product with less wasteful features might be expected to mention issues such as "over-packaging", "easy-to-dispose-of" food, food "packaging waste", and food waste that does not pollute household rubbish through "bad smells". All of these are ambiguous in technical terms, needing to be interpreted by technical staff and verified by the consumer before any actual development work can be completed. What does "over"-packaging actually mean? Too much of what is there, and in whose view is it too much?

There are a number of possible approaches to classifying innovation and new product development. The approach taken here discusses the issue from three perspectives, with some overlap between them. They are consumer behavior, food label, and product design (Figure 16.5). The first perspective is on behavioral

change in the sense that it is possible to "innovate the consumer" as much as it is to develop a new product. In essence, the consumer is asked to change behavior towards a given target. This obviously draws on the previous section of this chapter, which provided a generic glimpse of the subject and gave examples of behavioral patterns being promoted towards consumers to help them to turn intentions in respect of more sustainable practices into actions. The second perspective is concerned with innovating communications between producers and consumers, and here the food label, in particular, comes into play. The third perspective considers the more technical aspects of some actual product and process developments. It also considers the degree of novelty that a consumer might perceive. Novelty may be high, as when something is new relative to everything that has been experienced before, or it may be minor (Bianchi, 1998; Schröder, 2003:76), and consumers will respond differently in each case. Consumer neophobia towards unfamiliar foods may be reduced by various means, such as message framing, use suggestions, and appropriate supply and communication channels (Schröder, 2003:220). Social marketing (Kotler and Lee, 2006) again merits a mention here as this is almost entirely focused on innovating consumer behaviors, in this case socially desirable behaviors. A recent document by the OECD (2011) suggests that for environmentally friendly household practices to thrive, appropriate government incentives would be particularly helpful. And then there is "choice editing" (NCC, 2006), a concept much mentioned in debates around sustainable consumption. It is an approach that presents the opportunity to create tight articulation between producer and consumer values (i.e., where a producer understands their market thoroughly). The consumer finds certain choices removed, but these are choices they would not, in any case, wish to make. In the UK, until a few

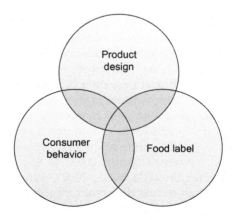

FIGURE 16.5   Three perspectives on innovation for sustainable consumption.

decades ago, chickens were usually sold with their giblets packed into the bird's cavity, and pork chops with kidneys attached. These choices have now been almost completely edited out, and this must surely have made a positive impact in terms of household waste.

If less wasteful consumption is to be promoted successfully by government and industry, effective labeling is a key requirement. Labels are designed primarily to inform, but they also serve to cue appropriate consumer behavior at the point of purchase or use. Date marks inform consumers about the shelflife and storability of a food at the point of sale, and about its quality/safety in the use phase. In the UK, date marking prepackaged food with a "use by" date for perishable foods and a "best before" date for stable foods is a statutory requirement, even though only just over half of UK consumers know what either of the marks means (FSA, 2008). Unfortunately, rather than promoting consumer understanding of these marks, some retailers have undermined the statutory labels by supplementing them with other marks, ostensibly directed towards shop staff rather than consumer facing, for example, "sell by" dates. Defra (2011a) has now issued guidance on the application of date labels to food. Food companies increasingly reference sustainability as a point for differentiation. Many are taking steps to establish and reduce their carbon footprint and they have also been assessing their sourcing strategies. In some cases, this has led to the elimination of ingredients linked to deforestation, including beef, soy, and palm oil that are directly linked to deforestation (Garnett, 2011). Value-based labeling (VBL) communications are increasingly deployed to signal and build trust in some of the credence attributes of food. Such communications create value for consumers as demonstrated by McEachern and Schröder (2004) for the fresh-meat sector in the UK, where VBL fall into several categories such as producer-led, independent, and retailer-led. The level of quality assurance inherent in VBL varies and, in combination with the proliferation of such labels, this introduces yet another element to the many barriers to effective behavioral change. Government guidance is increasingly available for consumers. Defra's (2007) "A Shopper's Guide to Green Labels" includes food labels (LEAF Marque, Marine Stewardship Council logo), organic labels, "the wider world" (FAIRTRADE mark, Rainforest Alliance certified logo), eco-labels, and symbols with a specific meaning such as the Mobius Loop indicating recyclability. A comprehensive review and analysis of global approaches to the environmental labeling of food has recently been published for Defra by the University of Hertfordshire (2010). For people to successfully adopt low-carbon lifestyles, they need to be able to understand how the different aspects of their consumption contribute to their personal and household footprints. UK-based food retailer Tesco (2011:28) committed in 2009 to "help our customers to halve their carbon footprints by 2020". To this end, the company developed a "universal carbon footprint label that describes the emissions associated with each product". To date, they have carbon labeled 525 products in the UK. Other retailers are undertaking similar initiatives. Footprint labels cleverly combine informational and emotional cues; quantification of impact sits alongside a simple but powerful symbol and reminder to "walk lightly" on the earth, not leaving a footprint or, at worst, only a very shallow and temporary one.

It would appear that human ingenuity may have always been able to conceptualize the by-products from food processes as resources rather than wastes, thus redefining them as co-products. Such co-products have, on occasion, overtaken the original core product in value terms, especially where the line between food and pharmaceutical has been crossed. The UK Institute of Food Research (IFR, 2011) is actively engaged in this kind of innovation, which focuses on the "total transformation of food-processing-derived plant-based organic waste products". In fact, wastes from fruit and vegetable processing are prime candidates for this, not only because of the high-value components present, but also because these components would find favor with consumers and would be unlikely to be met with neophobia. A particularly promising class of components would be bioactive phytochemicals, such as fruit pigments, which could then be sold as fine chemicals (Pap et al., 2004). These could be incorporated into functional foods, for example "functional" snack bars. This picture in some respects mirrors the history of whey utilization. For many years, surplus cheese whey was commonly fed to pigs and, based on this approach, complementary industries became established in various geographic areas. This arrangement survives today, with Parma ham and Parmigiano-Reggiano cheese, two prime food products whose designation is protected in law (Parma Consortium, 2007). Modern processing methods have enabled the production of whey powders, flavoured beverages, and other conventional foods, and whey proteins have been retained in cheese curds (Hinrichs, 2001). However, more sophisticated and profitable approaches sought to maximise consumer benefits from the specific qualities of whey proteins. Processes were developed that isolated whey proteins and these isolates are being used as ingredients in the formulae of niche, high-value, "functional" foods as illustrated by infant formulae and sports supplements (Hoffman and Falvo, 2004; Marshall, 2004; NESTLÉ BABY, 2011). Whey proteins are also under consideration as coatings for food packaging materials, where one of their desired properties is biodegradability (Schmid et al., 2011). The technology of protein recovery and utilization from wastes such as blood and offal has also been known for some time. For example, proteins may be spun into fibers for the production of meat analogues (Ledward and Lawrie, 1984). However, in common with much food research conducted up until the early 1980s, little attention was given to what consumers might regard as palatable. The focus tended to be on product performance rather than provenance. This attitude was fundamentally challenged as a series of food scares unfolded from the late 1980s. The most notable of these was of course BSE, where low-grade and unsafe meat, sometimes mechanically recovered, had entered the food chain via diseased cattle (Smith et al., 1999).

The UK government identified the less popular cuts of meat, including offal, as a target for innovation in terms of the upgrading of "waste" (MAFF, 1999). Offal remains popular in certain "cultural" dishes, in particular, haggis in Scotland, but few consumers would want to know too much about its recipe—a degree of mystique will be appreciated as part of the ritual, and perhaps the spiritual connection to the ancestors plays some role in overcoming negative feelings when eating such foods. Reformed meat using enzyme transglutaminase to cause pieces of meat to adhere together is permitted according to the EU-Ökoverordnung but is not acceptable to German organic food producers such as Bioland and Demeter (Anon, 2011). However, animal by-products (ABPs), such as animal carcasses, parts of carcasses, or products of animal origin not intended for human consumption must by law be disposed of safely, whether as feed, fertilizer, or technical products (Defra, 2011b). The Animal By-Products Regulations (ABPR) also apply to the disposal of dead fish and shellfish. The mycoprotein-based meat substitute Quorn is of interest to the current debate for a number of reasons. It is a highly novel food yet did not encounter much, if any, neophobia, and it is a food that does not generate waste in the way that meat production does. Building on the success of Quorn since its introduction to the vegetarian market in 1994, Finnigan (2010) suggests that a better understanding of the potential contribution of mycoprotein is needed, both in nutritional and environmental terms. LCA shows that Quorn mince may be 5—10 times lower in its embedded GHG content than meat and similarly better in terms of its water footprint.

Packaging innovation is relevant to food waste reduction in a number of ways. Its role as carrier of food labeling has already been mentioned, and the most important role that packaging has for any food in transit is, of course, that it protects that food from damage and deterioration. As it is an integral part of any food product, the type and amount of packaging used will also influence the perception that the consumer has of the food itself. A food that is being marketed as sustainable in some way therefore requires suitable packaging in order to preempt the possibility of any cognitive dissonance being caused to consumers. Indeed, there may be an expectation to find certain claims, such as "suitable for composting", "biodegradable", as well as the appropriate recycling label (BRC, 2009). While packaging contributes to a food's LCA, Butler (2011) states that, in sustainability terms, the environmental impact of packaging is minimal compared with the food it protects. Nevertheless, the voluntary Courtauld Commitment 2 targets have been widely adopted (FDF, 2011). These require the UK food supply chain to reduce weight, increase recycling rates, and increase the recycled content of all grocery packaging. However, sustainable food packaging presents a major topic in its own right and has to remain outside the scope of this chapter. A packaging research listing, available from WRAP (2011b), illustrates current activity in the field. Butler (2011) argues that better and smarter food-saving packaging would help to reduce consumer food waste: for example, portion-controlled packaging, re-sealable packaging, and "easy out" packaging for viscous sauces and extended shelf life, for example modified atmosphere packaging (MAP). Finally, a widespread practice and one very much in the spotlight is "bundling", especially when it comes in the format of offers, such as "buy-one-get-one-free" (BOGOF). UK shoppers are divided on the possible effects of BOGOF offers. While nearly three in ten think that BOGOFs create food waste, nearly a quarter use them to try new products (IGD, 2009).

5. CONCLUSIONS

Inefficiencies in the food supply chain today lead to significant food waste, and this implies wasted resources in producing such food. Households are by far the weakest link in this chain, creating the overwhelming portion of the overall waste. However, while households regularly throw away large amounts of food, it is still uncertain whether such behavior is entirely irrational. The need for a more sophisticated understanding of what causes household food waste therefore remains and should be pursued so that behavioral innovations have a better chance of bearing fruit. Consumers also have an important role as adopters of product innovation. But it is for the food industry to ensure that the consumer voice is accurately captured and translated when product developments are undertaken. This applies both to the product *per se* and to any product communications, especially labels.

References

Addy, R., 2011. Food firms wage war on waste. Food Manuf. Available at: <www.foodmanufacture.co.uk/content/view/print/559253> (accessed 28.09.11.).

Ajzen, I., 1985. From intentions to actions: a theory of planned behaviour. In: Kuhland, J., Beckman, J. (Eds.), Action Control: From Cognition to Behaviour. Springer Verlag, Heidelberg, pp. 11–39.

Ajzen, I., Fishbein, M., 1980. Understanding Attitudes and Predicting Social Behaviour. Prentice-Hall, Inc, Englewood Cliffs, NJ.

Allan, T., 2011. Virtual Water: Tackling the Threat to Our Planet's Most Precious Resource. I.B. Tauris, London.

Anon, 2009. Ghent goes veggie once a week. Food&Drink. Available at: <www.foodanddrinkeurope.com/content/view/print/246772> (accessed 16.05.11.).

Anon, 2011. Bio-verbände. Kein klebefleisch. ÖKO-TEST 2011, 35.

Ansoff, I.H., 1957. Strategies for diversification. Harv. Bus. Rev. 35, 113–124.

Bauman, Z., 2007. Consuming Life. Polity, Cambridge.

Bettman, J., Luce, M., Payne, J., 2006. Constructive consumer choice processes. In: Lichtenstein, P., Slovic, P. (Eds.), The Construction of Preference. Cambridge University Press, Cambridge, pp. 323–341.

Bianchi, M., 1998. Introduction. In: Bianchi, M. (Ed.), The Active Consumer. Novelty and Surprise in Consumer Choice. Routledge, London, pp. 1–18.

BRC (2009). On-Pack Recycling Label. Available at: <www.oprl.org.uk> (accessed 30.06.11.).

Butler, P., 2011. Smart packaging solutions to food waste reductions. Food Sci. Technol. Today 25, 36–38.

Buzby, J.C., Guthrie, J.F., 2002. Plate Waste in School Nutrition Programs: Final Report to Congress. Economic Research Service/USDA (E-FAN-02-009, March 2002).

Cabinet Office, 2008. Food Matters: Towards a Strategy for the 21st Century. Cabinet Office, London.

Cabinet Office, 2010a. MINDSPACE: Influencing Behaviour through Public Policy. Cabinet Office and Institute for Government, London.

Cabinet Office, 2010b. Applying Behavioural Insight to Health. Cabinet Office, London.

Cabinet Office, 2011. Better Choices: Better Deals. Consumers Powering Growth. Cabinet Office, London.

Casimir, G.J., Tobi, H., 2011. Defining and using the concept of household: a systematic review. Int. J. Consum. Stud. 35, 498–506.

Chatzidakis, A., Hibbert, S., Smith, A.P., 2007. Why people don't take their concerns about fair trade to the supermarket: the role of neutralization. J. Bus. Ethics. 74, 89–100.

Chin, K.-S., Lam, J., Chan, J.S.F., Poon, K.K., Yang, J., 2005. A CIMOSA presentation of an integrated product design review framework. Int. J. Computer Integr. Manuf. 84, 260–278.

Co-operative Bank, 2010. Ethical Consumption Report 2010. Ethical Shopping through the Downturn. Co-operative Bank, Manchester.

Defra, 2007. A Shopper's Guide to Green Labels. Understanding Environmental Labels on Products. Defra, London.

Defra, 2008. "The Milk Roadmap". Sustainable Consumption & Production Taskforce, Dairy Supply Chain Forum. Defra, London.

Defra, 2010. Household Food and Drink Waste Linked to Food and Drink Purchases. Defra, London.

Defra, 2011a. Guidance On the Application of Date Labels to Food. Defra, London.

Defra, 2011b. Control of Animal By-products. Available at: <http://defra.gov.uk/food-farm/byproducts/> (accessed 09.09.11.).

Department of Health, 2003. DoH Five-a-Day Programme. Department of Health, London.

Drucker, P.F., 2007. The Essential Drucker. Elsevier, Oxford.

DTI, 2003. Innovation Report. Competing in the Global Economy: The Innovation Challenge. DTI, London.

EU, 2008a. Directive 2008/98/EC of the European Parliament and of the Council of 19 November 2008 on Waste and Repealing Certain Directives, Official Journal of the European Union L 312/3.

EU, 2008b. Food: From Farm to Fork Statistics, Eurostat Pocketbooks. European Commission, Brussels.

EU, 2010. OPEN: EU Scenario Scoping Report. One Planet Economy Network: EU. European Commission, Brussels.

EU, 2011. A Resource-Efficient Europe—Flagship Initiative Under The Europe 2020 Strategy, Com(2022)21 Final. European Commission, Brussels.

Fairtrade Foundation, 2011. Sales of Fairtrade certified products in the UK. Available at: <www.fairtrade.org.uk/what_is_fairtrade/facts_and_figures/> (accessed 23.02.11.).

FAO, 2006. World Agriculture: Towards 2030/2050—Interim Report. Global Perspectives Studies Unit, Food and Agriculture Organisation of the United Nations, Rome.

FDF, 2011. Reducing Packaging. Food and Drink Federation, London. Available at: <www.fdf.org.uk/environment/reduce_packgaing.aspx> (accessed 16.05.11.).

Featherstone, M., 2007. Consumer Culture and Postmodernism, second ed. Sage, London.

Festinger, L., 1957. A Theory of Cognitive Dissonance. Stanford University Press, California.

Finnigan, T.J.A., 2010. Food 2030. Life Cycle Analysis and the Role of Quorn Foods within the New Fundamentals of Food Policy, Summary Document. January 2010. Available at: <www.mycoprotein.org/assets/timfinnigan2030.pdf> (accessed 27.09.11.).

Foresight, 2011. The Future of Food and Farming: Challenges and Choices for Global Sustainability. Final Project Report. Government Office for Science, London.

Friel, S., Dangour, A.D., Garnett, T., Lock, K., Chalabi, Z., Roberts, I., et al., 2009. Public health benefits of strategies to reduce greenhouse-gas emissions: food and agriculture. Lancet 374, 2016–2025.

FSA, 2008. Consumer Attitudes to Food Standards. Wave 8. Food Standards Agency, UK.

Garnett, T., 2008. Cooking up a Storm: Food, Greenhouse Gas Emissions and Our Changing Climate. Food Climate Research Network, Centre for Environmental Strategy, University of Surrey.

Garnett, T., 2011. Where are the best opportunities for reducing greenhouse gas emissions in the food system (including the food chain)? Food Policy 36, S23–S32.

Gerwin, D., Barrowman, N.J., 2002. An evaluation of research on integrated product development. Manag. Sci. 48, 938–953.

Gigerenzer, G., 2000. Adaptive Thinking. Rationality in the Real World. Oxford University Press, Oxford.

Gladwell, M., 2002. The Tipping Point. How Little Things Can Make a Big Difference. Little, Brown and Company, Boston.

Green Alliance, 2011. Designing out waste consortium. Available at: <www.green-alliance.org.uk/consortium2/> (accessed 07.10.11.).

Hart, S.L., 2007. Beyond greening: strategies for a sustainable world. Harv. Bus. Rev. Green Bus. Strategy, 99–123.

Helm, D., 2009. EU Climate-Change Policy—A Critique, Smith School Working Paper Series. University of Oxford, Oxford.

Hennel, M., 2011. Food manufacturers turn to demand forecasting to gain a competitive edge. Food Manuf. Available at: <www.food-manufacturing.com/scripts/ShowPR~RID~7171.asp> (accessed 11.11.11.).

Hertwich, E.G., 2005. Life cycle approaches to sustainable consumption: a critical review. Environ. Sci. Technol. 39, 4673–4684.

Hertwich, E.G., Peters, G.P., 2009. Carbon footprint of nations: a global, trade-linked analysis. Environ. Sci. Technol. 43, 6414–6420.

Hinrichs, J., 2001. Incorporation of whey proteins in cheese. Int. Dairy J. 11, 495–503.

Hoffman, J.R., Falvo, M.J., 2004. Protein—which is best? J. Sports Sci. Med. 3, 118–130.

Hofmeister, K.R., 1991. Quality function deployment: market success through customer-driven products. In: Graf, E., Saguy, I.S. (Eds.), Food Product Development. From Concept to the Marketplace, pp. 189–210.

Holbrook, M.B., 1999. Introduction to consumer value. In: Holbrook, M.B. (Ed.), Consumer Value: A Framework for Analysis and Research. Routledge, London, pp. 1–28.

IFR, 2011. Sustainability in the Food Chain. Research. Institute of Food Research, Norwich. Available at: <www.ifr.ac.uk/SFC/research/default.html> (accessed 26.09.11).

IGD, 2007. Supply Chain Food Waste. Institute of Grocery Distribution, Watford.

IGD, 2009. Promotions and Customer Loyalty. Institute of Grocery Distribution, Watford.

Ironmonger, D., 2000. Household Production and the Household Economy. Household Research Unit, University of Melbourne. Available at: <http://www.economics.unimelb.edu.au/Household/Papers/2001/Household%20Production%20Dept%20of%20Eco%20Research%20Paper.pdf> (accessed 04.08.11).

Jack, F.R., O'Neill, J., Piacentini, M.G., Schröder, M.J.A., 1997. Perception of fruit as a snack: a comparison with manufactured snack foods. Food Qual. Prefer. 8, 175–182.

Jackson, T., 2006. Readings in sustainable consumption. In: Jackson, T. (Ed.), The Earthscan Reader in Sustainable Consumption. Earthscan, London, pp. 1–23.

Jones, A., Nesaratnam, S., Porteous, A., 2008. Non-Municipal Waste Disposal Routes. The Open University, Milton Keynes, Factsheet No. 9.

Just, D., Mancino, L. and Wansink, B., 2007. Could Behavioural Economics Help Improve Diet Quality for Nutrition Assistance Program Participants? United States Department of Agriculture, Economic Research Service, Economic Research Report Number 43.

Kotler, P., Lee, N., 2006. Marketing in the Public Sector. A Roadmap for Improved Performance. Wharton School Publishing, Upper Saddle River, NJ.

Ledward, D.A., Lawrie, R.A., 1984. Recovery and utilisation of by-product proteins of the meat industry. J. Chem. Technol. Biotechnol. 34, 223–228.

Lehtinen, U., Torkko, M., 2005. The lean concept in the food industry: a case study of a contract manufacturer. J. Food Distr. Res. 36, 57–67.

Lewin, S.B., 1996. Economics and psychology: lessons for our own day from the early twentieth century. J. Econ. Lit. XXXIV, 1293–1323.

Macdiarmid, J., Kyle, J., Horgan, G., Loe, J., Fyfe, C., Johnstone, A., et al., 2011. Livewell: A Balance of Healthy and Sustainable Food Choices. WWF-UK, Godalming.

MAFF, 1999. Working Together for the Food Chain: Views from the Food Chain Group. Ministry of Agriculture, Fisheries and Food, London.

Marshall, K., 2004. Therapeutic applications of whey protein. Altern. Med. Rev. 9, 136–156.

McEachern, M.G., Schröder, M.J.A., 2004. Integrating the voice of the consumer within the value chain: a focus on value-based labelling communications in the fresh-meat sector. J. Consum. Mark. 21, 497–509.

Munasinghe, M., Dasgupta, P., Southerton, D., Bows, A., McMeekin, A., 2009. Consumers, Business and Climate Change. Report by the Sustainable Consumption Institute at the University of Manchester, UK.

National Geographic, 2010. Greendex 2010: Consumer Choice and the Environment—A Worldwide Tracking Survey. National Geographic, US.

NCC, 2006. I Will If You Will. Towards Sustainable Consumption. National Consumer Council, UK.

NESTLÉ BABY, 2011. Explaining formula: make every start—the best start. Available at: <www.nestle-baby.ca/> (accessed 09.09.11).

Nuffield Council on Bioethics, 2007. Public Health: Ethical Issues. Nuffield Council on Bioethics, London.

Oakdene Hollins, 2010. Mapping Waste in the Food and Drink Industry. A report for Defra and The Food and Drink Federation.

OECD, 2002. Towards Sustainable Household Consumption? Trends and Policies in OECD Countries. OECD.

OECD, 2011. Greening Household Behaviour: The Role of Public Policy. OECD.

ONS, 2010. Living Cost and Food Survey 2009. Office for National Statistics, London.

Oxfam, 2009. 4-a-week, changing food consumption in the IL to benefit people and planet. Oxfam GB Briefing Paper.

Pap, N., Pongrácz, E., Myllykoski, L., Keiski, R., 2004. Waste minimization and utilization in the food industry: processing of arctic berries, and extraction of valuable compounds from juice-processing by-products. In: Pongrácz, E. (Ed.), Proceedings of the Waste Minimization and Resources Use Optimization Conference. Oulu University Press, University of Oulu, Finland, pp. 159–168.

Parma Consortium, 2007. Available at: <www.prosciuttodiparma.com/eng/info/pigs/> (accessed 11.11.11).

Porter, M.E., 1985. Competitive Advantage. Creating and Sustaining Superior Performance. The Free Press, New York.

Prelec, D., Herrnstein, R., 1991. Preferences or principles: alternative guidelines for choice. In: Zeckhauser, R. (Ed.), Strategy and Choice. MIT Press, Cambridge, MA, pp. 319–340.

Rodriguez, J.P., Beard, T.D., Bennett, E.M., Cumming, G.S., Cork, S.J., Agard, J., et al., 2006. Trade-offs across space, time, and ecosystem services. Ecol. Soc. 11, 28.

Rogers, E.M., 2003. Diffusion of Innovations. Free Press, New York, fifth ed.

Schmid, M., Held, J., Wild, F., Noller, K., 2011. Thermoforming of whey protein-based barrier layers for application in food packaging. Food Sci. Technol. 25, 34–35.

Schröder, M.J.A., 2003. Food Quality and Consumer Value. Delivering Food That Satisfies. Springer, Berlin.

Schröder, M.J.A., McEachern, M.G., 2004. Consumer value conflicts surrounding ethical food purchase decisions: a focus on animal welfare. Int. J. Consum. Stud. 28, 168–177.

Schwartz, B., 2004. The tyranny of choice. Sci. Am. 290, 70–75.

Sen, A., 1999. Development as Freedom. Oxford University Press, Oxford.

Smith, A.P., Young, J.A., Gibson, J., 1999. How now, mad cow? Consumer confidence and source credibility during the 1996 BSE scare. Eur. J. Mark. 33, 1107–1122.

Stern, N., 2007. The Economics of Climate Change: The Stern Review. Cambridge University Press, Cambridge.

Sykes, G.M., Matza, D., 1957. Techniques of neutralization: a theory of delinquency. Am. Soc. Rev. 22, 664–670.

Szmigin, I., Carrigan, M., McEachern, M.G., 2009. The conscious consumer: taking a flexible approach to ethical behaviour. Int. J. Consum. Stud. 33, 224–231.

Tesco, 2011. Tesco PLC Corporate Responsibility Report 2011. Tesco, UK.

Thaler, R., Sunstein, C., 2008. Nudge: Improving Decisions about Health, Wealth and Happiness. Yale University Press, London.

Thøgersen, J., Ölander, F., 2003. Spillover of environment-friendly consumer behaviour. J. Environ. Psychol. 23, 225–236.

UN, 1945. Charter of the United Nations and Statute of the International Court of Justice. San Francisco.

UN, 1972. Declaration of the United Nations Conference on the Environment. Stockholm.

UN, 1992. Declaration of the United Nations Conference on Environment and Development. Rio de Janeiro.

UN, 1998. Kyoto Protocol to the United Nations Framework Convention on Climate Change. United Nations.

UN, 2005. Johannesburg Plan of Implementation. UN Department of Economic and Social Affairs, Division for Sustainable Development.

UN, 2010. Trends in Sustainable Development, Towards Sustainable Consumption and Production. UN Department of Economic and Social Affairs, Division for Sustainable Development.

UN, 2011. Durban Platform for Enhanced Action. United Nations.

University of Hertfordshire, 2010. Effective approaches to environmental labelling of food products. Appendix A: Literature review report. University of Hertfordshire, Report for Defra 15 November 2010.

WCED, 1987. Our Common Future, Report of the United Nations World Commission on Environment and Development (Brundtland Report).

Which, 2005. In-store tricks. Which? August 2005, pp. 12-13.

Wiltshire, J., Tucker, G., Fendler, A., 2008. Carbon footprint of British food production. Food Sci. Technol. Today 22, 23–26.

Worldwatch Institute, 2010. State of the World 2010. Transforming Cultures: From Consumerism to Sustainability. Washington DC.

WRAP, 2007. We Don't Waste Food! A Householder Survey. WRAP, Banbury.

WRAP, 2008. Helping Consumer Reduce Fruit and Vegetable Waste: Final Report. WRAP, Banbury.

WRAP, 2009. Household Food and Drink Waste in the UK. WRAP, Banbury.

WRAP, 2010a. Waste Arisings in the Supply of Food and Drink to Households in the UK. WRAP, Banbury.

WRAP, 2010b. Reducing Food Waste Through the Chill Chain. WRAP, Banbury.

WRAP, 2011a. The Water and Carbon Footprint of Household Food and Drink Waste in the UK. WRAP, Banbury.

WRAP, 2011b. Packaging Research Listing. WRAP, Banbury.

WWF, 2010. Living Planet Report: Biodiversity, Biocapacity and Development. WWF International.

Concluding Remarks and Future Prospects

Maria R. Kosseva and Colin Webb

1. PREVENTION OF FOOD LOSSES AND WASTE

Food waste is a key environmental, social, and economic issue. Growth of consumption and wealth has put increased pressure on waste management and prevention strategies that aim to reduce the negative effects on the ecosystem and public health. Recent EU policy instruments and strategies in this area, such as the revised Waste Directive (2008/98/EC), the Thematic Strategy on the Prevention and Recycling of Waste, and the 6th Environmental Action Program (EAP), prioritize waste prevention and decoupling of waste generation from economic development and environmental impacts (Fischer and Kjaer, 2012).

A transition to sustainable consumption requires a strong effort to develop a cross-disciplinary agenda of research, linking together agricultural, environmental, social, and health concerns under the principle of sustainability and also applying them to daily consumption practices. Cross-disciplinary frameworks should look at the complexity of food production and consumption to elucidate invisible links, possible synergies, and key points for change. They may foster innovation from society itself to encourage good practices and facilitate their diffusion through support to learning processes, institutional/regulatory facilitation, and technology transfer. Lines of research could apply the criteria of sustainability to lifestyles, diets, consumption technologies, and food provision systems. They could also provide enterprises with necessary knowledge to redesign products and processes in a sustainable way and to communicate relevant messages to consumers and to the public (EC SCAR, 2011). The EC's Roadmap to a Resource Efficient Europe (EU, 2011) outlines a vision for 2050 and milestones for 2020. One of the main themes is in the area of "Treating waste as a resource". With growing global food demand and escalating food prices, the European target of 50% reduction of food waste by 2020 is of key importance in ensuring global food security and good environmental control. There is a pressing need to find new ways of reducing waste throughout the food supply chain. Novel research on how to minimize food wastes during food processing, transport, and retail is required, while options for recycling and reuse (e.g., composting and bioenergy) should also be further explored while taking into consideration health issues (FACCE, 2010).

In this context, many OECD countries recognize the importance of knowledge networks and have extensively embedded the support of such networks in their innovation policy. Today's environmental challenges require a new approach to policy making that fosters eco-innovation through partnerships and collaboration (OECD, 2009). The EC has several initiatives for establishing platforms and networks composed of expert stakeholders to help fulfill the Lisbon Strategy objectives of "a competitive Europe". For example, European Technology Platforms bring together stakeholders led by industry to define medium to long-term research and technological development objectives and to better align EU research priorities with industrial needs. In this frame, the draft Strategic Research and Innovation Agenda of the European Technology Platform *Food for Life* addresses the topic of FIW prevention as one of the essential issues required to achieve a sustainable food system (ETP, 2011).

A further example in the USA, The Green Suppliers' Network, was established by the EPA. Their "lean and clean" initiative aims at eliminating non-value-added activity to drive down costs and improve efficiency in the manufacturing process.

1.1 EU Measures to Reduce Food Industry Waste

Within the EU, management of FIW involves several policy areas including sustainable resource management, climate change, energy, biodiversity, habitat protection, agriculture, and soil protection. However, significant limitations accompany the work on FIW quantification, as there is a lack of robust data on how much food is wasted across the food supply chain. Methods for collecting and calculating the food waste data submitted to EUROSTAT differ among Member States (MS). Limitations in the reliability of EUROSTAT data, due to a lack of clarity on the definition and method, may be significant. According to the

EC FP7 Theme 2 work program 2012, improving the FIW reporting requirements at EU and MS level is seen as an essential step for the prevention of food wastage (EC, 2012). It will also enable the establishment of policy initiatives aiming at coherent food safety and hygiene regulation, labeling (best-before date), food distribution, and awareness/education campaigns to all players involved. There is no comprehensive EU-wide database on the advantages, drawbacks, and total cost of local versus global food production and supply systems, as recently stated in the orientation paper of the EC FP7 work program 2012, whose main objective is: "Building a European Knowledge Based Bio-Economy by bringing together science, industry and other stakeholders, to exploit new and emerging research opportunities that address social, environmental and economic challenges: the growing demand for safer, healthier, higher quality food and for sustainable use and production of renewable bio-resources, . . . etc.".

Although FIW constitutes a large proportion of bio-waste, the first overview of the situation in the European Union was published only recently (EC, 2010). Urgent actions are needed to reduce food waste particularly in the household sector (comprising ~42%

of total EU27 food waste). Improvement is needed in all the MS to provide reliable statistical data on FIW generation. The production and consumption of food products has shifted over the last thirty years as a result of rising per capita incomes, lifestyle changes, and demographic shifts, such as an increase in single-person households. Concentration and competition in the international food market has driven changes in the variety and availability of food products. Attitudes towards food safety, product labeling, and the impact of food consumption on the environment have had a broader impact (EC, 2010). The EC used typical FIW prevention measures, illustrated in Figure 1, as a part of the continuous improvement initiatives required to solve food waste problems.

The same EU document reported that manufacturing food wastes in the EU27 are estimated at almost 39% of the total. Thus, the food-manufacturing sector represents another target where reduction of discharged largely unavoidable and technically malfunctional products should be addressed. Manufacturing is a driver of productivity, innovation, R&D, and technological change. For the manufacturing and processing industry, innovation is seen as the key to future success.

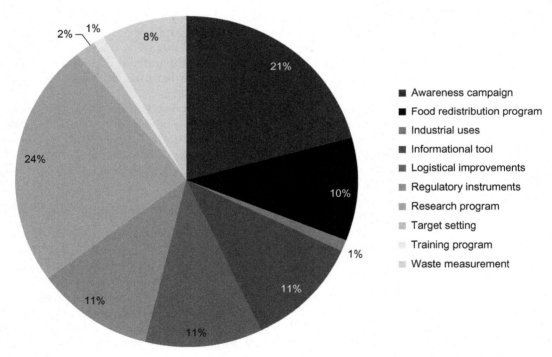

FIGURE 1　**Types of initiatives used to prevent food waste.** FW prevention initiatives were identified via literature research and through a stakeholder questionnaire. Typical initiatives and their percentage of total efforts are research program (development of specific prevention methodologies) (24%); awareness campaigns (e.g., WRAP's "Love Food Hate Waste") (21%); informational tools (e.g., sector-specific prevention guidelines) (11%); logistical improvement (e.g., stock management improvements) (11%); regulatory measures (e.g., requirement for separate collection of FW in Ireland) (11%); food redistribution programs to charitable groups (10%); waste measurement activity (8%); target setting (2%); training program (1%); development of industrial uses—turning FW into valuable products (1%). *Source: EC (2010).*

2. CHALLENGES FOR THE PROCESSING INDUSTRY

Society requires a sustainable industrial development framed by environmental responsibility, renewable energy use, and higher energy efficiency. These requirements are expressed in more stringent legislation regarding waste production, CO_2 emissions, and air and water pollution. On the other hand, economic pressures are escalating. An unstable economic situation is forcing industry to be more flexible and to be able to change production capacities quickly. Furthermore, frequent variations in feeds compositions, together with fast and sometimes unpredictable price oscillations, place much more emphasis on process dynamics. Combined, all the listed expectations are raising many scientific and technical issues for modern process technologies and their control (Nikačević et al., 2012). Looking at the chemical engineering success in coping with the challenges in the process industry, one could identify process intensification (PI) concepts as the most promising (Dudukovic, 2009), especially when linked with green chemistry (GC), product design, and process systems engineering (PSE). One could apply a similar approach to the food-processing industry.

Current societal requirements and the state of the environment call for accelerated development of more sophisticated and efficient process systems. Process intensification concepts have offered significant advantages in material and energy efficiency in diverse process applications, up to date mostly by using PI methods in the design of novel types of equipment. Further advances in economic and environmental performance can be achieved if operation and control of a whole process are considered simultaneously and systematically together with different PI design options (Nikačević et al., 2012).

Within the financial state of the process industry, to support "value preservation" and "value growth", three major future research challenges need to be addressed: product discovery and design, enterprise and supply chain optimization, and global life cycle assessment. One motivating opportunity might be process intensification as a way to improve industrial plants. Another opportunity could be stronger interaction between product and process design as part of a life cycle analysis of materials used (Grossmann, 2004).

2.1 Product Discovery and Design

Currently, traditional process design is expanding to include molecular product design in order to move towards the molecular level (Grossmann, 2004). A major challenge that remains is the need to develop more accurate predictive capabilities (e.g., applying molecular simulation models) for properties of compounds in order to apply optimization methods. At the macroscopic level, the emphasis in research is on the linking of new products to market needs and the systematic exploration of alternatives for developing new products, which normally must be accomplished in multidisciplinary teams composed of scientists, engineers of other disciplines, and business people (Cussler and Moggridge, 2001). An interesting problem here that has not received enough attention is the integration of product and process design. Process design will still involve significant research challenges. The interaction between design and control also continues to attract attention. Design and control of batch and biological processes have also become more important. Another related challenge is the area of process intensification that requires the discovery of novel unit operations that integrate several functions and that can potentially reduce the cost and complexity of process systems (Stankiewicz and Moulijn, 2000).

At the process level, significant progress has been made in the synthesis and optimization of water networks. Progress has also been made to better understand the implications of waste at the level of the synthesis and analysis of a process flow sheet. Little work, however, has been done to assess environmental implications at the level of product design and the integration with processing. More importantly, however, is the need to adopt broader approaches to the life cycle assessment of products and processes in order to predict more accurately their long-term sustainability (Grossmann, 2004).

Life cycle assessment (LCA), now an ISO-standardized methodology, is an important example of the effort to expand the traditional process boundary (Bakshi and Fiksel, 2003). LCA considers both the upstream and downstream processes associated with a given product in terms of energy use, material use, waste generation, and business value creation. However, careful consideration of life cycle implications can sometimes yield surprising results. For example, efforts to develop green plastics, such as polylactides, seem appealing because the feedstocks are renewable and the plastics are biodegradable. But it turns out that the extraction and processing of the plastics is extremely energy intensive, and the biological breakdown of the plastics releases greenhouse gases (Gerngross and Slater, 2000).

2.2 Sustainability and Eco-Innovation

The scope of sustainability is so broad that it requires integration of many disciplines including engineering, biology, medicine, economics, law, ethics, and social sciences. Sustainable development is among the most pressing and urgent challenges facing humanity today. The need for research and education to meet this challenge has been identified in virtually every recent study on engineering research needs (NRC-BCST, 2003; Pfister and AC-ERE, 2003). Achieving sustainability requires a new generation of engineers who are trained to adopt a holistic view of processes as embedded in larger systems. Engineering can no longer be performed in isolation and must consider interactions among industrial processes and human and ecological systems (Bakshi and Fiksel, 2003). Fiksel (2003) revealed the major pathways whereby sustainability contributes to shareholder value.

Pursuit of sustainability has resulted in the flourishing of a variety of innovative business practices. The following are examples of sustainable business practices that simultaneously benefit both an enterprise and its stakeholders (Bakshi and Fiksel, 2003).

- *Design for Sustainability* is manifested via development of green chemical routes, process intensification, and process redesign (Rittenhouse, 2003).
- *Eco-efficient Manufacturing* focuses on reducing the "ecological footprint" of a company's operations, including the inputs of materials and natural resources, including water and energy required to manufacture and deliver a unit of output (Verfaillie and Bidwell, 2000).
- *Industrial Ecology* is more than a practice—it is a framework for shifting industrial systems from a linear model to a cyclical model that resembles the flows of natural ecosystems, which some have called "biomimicry" (Benyus, 1997). However, the available tools for systematic design of industrial ecology networks are still in their infancy (Allen and Butner, 2002).

Many companies are now considering the environmental impact throughout the product's lifecycle and are integrating environmental strategies and practices into their own management systems. Some pioneers have been working to establish a closed-loop production system that eliminates final disposal by recovering wastes and turning them into new resources for production. Eco-innovation helps to make possible this kind of evolution in industry practices (OECD, 2009).

2.3 A Nature-Inspired Engineering Approach

Nature-inspired engineering researches the fundamental mechanism underlying a desired property or function in nature, most often in biology, and applies this mechanism in a technological context. In the framework of chemical engineering, we call this approach nature-inspired chemical engineering (NICE) (Coppens, 2009).

NICE aims to innovate, guided by nature, but it does not mimic nature and should be applied in the right context. Emphasizing reactor and catalysis engineering, it illustrates how mechanisms used in biology to satisfy complicated requirements, essential to life, are adapted to guide innovative solutions to similar challenges in chemical engineering. These mechanisms include: (1) use of optimized, *hierarchical networks* to bridge scales, minimize transport limitations, and realize efficient, scalable solutions; (2) careful *balancing of forces* at one or more scales to achieve superior performance, for example, in terms of yield and selectivity, and (3) emergence of complex functions from simple components, using *dynamics as an organizing mechanism* (Coppens, 2012). In this way, NICE complements an ongoing revolution in bio-inspired chemistry and materials science (Ozin et al., 2009).

What makes biological organisms especially interesting from the viewpoint of chemical reaction engineering is that efficiency, scalability, robustness, and adaptability are quintessential to both, yet nature uses an arsenal of tools barely touched in engineering. A tree can be viewed as a photosynthesis reactor, converting carbon dioxide and water into biomass (the growing tree) and oxygen. Biological structures are an excellent source of inspiration for engineering designs that bridge multiple length scales, maintaining efficiency under scale-up. At the smallest scales, the structure is very specific, and dependent on the intrinsic function. At intermediate scales, uniform arrays appear common. At larger scales, fractal interpolation is very powerful to preserve the desired functionality. Other examples of nature-inspired chemical engineering include membranes for separation processes that imitate the key features of protein channels crossing cell walls in order to achieve high flux and selectivity. Helping to create sustainable processes, nature-inspired designs unite the atomistic and the holistic approaches, using efficient mechanisms in natural systems as guidance for artificial designs (Coppens 2009, 2012).

3. VALORIZATION OF FOOD INDUSTRY WASTE

Among the EC initiatives whose implementation represents only 1% of total EC efforts in this area is the industrial uses of otherwise inedible food (as shown in Figure 1). Therefore, in this book we focus on the routes for recovery of valuable commodities and energy locked in food wastes, aiming to expand their industrial applications.

Food waste streams have a vast potential, which has been underestimated so far. Through the identification and isolation of valuable components present in the waste streams, new opportunities are created for both new and existing markets in the food industry, fine chemistry, cosmetics, pharmaceuticals, and others. Bulk components of FIW consist of carbohydrates, such as fibers and sugars, proteins, fats, and oils, while minor components include minerals, nutrients, vitamins, antioxidants, aromas, colorants, and so on. Determining the biowaste constituents is therefore a first step to valorization. Profound analytical expertise and specially designed tools are needed to efficiently differentiate the molecules of interest from the biological matrix. The importance of plants as a source of functional compounds or new drug molecules is illustrated by the fact that in the past 20 years 28% of new drug entities were either natural products or derived from them as semisynthetic derivatives. Usually, organic solvents are used for extraction, but the technique can be converted to a sustainable process by choosing solvents such as water and/or bioethanol with or without additives. The most advanced green extraction technique uses supercritical liquids with, for example, carbon dioxide (CO_2) as a solvent (D'Hondt and Voorspoels, 2012).

Recovery of commodities from agricultural and food waste products can be successfully accomplished using solid-state fermentation (SSF) technology. It provides many novel opportunities as it allows the use of the wastes without need for extensive pretreatment. Moreover, SSF can improve economic feasibility of the biotechnological processes, offering waste reduction in design and operation. The biorefinery presents a promising approach for FIW processing (Botella et al., 2009). The use of SSF in many of the biological processing steps in the biorefinery will help to minimize water use. However, in order to fulfill this potential it will be necessary to have reliable large-scale SSF bioreactors and strategies for optimizing their operation (Mitchell et al., 2011).

Once the high-value compounds are characterized and a process technology selected, it is essential to assess the economic viability. The price and volume of an end product determines the degree of technical complexity that is viable. For example, in the pharmaceutical industry prices can be up to €300 per mg. Natural medicinal drugs with high purity can therefore be produced in high-tech, kg-scale installations. For cosmetics, food, or food supplements, prices are lower, volumes are higher, and the technology is preferably simple. Depending on the degree of purity, or

proven functionality, prices range from €10 to €75 per kg for crude extracts and go up to €700 per kg for functional extracts. Thus a market-research-based business plan is essential. Generally, it is more sustainable to first recuperate as many materials and substances as possible from a waste stream and extract energy only after the economically viable substances have been removed (D'Hondt and Voorspoels, 2012).

3.1 Dry Anaerobic Digestion

The dry AD process (~20% solid content) has been regarded as a sustainable waste recycling approach to treat a wide range of solid feedstocks, including animal wastes, agricultural residues, organic fraction of municipal solid wastes, solid industrial and commercial effluents, energy crops, and sewage sludge. The advantages of the solid state anaerobic digestion process compared with wet AD are higher biogas yield per unit volume of digester, smaller digester, greater organic loading rate, lower energy requirements for heating, limited leachate, reduced nutrient runoff during storage and distribution of residues, and easier handling of digested slurry to farm. However, the process has suffered from having much longer retention time, incomplete mixing, the accumulation of volatile fatty acids, and the requirement of a larger amount of inocula. Thermophilic digestion is regarded as the more efficient method as it offers an enhanced hydrolysis process, shorter retention time, and pasteurization of wastes. The commercially available dry systems prove the capability of this process to effectively convert waste material into energy. Optimal thermophilic condition, inoculation, co-digestion, percolation, and the addition of additives can be used to enhance the process. It is essential to control pH, ammonia, buffering capacity, and volatile fatty acid levels to ensure optimal efficiency and maximize gas yield, while shortening the retention time (Nishio and Nakashimada, 2007; Jha, 2012). Furthermore, the successful mixing of different wastes results in a better digestion performance by improving the content of the nutrients and even reduces the negative effect of toxic compounds on the digestion process.

3.2 Thermophilic Aerobic Bioremediation

Thermophilic bioremediation technology for treatment of high-strength organic wastewaters appears to combine the advantages of low biomass yields and rapid kinetics associated with high-temperature operation and the feasible process control of aerobic systems. It also has the potential to both produce pathogen-free products and generate thermal energy from the process. Furthermore, the average velocity of thermophilic aerobic bioremediation was almost twice as high as that under mesophilic conditions. This promising technology could be extensively used for the recovery of valuable products, such as animal feed, fertilizer, biopolymers, biosurfactants, and organic acids, and calls for further investigation of the opportunities.

The aerobic technologies adapted by many dairy industries for processing of their wastewaters are usually highly energy intensive and may lead to uncertainty regarding stabilized performance due to factors such as overloading and bulking sludge. In contrast, anaerobic technologies are simpler, require a lower budget to operate, and have the potential of producing biogas with a high methane content while utilizing the waste products (Kosseva, 2011a).

3.3 Hydrogen Production

To consolidate the benefits of using hydrogen as a fuel or energy carrier, alternative cleaner processes that rely on renewable feedstock must be developed (Ferchichi et al., 2005). Fermentative Hydrogen Production (FHP) has been reported from numerous waste and wastewater sources, including bean curd manufacturing waste, brewery and bread wastes, rice and wheat bran, rice winery wastewater, molasses and sugary wastewater, waste-activated sludge, municipal solid waste, starch wastewater, food waste from cafeteria, peptone degradation, and lignocellulose materials such as rice straw, coir, and sugar bagasse. These studies have shown that FHP can rely on carbohydrate-rich wastewater and waste as feedstock, thereby providing a prospect of integrating pollution reduction with energy generation. However, not only does the sustainability of FHP depend on the availability of locally abundant renewable feedstock but also the establishment of fermentation conditions that increase both the rate and the yield of hydrogen production from these materials. Thermodynamic and metabolic constraints suggest that it would be impossible to find an organism capable of the complete conversion of sugar-based substrates to hydrogen by fermentation and that human intervention is needed to solve this problem. For a hybrid system using photofermentation, the key questions are about efficient photosynthetic bacteria and materials, developed by scientists, with sufficiently low-cost transparent and hydrogen impermeable photobioreactors. For a hybrid system using MECs[1], a different outstanding question arises: can

MECs be developed that have sufficient current densities, require lower voltages, and use inexpensive cathodes? Solving the outstanding problem of increasing hydrogen yields could lead to the development of systems capable of the complete conversion of waste streams and energy crops to hydrogen, a potential sustainable fuel of the future (Hallenbeck and Ghosh, 2009).

3.4 Fuel Cells

Because of their elegant functional principle, fuel cells (FCs) remain an extremely attractive option for the direct transformation of chemical into electrical energy. Microbial fuel cells (MFCs) have gained a lot of attention in recent years as a mode of converting organic waste, including low-strength wastewaters and lignocellulosic biomass, into electricity. FCs exploiting isolated redox enzymes (an example is the enzyme hydrogenase, which shows similar activity for hydrogen oxidation as platinum) are termed enzymatic FCs (EFCs), as distinct from MFCs, where whole organisms are utilized. The basic working principle of an EFC mimics the cellular respiration of living cells. When designing biomimetic energy conversion systems based on redox enzymes, the following points have to be considered: (1) enzyme immobilization, (2) contact between enzyme and electrode surface, (3) enzyme kinetics, (4) enzyme electrode architecture, and (5) integration of electrodes into the overall system. The first three aspects have been extensively studied in the past, mainly in the framework of the development of biosensors. But not all of these methods can be transferred directly to EFCs, because they will lead to electrode kinetics unfavorable for FC operation. For example, for FC applications, current densities have to be improved significantly. This goal can be achieved possibly by improvement of the electrode structure and also by engineering of the reaction system for improved catalytic efficiency (Sundmacher et al., 2012).

3.5 Progress in Immobilization of Enzymes

The immobilization of enzymes significantly increases their stability and reduces cost; therefore, it is widely pursued as efficient, selective, and environmentally friendly catalysis. Recent achievements regarding the immobilization of enzymes in inorganic mesoporous materials and the modifications of those materials are summarized by Tran and Balkus (2011). Enzymes immobilized in/on fibrous membranes provide high surface area for high-throughput biocatalysis. These membrane bioreactors also allow for

---

[1]MEC (microbial electrohydrogenesis cell): a bioelectrochemical cell, basically a modified microbial fuel cell (MFC), in which an applied voltage drives H_2 evolution.

biotransformations to be carried out within a continuous flow process while maintaining enzyme stability under operating conditions as a result of the immobilization.

3.6 Sustainable Packaging

Examples of some developments in nanotechnology include packaging materials with improved barrier properties and increased resistance to high temperature and mechanical stresses, as well as nutrient delivery systems that enable targeted delivery. These applications exemplify the use of nanotechnology to achieve products with improved control, selectivity, security, functionality, bioavailability, and product targeting (Augustin and Sanguansri, 2009). Traditionally, the food to be coated is dipped in a polysaccharide, protein-based solution, or emulsion, and a thin layer of the coating material is formed around the surface of the food product. Multilayered coatings may be obtained using layer-by-layer electrodeposition (Weiss et al., 2008). Coatings may be used as carriers of functional ingredients (e.g., antimicrobial agents) by using microencapsulation or nanoencapsulation techniques (Vargas et al., 2008). Edible films based on chitosan, with improved barrier and mechanical properties, may be obtained by incorporating nanoparticles. Naturally occurring biodegradable materials have been used as matrixes for nanoparticle synthesis to replace generally toxic, environmentally unfriendly organic surfactants, additives, and solvents. The selection of these materials is based on principles of green chemistry and chemical engineering, and it requires understanding of surface interaction between the synthesized nanomaterials and their surrounding chemical environments. Several green chemical pathways, which use biodegradable matrix reagents, including plant polyphenols, agricultural residues, and vitamins, have proved to be environmentally benign (Zeng, 2012).

Although global demand for biodegradable or plant-based plastic will quadruple by 2013, there is no good infrastructure in place to get optimum benefit from sustainable packaging. Europe leads the regulatory reform process, with the USA trailing behind. Asia and Australia are making headway, but Asia (and in particular China) suffers from a confused approach. Consumers are driving demand, with major retailers such as Coca-Cola and Walmart getting on board. However, despite increased demand, the key cost barriers for suppliers are R&D costs, production costs, and economies of scale. SMART, active, and intelligent packaging has been focused to date on retailer benefits, namely spoilage reduction and extending shelf life. In the future, the expectation is to be more consumer focused: delivering enhanced freshness and better information to elevate product quality for consumers, thus reducing the amount of food wasted (www.just-food.co.uk, 2010).

3.7 Progress in Encapsulation

Nanoliposomes are microscopic vesicles composed of phospholipid bilayers entrapping one or more aqueous compartments. Because of their biocompatibility and biodegradability, liposomes are being used in applications ranging from drug and gene delivery to diagnostics, cosmetics, long-lasting immunocontraception, and food nanotechnology. There are many potential applications for liposomes in the food industry, ranging from the protection of sensitive ingredients to increasing the efficacy of food additives. In order to extend the degree of utilization of liposomes, future research has to focus on the production of the lipid vesicles through safe, scalable methods by using low-cost ingredients (perhaps derived from dairy by-products or other FIW) (Kosseva, 2011b). Another research and development area, which has remained relatively unexplored, is that of encapsulation of antimicrobials for the protection and preservation of foodstuffs. The goal is to demonstrate the true potential for antimicrobial-loaded liposomes and nanoliposomes to improve the quality and safety of a wide variety of food products (Mozafari et al., 2008).

The future challenge in food processing will be to decide on what application of nanotechnology in food would be of most benefit to the consumer, environment, and industry. This requires further research into biopolymer assembly behavior and applications of nanomaterials in the food industry. It is important to educate the consumer about the implications of applying nanotechnology in food and for regulatory bodies to take an active role in approvals of new products made as a result of nanotechnology. This would include evidence that food products made using nanotechnology are safe and have benefits that would otherwise not be possible using current practices (Augustin and Sanguansri, 2009).

4. CONCLUSIONS

Across both developed and developing societies, food is produced, processed, transported, sold, driven home, and then, roughly one third of the time, thrown into the bin for landfill. Then in landfill, methane gas is given off, which is far more destructive than CO_2. Wasting food while millions of people around the world suffer from hunger raises moral questions and could lead to a future food crisis. There are also environmental impacts associated with the inefficient use of natural resources such as water, energy, and land. Apart from the environmental challenges posed, such food waste streams represent considerable amounts of potentially reusable materials and energy.

Food intake is a vital source of energy for human beings. In the same way, food wastes should be used as a reservoir of energy and commodities or a pool of valuable ingredients for novel manufacturing processes. In this book, we have provided a comprehensive state-of-the-art literature review on food waste assessment, management techniques, and processing technologies. Based on our own research achievements in the recovery of commodities from food industry wastes, we have proposed several routes for industrial applications of this waste. It is vital to apply green production principles, criteria, and upgrading concepts in order to develop sustainability in the global food and drink production industry. In a similar fashion, it is vital to focus our efforts on providing renewable energy sources for clean energy.

References

Allen, D.T., Butner, R.S., 2002. Industrial ecology: a chemical engineering challenge. Chem. Eng. Prog. 98 (11), 40.

Augustin, M.A., Sanguansri, P., 2009. Nanostructured materials in the food industry. Adv. Food Nutr. Res. 58 (5), 184–207.

Bakshi, B.R., Fiksel, J., 2003. The quest for sustainability: challenges for process systems engineering. AIChE J. Perspect. 49 (6), 1350–1358.

Benyus, J., 1997. Biomimicry. William and Morrow, New York.

Botella, C., Diaz, A.B., Wang, R., Koutinas, A., Webb, C., 2009. Particulate bioprocessing: a novel process strategy for biorefineries. Proc. Biochem. 44 (5), 546–555.

Business Insights, Reference, 2010. Available at: <www.just-food.co.uk/> Reference 116311, p.110.

Coppens, M.O., 2009. Multiscale nature inspired chemical engineering. In: Fish, J. (Ed.), Multiscale Methods—Bridging the Scales in Science and Engineering. Oxford University Press, Oxford, pp. 536–560.

Coppens, M.O., 2012. A nature-inspired approach to reactor and catalysis engineering. Curr. Opin. Chem. Eng. 1, 1–9.

Cussler, E.L., Moggridge, G.D., 2001. Chemical product design. Cambridge University Press.

D'Hondt, E., Voorspoels, S., 2012. Maximising value: making the most of biowaste. Waste Management World Jan.–Feb., 19–21.

Dudukovic, M.P., 2009. Frontiers in reactor engineering. Science 325, 698–701.

EC, 2010. European Commission, Final report—Preparatory Study on Food Waste Across EU 27, October 2010, European Commission, Brussels.

EC, 2012. FP7 Work Program; Orientation paper; Cooperation Theme2, Food, Agriculture, Fisheries, and Biotechnology, European Commission, Brussels.

EC SCAR, 2011. European Commission Standing Committee on Agricultural Research. Sustainable Food Consumption and Production in a Resource-Constrained World, European Commission, Brussels.

ETP, 2011. Food for Life. Draft Strategic Research and Innovation Agenda. European Technology Platform.

EU, 2011. Roadmap to a Resource Efficient Europe. Communication from the Commission. Available at: <http://ec.europa.eu/environment/resource_efficiency/pdf/com2011_571.pdf>.

FACCE, 2010. Scientific Advisory Board on a Strategic Research Agenda within the JPI Agriculture, Food Security and Climate Change (FACCE), Preliminary Report.

Ferchichi, M., Crabbe, E., Gil, G., Hintz, W., Almadidy, A., 2005. Influence of initial pH on hydrogen production from cheese whey. J. Biotechnol. 120, 402–409.

Fiksel, J., 2003. Revealing the Value of Sustainable Development, Corporate Strategy Today, VII/VIII (2003).

Fischer, C., Kjaer, B., 2012. Recycling and sustainable materials management. Copenhagen Resource Institute (CRI).

Gerngross, T.U., Slater, S.C., 2000. How green are green plastics? Sci. Am. (Aug.).

Grossmann, I.E., 2004. Challenges in the new millennium: product discovery and design, enterprise and supply chain optimization, global life cycle assessment. Comp. Chem. Eng. 29, 29–39.

Hallenbeck, P.C., Ghosh, D., 2009. Advances in fermentative biohydrogen production: the way forward? Trends in Biotechnol. 27 (5), 287–297.

Jha, A.K., 2012. Biowaste: dry advice. Waste Manage. World Jan.–Feb., 23–24.

Kosseva, M.R., 2011a. Management and processing of food wastes. In: second ed. Moo-Young, Murray (Ed.), Comprehensive Biotechnology, vol. 6. Elsevier, pp. 557–593.

Kosseva, M.R., 2011b. Immobilization of microbial cells in food fermentation processes. Food Bioprocess Technol. 4, 1089–1118.

Mitchell, D.A., de Lima Luz, L.F, Krieger, N., 2011. Bioreactors for solid-state fermentation. In: Murray Moo-Yong (Ed.), Comprehensive Biotechnology, second ed. v. 2. Elsevier, pp. 347–360.

Mozafari, M.R., Khosravi-Darani, K., Borazan, G.G., Cui, J., Pardakhty, A., Yurdugul, S., 2008. Encapsulation of food ingredients using nanoliposome technology. Int. J. Food Prop. 11, 833–844.

Nikačević, N.M., Huesman, A.E.M., Van den Hof, P.M.J., Stankiewicz, A.I., 2012. Opportunities and challenges for process control in process intensification. Chem. Eng. Process. Int.

Nishio, N., Nakashimada, Y., 2007. Review. Recent development of AD processes for energy recovery from wastes. J. Biosci. Bioeng. 103 (2), 105–112.

NRC-BCST, 2003. Beyond the molecular frontier: challenges for chemistry and chemical engineering, Committee on Challenges for the Chemical Sciences in the 21st Century, National Research Council.

OECD, 2009. Eco-innovation in industry. Enabling green growth. OECD.

Ozin, G.A, Arsenault, A.C, Cademartiri, L., 2009. Nanochemistry—a Chemical Approach to Nanomaterials, second ed. Royal Society of Chemistry.

Pfister, S., and the AC-ERE, 2003. Complex Environmental Systems: Synthesis for Earth, Life and Society in the 21st Century, Report, National Science Foundation.

Rittenhouse, D., 2003. Piecing together a sustainable development strategy. Chem. Eng. Prog. 99 (3), 32.

Stankiewicz, A., 2008. What is process intensification? European Roadmap of Process Intensification.

Stankiewicz, A., Moulijn, J.A., 2000. Process intensification: transforming chemical engineering. Chem. Eng. Prog. 96 (1), 22–34.

Sundmacher, K., Hanke-Rauschenbach, R., Heidebrecht, P., Rihko-Struckmann, L., Vidakovic-Koch, T., 2012. Some reaction engineering challenges in fuel cells: dynamics integration, renewable fuels, enzymes. Curr. Opin. Chem. Eng. 1, 1–8.

United Nations, 2011. World economic and social survey 2011. The Great Green Technological Transformation.

Vargas, M., Pastor, C., Chiralt, A., McClements, D.J., Gonzalez-Martinez, C., 2008. Recent advances in edible coatings for fresh and minimally processed fruits. Crit. Rev. Food Sci. Nutr. 48, 496–511.

Verfaillie, H.A., Bidwell, R., 2000. Measuring Eco-Efficiency: a Guide to Reporting Company Performance. WBCSD, Geneva.

Weiss, J., Decker, E.A., McClements, D.J., Kristbergsson, K., Helgason, T., Awad, T., 2008. Solid lipid nanoparticles as delivery systems for bioactive food components. Food Biophys. 3, 146–154.

World Economic Forum, 2009. Driving sustainable consumption, value chain waste.

Zeng, H.C., 2012. Nanotechnology for emerging applications, Editorial overview. Curr. Opin. Chem. Eng. 1, 1–2.

Food Science and Technology International Series

Amerine, M.A., Pangborn, R.M., and Roessler, E.B., 1965. Principles of Sensory Evaluation of Food.

Glicksman, M., 1970. Gum Technology in the Food Industry.

Joslyn, M.A., 1970. Methods in Food Analysis, Second Ed.

Stumbo, C. R., 1973. Thermobacteriology in Food Processing, Second Ed.

Altschul, A.M. (Ed.), New Protein Foods: Volume 1, Technology, Part A—1974. Volume 2, Technology, Part B—1976. Volume 3, Animal Protein Supplies, Part A—1978. Volume 4, Animal Protein Supplies, Part B—1981. Volume 5, Seed Storage Proteins—1985.

Goldblith, S.A., Rey, L., and Rothmayr, W.W., 1975. Freeze Drying and Advanced Food Technology.

Bender, A.E., 1975. Food Processing and Nutrition.

Troller, J.A., and Christian, J.H.B., 1978. Water Activity and Food.

Osborne, D.R., and Voogt, P., 1978. The Analysis of Nutrients in Foods.

Loncin, M., and Merson, R.L., 1979. Food Engineering: Principles and Selected Applications.

Vaughan, J. G. (Ed.), 1979. Food Microscopy.

Pollock, J. R. A. (Ed.), Brewing Science, Volume 1—1979. Volume 2—1980. Volume 3—1987.

Christopher Bauernfeind, J. (Ed.), 1981. Carotenoids as Colorants and Vitamin A Precursors: Technological and Nutritional Applications.

Markakis, P. (Ed.), 1982. Anthocyanins as Food Colors.

Stewart, G.G., and Amerine, M.A. (Eds.), 1982. Introduction to Food Science and Technology, Second Ed.

Iglesias, H.A., and Chirife, J., 1982. Handbook of Food Isotherms: Water Sorption Parameters for Food and Food Components.

Dennis, C. (Ed.), 1983. Post-Harvest Pathology of Fruits and Vegetables.

Barnes, P.J. (Ed.), 1983. Lipids in Cereal Technology.

Pimentel, D., and Hall, C.W. (Eds.), 1984. Food and Energy Resources.

Regenstein, J.M., and Regenstein, C.E., 1984. Food Protein Chemistry: An Introduction for Food Scientists.

Gacula Jr. M.C., and Singh, J., 1984. Statistical Methods in Food and Consumer Research.

Clydesdale, F.M., and Wiemer, K.L. (Eds.), 1985. Iron Fortification of Foods.

Decareau, R.V., 1985. Microwaves in the Food Processing Industry.

Herschdoerfer, S.M. (Ed.), Quality Control in the Food Industry, Second Ed. Volume 1—1985. Volume 2—1985. Volume 3—1986. Volume 4—1987.

Urbain, W.M., 1986. Food Irradiation.

Bechtel, P.J., 1986. Muscle as Food.

Chan, H.W.-S., 1986. Autoxidation of Unsaturated Lipids.

Cunningham, F.E., and Cox, N.A. (Eds.), 1987. Microbiology of Poultry Meat Products.

McCorkle Jr. C.O., 1987. Economics of Food Processing in the United States.

Japtiani, J., Chan Jr., H.T., and Sakai, W.S., 1987. Tropical Fruit Processing.

Solms, J., Booth, D.A., Dangborn, R.M., and Raunhardt, O., 1987. Food Acceptance and Nutrition.

Macrae, R., 1988. HPLC in Food Analysis, Second Ed.

Pearson, A.M., and Young, R.B., 1989. Muscle and Meat Biochemistry.

Penfield, M.P., and Campbell, A.M., 1990. Experimental Food Science, Third Ed.

Blankenship, L.C., 1991. Colonization Control of Human Bacterial Enteropathogens in Poultry.

Pomeranz, Y., 1991. Functional Properties of Food Components, Second Ed.

Walter, R.H., 1991. The Chemistry and Technology of Pectin.

Stone, H., and Sidel, J.L., 1993. Sensory Evaluation Practices, Second Ed.

Shewfelt, R.L., and Prussia, S.E., 1993. Postharvest Handling: A Systems Approach.

Nagodawithana, T., and Reed, G., 1993. Enzymes in Food Processing, Third Ed.

Hoover, D.G., and Steenson, L.R., 1993. Bacteriocins.

Shibamoto, T., and Bjeldanes, L., 1993. Introduction to Food Toxicology.

Troller, J.A., 1993. Sanitation in Food Processing, Second Ed.

Hafs, D., and Zimbelman, R.G., 1994. Low-fat Meats.

Phillips, L.G., Whitehead, D.M., and Kinsella, J., 1994. Structure-Function Properties of Food Proteins.

Jensen, R.G., 1995. Handbook of Milk Composition.

Roos, Y.H., 1995. Phase Transitions in Foods.

Walter, R.H., 1997. Polysaccharide Dispersions.

Barbosa-Canovas, G.V., Marcela Go'ngora-Nieto, M., Pothakamury, U.R., and Swanson, B.G., 1999. Preservation of Foods with Pulsed Electric Fields.

Jackson, R.S., 2002. Wine Tasting: A Professional Handbook.

Bourne, M.C., 2002. Food Texture and Viscosity: Concept and Measurement, Second ed.

Caballero, B., and Popkin, B.M. (Eds.), 2002. The Nutrition Transition: Diet and Disease in the Developing World.

Cliver, D.O., and Riemann, H.P. (Eds.), 2002. Foodborne Diseases, Second Ed.

Kohlmeier, M., 2003. Nutrient Metabolism.

Stone, H., and Sidel, J.L., 2004. Sensory Evaluation Practices, Third Ed.

Han, J.H., 2005. Innovations in Food Packaging.

Sun, D.-W. (Ed.), 2005. Emerging Technologies for Food Processing.

Riemann, H.P., and Cliver, D.O. (Eds.), 2006. Foodborne Infections and Intoxications, Third Ed.

Arvanitoyannis, I.S., 2008. Waste Management for the Food Industries.

Jackson, R.S., 2008. Wine Science: Principles and Applications, Third Ed.

Sun, D.-W. (Ed.), 2008. Computer Vision Technology for Food Quality Evaluation.

David, K., and Thompson, P., (Eds.), 2008. What Can Nanotechnology Learn from Biotechnology?

Arendt, E.K., and Bello, F.D. (Eds.), 2008. Gluten-Free Cereal Products and Beverages.

Bagchi, D. (Ed.), 2008. Nutraceutical and Functional Food Regulations in the United States and Around the World.

Singh, R.P., and Heldman, D.R., 2008. Introduction to Food Engineering, Fourth Ed.

Berk, Z., 2009. Food Process Engineering and Technology.

Thompson, A., Boland, M., and Singh, H. (Eds.), 2009. Milk Proteins: From Expression to Food.

Florkowski, W.J., Prussia, S.E., Shewfelt, R.L. and Brueckner, B. (Eds.), 2009. Postharvest Handling, Second Ed.

Gacula Jr., M., Singh, J., Bi, J., and Altan, S., 2009. Statistical Methods in Food and Consumer Research, Second Ed.

Shibamoto, T., and Bjeldanes, L., 2009. Introduction to Food Toxicology, Second Ed.

BeMiller, J. and Whistler, R. (Eds.), 2009. Starch: Chemistry and Technology, Third Ed.

Jackson, R.S., 2009. Wine Tasting: A Professional Handbook, Second Ed.

Sapers, G.M., Solomon, E.B., and Matthews, K.R. (Eds.), 2009. The Produce Contamination Problem: Causes and Solutions.

Heldman, D.R., 2011. Food Preservation Process Design.

Tiwari, B.K., Gowen, A. and McKenna, B. (Eds.), 2011. Pulse Foods: Processing, Quality and Nutraceutical Applications.

Cullen, PJ., Tiwari, B.K., and Valdramidis, V.P. (Eds.), 2012. Novel Thermal and Non-Thermal Technologies for Fluid Foods.

Stone, H., Bleibaum, R., and Thomas, H., 2012. Sensory Evaluation Practices, Fourth Ed.

Index

Note: Page numbers followed by "*f*" indicate a figure; page numbers followed by "*t*" indicate a table.

Printed in the United States
By Bookmasters